Structure and Fabric
Part 2

MITCHELL'S BUILDING CONSTRUCTION

The volumes in this series of standard text books have been thoroughly re-written, re-illustrated, and amplified in order to bring them into line with the present-day needs of students of architecture, building, and surveying. All quantities are expressed in SI units.

There are five related volumes:

ENVIRONMENT AND SERVICES
Peter Burberry Dip Arch ARIBA
new edition 1975

MATERIALS
Alan Everett ARIBA

STRUCTURE AND FABRIC Part 1
Jack Stroud Foster FRIBA

STRUCTURE AND FABRIC Part 2
Jack Stroud Foster FRIBA
Raymond Harington Dip Arch, ARIBA, ARIAS

COMPONENTS AND FINISHES
Harold King ARIBA
Alan Everett ARIBA

Structure and Fabric Part 2 is an extension and development of Part 1 and with it forms a complete work in the area of its subject title.
The chapters are:
1   Contract planning and site organisation
2   Contractors' mechanical plant
3   Foundations
4   Walls and piers
5   Multi-storey structures
6   Floor structures
7   Chimney shafts, flues and ducts
8   Stairs, ramps and ladders
9   Roof structures
10 Fire protection
11 Temporary works

The book forms a suitable text for courses in architecture, building technology, quantity surveying, structural engineering, and environmental engineering.

Jack Stroud Foster is Principal Lecturer in Architecture at the Polytechnic of Central London in which he is responsible for the co-ordination and development of Constructional Studies.

Raymond Harington Dip Arch, ARIBA ARIAS, is Director of Studies, B Arch Course, Mackintosh School of Architecture, Department of Architecture of the University of Glasgow and the Glasgow School of Art.

MITCHELL'S BUILDING CONSTRUCTION

Structure and Fabric

Part 2

Jack Stroud Foster *FRIBA*

Raymond Harington *Dip Arch, ARIBA, ARIAS*

B T Batsford Limited
London

© *Text Jack Stroud Foster and
Ray Harington 1976*
© *Illustrations Jack Stroud Foster and
B T Batsford Limited 1976*
First printed 1976
ISBN 0 7134 0522 8 (hardcover)
ISBN 0 7134 0523 6 (paperback)

*Printed in Great Britain by
The Anchor Press, Tiptree, Essex
for the publishers
B T Batsford Limited
4 Fitzhardinge Street
London W1H 0AH*

Contents

Preface

This volume in Mitchell's Building Construction is a revision and expansion of *Mitchell's Advanced Building Construction – The Structure* by Jack Stroud Foster, and is intended to extend and develop the material in *Structure and Fabric* Part 1.

Much of the preface to the original work is appropriate to this volume and it, therefore, forms the text of this preface.

The subject of the book has been treated basically under the elements of construction. Most of these are interrelated in a building and, as far as possible, this has been borne in mind in the text. As in Part 1 ample cross-references are given to facilitate a grasp of this interrelationship of parts. Contract planning and site organization, and the use of mechanical plant, are both subjects relevant to constructional techniques and methods used on the site and to the initial design process for a building. These have been considered in the first two chapters. The subject of fire protection by its nature is extremely broad but it is so closely linked with the design and construction of buildings that it has been covered on broad lines in order to give an understanding of those factors which influence the nature and form of fire protection as well as to give detailed requirements in terms of construction.

In view of the continual production of new and improved materials in various forms and the continuous development of new constructional techniques, using both new and traditional materials, the designer can no longer be dependent on a tradition based on the use of a limited range of structural materials, but must exercise his judgment and choice in a wide, and ever-widening, realm of alternatives. This necessitates a knowledge not only of the materials themselves but of the nature and structural behaviour of all the parts of a building of which those materials form a part. Efficiency of structure and economy of material and labour are basic elements of good design. They are of vital importance today and should have a dominating influence on the design and construction of all buildings.

In the light of this something is required to give an understanding of the behaviour of structures under load and of the functional requirements of the diff-

erent parts; to give some indication of their comparative economics and efficient design, their limitations and the logical and economic application of each. In writing this volume and its companion volume it has been the aim to deal with these aspects. This book is not an exemplar of constructional details. Those details which are described and illustrated are meant to indicate the basic methods which can be adopted and how different materials can be used to fulfil various structural requirements. The illustrations are generally not fully dimensioned; such dimensions as are given are meant to give a sense of 'scale' to the parts rather than to lay down definite sizes in particular circumstances. The function of the book is not primarily to give information on *how* things are done in detail, as this must be everchanging. Rather, the emphasis is on *why* things are done, having regard particularly to efficiency and economy in design. An understanding of the function and behaviour of the parts and of the logical and economic application of material, should enable a designer to prepare satisfactory constructional details in the solution of his structural problems.

This volume is intended primarily as a textbook for architectural, building and surveying students, but it is hoped that students of civil and structural engineering will find it useful as a means of setting within the context of the building as a whole their own studies in the realm of building structures.

In a book of this nature there is little scope for original work. The task consists of gathering together existing information and selecting that which appears to be important and relevant to the purpose of the book. The authors acknowledge the debt they owe to others on whose work they have freely drawn, much of which is scattered in the journals of many countries. An endeavour has been made to indicate the sources, either in the text or in footnotes. Where this has not been done it is due to the fact that over a period of many years of lecturing on the subject much material has been gathered, both textual and illustrative, the sources of which have not been traced. For any such omissions the authors' apologies are offered.

1975   JSF and RKH

Acknowledgment

The authors are indebted to many people and organisations who have given help and guidance in the preparation of this book and from whom they have received much information. The following, to whom thanks are especially due, were named in the original *Advanced Building Construction* where a note of their particular contribution was made: J E Crofts; Kenneth A Lock; A G Stone: Ivan Tomlin; Professor Z S Makowski; Leonard R Creasy; Kenneth W Dale; Peter Dunican; J W Tiller.

We are grateful to Concrete Limited; British Lift Slab Limited; Dollery and Palmer Limited and Putzmeister Limited; How-Kinnell Limited; Omnia Constructions Limited and many other firms and organisations who have freely given information and permission to base illustrations on material which they have readily provided.

With the permission of the Controller of HM Stationery Office we have drawn freely on *Principles of Modern Building,* Volumes 1 and 2 and on *Post-War Building Studies, Building Research Station Digests* and *Current Papers,* and have quoted from *The Building Regulations, 1972.* We have also drawn on *British Standard Specifications* and *Codes of Practice* with the permission of the British Standards Institution, from whom official copies may be obtained. Extracts from the *London Building (Construction) By-laws, 1972* have been made with the permission of the Director General and Clerk to the Greater London Council. We owe much in Chapter 3 to the reading of Capper and Cassie's book *The Mechanics of Engineering Soils* and much of the material on the economic aspects of Chapters 5 and 9 has been gathered from Leonard R Creasy's excellent paper on the subject which is referred to in the text.

We acknowledge with thanks the following for permission to quote from books or papers, to reproduce tables or to use drawings as a basis for illustrations in this volume: Architectural Press: *Building Elements,* R Llewellyn Davies and A Petty; *Guide to the Building Regulations, 1972,* A Elder (for tables 17 and 18); *Guide to Concrete Block-work,* Michael Gage (for figure 108 *B*); *Principles of Pneumatic Architecture,* Roger N Dent (for figures 204 *A,C,F* and 231 *A,B,E,F,G,H*); *Structure in Building,* W Fisher Cassie and J H Napper (for part of figures 38 and 44 *A,B*). Cement and Concrete Association: *Concrete Block Walls* (for figure 108 *C,D*). Concrete Publications: *Reinforced Concrete Chimneys,* C Percy Taylor and Leslie Turner. Crosby Lockwood Staples: *Practical Problems in Soil Mechanics,* H R Reynolds and P Protopapadakis (for figures 33 *A,* and 37); *Design Problems of Heating and Ventilating,* A T Henley; *Oil Fuel Applications,* A T Henley. HMSO: *Air Structures – A Survey,* F Newby (for figures 231 *C,D* and 204 *D,G,H,L*); DoE *Construction 3* and 7 (for figures 81 and 78 respectively). Newnes-Butterworth: *Structural Steelwork for Students,* L V Leech (for parts of figures 115, 120 and 207). Pitman: *Heating and Air Conditioning Equipment for Buildings,* F Burlace Turpin; *Soil Mechanics Related to Building,* J H G King and D A Creswell (for figures 33 *B,C,D,F* and 39); *The Fabric of Modern Buildings,* E G Warland. Spon: *Mechanics of Engineering Soils,* P L Capper and W F Cassies (for parts of chapter 3); *Walls and Wall Facings,* D N Nield. *Acier-Stahl-Steel 6/1974,* Centre Belgo-Luxembourgeois d'Information de l'Acier (for figure 117 *B*). *Architects' Journal,* Architectural Press, 19 February 1964 (for figure 158), 18 February 1971 (for tables 20, 21), 3 June 1970 (for figure 100 top left), 15 September 1971 (for figure 80). *Architectural Design,* Standard Catalogue Co Ltd. *BDA Technical Notes* Volume 1 no 4 September 1971 (for figure 76), The Brick Development Co Association. *Build International,* October 1969, Applied Science Publishers Ltd (for figure 121). *Building Specification,* February 1970 (for figure 100 right). *Building Research Congress 1951: Papers,* Division 1 (for figure 36). British Gas Corporation: 'Flexibility with Flues' (for figure 168), 'Fan diluted flues' text page 277; 'Gas Handbook for Architects and Builders (for figure 19). Constrado: *Structural Steelwork Simplified* (for figures 115 *F,G,H,I,O,P* and 120 *F,G,J*). *Proceedings of the Institution of Civil Engineers,* Institution of Engineers, *RIBA Journal,* October 1973 (for figures 16 and 161);

The Structural Engineer, Institution of Structural Engineers. Also thanks to the Australian Department of Labour and National Service — Industrial Services Division; the British Gas Corporation; and to Messrs Anthony Collins, H J B Harding and R Glossop, F Kerr, Professor Z S Makowski, and H Werner Rosenthal.

We are grateful to George Dilks for his meticulous work in revising for this volume the original illustrations from *Advanced Building Construction* and for preparing the new illustrations. We must also express our appreciation to Thelma M Nye, of the publishers, for her help and patience in seeing the work through to press.

JSF and RKH

1 Contract planning and site organization

The planning and control of all resources necessary to the realisation of building production is of vital importance and at some point before commencing work on the site, thought must be given to the way in which the building operation will be organized. Most builders plan their work in some form or another but, in the past, only a few have done so in much detail and have committed their plan to paper. Buildings, and consequently their construction, have become increasingly complex and the proper management of a contract and the control of cost, on the part of the architect at design stage and the contractor during erection, are more than ever essential if building is to be carried out efficiently both in terms of time and money. Only by proper planning can aids to productivity, such as mechanical plant, incentives, and efficient use of labour, become fully effective. With the greater mechanization of building operations and the increased use of expensive plant, the contractor must obtain maximum use of the plant and speed the construction of the job in order to keep his costs to a minimum. The design/erection continuum must be seen as a production process from inception to completion and there must be a programme on which the job may be organized, against which performance may be assessed and within which control may be exercised.

As pointed out in Part 1 the building and civil engineering industry is peculiar in that the contractor who will be responsible for carrying out the work usually plays no part in the design of a project, and has no opportunity at this point to contribute from his experience on matters of construction, planning and the nomination of subcontractors and thus to assist in the work being carried out efficiently, quickly and economically. Although the negotiated contract is often suggested as a means of overcoming this lack of collaboration at the design stage, it is not always suitable or acceptable since the element of competitive tendering is absent.[1] In these circumstances it is, therefore, essential that the architect should have sufficient knowledge of contract planning and of its implications to ensure a well-organized job. Reference is made later to ways in which the architect can contribute to this end at the design stage.

Whether or not such contribution is made by the architect, the responsibility for actually carrying out the job in all its aspects is that of the contractor. In order to enable him to do this efficiently management methods common to other industries have become widely used in building[2]. The contemporary view of construction management embraces a great number of interrelated activities drawing on a vast range of resources: professional, manufacturing, different categories of contracting and supplying, off-site production of components involving an increased use of transport, and specialized assembly methods.

The subjects of this chapter, contract planning and site organization, together with general control, are the construction aspects of production management which itself is a part only of overall management in building. *Planning* makes efficient and economical use of labour, machines and materials, *organizing* is the means of delegating tasks and *control* enables planning and organization to be effective. It is possible here to deal with them only in outline and for a more detailed consideration reference should be made to other works[3].

CONTRACT PLANNING AND CONTROL

This involves working out a plan of campaign or a programme for the contract as a whole and assembling the necessary data. The primary function of such a programme is to promote the satisfactory organization and flow of the various building operations during the course of erection, by planning in advance the times and sequences of all operations and the requirements in labour, materials and equipment. In order to fulfil this function and also to provide important information required during

[1] See footnote page 15, Part I on other developments

[2] See *The Industrialisation of Building* page 24 *et seq*, Part 1
[3] *Introduction to Building Management*, 3rd edition, R.E. Calvert (Newnes Butterworth): *The Construction Industry Handbook* 1971, ed R.A. Burgess and others (Medical & Technical Publishing Co): *Principles of Construction Management for Engineers and Managers* 1966, R. Pilcher (McGraw Hill)

1

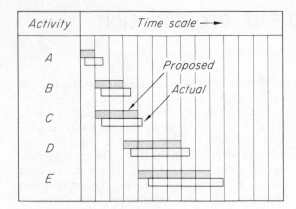

1 Gantt chart

the contract, a well planned programme will have certain clear objectives and *Building Research Station Digest* 91 states that it should:

(a) show the quickest and cheapest method of carrying out the work consistent with the available resources of the builder

(b) by the proper phasing of operations with balanced labour gangs in all trades, ensure continuous productive work for all the operatives employed and reduce unproductive time to a minimum

(c) provide an assessment of the level of productivity in all trades to permit the establishment of equitable bonus targets

(d) determine attendance dates and periods for all sub-contractors' work

(e) provide information on material quantities and essential delivery dates, the quantity and capacity of the plant required and the periods it will be on site

(f) provide, at any time during the contract, a simple and rapid method of measuring progress, for the builder's information, the architect's periodical certificate or the valuation of work for accounting purposes.

If a builder's tender for any sizeable job is to be realistic, planning must start at the estimating stage and the following considerations must be taken into account: the most economic methods to be used for each operation and the sequence and timing of the operations, having regard to the resources at the contractor's disposal; whether hand or mechanical methods will be most economical and the most suitable type of plant to be used in relation to the nature and size of the job; the space available and the best positions for the various machines to be used; the best methods of handling materials and

the most suitable places on the site for the storage of materials and for the placing of huts; suitable points of access to the site for lorries and machines. In deciding what methods to use for erection and the most suitable plant for different operations, the estimator would, when necessary, consult the contract planning and plant departments of the firm.

Traditional methods of production control have tended to be based on criteria of usefulness to site managers and indeed it must be kept well in mind that the site is where the building is finally assembled and where managerial control must be effective.

A typical site oriented control device is the Gantt Chart or Bar Chart which allows a fairly simple and easily read plan of operations to be made

(a)

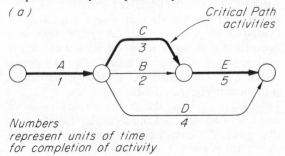

Numbers represent units of time for completion of activity

Dependencies:

E is dependent on activities
    C, B and A being completed
C and B are
    dependent on A being completed
D is dependent on A being completed

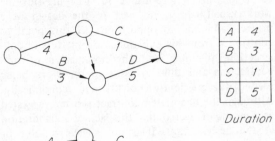

A	4
B	3
C	1
D	5

Duration

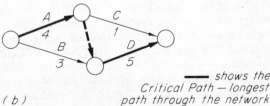

—— shows the
Critical Path — longest
path through the network

(b)

2 Network or Critical Path diagrams

available to all site personnel against which may be plotted actual performances (see figure 1). However, this excellent device only takes into account one of the resources, time, and unless further schedules of the resources needed for each operation are also available adjacent to the Bar Chart it does not inform on the critical relationships between the various activities depicted nor does it enable procedures involving a number of variables to be optimised since the complex interrelationships affecting the outcome of any plan (or alteration of plan) are not readily evident or quantifiable. This can, however, be achieved by means of a technique known as Network Analysis.

Network analysis

The essential difference between analysing a production problem by network as against linear or parallel linear methods lies in the identification of the dependency between operations. This approach leads to interrelated networks through which certain sequences can be seen to be 'critical' to the anticipated outcome in that they occupy the longest irreducible time necessary to execute the project (or part of the project) to which they are necessary. Figure 2 shows this in a simple set of five interrelated activities A, B, C, D, E of time values 1, 2, 3, 4 and 5 days.

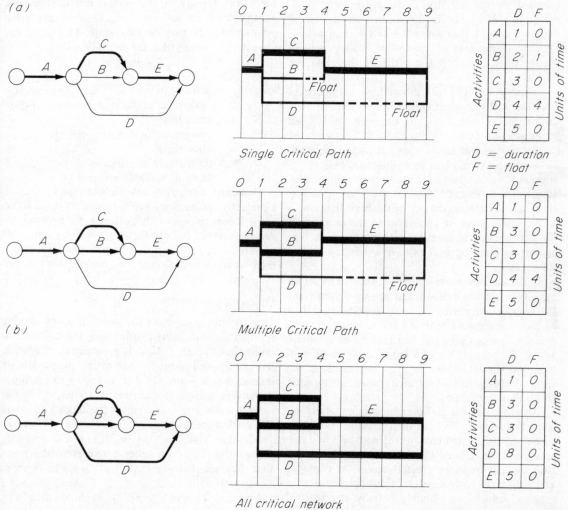

(a)

Single Critical Path

	D	F
A	1	0
B	2	1
C	3	0
D	4	4
E	5	0

D = duration
F = float

Multiple Critical Path

	D	F
A	1	0
B	3	0
C	3	0
D	4	4
E	5	0

(b)

All critical network

	D	F
A	1	0
B	3	0
C	3	0
D	8	0
E	5	0

NB Activities on Critical Path have zero "float"

(c)

3 Network or Critical Path diagrams

In preparing a network the project is broken down into its operational parts (this can be at a strategic level or in extreme tactical detail) termed 'activities' which are represented by linear arrows (*a*). The arrows are arranged to show the sequence of activities necessary to the occurrence of 'events' (shown by circles) which must precede further sets of activities. Activities (with the exception of 'dummy' activities) take up time, including waiting time, whereas events do not. An event cannot be said to have occurred until *all* the activities leading to it have been effectively completed.

Dummy arrows are used to show dependencies where activities are not directly sequential. For example in (*b*) activity *C* is dependent on *A* being completed. *D* is dependent on *A and B* being completed – shown by use of a dummy arrow. Dummy arrows have no time value but may still lie on the critical path if the activities they link are critical.

Figure 3 (*a*), (*b*), (*c*) shows how variations in time ascribed to the activities will result in different critical paths and how, in the case of (*c*), all sequences can become critical. In theory the most perfect plan would result in an 'all critical' network but in life this would lead to a wholly inflexible situation lacking any time to manoeuvre or rethink situations. It is in this context that the 'float' or difference in time between the non-critical and the critical activity times become important to the production planner working in changing circumstances, in that he will be given options as to how he may deploy resources.

It is often convenient for the network to be set up initially using non-scaled linear arrows to represent the sequence interrelationship of activities and then to plot the network against a linear time scale as a prelude to examining the distribution of resources and in the preparation of a bar chart form of presentation. The generalised network statements shown figure 3 (*a*), (*b*) and (*c*) are each shown developed in this way in time scale form.

Although time is the planner's main parameter, other factors will affect the final assessment of times to be ascribed to the constituent activities of a network, eg cost, labour and material availability and the demands of other projects under the planner's control. It is convenient to think in terms of a 'normal' time when planning initially and figure 4 shows a statistical method of arriving at the time envisaged for an activity, given some historical data,

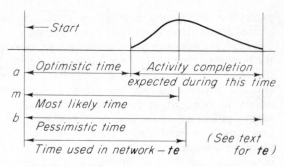

4   *Activity times*

using a Beta distribution in which the area under the curve represents the probability of the time expended on an activity lying between any given points along the baseline time scale. This particular distribution curve yields the formula:

$$t_e = \frac{a + 4m + b}{6}$$

where t_e = estimate of time to be used in network
   a = optimistic or shortest estimate of duration time
   b = pessimistic or longest estimate of duration time
   m = most likely duration of activity judged from all available evidence

However, there are occasions when the control of a project requires more sophisticated analyses to be carried out to allow the planner to respond to varying criteria for productivity: such as time/cost optimisation, levelling of resources and or deployment of resources amongst a number of projects.

Time/cost optimisation

This technique explores the possibilities of altering production time whilst optimising the costs of so doing. In building work it is generally the case that increased speed of production carries the penalty of increased cost – usually due to having to use more operatives and/or machinery or to paying higher rates. Clearly any reduction of the activity times on the critical path or paths will reduce the overall production time but in so doing will probably reduce the 'float' on other activities to the point that they also become critical as shown in figure 5. Given that activities *A, B, C, D, E* shown in case 1 can be carried out in 'normal' times shown in column X and that activities *A, C* and *E* are capable of being carried out by different means at 'crash'

Case 1

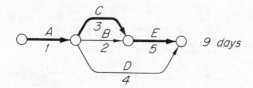

9 days

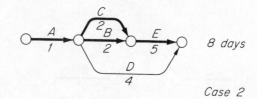

8 days

Case 2

Cost data

Activity	Dur'n	Cost	Dur'n	Cost	
A	1	120	$\frac{1}{2}$	200	CC
B	2	80	2	80	
C	3	100	2	150	CC
D	4	60	4	60	
E	5	200	3	300	CC

X Y (column headers above Dur'n/Cost pairs)

CC = crash costs

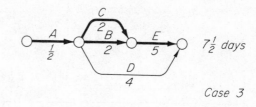

$7\frac{1}{2}$ days

Case 3

I.C. = rate of Indirect Costs — 70·00 per day

Activity	Duration (days)				
	9	8	$7\frac{1}{2}$	6	$5\frac{1}{2}$
A	120	120	200	120	200
B	80	80	80	80	80
C	100	150	150	150	150
D	60	60	60	60	60
E	200	200	200	300	300
I.C.	630	560	525	420	385
Totals	1190	1170	1215	1130	1175

Summary of cases 1–5

└ Best time for least cost

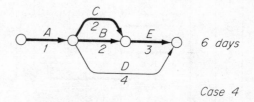

6 days

Case 4

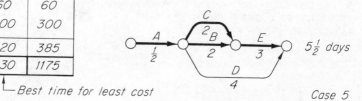

$5\frac{1}{2}$ days

Case 5

5 Time-cost optimisation

times for the increased rates shown in column *Y*, it is then possible to define three basic outcomes from the application of these figures: normal cost programme, all *crash* programme and best time/least cost programme ie an optimisation of time and cost. These alternatives are shown in the networks of cases 1, 5 and 4. The effect of the pro rata indirect or overhead costs which are added to the direct cost variations should be noted.

Resource levelling and control

This technique enables a planner to assess the requirements of various resources to serve any given network of activities and to utilise 'float' in uncritical activities to optimise his use of resources or to reduce imbalances of resource demand. The technique ascribes the various resources to each activity and by comparison with established norms identifies excessive demands. It is possible to re-position activities requiring excessive use of resources and to balance the total requirements within the resources available or at least to reduce the time of excessive demand. The repositioning of certain activities will often render them critical when they are taken together with fixed waiting periods nec-

5

A (i)

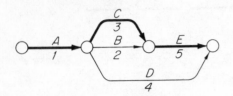

Network with Normal times (N)

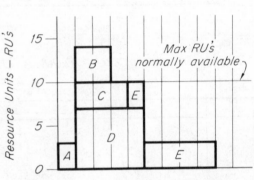

A (ii) Bar Chart based on Network (N)

Re-plan due to overload exposed by A (iii)

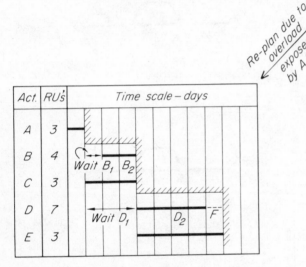

B (i) Bar Chart still based on Network (N), but activities B and D moved within time spaces available (F_1 and F_2)

A (iii) Resource Loading Chart for A (ii)

Re-check Network as to new critical activities

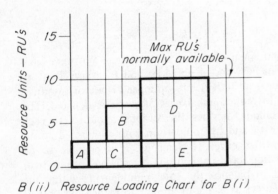

B (ii) Resource Loading Chart for B (i)

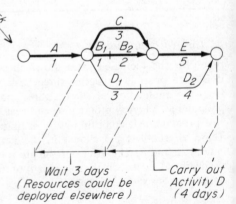

C

Wait 3 days (Resources could be deployed elsewhere)

Carry out Activity D (4 days)

6 Resource levelling

essary to the planned use of resources.

This is illustrated in figure 6 where the network shown in *A* (*i*) yields the scaled network in bar chart form shown in *A* (*ii*). By allocating the resource units, RUs, for each activity (shown in the second column) a resource loading diagram can be prepared as in *A* (*iii*) which in this case shows an excessive demand of four RUs above the resource units normally available during the second and third days, due to activities B and D coming together.

B (*i*) shows the repositioning of activities B and D in the excess times available for their execution and a resulting 'levelling' of the loading diagram to bring the requirements for resources within the limits of normal availability as in *B* (*ii*).

It will be noticed that this manoeuvre involves specific positioning of the waiting periods B_1 and D_1 and examination of the resulting network at *C* shows that these constraints on the commencement of activities B_2 and D_2 leads to the former becoming critical and reduces the float of the latter to one day only.

The foregoing brief description of some of the uses of network analysis has been based on a simplified description of the networks involved. Readers who wish to study these techniques in depth should consult one of the many books dealing specifically with techniques of presentation which allow discrete descriptions of activities by numbers which facilitate input statements into computors, for which many standard programmes exist for solving networks and analysing varying plans with a view to optimisation and control[1].

It has been said with some cause that one of the most useful aspects of network analysis lies in exercising the logic used to set up the basic network of activities since the planner has a full knowledge of the practical consequences of any sequence of activities and the importance of their relationships. This aspect of a network approach to planning is illustrated in the method of presentation of the logic known as a Precedence Diagram. Figure 7 (b) shows a typical network restated in this form which eliminates the need for dummy activities normally used in conventional networks to indicate dependency (a).

The overall programme

On acceptance of the tender, contract planning commences and a working, or overall, programme is prepared by the contractor's planning staff together with the plant engineers and the site agent or foreman for the job. As already indicated this will be used as a guide for site activities, for detailed planning, for the buying and delivery of materials, for the co-ordination of sub-contractors' and main contractor's work and for assessing job progress. Assumptions made at the estimating stage are borne in mind. It is essential at this point for the contractor to have full information from the architect in the form of a site survey, a full set of working drawings including, preferably, all details and full-size drawings together with those of all specialists, a specification, a copy of the bills of quantities and a complete list of all nominated sub-contractors.

Ample time should be allowed for planning before commencement of work on the site. For most jobs, unless particularly small, at least four weeks should be allowed. The smaller jobs must be planned in detail at the outset, since there is no time for making adjustments during the course of a short contract. The larger jobs may be planned on

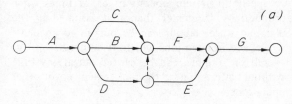

(*a*)

Dependencies:

C, B and D depend on A being completed
F depends on B, C and D being completed
E depends on D being completed
G depends on F and E being completed

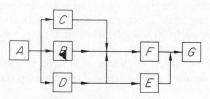

Precedence diagram (*b*)

7 *Precedence diagram*

[1] See *An Introduction to Critical Path Analysis* 3rd Ed, and *Critical Path Analysis: Problems and Solutions*, 1968, both by K.G. Lockyer (Pitman)

broader lines, since time will be available to carry out detailed planning as the job proceeds.

The preparation of the overall programme consists broadly of

(i) breaking the job down into a series of basic operations involving only one trade

(ii) establishing the quantities of work in each operation and the time content of each in terms of men and machines

(iii) arranging the operations in a sequence and balancing the size of gangs to give a maximum continuity of work for each trade and the minimum delay as one trade follows another

(iv) breaking down a large job into phases so that several operations may proceed simultaneously.

The programme is ultimately expressed in chart form which covers all the main operations throughout the contract, the phasing of the work on different parts where this is necessary and the duration of each operation, including the work of all sub-contractors and specialists. Together with this chart a written report or schedule is prepared, which includes a description of the methods to be used, schedules of plant giving the dates when each machine will be required, the labour requirements for each stage of the work, and information regarding site offices, storage huts, equipment and small tools. If, for any reason, the proposals in the contract plan differ from those assumed at the estimating stage, a written cost comparison is drawn up which shows the differences and the cost implications.

The overall programme shows the major operations and phasing of the job, but detailed short-term planning at regular intervals on the site is necessary to ensure the satisfactory allocation of labour and materials to each individual operation as the work proceeds. This is usually carried out in two stages: (i) a reasonably detailed programme is prepared at monthly intervals, to cover the four weeks ahead, and (ii) a detailed programme is prepared each week, to ensure that labour, materials and plant will be available when required. To enable the site foreman to give his full attention during the first few weeks to starting off the job, a detailed programme is prepared for him. This indicates in detail the materials and labour requirements of the first four weeks, together with the operational methods to be used.

The broad picture of the contract planning process given above will now be considered in greater detail.

Break down of job

For the purpose of the overall programme the job must be broken down into groups of basic operations, each of which involves only one trade. For example, in housing, the cutting and fixing of the carcassing timber in first floor and roof or the building of the brickwork or plastering throughout. For convenience in doing this the whole of the job is divided into stages which are commonly

(1) foundations and walling up to DPC
(2) carcase to completion of roofing-in
(3) finishings and all services
(4) drains and site works[1].

In larger jobs and multi-storey work the break-down stages can be

(1) sub-structure, or foundation work
(2) frame, or basic structure
(3) claddings, infillings, weather-proofing, etc
(4) finishings and services
(5) drains and site works.

Each stage is planned separately at first, to allow some flexibility in relating them on the site; delays due to bad weather or other causes can be provided for by varying the intervals between or overlapping the stages during the course of the job. Compensation for any variations from the programme arising within the stages can be made by increasing the gang sizes to speed up certain operations or, at times when productivity is greater than that assumed at the planning stage, labour can be put on to ancillary works and isolated jobs which, if omitted from the overall programme, can be carried out at any time without interfering with the sequence of other operations.

Quantities of work and time content

In order to relate the various operations throughout the job, it is necessary to define the work content of each by means of a schedule of basic quantities, from which the number of man hours and machine hours required to complete each operation can be ascertained by the application of output rates per man or machine hour, or, as they are called, labour and plant standards. These standards in each case are established on the basis of information fed back

[1] See *Building Research Station Digest* 91, 'The Programming of House-building' for complete schedule of operations on this basis

from previous contracts or from work studies[1], having regard to the type of labour which will be available and the likely demand on plant at the time of erection. The work content for each operation is then inserted on a schedule of basic operations which can be in the form of a series of *Data sheets*. These are lists of all operations in sequence under trade sections, each operation being numbered, against each of which is placed the quantity of work involved, the amount of labour, plant and materials in each, together with the estimated cost of each operation. Operations which can be carried out concurrently are noted.

These sheets together form a detailed analysis of the complete work and give information for all planning activities during the course of the contract. They provide a link between the overall programme and detailed work on the site and enable the site agent or general foreman to prepare the short term plans accurately. In addition, they provide the basis on which materials can be ordered and the correct amount of labour can be put on each operation. The sheets also provide definite operations against which operatives' time can be recorded for purposes of site bonus and costing procedures.

Sequence and timing of operations

In any section of work which contains two or more operations, one of the operations will govern the time required to complete the whole of the work. Similarly, in each stage into which a job may be divided, there will be one operation or a group of related operations governing the production time of the complete stage. This 'key operation' is the one which takes the longest time when the time cycles of all the operations are based on the use of the optimum size of gang for each. The longest of the key operations in each of the stages is termed the 'master operation' and fixes the rate of production for the whole job. The speed of the master operation is governed either by the time in which the work has to be completed, the size of the gang being fixed accordingly, or by the amount of labour available, in which case the size of the gang which can be put on it will fix the time required to complete the operation. In either case, it is essential to bring all other operations into phase with the master operation. This is necessary in order to ensure continuity of productive work for each trade or gang and to minimize unproductive time in preparation and clear-ing up at the beginning and end of each operation.

The time cycles of the operations in each stage are brought into phase by adjusting the size of the gangs so that the working time of each gang is, as far as possible, the same as that of the key operation. This avoids one trade being idle while another related trade completes its work[2]. Although some operations might be finished in a shorter period than the time cycle for the whole stage this would result in no difference in the overall building time, so that wherever possible gangs should be balanced as shown in figure 8 which illustrates the effect of the balancing of working times upon continuity of work and unproductive time in the erection of a pair of semi-detached houses. (*A*) indicates the result of haphazard selection of trade gang constitution, resulting in unbalanced working times. In (*B*) it can be seen that planned and phased production using balanced gangs results in balanced working times, a shortening of the complete time-cycle and a reduction in unproductive time.

However, it is not always possible to balance gangs because in some circumstances there may be physical limits to the size of gang which can be used on a particular operation, or in one trade there may be insufficient work to occupy even one man con-

[1] Work study is a tool of production management and is the name given to the study of work processes to find out if they are being done efficiently and, if not, to suggest means or alternative methods by which they may be carried out more efficiently. The process involves the examination of the way operations are performed, which is called *method* or *motion study*, and the time within which they are performed, which is called *time study*. Both of these studies are extensive but interdependent and are usually carried out concurrently by an executive trained in the technique of work study and called a 'Studyman' or 'Work Study Engineer'.

Although in normal building work a very large proportion of individual assembly operations are non-repetitive, in some of the trades there is considerable repetition in the work. In some cases 50 to 65 per cent of the work in the bricklayer and carpenter trades may be repetitive and some 40 per cent of work in concreting is repetitive. Work study, by establishing standard times and developing correct methods, gives considerable advantages in these spheres. In addition, methods for new work may be developed by this means and standards derived which are fair and which provide the same incentive to operatives to earn a bonus as the repetitive work.

For fuller details see *Work Study* by R.M. Currie (Pitman, London).

[2] In this respect see page 22 regarding the relationship between operations of plant and men.

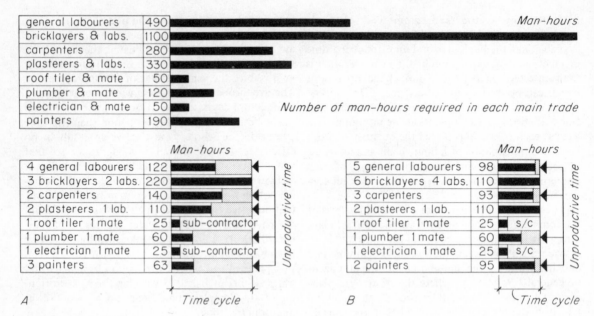

general labourers	490	Man-hours
bricklayers & labs.	1100	
carpenters	280	
plasterers & labs.	330	
roof tiler & mate	50	
plumber & mate	120	
electrician & mate	50	
painters	190	

Number of man-hours required in each main trade

Man-hours

4 general labourers	122	
3 bricklayers 2 labs.	220	
2 carpenters	140	
2 plasterers 1 lab.	110	
1 roof tiler 1 mate	25	sub-contractor
1 plumber 1 mate	60	
1 electrician 1 mate	25	sub-contractor
3 painters	63	

Man-hours

5 general labourers	98	
6 bricklayers 4 labs.	110	
3 carpenters	93	
2 plasterers 1 lab.	110	
1 roof tiler 1 mate	25	s/c
1 plumber 1 mate	60	
1 electrician 1 mate	25	s/c
2 painters	95	

Unproductive time

A   Time cycle   B   Time cycle

8 *Balancing of trade gangs*

tinuously throughout the complete time cycle. These operations, which must be carried out intermittently, are usually in the services installation and finishing trades which are quite often sub-contracted so that arrangements can be made for the work to be done at intervals within the main cycle of operations (see figure 8 *B*).

Phasing of work

Where the job is extensive or consists of a number of blocks, it is usual to phase the job as a whole by dividing it into a number of sections, each of which is planned on the lines indicated above and so related to the other sections that trade gangs can proceed from one to another in a continuous progression.

Each operation should commence as soon as possible without necessarily waiting until the whole of the preceding one is complete, and each should have the largest practicable gang on it. In each stage and in all phases every operation should continue without a break to completion and each gang should be able to work continuously until it can leave the site altogether. Maximum production results when each member of a balanced gang is continuously engaged on the same work. It has been shown[1] that in such circumstances a definite increase in production takes place as the contract proceeds, up to a certain point, after which it tends to fall off

slightly. In order to assist gangs special instruction is sometimes given on working methods by means of large scale or full-size mock-ups of parts of the structure, particularly when new systems are involved.

The programme chart

The final step is to prepare a working schedule on the basis of the balanced production in each stage, from which programmes for the various stages are drawn up. The stage programmes are combined to give the final overall programme based on the methods and plant to be used and on the balanced production of work. A short interval may be left between the stages to provide for delays due to bad weather or other causes. The extent of these intervals will usually be governed by seasonal conditions and local circumstances. If the job is extensive and allows some freedom for the redisposition of gangs in the event of delay at one point, no interval need be left between the stages.

The overall programme is usually expressed as a programme chart in the form of a Gantt or bar chart, on which the sub-division of work is shown on

[1] See *routine effect* in Part 1 and page 36 *National Building Studies Special Report* no. 29, 'Organization of Building Sites', HMSO.

horizontal lines and of time on vertical lines (figure 1). Sometimes this is called a progress chart, since it is a useful means of recording the progress of the work. This overall or working programme is intended only as an outline of the site operations as a whole. It cannot be detailed because so many unknown factors which may affect the operations make it essential for detailed programming to be carried out at regular, short intervals during the course of the work. This gives flexibility and allows for rapid revision should progress fall seriously behind the overall programme.

Such a chart may be quite simple, or complex. As a simple progress chart it will consist of a list of the basic operations with a bar opposite each to indicate the length of time the operation is planned to take and at what point relative to the other operations. When the job as a whole is phased, the bars are hatched in sections or lettered to indicate the work in each phase. Usually, the chart also indicates the dates on which orders for materials must be placed, the dates for the delivery of the various pieces of plant and the total number of men required each week, with or without a breakdown into trades[1]. A typical chart is shown on page 12.

In addition to the data sheets and the overall programme, a *schedule of contract information* is prepared giving the recommended labour force for each stage of the contract under trades, details regarding the sequence of operations given on the data sheets, and details of equipment and methods of construction to be used. This schedule will also include full details concerning all sub-contractors. A site layout plan and a site preparation programme (see page 17) will also be prepared at this stage, as well as the detailed programme for the first four week period of the contract (page 13).

Planning considerations

A number of factors which have a bearing on the decisions made during the contract planning stage are briefly considered here.

Site conditions and access

Site conditions will limit the type of plant that may be used. On wet sites it will be necessary to use tracked machines in the case of excavators and mobile cranes, and dumpers for transport. Sloping sites may make the use of rail mounted cranes un-

suitable or uneconomical. On confined sites there may be insufficient room for a mixer or mixing plant and it may be necessary to use truck mixed concrete. Limitations of access may fix the maximum size of plant which can be brought on the site for use on the job. A site closely surrounded by tall adjoining buildings may dictate the use of a derricking jib crane rather than a horizontal jib crane in order to be able to rise and clear the buildings as it turns from one position to another.

Nature of job

The type of structure and the general form, size and detailing of the building will all have an effect upon the way in which the contract is planned. Reference is made on page 16 to the significance of decisions made by the architect at the design stage. As far as the contractor is concerned, he must consider the nature of the structure in relation to the site so that he can decide where best to place his equipment and materials. It is desirable that all plant should be so placed on the site that the structure can be erected without moving the plant until most of it is completed. Plant should also be so placed that it can be removed easily at the completion of the job.

In some circumstances the contractor may request the adjustment of the structure in some way, in order to permit the most efficient planning of the contract. For example, it may be desirable to enlarge a lift shaft slightly in order that a climbing crane may be accommodated within it, or for certain parts designed originally as *in situ* cast work to be carried out as precast work, or vice versa, in order fully to utilize à crane on the job. In addition to conditions round the site, the height and width of the building will influence the choice of the type of crane. As shown in chapter 2, many mobile cranes have limitations of height and reach which make them unsuitable for buildings of three storeys and upwards.

Plant

The choice of the most suitable plant for any particular operation necessitates a consideration of the

[1] See *Building Research Station Digest* 91, 'The Programming of House-building', for further details of the technique of programming. See also *National Building Studies Special Report* No. 29, 'Organization of Building Sites', HMSO.

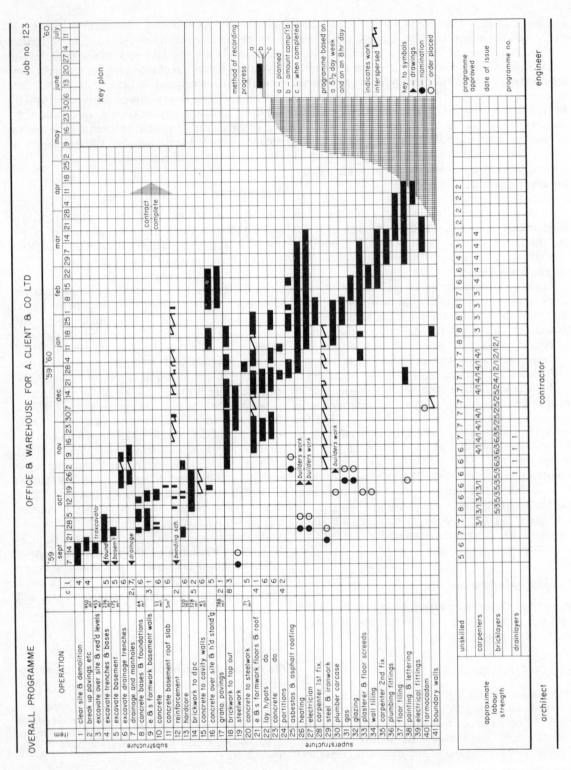

contract: A. Client & Co. Ltd.
job no. 123
period 7·9·59 to 5·10·59

MONTHLY PERIOD PROGRAMME

method of recording progress
a: planned
b: amt. completed
c: when completed

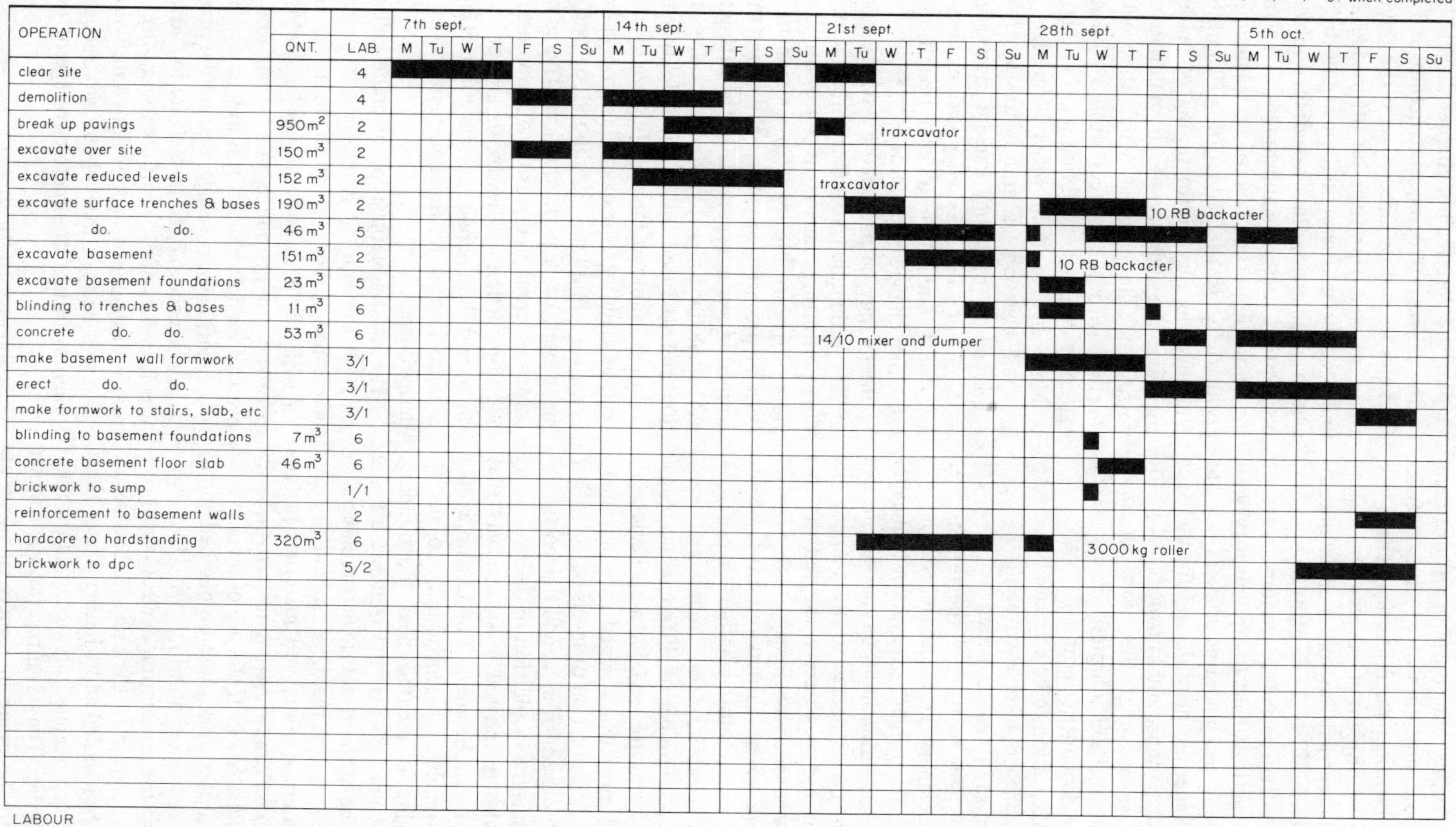
OPERATION
QNT.
LAB.
7th sept.
14th sept.
21st sept.
28th sept.
5th oct.
M Tu W T F S Su
clear site 4
demolition 4
break up pavings 950m² 2
excavate over site 150 m³ 2
excavate reduced levels 152 m³ 2
excavate surface trenches & bases 190 m³ 2
do. do. 46 m³ 5
excavate basement 151 m³ 2
excavate basement foundations 23 m³ 5
blinding to trenches & bases 11 m³ 6
concrete do. do. 53 m³ 6
make basement wall formwork 3/1
erect do. do. 3/1
make formwork to stairs, slab, etc 3/1
blinding to basement foundations 7 m³ 6
concrete basement floor slab 46 m³ 6
brickwork to sump 1/1
reinforcement to basement walls 2
hardcore to hardstanding 320m³ 6
brickwork to dpc 5/2
traxcavator
traxcavator
10 RB backacter
10 RB backacter
14/10 mixer and dumper
3000 kg roller

LABOUR	7th sept.							14th sept.							21st sept.							28th sept.							5th oct.						
	M	Tu	W	T	F	S	Su	M	Tu	W	T	F	S	Su	M	Tu	W	T	F	S	Su	M	Tu	W	T	F	S	Su	M	Tu	W	T	F	S	Su
unskilled	4	4	4	4	4	4		5	5	5	5	5	5		6	6	6	6	6	6		6	6	6	6	6	6		6	6	6	6	6	6	
carpenters																						3/1	3/1	3/1	3/1	3/1	3/1		3/1	3/1	3/1	3/1	3/1	3/1	
bricklayers																								1/1							5/2	5/2	5/2	5/2	
drivers					1	1		1	1	1	1	1	1		1	1	1	1	1	1		1	1	1	1										
steelfixers																																	2	2	

capabilities, limitations and outputs of different types of plant, some indication of which is given in chapter 2. The following paragraphs deal with a few aspects of the choice of plant at the planning stage[1].

Excavation can be carried out either mechanically by a number of different types of plant or by hand, the spoil can be transported in various types of vehicle and the length of haul to tip will vary with the job, so that many combinations of excavator and transporting machines are possible and the contract planner, from his experience and knowledge of plant, must in each case arrive at the most economic combination. The method adopted for excavating operations will be dependent upon

(a) the type of excavation to be carried out
(b) the nature of the soil to be excavated
(c) the volume of soil to be excavated
(d) the length of haul to tip and the terrain over which the machinery has to dig and travel.

For small quantities, hand excavation is cheaper than mechanical excavation and the type of transport will depend on the distance to be hauled, the nature of the ground to be traversed and the cost of temporary roads, where necessary.

The total work to be carried out must be reviewed in order to establish whether or not it is possible to use one machine for a number of operations rather than a number of different machines. For example, an excavator rigged as a backacter will dig trenches and also, when rigged as a skimmer, could carry out the reduction of levels on the site if these were not too great in area, and thus avoid the use of another machine in addition to the excavator.

Work must also be phased in such a way that mechanical plant can be used. For example, if a run of drain trench sufficiently long to justify mechanical digging is situated near the building, work must be planned so that the trenches are dug before the building of the structure is commenced in order to provide room for the digger to work.

Handling of structural units and materials in fabrication and erection can be carried out satisfactorily by crane or forklift truck. If the use of a crane is to prove economical the work must be planned round the crane, the influence of which will, to a large extent, determine the production cycle. Careful consideration must be given to the quantity and nature of materials to be handled and whether or not there is sufficient to keep a crane fully occupied throughout the working day. The advantages of the forklift truck relative to the crane in respect of materials

handling are described in chapter 2.

The delivery of incoming structural and fabricated elements should be phased with the building operations so that they can be off-loaded and placed immediately in their final positions by the crane wherever possible, thus avoiding double handling. In addition to the establishment of balanced gangs the most important factor in planning for high productivity is the reduction of double-handling. This involves the careful timing of materials deliveries and the delivery of all materials as near as possible to the point at which they will be used, together with the correct siting of hoisting plant, materials dumps and mixing plant in relation to the building and to each other. Materials should be grouped near cranes and hoists where these are employed so that they can be moved in order of requirement, and in such positions that the crane can hoist and place in one operation with the minimum change of position.

Bricks may be packaged by straps into multiples of fifty, or handled, together with blocks, on pallets, thus permitting the use of a forklift truck or providing a reasonable load for a crane to hoist. The sizes of precast elements or shuttering units should be related as far as possible to the lifting capacity of the crane in order to avoid excessive numbers of lifts and to speed up erection. In this respect the architect is able to consider the size of precast units, such as claddings for example, on broad lines only; bearing in mind the likelihood or not of a crane being used on a particular job. This is because the choice of crane depends on other considerations in addition to that of the size of any precast concrete units, and these, under normal contracting methods, become known to the contractor only after the design stage.

The most suitable type of crane will depend not only on the work it is required to perform but, as mentioned earlier, also on the nature of the building on which it is to be used. On large sites it is often necessary to introduce more than one crane and even on a single block if it is long, a single crane may give insufficient coverage. During the planning stage, it may be necessary to investigate the advantages and disadvantages of different combinations of cranes by means of diagrams such as shown in figure 9.

It will be seen that when two cranes are used in order to obtain complete coverage of the building,

[1] See also page 27, Part 1.

14

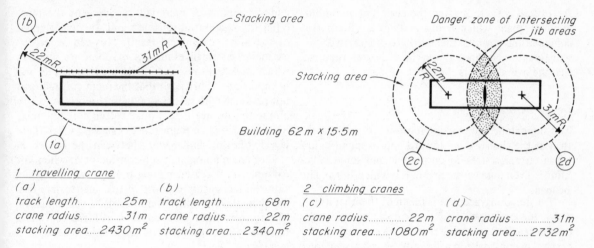

1 travelling crane

(a) | (b)
track length............25m | track length.............68m
crane radius............31m | crane radius............22m
stacking area....2430 m² | stacking area....2340 m²

2 climbing cranes

(c) | (d)
crane radius............22m | crane radius...........31m
stacking area......1080 m² | stacking area.....2732 m²

9 Crane coverage

the arcs of the jibs must intercept and adequate precautions on the job must be taken to avoid collision of the jibs and hoist ropes as the cranes slew. There is less likelihood of such a collision when derricking jib cranes are used, and when horizontal jib cranes are employed the booms should be set at different levels.

It will also be seen from these diagrams that as well as the amount of coverage given to the building, the stacking area for materials covered by the arc of the jibs is a significant factor, In 1a, with a short track length and long jib, the stacking area is mainly on one side. With the shorter jib but longer track length in 1b, stacking area is provided on one side and at the ends of the building. 2d, with the longer jibs, provides a greater area than the four small areas of 2c but results in a very large intersecting jib area.

Mixing Type and size of concrete mixer are dictated to a large extent by the quality and quantity of concrete required. The suitability of the various types of mixers in these respects is described in chapter 2. To obtain highest efficiency, the concreting equipment must be carefully combined according to the kind of work to be done. When small to medium quantities, say up to 23 m³ per day are required, a mixer together with hand loading of the aggregate skip, some form of weigh batching and hand barrow delivery can be economical, although scraper loading, being so cheap, is generally used except for very small quantities. If mechancial del-

ivery is used, then the labour requirement is reduced. When steady outputs of not less than 30 m³ per day are required, complete mechanization is best and this would involve a mechanical scoop or gravity loading of the mixer skip, gravity fed bulk cement and, for delivery, a crane carrying a full batch skip or, alternatively, a pneumatic concrete placer.

When a job incorporates some *in situ* concrete work and some precast work, it is sometimes possible to arrange site precasting to occupy the idle time of a mixer and its associated handling equipment.

The decision on whether or not to set up a central mixing plant will depend on the amount of concrete to be produced, the relative costs of setting up a central plant and a number of individual mixers, and the relative costs of mixing by the two methods. In addition, the cost of transport from the central mixer to the various points of placing will also affect the decision.

Type of plant In considering alternatives for the same operation, it must be borne in mind that the cost of mechanical operations is influenced by the nature of the particular job, the sequence of work dictated by the design of the structure and the methods used for erection, and by the amount of work to be done. The efficiency of any mechanized system cannot, therefore, be judged solely on the cost of equipment, but on the influence the system has upon all the related operations and thus upon

15

the overall cost. It is often the case that a combination of plant which is dearer than an alternative combination permits the work to be carried out with greater continuity and in a shorter time, so that the overall cost is cheaper[1].

Design factors

The importance of a contribution from the architect at the design stage toward improvement in job organization is stressed on page 1, and some indication is given here of the manner in which this might be made.

If, in designing, account is taken of the operations which the craftsmen must perform in carrying out the work, and unnecessary labours are avoided, greater speed in construction will result. Simplicity of construction and detailing leads to economies by enabling work to proceed quickly, thus reducing the contract period. Site operations can be simplified in many ways such as by reducing variations in the widths of foundations so that changes in trench-digger shovels are kept to a minimum; by maintaining floor slabs at a uniform level and allowing for variations in finishes by different thicknesses of screed rather than by variations in the floor slab; by designing openings and lengths of wall to brick dimensions in order to avoid the cutting of bricks. By avoiding breaks and returns in walls as far as possible, in order to simplify brick and block work, or shuttering in the case of concrete work, and to keep foundation runs straight to simplify digging.

Interference and delay arise on the job when more than one trade has to work on one item at the same time, or one trade has to wait on another before completing work it has partly finished. This can be avoided by the separation of trades at the design stage[2]. For example, by designing the brick walls to a single-storey building as panels running from floor slab to roof, with no openings in them, and the doors and windows as units between them running from floor to roof with no brickwork over, the bricklayer and carpenter can carry out their operations independently. If the roof is of timber construction, since no concrete lintels are required over openings, the concretor does not have to wait after having formed the floor slab in order to follow the carpenter and place the lintels. Finishing work is also simplified as the plasterer has only to work on plain surfaces.

Detailing which results in the division of work

and the mixing of materials can prevent the most efficient organisation of a job and the achievement of the most economical result. This can occur, for example, on small contracts of domestic work in which cross-walls in some blocks are of concrete and in others are of brick, so that the main contractor is not provided with sufficient concrete work to justify him setting up an efficient mechanized concrete mixing plant. Similar circumstances can arise when there is a division of similar work between the main contractor and a nominated sub-contractor or when, for example, in precast concrete multi-storey construction the intermediate floors of maisonettes are designed in timber rather than in reinforced concrete. The former, although cheaper as a 'measured' item, disturbs the continuity of work and results in a greater total cost[3].

The absence of continuity of work is an important factor which reduces the value of the mechanization of most building operations because so much time is spent on preparatory work between operations rather than on productive work. Continuity is more likely if the operations are simple and repetitive. This should be dealt with at the design stage by detailing which results in (i) as few operations as possible for each aspect of the work, and (ii) the separation of the work of fabrication from that of erection, since the problems of each are so different that any mechanization must be independent[4]. The detailing of elements which must be moved into position should be such that the loads to be handled closely approach the capacity of the plant likely to be used (see also page 23).

SITE ORGANIZATION

Site planning

As described earlier, a programme covering operations during the first four weeks will have been drawn up at planning stage, in the preparation of which, where possible, the general foreman or agent respon-

---

[1] See *The Structural Engineer*, February 1958: 'The Problem of Mechanical Handling in Building Operations' by J.F. Eden and D. Bishop, where this is discussed and illustrated by comparative priced programmes.

[2] See also *Separation of Work*, Part 1.

[3] See also Part 1, page 26 on the use of prefabricated components in otherwise traditional construction.

[4] See also Part 1, page 26, on continuity of work.

sible for the contract will have assisted, so that he is in agreement with the proposals laid down. This programme will be generally in two parts:

(a) Site preparation programme, which will cover the demolishing of any existing buildings, the setting out of the site and marking out of storage areas, the erection of huts and the construction of temporary access roads where necessary

(b) Period 1 programme, on the lines of that shown on page 13, which will cover work during the first four weeks or so of the contract.

Together with these will be provided site layout plans to show (a) traffic routes, on which will be indicated any areas requiring particular attention, such as levelling-off or covering with temporary Summerfield track, and any direction signs required; (b) the location of offices, huts and stores; (c) the position of bulk storage areas both during and after excavation together with the location of any equipment.

The general foreman or agent will in addition also be provided with copies of the overall programme, schedule of contract information and data sheets as well as all other necessary documents such as bills of quantities, specification, set of contract drawings, details of all material orders placed and to be placed at various dates during the contract, and finally details of the type and quantity of equipment to be used and the approximate periods when they will be required on the site.

Period planning

Work on the site will commence on the basis of the first monthly programme and during the third week of this period, and all subsequent stages, the next monthly programme will be prepared on the basis of the overall plan and data sheets. In the preparation of the monthly plan consideration must be given to the labour force desirable and practicable in the circumstances at the time, to plant requirements and availability, to the phasing and overlapping of operations to ensure completion of the work in the minimum time and to the planning of labour to maintain group identities. Steps must be taken to give adequate warning to all sub-contractors when they will be required on site.

Weekly planning

Towards the end of each week progress will be reviewed and the next week's planned progress con-

firmed or modified if necessary. The following week's planned labour requirements will be reviewed and an estimate made of materials required for the next week but one and of any action required to be taken regarding equipment. This weekly review will be prepared by the general foreman in consultation with his trade foremen and any sub-contractors' foremen, and a written report will be submitted to the contractor's planning department.

In certain cases where close integration of fully mechanized operations is required over a short period, particularly in the case of reinforced concrete structures, a weekly programme would be drawn up in chart form by the planning department. Such a chart is illustrated on page 19. In addition to weekly planning, the general foreman will hold a brief meeting each day with his trade foremen and sub-contractor foremen to review the next day's work and to make the necessary preparations in regard to the placing of materials and equipment in readiness for the next day's operations.

The general foreman will, at the beginning of the job whenever possible, indicate to the local employment office his anticipated 'build up' of labour force during the course of the contract.

Progress control

Good site planning is a prior necessity to smooth and effective progress in construction work, but a regular review of the progress of all operations and its comparison with the programme or plan is essential.

Progress is maintained by the foreman, or on larger jobs by a progress engineer, by the proper organization of the delivery and placing of materials, by ensuring that all equipment and plant is in its correct position at the right time, and by adjusting the size of labour gangs when progress is likely to fall behind the programme because of unforeseen circumstances. Progress is checked during weekly planning by estimating or measuring the work completed, the percentage of each operation or group of operations completed being established and compared with the programme. Progress is marked on the charts as indicated in figure 1 and on page 12. When progress varies appreciably from the overall programme and where for this and any other reason it is considered desirable to alter the planned sequence of operations, the general foreman would consult the planning department before making such changes.

Close co-operation between the site staff and the planning department is often maintained by means of regular and formal site production meetings between the general foreman and the planning engineer responsible for the job. When considering any changes, the effect on the supply of materials must be borne in mind, and when progress is faster than planned, the supply of materials in time for the work becomes the predominating factor. All sub-contractors must be notified immediately of any changes in the planned programme of work.

The general foreman should maintain a record of current and planned labour strength in the form of a schedule or chart on which the following week's planned labour requirements will be entered during weekly planning. In addition, all incoming material will be recorded on a form, one for each main item, which should show amongst other information dates of order and receipt, quantity delivered and the balance of material outstanding.

As an aid to progress control on a job of any size, in addition to the site production meetings mentioned above, other regular site meetings should be held at which should be present the contract manager, site agent or general foreman, architect, clerk of works, quantity surveyor and any sub-contractors when necessary. At these meetings all aspects of the job requiring attention are discussed and decisions for future action made. To ensure that maximum benefit is obtained from such a site meeting, an agenda should be prepared and circulated some days prior to the meeting and minutes should be prepared and circulated as soon as possible after the meeting.

Site layout

The need for an efficient layout of site has been referred to on page 26 of Part 1.

The layout of every site may be divided into an administrative area and a construction area. In the former will be located stores, offices, sub-contractors' huts and canteen and similar accommodation if this is provided, and in the latter, which will be the actual site of the buildings being constructed, will be located consumable stores adjacent to the various buildings and all equipment required for construction purposes. The layout of both these areas forms an essential part of early planning in every contract, the neglect of which will lead to delay in the initial progress of the job, the tying up of more capital than necessary in materials and financial loss on the part of the contractor.

Proper access and departure routes for lorries should be provided and these should be clearly signposted. In determining the traffic routes attention must be paid to the position of all main services, such as water, gas and electricity, and to drains and excavations. Temporary roads must be positioned with sufficient distance between them and future buildings to allow for the movement or positioning of all mechanical plant.

The *administrative area* should be located to give quick access to that area of the site which will require maximum labour control and the main storage area, sub-contractors' huts and canteen, should be so located that accessibility for unloading materials is good and so that they are a minimum distance from the construction areas. The site office should be sited on the route into the administrative area and with as good a view as possible of the construction areas. The size of the site office will vary with the accommodation to be provided, which will depend in turn upon the size of the contract. It should provide accommodation for all or any of the following: agent, general foreman, quantity surveyor; timekeepers, bonus surveyor, checker, clerk of works, resident engineer, small lock-up store and conference room.

All contracts of any size in the present day require adequate telephone facilities for communication, electricity for power, compressed air for equipment, and lighting and heating facilities for office huts. At a very early point in the planning stage the necessary arrangements for these services will be put in hand, particularly requests for telephone facilities and electricity.

The stores area should be situated near the site office and will consist of covered huts for valuable or non-weatherproof stores, such as paint and ironmongery, and a locked pen for larger valuable stores which are weatherproof, such as metal window frames and pipes. Areas for sub-contractors' stores will be located near the sub-contractors' huts and sometimes they will be situated within the main stores area.

Where possible the moving of the administrative area to another part of the site during the course of a contract should be avoided by careful initial site planning, but where this is essential because of the nature of the site or the job, a further site layout

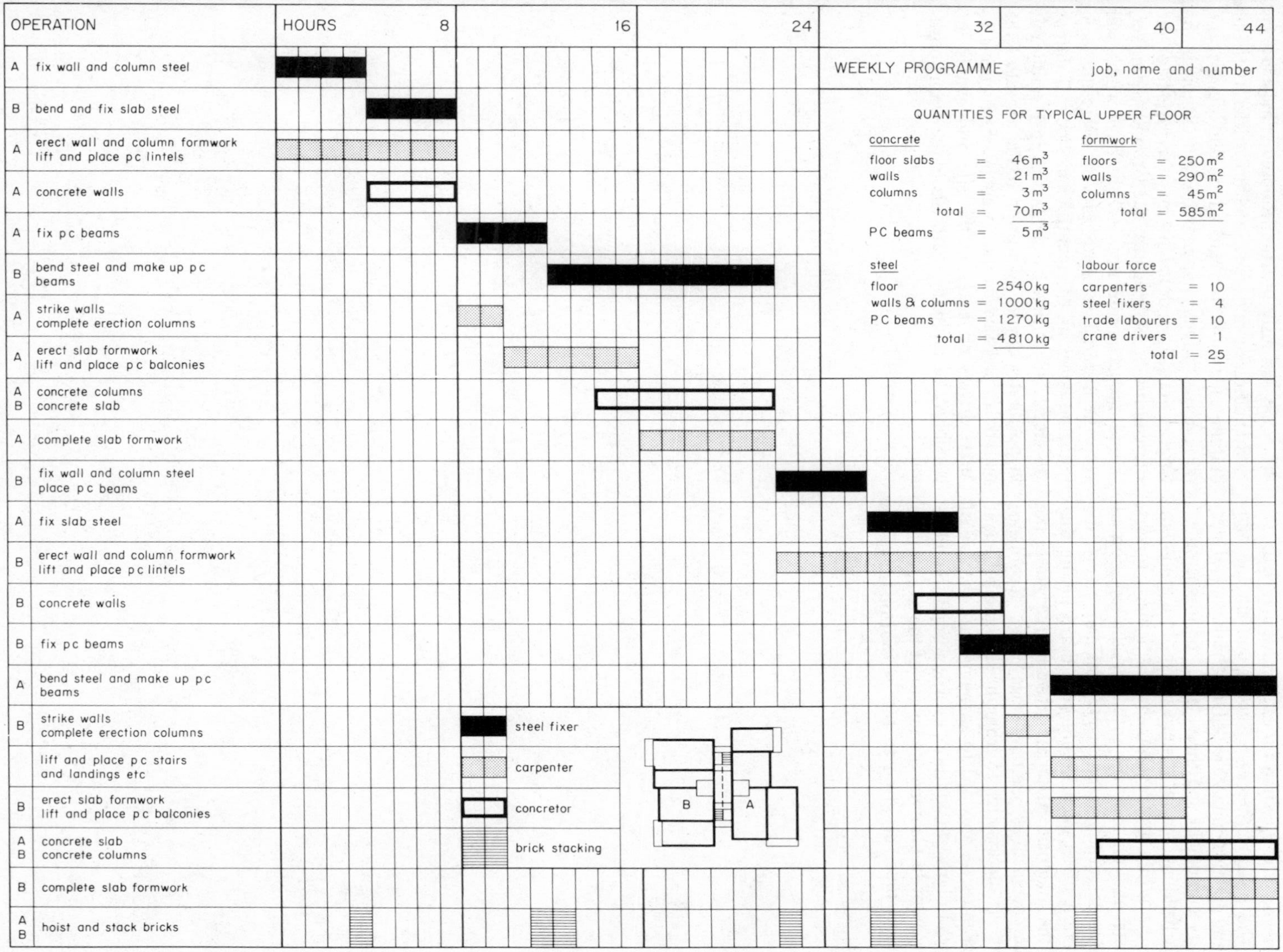

WEEKLY PROGRAMME

job, name and number

QUANTITIES FOR TYPICAL UPPER FLOOR

concrete			formwork		
floor slabs	=	46 m³	floors	=	250 m²
walls	=	21 m³	walls	=	290 m²
columns	=	3 m³	columns	=	45 m²
total	=	70 m³	total	=	585 m²
PC beams	=	5 m³			

steel			labour force		
floor	=	2540 kg	carpenters	=	10
walls & columns	=	1000 kg	steel fixers	=	4
PC beams	=	1270 kg	trade labourers	=	10
total	=	4810 kg	crane drivers	=	1
			total	=	25

	HOURS	8	16	24	32	40	44

OPERATION	
A	fix wall and column steel
B	bend and fix slab steel
A	erect wall and column formwork lift and place pc lintels
A	concrete walls
A	fix pc beams
B	bend steel and make up pc beams
A	strike walls complete erection columns
A	erect slab formwork lift and place pc balconies
A	concrete columns
B	concrete slab
A	complete slab formwork
B	fix wall and column steel place pc beams
A	fix slab steel
B	erect wall and column formwork lift and place pc lintels
B	concrete walls
B	fix pc beams
A	bend steel and make up pc beams
B	strike walls complete erection columns
	lift and place pc stairs and landings etc
B	erect slab formwork lift and place pc balconies
A	concrete slab
B	concrete columns
B	complete slab formwork
A B	hoist and stack bricks

steel fixer

carpenter

concretor

brick stacking

19

plan would be prepared for each move involved.

The *construction area* should contain the minimum practical quantities of materials and of necessary equipment and these should be so positioned that handling and movement is kept to a minimum. As the position of equipment, particularly mixers, hoists and cranes, will influence the position of materials such as sand, aggregates and bricks, the position of all plant should be planned before that of the materials. Materials arrive on the site in the order decided at planning stage, or in accordance with instructions issued from the site, and sufficient area must be provided to accommodate the size of batch ordered. In addition, overflow areas should be allocated. In planning the layout of the site, consideration must be given to the excavation stages as these may seriously restrict proposed storage areas.

Standardized materials, such as bricks, tiles and drainpipes, should be stacked in unit dumps, the numbers in which remain constant although the length, breadth and height may be varied to suit site conditions. Aggregates, sited round the mixer, should be accommodated in light bays which will keep the materials separate, prevent waste and facilitate the checking of quantities in stock. If dumps are clearly identified by signboards, preferably put in position a day or two prior to the arrival of the materials, this saves time in giving instructions to lorry drivers on entering the site and avoids the possibility of dumping materials in the wrong position.

2 Contractors' mechanical plant

In its widest sense 'contractors' plant' implies the machinery, tools (other than craftsmen's personal tools) and other equipment used in the contractor's yard and workshop, and on the site. In this chapter contractors' mechanical plant and power tools only will be discussed, in particular those used on the building site. Machinery used solely in the workshop, such as woodworking machinery,[1] will not be considered nor will contractors' general equipment such as ladders, lifting tackle and wheelbarrows. Scaffolding, the erection of which often presents minor structural problems, is discussed in chapter 11.

The machines and power tools which are the subject of this chapter are divided into three classes according to their degree of mobility: (i) fixed, (ii) portable, (iii) mobile. The first group includes machines which operate from a fixed position on the site, the second group, machines and tools which can be moved about by pulling, pushing or carrying by hand, and the third group, those which can move from one place to another under their own power. They may further be divided into classes according to their function and are later discussed on this basis.

THE MECHANIZATION OF BUILDING OPERATIONS

As shown in chapter 2 of Part 1 mechanization is one aspect of the industrialisation process and is one of the rationalising methods by means of which this process achieves increased productivity in building operations.

Modern off-site production of most building materials and components is generally highly mechanised; the mechanisation of site operations has been far less general and for this there are a number of reasons, some of which have been referred to in Part 1. Compared with other industries the problem of applying machines to operations on the building site is far more complicated. Building work is less repetitive, is less continuous and involves the movement of plant from one place to another as one job is completed and the next commenced. Moving a machine and setting it up can be expensive and in the case of the heavier machines the cost of this can form a large proportion of the total cost of the operations they perform. In addition, most jobs present different problems and site conditions vary. Furthermore, as pointed out in Part 1, the application of mechanised plant to a construction process which is primarily one of assembly is limited. In relation to this perhaps the greatest hindrance lies in the fact that the greater majority of building designs, building techniques and sequences of operations on the site have been based on manual methods. Unless a building is so designed and the contract work so organised that machines can be operated for continuous periods at full capacity their use will not be economic. The operations to which mechanisation can most readily be applied are those related to earth-moving and excavating and to the structure of the building, such as the mixing and transport of concrete, and to the handling of materials.

Nevertheless for reasons given under the section on *The Industrialisation of Building* in Part 1 there has been in Great Britain over recent years a marked trend towards a greater degree of mechanisation on site. The introduction of suitable mechanical aids, by replacing unskilled labour in particular, can result in reduced labour costs and, by increasing the speed of construction, can result in earlier completion, enabling the building owner to occupy the building and recover his capital outlay at an earlier date. Mechanical plant may also be introduced to carry out operations for which manual labour is not available or is in short supply or to carry out operations which cannot be done either economically or physically by manual labour. Its introduction can reduce the effort required to be made by the operative in carrying out his work and can improve his working conditions.

A wide range of mechanical aids is now available many of which are the outcome of work done over the years by the Building Research Station, especially those designed for use on housing and other small scale work. The most marked development is

---

[1] For this see *Mitchells Building Construction: Components and Finishes*, chapter 2

probably to be seen in the use of the crane, originally employed in this country primarily for erection and fixing and now widely used for the handling and movement of materials. This became possible because of the introduction of tall, long jib cranes from the continent, where they had long been used for this purpose, followed by the development and production of similar cranes in Britain.

The high cost of the more expensive plant can be justified only if the plant is kept in more or less continuous use. Many firms, especially the smaller contractors, experience difficulty in maintaining the necessary sequence of operations over a period long enough to justify the high initial cost of the plant. There are now, however, a considerable number of specialist firms with the necessary equipment who can be hired to carry out a particular operation, as well as firms who only stock plant for hire, so that with adequate and careful planning it is possible for even the smaller contractor to mechanise those operations which can thereby be performed more cheaply.

Planning for the use of mechanical plant

The efficient employment of mechanical plant depends on a number of factors which must be given careful consideration at the outset of each job. Haphazard mechanization will not necessarily either reduce costs or reduce the total construction time. Careful planning of the work throughout is essential.

Nature of job and site

While it is true that on large contracts, provided the work is satisfactorily planned, mechanisation will usually be advantageous, this is not necessarily so on smaller contracts. The latter are likely to be carried out by traditional building methods in which a large proportion of the work is skilled craftsmen's work[1] which is difficult to mechanise. Apart from this the problem of introducing mechanical plant into traditional building work arises from the fact that machines are relatively highly specialised whereas the operatives they replace are extremely versatile and can readily transfer from one type of operation to another within their competence with little loss of productivity[2]. A large contract normally has sufficient work of various types to justify the introduction of specialised machines and to enable them to be used economically. However, as indicated in

Part 1 (page 27), there is now available a number of small pieces of plant designed primarily for the small builder which make it economic for him to employ them on many contracts. On the remainder other means of rationalization alone may be more effective, such as careful programming of the work, flexible methods of working, efficiently planned site organization and the use of production aids to normal manual methods. These include ready dry-mixed mortars, building jigs and frame scaffolding.

Site considerations have a considerable effect upon the choice of mechanical plant and site conditions must be suitable for its safe, efficient and economic use. Some of the limitations of site conditions in this respect are mentioned on page 11.

Relationship between operations of plant and of men

It is essential, particularly in mechanical handling, that the number of men working on any operation should be correctly related to the output of the mechanical plant serving them. This is necessary in order to avoid the plant being idle from time to time while the men use the material already delivered to them. Concreting gangs, for example, must be related in size and number to the size of concrete mixer used so that each load of concrete can be received and placed by the time the next load is ready for delivery. The number of men who can work efficiently on any one site, is of course, limited by the size of the job, the nature of the structure and other considerations, so that this will set a limit to the size of mechanical plant capable of being used to advantage (see also page 9).

Careful planning and programming of the contract as a whole

This is essential in order that the plant, which is expensive to hire or to purchase and maintain, is occupied to the maximum extent while on the site. As indicated under *Post-traditional building* in chapter 2 of Part 1, the introduction of expensive plant does, in fact, result in an all-round improvement in planning and organization arising from the need to plan the sequence of operations in a rational

---

[1] See Table 2, page 20, Part 1
[2] See also Part 1, pages 21 and 27 on this topic

way so that the cost of the machines may be spread over the maximum amount of work. This in itself results in greater overall productivity.

Ideally, the sequence of all operations throughout the job, whether mechanised or manual, should be so arranged that no plant on the site is ever idle. This need for continuity of work, which applies to men as well as machines, is stressed in Part 1[1] and has already been referred to on page 16 of this volume in relation to the effect of the architect's design details. In respect of the plant itself this requires careful consideration in the choice of the most appropriate plant for each particular set of operations to ensure that it will always be working to its full capacity, yet neither too quickly nor too slowly for the manual work to which it is related (see pages 9 and 11).

Suitability of the design of the building

As pointed out in Part 1[2] the design of a building determines the work to be done in its erection and, to a large extent, the sequence in which that work is carried out. If, therefore, full benefit is to be derived from the use of mechanical plant all aspects of mechanization and building methods must be considered at the design stage, so that when the contractor takes over the work can be arranged in a manner suited to mechanical rather than to wholly manual operations. It is essential that the architect should be aware of the advantages of mechanized methods, of the nature and capabilities of various types of mechanical plant and of the factors which are likely to make their use of greatest value. These include continuity of operations and the use of plant at maximum capacity every time it is operated. For example, precast concrete cladding panels can be designed as large or small units. A crane will be more economically employed in hoisting them if the panels are designed as large units approaching in weight the maximum lifting capacity of the crane likely to be used on the job, rather than if they are smaller in size and weight. The cost of hoisting one large or one small panel will be much the same so that the relative cost of hoisting a number of the small panels equivalent in area to one of the larger panels will be high.

Simplification of design, particularly in constructional detailing, is necessary to assist the mechanization of building, and must become a normal process in architectural practice if greater advantage is to be

taken of plant commonly used and if the development of new plant is to be encouraged (see also page 16).

MECHANICAL PLANT

In addition to the classification according to mobility already given on page 21, most mechanical plant may be classified according to function and will be described under the following headings:

Excavating – Hoisting – Transporting – Mixing.

Excavating

The process of excavating covers both surface and deep excavation of soil and often involves the movement of the excavated soil by the excavating equipment itself.

Excavating plant, apart from a few special machines such as trenchers, may be broadly divided into two types (i) equipment based on an *excavator*, which is a tracked or wheeled self-propelled machine consisting of a chassis carrying a revolving platform and a power-operated jib or boom controlled by wire ropes, together with a driver's cabin, and which is designed to be used for a number of different excavating operations by changing the booms and buckets, and (ii) that based on a *tractor*, either tracked or wheeled.

The nature of the excavation to be performed, the type of soil to be excavated, the distances which the excavated soil must be carried to transport and the condition and gradients of the site are among the factors which must be borne in mind in selecting the equipment for any particular job. Excavating plant cannot have much influence on architectural design, but the machines are expensive if kept idle or are only partly worked, so that careful planning of the work relative to the job and the plant to be used is essential especially in view of the fact that the excavation/foundation stage of construction can be such a significant factor in a job as a whole that it can result in either the financial success or failure of a contract.

---

[1] See chapter 2, pages 26 and 46
[2] See chapter 1, page 13

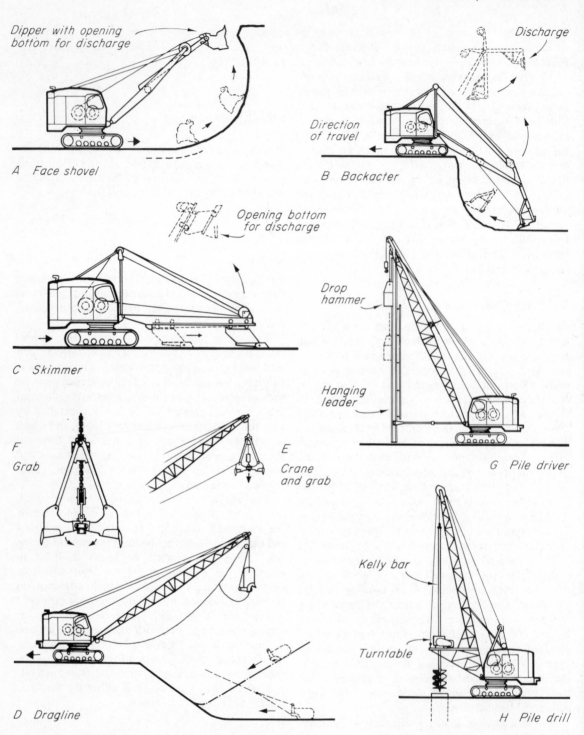

Dipper with opening bottom for discharge

A Face shovel

Discharge

Direction of travel

B Backacter

Opening bottom for discharge

C Skimmer

Drop hammer

Hanging leader

G Pile driver

F Grab

E Crane and grab

Kelly bar

Turntable

D Dragline

H Pile drill

10 Excavator equipment

24

Excavators

An excavator may be rigged as a face shovel, backacter, skimmer, dragline or crane and grab as shown in figure 10, all of which can be used to load the spoil into transporting equipment. An excavator crane may be used for pile driving and may also be rigged for drilling cylinder piles.

Excavators are generally fitted with crawler tracks rather than pneumatic tyred wheels as the former, although slower in travel, spread the load and enable the machine to work in comparatively bad conditions.

Face shovel Sometimes called a mechanical shovel or navvy. This is a widely used machine for excavating against a face or bank, consisting of an open-top bucket or dipper with a bottom opening door, fixed to an arm or dipper stick which slides and pivots on the jib of the crane (figure 10 *A*). The bucket, which has renewable cutting teeth on its cutting edge, is filled by driving it forward and upward into the material to be excavated and is emptied by placing it over the required point of discharge and opening the bottom door. The working of the dipper stick and bucket and the operation of the crane are all under the control of the one operator.

Face shovels excavate from just below ground level to heights of up to 6.10 m above ground level depending on the machine size. A face shovel is able to discharge at any point within its radius as it rotates through 360 degrees.

This type of machine is suitable for excavating all clay, chalk and friable materials and for handling rock and stone which has been loosened previously by other methods. It is particularly suitable for digging into a face such as a bank and for excavation in large quantities where the excavated material is to be transported and loaded into lorries or dumpers. It is the best piece of equipment for loading broken rock and also for loading gravel and similar loose excavated material. It is not suitable for surface excavation for which a skimmer is used.

Backacter Sometimes called a drag shovel, back shovel, or trench hoe. This is similar to the face shovel except that the dipper stick pivots on the end of the jib and the dipper or bucket works towards the chassis and normally has no bottom door but is emptied by swinging away from the chassis to invert the bucket (figure 10 *B*). It

excavates from ground level or slightly above, to a depth of about 6 m, depending on the size of the machine. Large backacters have a reach of about 7.6 m in front of the machine.

The backacter is mainly used to excavate trenches. For this purpose it is not so fast as an endless bucket trencher nor does it excavate such a clean trench as the movement of the bucket tends to roughen the sides of the trench; it can however deposit the spoil at a greater distance than the trencher. As the machine excavates it moves backwards away from the excavation so that the crawler tracks do not force in the sides of the newly excavated trench. Working from the side of the area to be excavated the backacter is occasionally used for the excavation of open areas such as small basements and loads the spoil into lorries or dumpers for transport.

Skimmer This arrangement is similar to the face shovel except that in this case the bucket slides on rollers directly along the jib and thus has a more restricted movement (figure 10 *C*). It cuts to a depth of 230 to 300 mm and is used for surface excavation and levelling in conjunction with transport to haul away the excavated material. The surface of the ground left after the skimming operation requires some trimming to be carried out by hand. It may also be used for loading loose excavated material and materials such as sand and gravel: when the bucket reaches the end of the jib, the latter is raised, the bucket turned over the transport and the spoil tipped through its bottom opening door.

Dragline For this purpose the machine is usually fitted with a long slender boom or jib and the bucket, which in operation faces towards the machine and has no door, is supported by cable only as on a crane (figure 10 *D*). The bucket is 'thrown' out by slewing and releasing the supporting cable like a fishing line and is then filled by being dragged towards the machine by another cable. This cable is so arranged over pulleys at the bucket end that the latter may be kept in such a position when hoisted and the jib raised that it contains the spoil until it is over the transport, where it is then inverted in order to empty it. The dragline excavates to a depth of 6.10 to 9.10 m with a reach of 15.20 to 21.30 m depending on the length and angle of the jib.

25

A dragline works from the side of the excavation at normal ground level and is used for excavating large open excavations such as basements when the depth is beyond the limit of the boom of a backacter; it loads into lorries or dumpers for removal of the spoil. If the total depth of the excavation is greater than that which may be safely carried out without the use of side timbering, the dragline would be used up to the point where timbering becomes necessary after which other methods would be required depending on the job in hand. In the case of extensive excavations of this nature ramps would be formed in order to introduce face and drag shovels and skimmers at the bottom of the excavation. Care must then be given to the maintenance of the ramps until these machines have been removed.

Crane and grab   Alternatively called a grabbing crane or clamshell. The grab, of which there is a large variety designed for different purposes and soils, consists of two hinged half-buckets or jaws pivoted to a frame which is suspended by cable from a long jib of an excavator, such as used for the dragline, or from a crane (figure 10 *E*). Grabs vary in weight as digging capacity depends on this; for hard soils heavy grabs with long sharp tines are used. The excavating action is vertical, the grab being dropped in an open position into the soil to be excavated and closing under the soil as it is hoisted up (*F*). The jaws open for discharge either by means of a trip cord or automatically as the grab is lowered to rest. The reach depends on the length and angle of the crane jib.

The grab is used for deep excavations of limited area on all types of soil except rock.

Pile driving and drilling   It is possible to use an excavator crane for pile driving by equipping it with hanging leaders which guide the pile and the hammer during driving (figure 10 *G*). The leaders are suspended from the top extremity of the jib and are strutted back at the bottom to the excavator in order to maintain them in a vertical position. Hanging leaders are not suitable for long piles because of the great length of crane jib required and are used for driving sheet piling and timber and precast concrete piles of lengths less than about 12 m. This method avoids the use of guide trestling when driving sheet piling[1].

In addition to pile driving excavators may also be specially rigged for drilling wide diameter cylinder piles (see page 103). At the base of the jib a platform or boom is constructed carrying a turntable through which the square drilling rod or kelly bar passes, the bottom end of which carries the drilling auger and the top of which is supported by the cable over the jib (figure 10 *H*).

Tractor-based equipment

Tractor-based equipment is designed either as attachments to normal tracked or wheeled tractors, or as machines in which the earth-moving attachments and the tractor are designed as a single integrated unit. Loading shovels, trench diggers and light bulldozers designed as attachments to tractors are primarily used for small scale work. All operations are hydraulically controlled.

Tractor shovel   This consists of a tipping bucket at the front attached by strong pivoted arms or booms to the frame of the machine (figure 11 *A*). The bucket is loaded by lowering it by the arms on to the ground, the open side facing forward, and driving the machine forward into the soil to be excavated or into a heap of material to be moved. The arms are then raised, the bucket being levelled to prevent the material falling out, and the machine is reversed and driven to the point of discharge, either transport or a tip, where the bucket is then tilted down to permit the contents fo fall out; mechanism ensures that the bucket is locked level at all heights while the machine is moving. Raising and tipping is by hydraulic rams.

Buckets of various sizes may be attached, the larger ones being used for dealing with lighter materials, and teeth can be fitted to the front edge; most machines can be fitted with a bulldozer blade in place of a bucket. The bucket excavates to a depth of about 150 mm below the surface of the ground and the maximum height over which it can discharge varies with each make, but some with high pivoted side arms are designed to load at heights as much as 4.90 m. On some machines, called *overloading shovels,* the bucket, after being loaded, is carried on an inclined frame, or by arms, over the tractor to discharge at the rear, and this minimises travel for each cycle of digging and loading (figure 11 *B*).

The term 'tractor shovel' is applied to the heavier machines, particularly tractor crawler shovels and

---

[1] See also page 98 with reference to pile driving

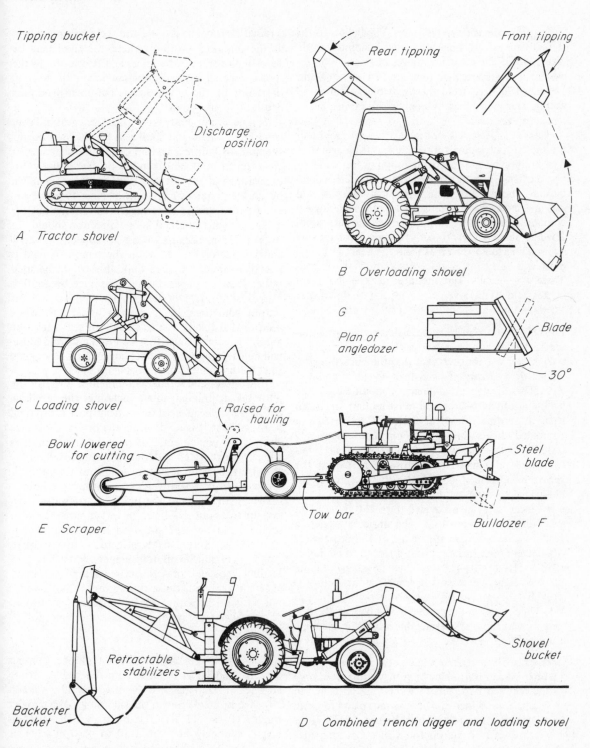

Tipping bucket

Discharge position

A   Tractor shovel

Rear tipping

Front tipping

B   Overloading shovel

C   Loading shovel

G

Plan of angledozer

Blade

30°

Raised for hauling

Bowl lowered for cutting

Steel blade

Tow bar

E   Scraper

Bulldozer   F

Retractable stabilizers

Shovel bucket

Backacter bucket

D   Combined trench digger and loading shovel

11   Tractor based equipment

the larger pneumatic tyred models. The lighter forms of this type of machine are called 'loading shovels (figure 10 C). The main advantages of tractor shovels over excavators are that they are fast working and can travel quickly from the point of excavation to waiting transport, thereby enabling the transport to remain on the road.

Tractor shovels, either track or wheel mounted, are used for stripping top soil, excavating against a face, bulldozing and for loading spoil or loose materials, particularly heavy, coarse material such as demolition rubbish, into transport. Most types of soil can be excavated, except rock and waterlogged ground, especially when crawler tractors are used. Loading shovels are primarily designed for moving and loading loose materials from the heap, although most will carry out surface excavation of top soil. These are usually tyre-mounted as the loads are lighter and tyres give greater speed than crawler tracks and greater mobility from site to site.

Trench digger This may be called a tractor mounted backacter and operates on the same principle as a backacter excavator except that the bucket is controlled by hydraulic rams instead of cables and pulleys. The digging attachment is mounted on the back end of the tractor and most models are fitted with the normal shovel at the front to produce a combined digger and shovel machine, commonly called a loader digger (figure 11 D). In some types the digger attachment is removable to permit the shovel to operate alone. As with the shovels the discharge height varies with different makes.

Trenches may be excavated from 250 to 914 mm wide and up to 4 m deep. The digger arm gives a reach of about 5.50 m and an arc of 180 degrees for discharge. Some makes permit a trench to be dug at right angles to the tractor. When digging is taking place at right angles to the tractor side stability is provided by projecting legs or outriggers on each side which rest on the ground at the base of the digging boom; these are retractable when the machine is moved from the site.

Scraper The scraper is a large box or bowl with an open front and bottom cutting edge, supported on a frame between two pairs of wheels and attached to a tractor from which the bowl is raised and lowered by cable or hydraulic power (figure 11 E). It is towed forward by the tractor with the cutting edge of the bowl lowered and is filled as the cutting edge digs into or scrapes the ground. When full the bowl

is raised clear of the ground and the scraper is towed to the dumping position where the spoil may be deposited or spread over an area. The ejection of the spoil is generally by tail-gate drawn from the back to the front of the bowl. Soil is cut to a maximum depth of about 300 m.

Scrapers are used for surface excavation over large areas where the spoil can be disposed on the site and for bulk excavation over somewhat smaller areas under the same conditions, particularly for excavations of the cut and fill type. Scrapers drawn by crawler tractors can work in most types of soil except rock and in waterlogged conditions but the maximum length of haul in one direction should not exceed 450 m because of the slow speeds at which crawler tractors move. When the length of haul is less than about 70 m a bulldozer would be used rather than a scraper. Pusher assistance by another tractor at the rear of the scraper can increase the output, sometimes by as much as 30 per cent. When used on hard soils, such as chalk or soft rock, the ground generally must first be broken up and loosened by a tractor-drawn ripper consisting of hardened steel teeth mounted on a wheeled frame.

The tractor and scraper equipment is ideal for the purposes mentioned above including site levelling, and is very cheap because no transport is required provided that it is used within the limits of distance given. The equipment can only be used on sites where the excavated material can be disposed on the site itself, because it cannot load into transport. It is capable of comparatively large outputs per hour.

Where the haul to the point of disposal is long, say up to about 1½ miles in one direction, motor scrapers can be used provided that the ground is firm and not unreasonably affected by the weather. Apart from smaller motor scrapers which are self-loading, most motor scrapers require pusher assistance for loading because their grip on the ground is not so good as with a tracked machine. Motorised scrapers have a travelling speed in the region of 20 mph.

Bulldozer and angle-dozer The bulldozer consists of a rectangular steel blade with renewable cutting edge set at right-angles to the direction of travel and attached by steel arms to the sideframes of a crawler tractor (figure 11 F). The blade, which varies in height and may be from 1.50 to 4.30 m wide, is raised and lowered by hydraulic power and operates between 450 mm below to 1.00 m above the tracks.

28

The angle-dozer (G) is a variation of the bulldozer with a blade which can be set at an angle of about 30 degrees to the direction of travel so that when operated the material is pushed to one side in a similar way to the action of a snow plough. The blade may also be pivoted slightly in a vertical plane.

The bulldozer may be used for excavating natural soil or for moving loose soil or debris which it pushes forward as the tractor forces it ahead. The depth of soil or the amount of material moved is governed by the raising and lowering of the blade. As in the case of the scraper, the bulldozer is unable to load material into transport so that it must be left on the site. It is capable of pushes not exceeding 90 m and is suitable for the same type of surface and bulk excavation as the tractor and scraper. It may be used also for backfilling, for spreading material from dumps and spreading such materials as gravel, sand, clinker and hardcore or for piling materials ready for loading by other equipment.

The angle-dozer is used for backfilling all forms of open excavations and trenches, the filling being pushed in at an angle, and for moving earth in line parallel to the direction of the tractor.

The following pieces of equipment are designed as single special purpose machines

Grader

The action of the grader is similar to that of the angle-dozer and consists of a wheeled frame under which the blade is suspended between the wheels. It is generally designed as a self-propelled machine although it can be pulled and operated by a tractor. The blade can be pivoted over a wider angle, both vertically and horizontally, than in the case of the angle-dozer so that a steep bank may be graded. Some models are designed so that the wheels can be made to incline at an angle to the frame in order to follow the angle of the ground on which the grader is working. Both tractor-drawn and motorized graders can be designed as elevating graders, in which the excavated material is transferred to a conveyor belt and elevated to transport alongside.

As the name implies graders are used for grading, that is for the final levelling off of excavated areas, for maintaining haul roads, for trimming of banks and cutting ditches.

Trenching machine

Alternatively called a trencher or a trench cutter.

The trenching machine consists basically of a number of excavating buckets mounted either on a wheel or on a continuous chain mounted on a vertical boom similar to a dredger. The chain-type is capable of deeper digging than the wheel type and is more easily dismantled for transport (figure 12 A). Both types have a transverse belt conveyor immediately under the line of filled buckets which deposits the spoil on either side of the trench being excavated. Alternatively, the conveyor, if long enough, can raise the spoil to a sufficient height for direct discharge into transport.

Unlike a backacter the trenching machine is a single purpose tool, but because its speed in cutting is fast it is more economic when there is a large amount of trenching work to be executed. To permit the boom carrying the buckets to enter or leave the the cut it may be pivoted and pulled with cables to alter its angle of inclination, or it may be mounted on rollers running on an inclined rail. In the first case, during transport the boom will be elevated at an angle; in the second case, it will be drawn back into a position parallel with the ground, making transport easier. Trenching machines are fitted with

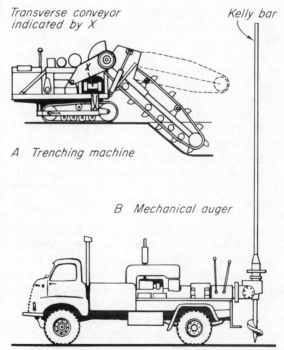

Transverse conveyor indicated by X

Kelly bar

A Trenching machine

B Mechanical auger

12 Special purpose machines

crawler tracks and move away from the trench being excavated.

The trencher is an alternative to the backacting shovel and is used for continuous runs of trenching up to 4.30 m deep and as much as 1.80 m wide, being much faster and cutting cleaner trenches than the backacter. As in the case of the backacter, the ground must be reasonably firm and dry, particularly where the trenches are deep. It is a particularly useful machine when a large amount of trenching is to be carried out on a fairly level site on which little other excavation is required.

Mechanical auger

This is a boring tool commonly used for forming short bored piles 2.10 to 3.00 m deep, but it can drill up to a depth of about 6.10 m with diameters of 230 to 914 mm drilled vertically or at an angle. The machine itself, consisting of a power unit, usually diesel-driven, the necessary controls and a drilling arm carrying the auger and its guide, is self-contained and is mounted on the platform of a lorry or similar means of transport (figure 12 B).

Hoisting

This section and the next on Transporting is concerned with the handling of materials and components.

Most materials used in the erection of buildings are handled several times during the course of construction. Various forms of wheeled transport are used for handling the materials at ground level and various forms of hoisting equipment are used for raising them to upper levels. The direct effect of mechanically handling materials is on the cost of the handling operation, but it has an indirect effect in that the speed of the whole contract is increased thus saving overhead costs and, by improving the phasing of various operations, non-productive time is reduced.

The plant used for hoisting, which consists of cranes, hoists and elevators, primarily performs vertical movement although, apart from the hoist, some horizontal movement is also involved and is functionally desirable. In this respect hoists give only a one-dimensional vertical movement, elevators give a two-dimensional vertical movement, and cranes give three-dimensional movement. Cranes are, therefore, particularly useful in solving handling problems[1].

Cranes Cranes may be divided into two broad groups: mobile cranes and stationary cranes. Travelling or rail-mounted cranes are included in the category of stationary cranes since their degree of mobility is limited to their working position on the site.

Mobile cranes

These may be either self-propelled or truck-mounted, the former being capable of travelling with a suspended load. Most types are fitted with a low-pivot derricking jib, either of strut type, held at the upper end by a suspension cable (figure 13 A, B) by means of which it is derricked, or of cantilever type (C, D, F) derricked by cable or hydraulic ram.

Mobile cranes can slew or turn a full cycle. In some the jib, power unit and cabin are power-turned above the tracks or wheels (figure 13 A, C, for example) or, as in the case of some wheeled cranes, the whole machine slews in its own length by means of four-wheel steering and requires no slewing mechanism. (F).

Self-propelled cranes include those mounted on pneumatic tyres which can be driven at slow speed from site to site on normal roads but they need hard ground on which to work on the site (figure 13 A). On most sites a crawler-mounted crane is necessary if the crane is to be used on unprepared ground and this is a widely used type of building crane, even though it cannot travel on the road and its travelling speed on site is slower than a tyre-mounted crane (B).

It is frequent practice to rig an excavator to produce a crawler-mounted crane by fitting the long jib used for dragline or grab-work, although in many types the controls do not allow a sufficiently sensitive control of the load since they are designed for rapid digging rather than handling of materials. It should also be borne in mind that it is illegal to rig an excavator as a crane unless it is calibrated with a load-radius arc which indicates the radius of jib at any point and the permissable load at that radius, and is fitted with an automatic warning in the cab. This causes a bell to ring and a light to go on when the maximum load is being approached and these continue to operate when the maximum load is being carried.

[1] See, however, section on forklift trucks, page 43.

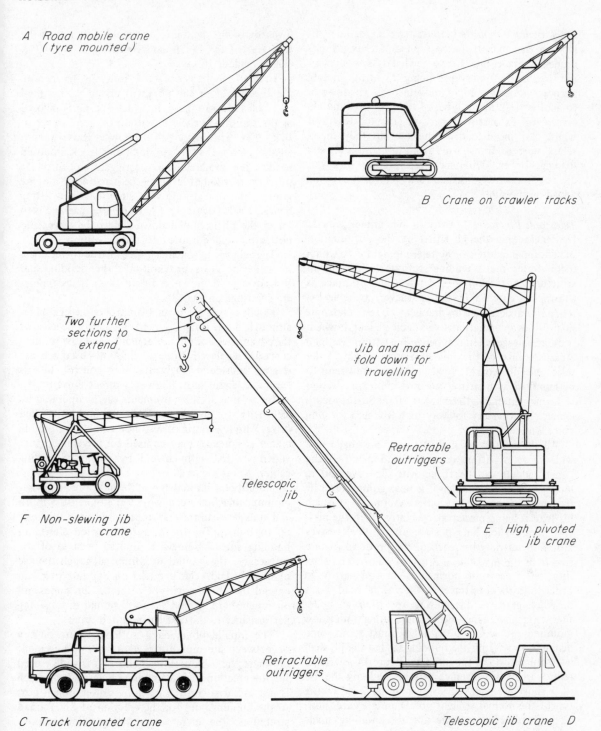

A   Road mobile crane
    (tyre mounted)

B   Crane on crawler tracks

Two further
sections to
extend

Jib and mast
fold down for
travelling

Retractable
outriggers

Telescopic
jib

F   Non-slewing jib
    crane

E   High pivoted
    jib crane

Retractable
outriggers

C   Truck mounted crane

Telescopic jib crane   D

13   Mobile cranes

Truck or lorry-mounted cranes have a greater mobility on the road than even a pneumatic tyred self-propelled crane but are somewhat less mobile on the site, their primary use being for highly mobile purposes requiring rapid movement from one site to another (*C*). Many types of self-propelled mobile cranes and excavator cranes are designed to travel whilst carrying a load on the site but they themselves must be transported from job to job on low loading lorries. Truck-mounted cranes, however, are not frequently designed to travel whilst carrying a load on the site.

Telescopic jib cranes   Many mobile cranes provide for lengthening the jib either by the introduction of additional sections or by telescoping the jib which incorporates the extending sections within the jib structure. A truck-mounted telescopic jib crane is shown in figure 13 *D*, the cantilever jib being extended and derricked hydraulically. The telescopic jib provides an operationally-variable jib length which is an advantage when manoeuvering within confined areas. Because of the greater relative weight of the telescopic jib working loads at longer radii tend to be less than with lattice type jibs and it is restricted in its maximum length to about 30 m. Strut type jib cranes are necessary for very high lifts and for long range work.

All these types of mobile cranes are made in a wide range of lifting capacities and many of the larger models can be fitted with jibs over 30 m long. Since the crane driver is near ground level he frequently is not able to see the exact point at which his load is being placed, particularly when the jib is very long. When long jibs are used it is necessary either to restrict the reach of the jib or to fit outriggers to the machine in order to improve its stability, since they are normally light and narrow in order to fit them to travel on the normal road.

When used for distributing operations in building work these cranes, because the straight jib is low mounted and works at an angle, must stand some distance away from the building so that the jib shall not foul the top of the building (figure 14 *A*). Some reduction in this distance may be made by the use of a short swan-necked jib or a fly jib fixed to the top of the normal straight jib. These are extensions to the jib placed at an angle, and are sometimes made to derrick independently of the main jib (*B*). These jib extensions enable the radius of the jib to be extended in the case of light loads without the

lowering of the main jib, thus keeping it clear of the top edge of the building. Extensions of 3 to 9 m can be made in this way.

Greater benefit is obtained, however, by increasing the height of the jib pivot and high mounted jib cranes are produced in which the jib is pivoted at the top of a vertical extension or short tower so that they are really small mobile tower cranes (figure 14 *C*). Crawler-mounted and truck-mounted models are available and outriggers are fitted to provide a method of levelling the machine on sloping ground and also to provide stability when lifting heavier loads (figure 13 *E*). When the machine is to travel the jib is folded down on to the tower and both are dropped into a horizontal position.

The value of the smaller types of true tower crane in achieving close proximity to the building and full coverage is shown in figure 14 *D*. These cranes are described on page 38.

Mobile cranes are driven by petrol or diesel engines although diesel is rapidly replacing petrol because of the advantages both in maintenance and low running costs. Diesel electric drive is sometimes used because it gives considerable advantages in control. In this form the diesel motor drives a direct current generator so that all the movements can be operated by electricity to give a very smooth and controlled drive. Wholly electric drive is not suitable for mobile cranes although it may be used for travelling cranes which are fed with current by means of trailing cables.

Mobility is limited by site conditions. Crawler-mounted machines can work on slopes of 1 in 40 and, unladen, can traverse unprepared sites on slopes up to 1 in 5. The use of tyre mounted cranes on building sites is somewhat limited because of the softness of the ground in winter, although the use of quickly laid vehicle tracks on building sites can extend the use of this type of crane. On slopes and on rough, hard ground they are similar in travelling performance to the crawler-mounted types.

The mobile jib crane is suitable where on site or between-site mobility is a primary requirement or where the job-duration is short. It is widely used for the erection of low buildings where a long reach is not essential and the machine can approach near to the building, and for the erection of low framed structures, the crane often being able to move between the columns of the structure. It is also used for loading and unloading operations and for filling hoppers.

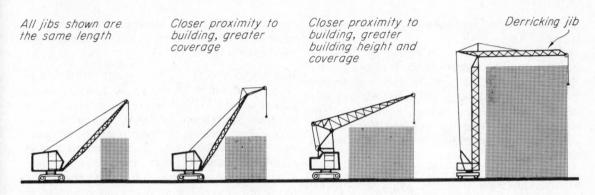

All jibs shown are the same length

Closer proximity to building, greater coverage

Closer proximity to building, greater building height and coverage

Derricking jib

A Mobile crane   B With fly jib   C High mounted jib   D Light tower crane

14 Cranes – relative amount of working areas and coverage

As indicated on page 30 some self-propelled mobile cranes are designed without slewing jibs so that they must be driven into the correct operating position. In order to give greater manoeuverability for this purpose, rear-wheel or four-wheel steering is usually employed (figure 13 *F*). In some forms a certain amount of flexibility is given by the provision of a derricking fly jib generally operated by a hydraulic jack. This type of crane is usually used in confined spaces for handling small loads.

Stationary cranes

The word 'stationary' here refers in particular to cranes of various types which are fixed firmly to some form of base at their working position. The term also covers cranes which have a limited lateral movement restricted to the immediate vicinity of their working position.

Derrick cranes   These are simple and inexpensive cranes used where heavy lifts at long radii have to be made and where the duration of the work justifies the installation of the crane. There are two types, the guy derrick and the Scotch derrick, the basic difference between them being the way in which the mast is supported. Both types handle their maximum loads over a greater range of radius than most other forms of crane of comparable capacity.

Guy derrick   This consists of a vertical guyed mast pivoted to the base and pivoted at the top to the cap or spider plate taking the ends of the guys. This permits the slewing of the crane. The jib is hinged to the base of the mast and both mast and jib are of lattice steel construction. Winches are provided at

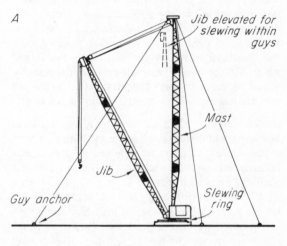

A

Jib elevated for slewing within guys

Mast

Jib

Guy anchor

Slewing ring

Guy derrick

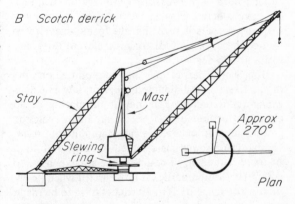

B Scotch derrick

Stay

Mast

Slewing ring

Approx 270°

Plan

15 Derrick cranes

33

the base for derricking the jib and for hoisting loads (figure 15 *A*).

The base occupies little space but the fixings of the guys must be widely spaced. The jib is shorter than the height of the mast and when it is elevated to an almost vertical position it passes under the guys and the crane will slew through a full circle. It is either hand or power operated and the winch is often mounted some distance away from the base of the mast.

A wide range of sizes is available but the normal capacities are 5 and 10 tonnes, although with some types loads of up to 150 tonnes can be handled. The jibs are about 30 m long.

The guy derrick may be erected on the ground or on the building under construction and it is very widely used for erecting steel frames. For this purpose it is set within the structure and is raised from time to time to rest on the steelwork of the frame as the latter rises, the jib being disconnected from the mast and raised by the derricking blocks and winch to the next level, where it is temporarily guyed in order to raise the mast; after which the two are reassembled to operate at the higher level[1]

Scotch derrick   This consists of a slewing mast and a derricking jib related in the same manner as in a guy derrick, except that only the jib is of lattice construction and the mast is considerably shorter than the jib which may be as long as 37 m (figure 15 *B*). Both sets of winches and motors are mounted on the mast itself. The mast is maintained in a vertical position by two latticed steel backstays which run from its top to the extremities of two base sleepers set to form an angle of 90 degrees with each other at the base of the mast, the ends being either loaded with ballast or kentledge or anchored to heavy concrete blocks. The junction of the mast with the sleepers is such that the former may freely rotate. The stays limit the practical slew of the crane to about 270 degrees.

The Scotch derrick may also be mounted on rails or on fixed or rail-mounted piers or towers called *gabbards* to increase the height of placing or to clear obstructions or to free the site below, since at ground level it occupies a large amount of space. Gabbards are sometimes erected to heights of 40 m or more, although, of course, these are expensive to erect. The Scotch derrick is most useful when heavy lifts in the range of 3 to 20 tonnes have to be made over a wide radius. Like the guy derrick, Scotch

derricks are used for the erection of steel frames and when both are used on one job they are employed to raise each other to higher levels.

Monotower derrick   This consists of a braced tower, which may be up to 60 m high, firmly secured to adequate foundations and surmounted by a derrick crane (figure 16 *A*). As there are no stays or guys to give stability, the mast of the crane extends down through the centre of the tower to a pivot bearing well down its height or even down to foundation level, passing through a slewing ring at the top of the tower. The derricking jib is pivoted to the mast above the slewing ring and a horizontal counter jib or boom projects on the opposite side, carrying a ballast counterweight at the end. The driver's cabin and controls are at the top of the tower. These cranes are full circle slewing and for construction purposes are from 5 to 20 tonnes capacity with jibs up to 45 m long.

The crane occupies a very small amount of ground area but has an exceedingly large coverage with the driver in an elevated position having a good view of the working area. It is, however, a costly crane to erect and the foundations are more expensive than those of a normal derrick. It is, therefore, suitable only for use on certain types of large multi-storey buildings where its erection is justified by its advantages of capacity and reach, although for most structures of this nature it has been replaced by the tower crane.

Tower crane   This is actually a form of monotower crane and some types have the derricking jib and ballast jib of the monotower derrick (figure 16 *D*). The tower crane is of continental origin and is capable of distributing materials over the whole area of a tall building, being particularly suitable for handling relatively light loads at great height and reach. They are electrically driven, are light in weight and can be easily transported and quickly erected. They are made with jibs up to 30 m long and some of the largest machines have a lifting capacity at this radius of 5 tonnes.

The lattice mast is built up in sections and may be fitted with a derricking jib or fixed horizontal or saddle jib. All types slew through a full circle

[1] For an illustration of this process and for other details of the erection of steel frames for buildings, see *The Erection of Constructional Steelwork* by Thomas Barron, Iliffe, 1963

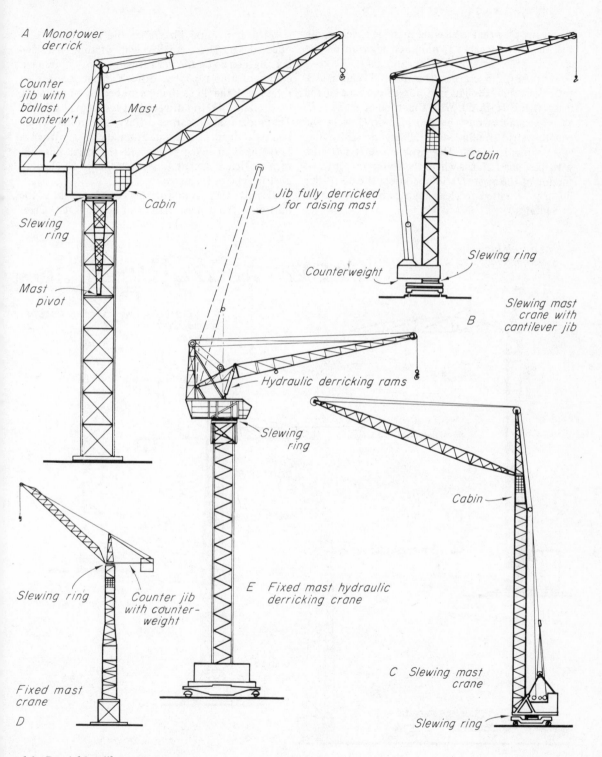

A Monotower derrick

Counter jib with ballast counterw't

Mast

Cabin

Slewing ring

Mast pivot

Jib fully derricked for raising mast

Cabin

Counterweight

Slewing ring

Slewing mast crane with cantilever jib

B

Hydraulic derricking rams

Slewing ring

Slewing ring

Counter jib with counter-weight

E Fixed mast hydraulic derricking crane

Cabin

C Slewing mast crane

Fixed mast crane

D

Slewing ring

16 Derricking jib tower cranes

by means of either a slewing mast or a slewing jib and the driver's cabin is mounted high up on the crane.

Derricking jib cranes often have slewing masts with the counterweight in a box at the base of the mast (figure 16 *B*, *C*). With this type of crane sufficient clearance must be provided to prevent the counterweight fouling the building or scaffolding when the mast slews. This problem does not arise with fixed mast cranes where the counterweight is at the top of the mast (*D*). In operation, the derricking jib may be raised to clear any nearby obstructions or buildings.

With horizontal jib cranes the mast is usually fixed, the jib and counterweight rotating on a slewing ring high up on the mast (figure 17 *B*, *C*). In most cranes of this type the driver's cabin is made to rotate with the jib so that, as in the case of a slewing mast crane (*A*), the driver is always facing his work. The horizontal jib carries a traversing trolley which takes the load in and out from the mast. When a horizontal jib crane is erected the mast must be of a sufficient height to permit the jib to clear all obstructions as it rotates.

Because there is no rotating counterweight at the bottom a fixed mast crane can be mounted nearer

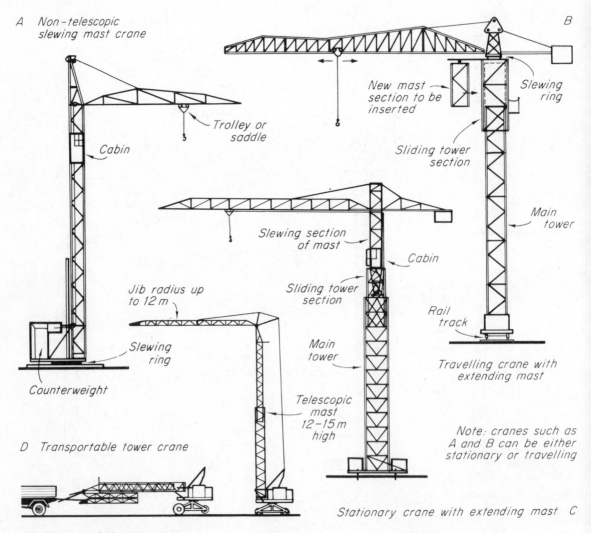

A Non-telescopic slewing mast crane

Trolley or saddle

Cabin

Jib radius up to 12 m

Slewing ring

Counterweight

D Transportable tower crane

New mast section to be inserted

Slewing ring

Sliding tower section

Main tower

Slewing section of mast

Cabin

Sliding tower section

Main tower

Telescopic mast 12–15 m high

Rail track

Travelling crane with extending mast

Note: cranes such as A and B can be either stationary or travelling

Stationary crane with extending mast C

B

17 Horizontal jib tower cranes

to the building than one with a slewing mast, thus producing a greater effective reach.

The derricking jib crane can usually be erected more quickly than a horizontal jib type of similar capacity, but the latter has the advantage of ease of operation and speed. In addition, the horizontal jib crane has the advantage that the load can be brought to within about 1.5 m of the mast, whereas with most derricking jibs it cannot be brought so near. The height of the building which a derricking jib tower crane can serve effectively is, as in the case of a low mounted jib mobile crane, limited by the inclined jib and unless designed with an extending mast they are, therefore, most often used for buildings of limited height.

Most cranes are not sufficiently stable with masts higher than about 30 m but with fixed mast types erected close to the building and designed to be built up in sections as the structure rises, the crane can be built up to greater heights by tying the mast back to the building to give the necessary stiffness and stability. Although not common, it is possible to tie back a slewing mast crane in this manner by the use of an encircling ring to which ties are attached.

In order to permit the height of the mast to be increased without dismantling the jib, cranes, particularly the horizontal jib types, may be designed with extending masts or telescopic masts. In the former the jib is carried on a short length of mast moving inside the top of the main mast, the two being secured together when the crane is in operation (figure 17 C). In order to increase the height, a section with one side temporarily open to pass round the inner mast is hoisted by the crane's own trolley and added to the top of the main mast. The inner mast, together with the jib, is then raised to its new position within it. Alternatively, the short supporting mast may slide outside the main mast and the extending operation then consists of raising the short outer section almost to the top of the main mast and temporarily opening up one of its sides to permit the introduction and fixing of a new section to the top of the main mast, the new section being hoisted and brought into position by the travelling trolley (B). One form of derricking jib crane with the slewing action at the top and with hydraulic derricking which enables it to be elevated to an extremely steep angle, is designed in a similar manner, the jib and driving unit being attached to a short section which slides outside the head of the

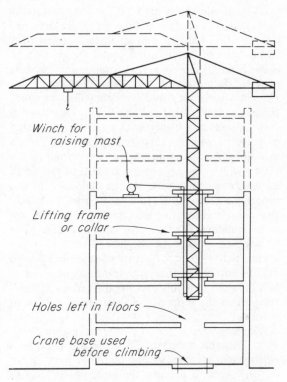

Winch for raising mast

Lifting frame or collar

Holes left in floors

Crane base used before climbing

18 *Climbing crane*

main mast (figure 16 E). The extension of the mast is carried out by elevating the jib to an almost vertical position, clear of the head of the mast, so that a new section already hoisted by the jib may be dropped down on to the mast head. When fixed in position the short outer section with the jib is then raised to its new, higher position. The jib of this particular crane pivots well outside the axis of the mast to give a small minimum radius which permits this operation to take place.

In cranes with telescopic masts the mast is in two or more sections nesting into each other, which permits the height to be altered without partial dismantling and re-erection. This type of mast is usually of the slewing form and is most common on mobile or transportable cranes (figure 17 D).

Climbing crane   An alternative to the extending or telescopic mast as a means of increasing the height of the jib is the climbing crane set inside the building. This has a short fixed mast which is raised from floor to floor as the building rises (figure 18). At the beginning of the job the crane is securely anchored to the base of the building to act as a fixed crane and this is used for the construction of the lower parts of the building. When the latter has reached the height of two or three storeys, holes having been left in the floors round the mast of the crane, lifting frames or collars fitted round the mast are fixed in position on two adjacent upper floors. Flanged wheels are fitted to these collars in which the angle members of the mast run. The tower is then hoisted to a higher level by means of a hand or powered winch, or by re-arranging the crane's own hoisting ropes, and wedged in position to the collars which transfer the crane loads to the floors on which the collars rest. As the building rises this process is repeated until at the completion of the job, the crane is dismantled and lowered over the side of the building. The floor structure, of course, must be designed to take the weight and loads from the crane which is mounted on steel joists spanning on to main beams. Alternatively, the crane may be mounted in a lift shaft, but this presents difficulties in shuttering when the walls of the shaft are of reinforced concrete.

Climbing cranes may have derricking or horizontal jibs, the derricking jib being useful on sites where there are surrounding buildings since it can be derricked up to clear the buildings as it turns and then be lowered to pick up or drop its load. Climbing cranes are useful for the construction of tall buildings and on congested sites where no space is available outside the building for the operation of a travelling or a stationary crane. In order to obtain the smallest minimum radius some climbing cranes with derricking jibs have the jib pivoted well behind the centre of rotation, as in the case of the hydraulically derricked crane mentioned earlier.

An inside mounted crane requires a shorter jib to cover a given area than one mounted outside. This, together with the fact that a short mast only is needed, makes the climbing crane a cheap crane for tall buildings if the structure is strong enough to take the load, although it may involve expense in other directions if it obstructs work internally.

The driver's cabin on a climbing crane must be right at the top of the mast under the jib, but these cranes as well as other types of tower cranes can be remotely controlled by means of a portable control panel connected to the crane by a wandering lead. This permits the driver to position himself at the most convenient point for observing the hook of the crane and its load. This form of control, however, is not suitable when two cranes are in use, close to each other, because the drivers have a poor view of the jibs as they slew across each other's arc.

Rail-mounted or travelling crane   As already shown all types of tower crane can be mounted on a fixed base adjacent to or inside the building being erected. Where conditions permit, however, a rail-mounted or travelling crane moving along the side of the building has the advantage of increased coverage due to the movement along the rails (see figure 9). Both horizontal and derricking jib cranes may be mounted in this way (figures 17 *B* and figure 16); travelling action is electrically operated.

The crane is mounted on a chassis supported on four double-flanged rail wheels or on four double-wheel bogies. These may be mounted to permit curves to be negotiated. A firm level track is essential and this sometimes presents difficulties on sloping sites. Curves in the track should be avoided where possible, but are preferable to the use of turntables. When the shape of the building necessitates a change on to different tracks a trolley on a sunk track at right angles to the main tracks is preferable to turntables for moving the crane from one track to another.

A height of about 30 m is considered to be the maximum at which these cranes can travel with a load.

Transportable tower crane   Apart from the mobile cranes with high mounted jibs referred to earlier, a number of transportable tower cranes are available with smaller heights and capacities than the standard types of crane described above. These are designed for work on small low buildings such as houses, small blocks of flats and small factories. Those intended primarily for two-storey house building have a height of 6 to 9 m to the derricking jib pivot and carry about 350 kg at a maximum radius of 10 m. Other models with mast heights of 12 to 15 m with slightly greater capacities have 12 m radius horizontal jibs with traversing trolleys which, however, may be derricked if necessary in order to work at greater heights (figure 17D). For towing from site to site these cranes are folded in the same way as a high-mounted jib mobile crane, the jib folding down on to the mast and then the two together folding down to a horizontal position. The base, fitted with tyred wheels, acts as a trolley or trailer. The masts of the taller types are made in two halves, the upper half telescoping into the lower half for transport.

Stabilisers, or outriggers, are required which raise

the wheels clear of the ground when the crane is in operation. Control is remote by wandering lead and portable control panel. The greater height and coverage possible with these cranes compared with mobile cranes with low-pivot jibs is indicated in figure 14 *D*.

Tower cranes may also be truck- or crawler-mounted with telescopic towers to facilitate transport and erection. Truck-mounted types require extended outriggers for stabilising when in operation, as do cranes mounted on one pair of crawler tracks. Those mounted on four widely spaced crawler tracks, each of which can be adjusted for height, are stable.

Fixed jib slewing crane　A *mast crane* or a *scaffold crane* can be used for placing materials on a scaffold. In both types the jib is short and non-derricking and the load can only be moved laterally in a circle of fixed radius. The mast crane or fixed jib slewing crane is a simple piece of apparatus consisting of a short jib fixed to the top of a telescopic slewing mast about 10.5 m maximum height mounted on a wheeled base (figure 19 *A*). The powered hoisting will lift about 400 kg. The scaffold crane is a simple fixed slewing jib fitted with a lightweight powered winch and can be attached in a few minutes to a

specially braced scaffold (*B*). It can lift up to 250 kg, can be carried by two men and is particularly useful as an adjunct to manual handling on small construction jobs.

Portal crane　This is a travelling crane but of fundamentally different design to that of the tower crane, being derived from the factory or workshop overhead gantry crane.

It consists basically of a horizontal beam supported at each end on a pair of legs, the load being suspended from a trolley which traverses the beam. The beam may cantilever at one or both ends beyond the leg connections (figure 20 *A*) in which case the pair of legs may be set wider apart to permit the handling of wide units between them. The crane is mounted on four twin-wheeled bogies, two of which are powered, the bogies running on a pair of rails, one situated beneath each pair of legs. The bogies incorporate hydraulic jacks which enable the crane to be raised off the rails and turned through 90 degrees, thus increasing its manoeuverability. In use the crane straddles the building being erected.

Post-war development of system building encouraged its use in the building industry, particularly in the field of heavyweight component systems, since its inherent stability makes it possible for a high lifting-capacity crane to be produced at a lower capital cost than any comparable slewing crane.

Since the crane load is spread over four widely spaced points the individual wheel loading is relatively small. This is an important consideration, especially when ground conditions are poor. Compared with the portal a tower crane places a heavy concentrated load on the track. However, a considerable clear area is required for the erection and dismantling of a portal crane and this, together with the need to use two heavy mobile cranes to lift the beam into position, has led to development towards a self-erecting type. One model which has been produced consists of a pair of twin towers, made up of masts similar to those used for tower cranes, between which the beam is carried on lattice yokes, which slide over the masts (figure 20 *B*). A lifting rig is mounted on top of each twin tower by means of which the beam is raised and lowered.

The normal type has a beam span around 18 m, a lifting capacity up to 10 tonnes and a height of lift of 36 m. The self-erecting model described above has a beam span and a height of lift of 40 m and a capacity of 8 tonnes. Multi-storey system buildings

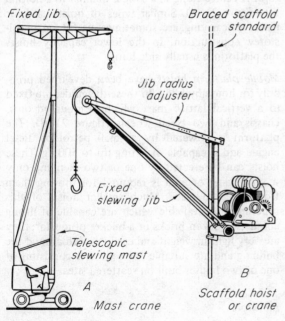

Fixed jib

Braced scaffold standard

Jib radius adjuster

Fixed slewing jib

Telescopic slewing mast

A

Mast crane

B

Scaffold hoist or crane

19　Fixed jib slewing cranes

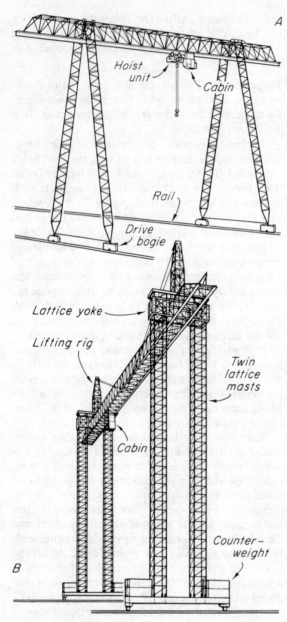

A

Hoist
unit

Cabin

Rail

Drive
bogie

Lattice yoke

Lifting rig

Twin
lattice
masts

Cabin

Counter-
weight

B

20 *Portal cranes*

up to 14 storeys in height can thus be erected under these cranes. As indicated above their low capital cost and high lifting capacity make this type of crane eminently suitable for heavy component system construction.

40

Hoists

A hoist consists of a horizontal platform which is moved up and down vertical guides by a powered winch and is usually termed a platform hoist, although the platform is sometimes replaced by a tipping skip to carry concrete. The guides are built up in sections so that they may be extended as the structure rises and are tied back to the structure or to scaffolding in order to provide stability.

Platform hoists are commonly made in sizes up to about 1.5 tonnes capacity and are used for lifting bricks, blocks and general building materials, but they provide vertical movement only since materials must be brought to the bottom of the hoist and taken away from the top for placing in the final positions. Generally speaking, a hoist is doing useful work for only about 10 per cent of its time on the site; this, however, can be increased considerably by palleting smaller materials and using packaged bricks. It is usually uneconomic to carry very heavy loads or long lengths of materials such as timber and scaffold tubing, because of the time expended in loading and unloading them off the hoist. In some circumstances, however, it may be essential to use a hoist for heavy loads and larger models up to a capacity of about 3 tonnes are made in which the platform is centre slung in a similar manner to a normal lift (figure 21 *A*). Similar types of hoists used for passenger-carrying are sometimes useful on multi-storey construction. In the lower capacity hoists the platform is usually side hung.

Mobile platform hoists have been developed primarily for house building. In these the guides are fixed to a vertical lattice mast which is mounted on a chassis and two tyred wheels (figure 21 *B*). The platform is powered by a small petrol or diesel engine and is capable of lifting up to 500 kg. These hoists can be erected by one or two men and only a small gang of men is required for moving them from place to place. Much smaller hoists of this type are also available which are capable of lifting about two dozen bricks or a bucket of mortar; they are very light in weight and easily moved around the building and are suitable for contracts consisting of one or two houses built on scattered sites.

Elevators

These consist of a series of buckets fixed to a rot-

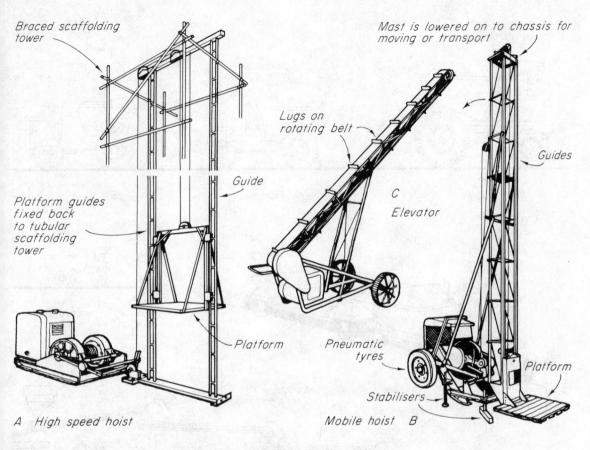

Braced scaffolding tower

Mast is lowered on to chassis for moving or transport

Lugs on rotating belt

Guide

C Elevator

Guides

Platform guides fixed back to tubular scaffolding tower

Platform

Pneumatic tyres

Platform

Stabilisers

A High speed hoist

Mobile hoist B

21 *Hoists and elevators*

ating belt or chain and are used for raising aggregates into the bins of weightbatchers and for similar purposes. If the buckets are replaced by lugs or prongs, the elevator can be used to raise bags of cement, bricks, tiles and similar components, which are fed on to the bottom by hand (figure 21 *C*). Elevators can work vertically but are usually set at an angle, according to the height of lift. They are powered by a small engine and are mounted on two wheels so that they can easily be moved around the site.

Transporting

The term transporting implies horizontal movement primarily but it can involve some vertical movement as, for example, in forklift trucks, conveyors and concrete pumps. The plant used for this purpose varies widely in its nature.

Dumpers

These are vehicles designed for the transport of materials which previously were usually carried by wheelbarrows, such as excavated spoil, hardcore and concrete. For round hauls longer than 36 m and, generally, to serve concrete mixers larger than 14/10 size, dumpers are more often used than wheelbarrows. They are faster and more economical than hand barrows and consist basically of a shallow tipping hopper or skip mounted on a wheeled chassis. They are powered by a diesel engine or, on the smaller types, by petrol engines.

Power barrow This is the name given to the smaller types of dumper which are three-wheeled, driven and steered on the back wheel by a pedestrian driver (figure 22 *D*) with capacities ranging from 0.14 to 0.40 m³, depending on ground con-

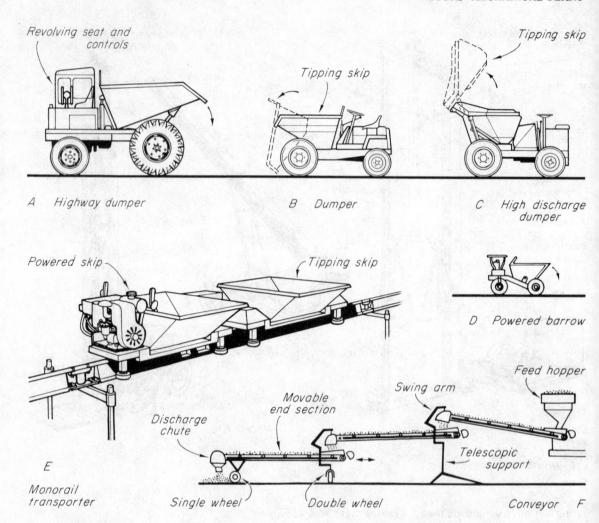

Revolving seat and controls

Tipping skip

Tipping skip

A   Highway dumper

B   Dumper

C   High discharge dumper

Powered skip

Tipping skip

D   Powered barrow

E

Monorail transporter

Discharge chute

Movable end section

Swing arm

Feed hopper

Telescopic support

Single wheel

Double wheel

Conveyor   F

22 Transporting equipment

ditions. On firm ground they can mount slopes up to about 1 in 7 or 8 but on soft ground excessive wheel spin tends to take place. Carrying about 0.20 m³ of wet concrete they travel at a maximum speed of 3 mph and a round haul of 45 to 55 m is about the practical maximum for this type of dumper. They are too large to be carried by most hoists.

Dumper   These have four tyred wheels with front-wheel drive and rear-wheel steering, with a driving seat or saddle placed so that the hopper or skip is in front to give a view for placing the load (figure 22 *B*). In most types the skip discharges at the front and is so balanced that discharge takes place

by gravity on the operation of a hand lever; after discharge the skip is automatically reset. Some types have hydraulic operation of the skip which permits close control of tipping. Side tipping skips which are mounted to swivel on a vertical axis enable the load to be discharged sideways at confined positions where manoeuvering for front discharge is difficult or impossible.

Multi-skip dumpers have interchangeable free standing skips which can be dropped when empty for reloading by a concrete mixer, for example, while a full skip is picked up and taken to the point of discharge or to within the orbit of a crane.

High discharge dumpers These have skips with high fronts at the top edge of which they are pivoted (figure 22 *C*). The high point of pivoting results in load discharge at a much greater height than with a normal dumper and this is particularly useful for transporting concrete from mixer and discharging into a crane skip. Discharge heights range up to 2 m and tipping is hydraulically controlled.

Small four-wheeled dumpers have capacities up to 1.5 m^3 and a maximum speed of about 8 to 10 mph when loaded. They can tackle similar slopes to a power barrow and although firm ground is best, light dumpers of this type can negotiate muddy and uneven ground. The skip capacities of the larger type of four-wheel dumpers range from 1.5 to 5 m^3 and may be used for round hauls of up to 2 miles. These are designed to travel on roads, for which purpose the driver's seat swivels through 180 degrees and a duplicate set of controls is provided (figure 22 *A*).

Dump truck This is a much larger capacity vehicle ranging from 5 to 22 m^3. It looks like a lorry and for capacities up to 7.5m^3 is generally for on- or off-highway use and above this capacity for off-highway use only.

Forklift trucks

A forklift truck is essentially a powered mobile chassis on the front of which is a vertical frame or mast on which a pair of 'forks', that is a pair of projecting tines, may be raised and lowered (figure 23 *A*). Loads can be picked up on the forks and be carried and deposited at any height within the limits of the mast. Those used on the uneven surfaces of building sites are known as rough terrain forklift trucks to distinguish them from those designed for use on the hard level surfaces of factories and warehouses.

The forklift truck is a machine for handling materials and its use for this purpose on building sites is of relatively recent origin. Much manual handling is involved in the movement of materials from their sources to the point of use. As already indicated dumpers, elevators, hoists and cranes are commonly employed to reduce this manual handling but their use still necessitates manual operations in loading and unloading dumpers, elevators and hoists and at the hook of a crane. The forklift truck and its driver, however, without assistance is able to unload from delivery lorries, stack in storage areas, distribute

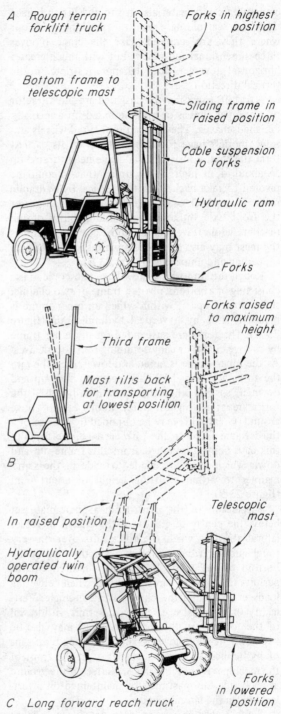

A Rough terrain forklift truck

Forks in highest position

Bottom frame to telescopic mast

Sliding frame in raised position

Cable suspension to forks

Hydraulic ram

Forks

Third frame

Forks raised to maximum height

Mast tilts back for transporting at lowest position

B

In raised position

Telescopic mast

Hydraulically operated twin boom

Forks in lowered position

C Long forward reach truck

23 Forklift trucks

components and materials on site and lift and place them on scaffolding[1] near the working positions where they are required. Like the crane it gives three-dimensional movement but with much greater horizontal movement although more restricted in the vertical direction.

A forklift truck may be designed around a tractor or a dumper chassis or may be based on a specially designed chassis. They run on four tyred wheels and some models have four-wheel drive. In its earliest form the mast consisted of a single height frame up to about 4 m high, so mounted that it could be pivoted to rake back from the vertical by hydraulic rams. Tilting back of the mast brings the load nearer the front axle thus increasing the stability of the machine while travelling. On most current machines the mast may also be tilted forward slightly for ease of loading and unloading (figure 23 B).

Rough terrain forklifts now have telescopic masts consisting of a bottom, pivoted frame of two channel slides braced apart, within which an inner frame is made to slide by a vertical hydraulic ram (figure 23 A). The forks are suspended on the inner frame by captive chains or cables running over top sheaves. As the inner frame is raised and lowered by the ram the forks are caused by this means to slide up and down it, so that when the frame is fully raised the forks are at its highest point, about 3.5 m above ground. Greater lift may be obtained by the use of a third frame fixed to the fork carriage, the forks in this case being suspended from this frame, up and down which they are made to slide in the same manner to attain a lifting height of about 6 m (figure 23 B).

Suspension of the mast from a pivot plate set below the centre of gravity of the chassis increases lateral stability when the truck tilts over on very rough ground when travelling with the load at the bottom of the mast. It also, to a limited extent, permits the mast to remain vertical when delivering loads even though the chassis is at an angle. Verticality of the mast is essential to permit withdrawal of the forks from the load. The mast may also be kept upright while loading and unloading by means of hydraulically operated outriggers at the front of the machine. By varying the pressures on the ground at each side the mast may be maintained on a vertical axis in the lateral plane.

Some trucks are designed to provide a forward reach of about 2 m. This is useful when unloading lorries off which all loads may be picked up from one side only. These machines also give a greater lifting height of about 6.5 m (figure 23 C).

The forklift truck is designed basically for handling unit or packaged loads, that is large individual components or smaller components packaged into suitable units, rather than loose materials such as sand and gravel, although the latter can be handled by the use of skip attachments. It is necessary for the loads to be on pallets or to be separated by spacing battens in order to permit the entry of the forks under the load. Manufacturers and building materials suppliers are increasingly prepared to palletise and to make up into packs materials which are normally delivered in bulk form, for example, bricks and roofing tiles. Indeed, many do so to permit the use of forklift loading at works or the use of mechanical handling devices fitted to their own delivery lorries.

As mentioned above, loose materials may be handled by general purpose skips fitted to the forks. These can replace dumpers for this purpose with the added advantage that when required the load can be lifted and discharged at the working height of the forklift. Manually or hydraulically operated concrete/mortar skips may be similarly fitted to the forks, permitting concrete or mortar to be distributed from the mixer directly to points of use. Large and awkward components such as baths and batches of prefabricated roof trusses can be carried and lifted to required levels.

In addition to skips the mast may be fitted with a crane jib. Some models are fitted with a back-hoe to dig small foundation and service trenches. The truck can thus become a versatile machine particularly suitable for smaller sites where the amount of unloading and distributing operations would not justify its use solely as a forklift. With maximum lifting heights around 6 m these machines can be put to greatest use on low-rise building work. Most machines can lift 1 to 2 tonnes to this height and some can carry these loads at 600 mm from the heel of the forks. Thus, with forks of appropriate length, unit loads can be 1200 mm in one direction which is about half the width of a lorry deck. Travelling speeds are from 10 to 13 mph.

[1] Since loads are often greater than the maximum recommended load of 736 kg per bay of the strongest steel scaffolds (CP 97, Part 1 *Common Scaffolds in Steel*) strengthening of the scaffolding at delivery points may be necessary.

Monorail transporter

This is a powered wagon or skip running on a single easily laid rail and is intended primarily to carry concrete from the mixer to the point of placing (figure 22 *E*). The rails are supported at the joints on two-legged stands and the wagons can negotiate curves. The wagon consists of a tipping skip carried by a rectangular frame chassis with a grooved pivoted wheel at each end; stability is provided by two idle rollers at each end which work on the bottom flange of the rail which, for this purpose, is fairly deep. The wagons, with capacities up to 0.6 m³, tip on either side of the track and are powered by a small petrol engine driving one of the grooved wheels. No driver is required as the wagon is despatched by engaging the clutch and at the end of the journey the clutch is disengaged either manually or by a trip device fitted to the rail. These transporters can travel up slopes of up to 1 in 8 and are generally used in otherwise inaccessible situations. The distance of travel should be related to the mixer cycle.

Conveyors

The usual type of conveyor used on building sites is the portable belt conveyor. They can be used to handle any small materials such as excavated spoil from the point of excavation to the boundaries of the site for loading into transport, for concrete placing or for filling up aggregate bins in weigh-batchers.

A conveyor system can be made up of units up to 19 m long but is usually made up of standard lengths 5 to 7.5 m long, with a feed hopper at one end and a discharge chute at the other (figure 22 *F*). These units are coupled by telescopic supports which position the discharge chute of one unit over the feed hopper of the next and at the same time permit articulation in a horizontal plane. To allow variation of the point of discharge at the end of the conveyor chain, the last two conveyor units are provided with wheeled support and the last suspension frame is fitted with rollers on which the final conveyor unit may telescope underneath the discharge chute of the previous conveyor. This arrangement, together with the single castor wheel support under the extreme discharge point, permits the final unit to be swung round in an arc, the radius of which may be varied as it is telescoped under the previous unit, thus permitting material to be fed into any position within the arc.

The conveyors are usually electrically driven with

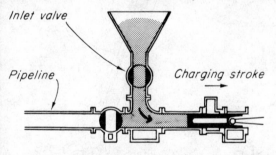

A   Concrete pump

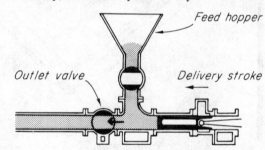

B   Concrete placer

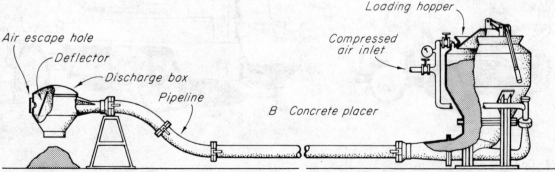

24   *Concrete pumps and placers*

a master switch at the discharge end controlling the whole conveyor chain.

Concrete pumps and placers

Pumps are mechanically operated by ram and placers are pneumatically operated by compressed air.

In the *mechanically operated pump* (figure 24 *A*) the concrete is pumped along a pipeline from the hopper by means of a ram pump and the pipe is always full of concrete, which comes out as a continuous discharge at the end. The concrete can be moved over a distance of 305 m horizontally, or vertically over a height of about 30 m. A re-mixer, or agitator, in the feed hopper is desirable to keep the concrete moving and to remedy any segregation which may occur in discharge from mixer to pump.

With a *pneumatically operated placer* (figure 24 *B*) the concrete is blown from the hopper in batches along the pipeline and the pipe is clear most of the time. With the right mix and with adequate air pressure and capacity it is possible to place concrete a distance of 395 m horizontally and 45 m vertically. It is not essential to use a very wet mix. The end of the pipeline on a pneumatic placer is fitted with a discharge box which permits the air driving a batch of concrete to escape but deflects the concrete down into the discharge opening.

The discharge ends of the pipes from pumps and placers may be arranged either to discharge concrete into its final position or to discharge it into storage hoppers for delivery into barrows where direct discharge is not convenient.

Mobile concrete pump   This is a lorry mounted

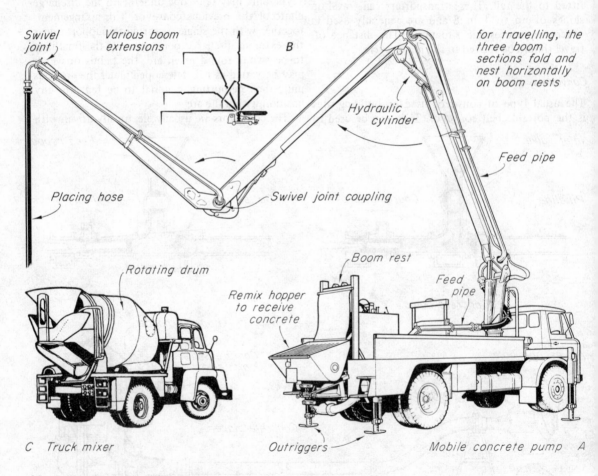

25 *Concrete pumps and mixers*

concrete pump together with a hydraulic folding boom which supports the first stages of the placing hose (figure 25 *A*). It is used mainly with ready-mixed concrete, the truck mixer delivering into a hopper at the rear of the mobile pump from which it is fed into the pump. When extended the folding boom facilitates quicker and more convenient movement of the pipeline and enables it easily to be elevated over pavement and site hoarding thus avoiding the need for the truck mixer to come off the road. The boom may be extended in various ways from the vertical to horizontal plane as shown in (*B*).

Mixing

In modern building construction a large amount of material must still be mixed with water, mainly concrete, mortar and plaster. Of these concrete is probably made in most quantities and the mixing of it has been most highly mechanized. Except on jobs requiring only an extremely small amount of concrete, a concrete mixer of some form is now used on every building site. The advantages of mechanical mixing over hand mixing, except for very small quantities, are greater economy, certainty of thorough mixing without loss of cement and accurate gauging of the water content.

Concrete mixers

Concrete mixers are made in various types and sizes and are broadly classified as (i) batch mixers and (ii) continuous mixers. The former deliver the mixed concrete in batches. The latter, which are not much used on building sites, provide a continuous and high output. There are five types of batch mixer

1 Tilting drum
2 Non-tilting drum
3 Reversing drum, a form of non-tilting mixer
4 Split drum
5 Paddle mixers (a) pan (b) turbo (c) trough

Types 1, 2 and 3, which are those most commonly used in building work, and type 5 are covered by BS 1305 and are designated according to size and type; size by the quantity in litres or cubic metres of mixed concrete delivered per batch, type by the letter: T, tilting drum; NT, non-tilting drum; R, reversing drum; P, pan mixer, giving designations for capacities up to 1000 litres such as 100T or 140NT. Mixers larger than 1000 litres capacity are designated by their output in cubic metres per batch, thus, T3.

In the first three types of drum mixer given above those smaller than 140 litres are tilting type only and those larger than 200 litres usually only non-tilting types. 140 and 200 litre models are made in all these types. The smaller sizes of tilting drum mixer are portable and in some cases are designed for towing. Most of the larger sizes are semi-portable, but in some the iron wheels are replaced by tyres in order to make them more mobile. Power is usually supplied to concrete mixers by diesel driven engine or electric motor.

Whereas the quantity of concrete required dictates the size of mixer to be used, the quality of concrete required dictates the type of mixer, since the different characteristics of the various types have an effect on the quality of concrete produced. Precast concrete work requires a high quality concrete; foundation work, say for normal strips, requires what is termed ordinary structural concrete; roads and similar areas require a lean concrete. Tilting and non-tilting drum mixers are suitable for ordinary structural quality concrete but not for dry-rich or dry-lean mixes. Split drum mixers produce good dry-lean concretes and pan mixers produce all qualities satisfactorily.

Tilting drum mixer   This consists of a pear-shaped drum open only at the narrow top end and revolving on a tilting axis which permits the drum to be tilted in one direction for loading into the open end, and in the opposite direction for discharging the batch of mixed concrete (figure 26 *B*, *C*). Blades are fixed to the inside of the drum and mixing takes place as these repeatedly pick up the materials and drop them to the bottom of the drum as it revolves.

These mixers are hand-loaded in the smaller sizes up to 100 litres. 140 litre machines are available with or without power loading skips, but very large tilting mixers are invariably mechanically loaded. The use of power loaders increases the output because the hopper or skip may be filled while the machine is actually mixing. These mixers generally discharge the concrete more cleanly and rapidly than the non-tilting mixers but mixing tends to be less thorough, resulting in a more variable concrete and in greater risk of segregation.

Non-tilting and reversing drum mixers   In these types the drum is cylindrical with partially closed ends and rotates in a vertical plane.

In the non-tilting type buckets are fixed to the inside of the drum which repeatedly pick up the

47

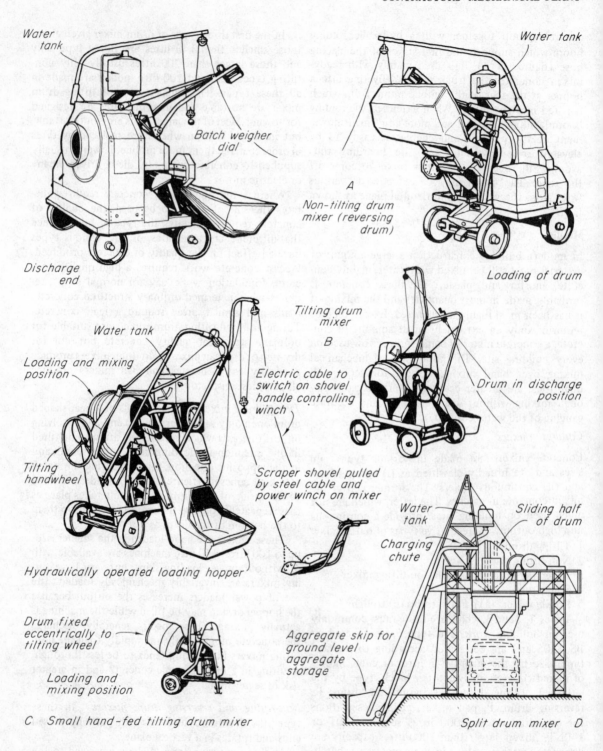

Water tank

Batch weigher dial

Discharge end

Water tank

A
Non-tilting drum mixer (reversing drum)

Loading of drum

Tilting drum mixer
B

Water tank

Loading and mixing position

Electric cable to switch on shovel handle controlling winch

Drum in discharge position

Tilting handwheel

Scraper shovel pulled by steel cable and power winch on mixer

Hydraulically operated loading hopper

Drum fixed eccentrically to tilting wheel

Loading and mixing position

C   Small hand-fed tilting drum mixer

Water tank

Sliding half of drum

Charging chute

Aggregate skip for ground level aggregate storage

Split drum mixer   D

26   *Concrete mixers*

materials and then let them fall to the bottom of the drum. Discharge is by means of a moveable chute projecting out and upwards from the discharge end during mixing. For discharge this is pulled down so that the concrete in its fall in the drum is caught on its inner end and slides down and out of the mixer.

In the reversing drum type blades instead of buckets are fixed to the inside of the drum. Discharge is achieved by reversing the direction of rotation causing the blades, which are spiral in form, to conduct the concrete to the discharge opening (figure 26 A).

Charging of the drum is carried out by a power loading side hopper or skip. The mixing action of these types repeatedly drops the materials within the drum a distance of one metre or so and, with rich, dry mixes in particular, the fine materials tend to stick on the sides and blades of the drum, producing a leaner mix than intended and increasing variability in concrete.

These mixers are easy to manufacture and to maintain and are popular in the 200 and 400 litre sizes, although larger ones are available. Some models are automatic in action producing one batch after another, only needing to be supplied with materials and with skips or other transport to take each batch as it is discharged.

Split drum mixer This mixer produces good results with dry-lean concretes and consists of a circular drum rotating on a horizontal spindle (figure 26 D). The drum is split vertically in two at right angles to the axis of the spindle and one half is arranged to slide along the spindle so that the drum opens to discharge the mix. With mixes other than dry-lean any leakage at the joint in the drum leads to loss of grout. Loading is through a chute into a feed-hole in one side of the drum.

Pan mixers These are of continental origin and consist of a shallow open topped drum or pan, 1 to 2.5 m in diameter, which may revolve on a horizontal plane or may be stationary. With the former a set of rotating paddles on a vertical spindle projects down into the drum in close contact with its side. These paddles rotate on the spindle but are fixed in position relative to the drum, the concrete mix being brought continually into contact with the paddles by the rotating action of the drum. With the latter two sets of rotating paddles project down from a horizontally rotating arm which brings the paddles into contact with the concrete.

An even distribution of water is obtained by feeding it through a pierced tube round the perimeter edge of the pan. Discharge is through the bottom at the centre (or with stationary pans sometimes at the side) so the mixer must be raised sufficiently to permit the entry of skips or dumpers, and loading must, therefore, be at high level. This makes it difficult to load and sometimes leads to a certain loss of dry cement in the mix through being blown away by wind. Should this occur there may be greater variations in the concrete than in that mixed in other types of machine. Discharge from the pan is very clean. Compared with drum mixers, pan mixers are expensive in initial and installation costs and are not readily portable, but they have a much higher rate of output and give rapid and positive mixing of all types of concrete. Capacities range from 200 litres to 1.5 m³.

A turbo-mixer is a form of closed, stationary pan mixer (figures 27A and 28). The paddles, set in two rows one above the other, project from the side of a rotating cylinder set within the pan, so that they

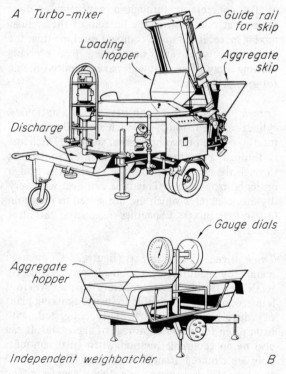

A Turbo-mixer

Guide rail for skip

Loading hopper

Aggregate skip

Discharge

Gauge dials

Aggregate hopper

Independent weighbatcher   B

27 Concrete mixing

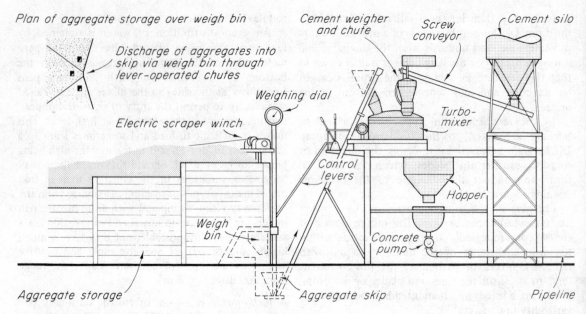

Plan of aggregate storage over weigh bin

Discharge of aggregates into skip via weigh bin through lever-operated chutes

Cement weigher and chute

Screw conveyor

Cement silo

Weighing dial

Turbo-mixer

Electric scraper winch

Control levers

Hopper

Weigh bin

Concrete pump

Aggregate storage

Aggregate skip

Pipeline

28 Central concrete mixing plant

move in the annular space formed between the 'paddle cylinder' and the side of the pan. Discharge of the concrete is through a semi-circular door situated at one point at the bottom of the annular mixing space. It is a more compact machine than the other pan types and the closed pan prevents blowing away of cement and makes it more suitable for general site use.

Trough mixer This consists of a stationary cylindrical open trough in which are mixing paddles rotating on one or two horizontal spindles. Discharge is through a bottom or side door or sometimes through the charging opening by rotating the trough on its horizontal axis. This type can deal with very dry-mix concretes which are not suited to the drum or pan type mixers. Capacities range from 280 litres to 2.25m^3.

Truck mixer Truck mixers (figure 25 *C*) are used mainly by manufacturers of ready-mixed concrete to deliver freshly mixed concrete to sites where it is uneconomic or impossible to install a mixing plant or where the site is particularly congested, with little room for mixer or storage of aggregates. It can also be an economic method when small amounts only are required from time to time. There are two types of truck mixers: one takes aggregate and cement batched in a central plant to which the water

is added and mixing commenced about five minutes away from the site; the other, which is really an agitator, takes a load of concrete which has already been batched and mixed in a central plant, and keeps it slowly turning during transit to the site. There is of course a maximum time limit of about one hour for the journey with such an agitator. This method enables the load of the mixer to be increased by about one third. It is possible to increase the load in the first type of truck mixer by means of what is known as shrink mixing. This can be done either by pre-mixing the dry materials at the central plant prior to loading the truck mixer, which permits a larger batch to be put into the mixer drum, or by dry batching a nominal full batch into the truck mixer, mixing it dry for a few moments and thus shrinking it, and then topping with more.

Figure 25 *C* shows the common type of truck mixer which discharges at the rear. In others the concrete is discharged from the front, above the driver's cab, so that the driver does not have to reverse to the point of delivery.

Good planning and progressing on the site is essential when ready-mixed concrete is used and one of the disadvantages of the method is that delivery of the concrete cannot be stopped immediately should a delay or mechanical breakdown occur on the site.

Weighbatchers

The batching of materials for concrete may be by volume or by weight but it is accepted that the careful control of water/cement ratio and proportioning by weight produces a higher quality concrete, which is much more consistent because of the greater control of the fractional quantities which the method makes possible.

Although volume batching is still used on small jobs and when normal quality concrete is mixed in hand loaded mixers, weight batching is invariably used when high quality concrete is required. When batching by weight is carried out using mixes based on the 50 kg bag of cement the aggregates only are weighed by means of a weighbatcher. Frequently, however, the cement is stored in bulk in a silo with its own weighing mechanism (figure 28).

Weighbatchers are available for use with a wide variety of mixers. Weighing may be by beam and jockey weight, by pendulum or by spring balance, or it may be hydraulic. Generally the pendulum or beam and jockey weight types give greater accuracy than spring balances. The hydraulic method uses an oil filled capsule in the bottom of the hopper; when the hopper is loaded with aggregate, the oil is pressed out of the capsule through a flexible pipe and operates a gauge dial.

Weighbatcher incorporated with mixer Some mixers up to 750 litre capacity incorporate a hydraulic weighbatcher on which the loading hopper rests, the weight of the aggregate in the hopper being registered on a gauge dial (figure 26). This may be loaded by hand or with a drag scoop.

Independent weighbatcher This is primarily for use with small portable mixers and consists of a pair of hoppers each connected by weighing mechanism to its own calibrated dial (figure 27 *B*). These batchers are usually hand-loaded and are therefore slower in use than other types which may be loaded by small drag scrapers.

Mobile and semi-mobile weighbatcher A mobile weighbatcher is mounted on wheels and jacked up for use and is usually large enough to supply one or two large mixers. Semi-mobile weighbatchers are designed to supply one or two mixers up to a total capacity of 750 litres. These batchers are usually fed by gravity from overhead mechanically loaded bins and equipment for handling and weighing cement in bulk is generally incorporated.

Cement silos

The storage of bulk cement in special silos for use with individual mobile mixers and central mixing plants is now quite common (figure 28). The silo is a tall metal cylinder, terminating at the bottom in a discharge hopper, which in most silos is fitted with weighing mechanism. The greatest advantage of bulk cement storage is that it ensures a full batch of cement being obtained by the mixer. In addition, it is easy to handle, it is cheap and it avoids the unloading of bagged cement. In most jobs of any size the economic result of these advantages will outweigh the initial cost of the silo, but the use of a silo, which is not mobile, does have the effect of making mobile mixers with which they are associated become static, so that in some cases the distances which the mixed concrete must be transported will be increased.

Central mixing plant

The concentration of batching and mixing operations in a single static plant instead of by a number of mobile mixers is often used on large, extensive sites and when large amounts of concrete per hour are required, such as on contracts involving mass concrete walls and large foundations. Such a concentration is called a central mixing plant (figure 28). Whether or not a central plant would prove the most economical method must be established for each job by a full appraisal at the contract planning stage of all the concreting requirements throughout the job.

The main advantages of central mixing are:
(i) An increased output by fewer machines and men is possible.
(ii) In suitable circumstances it is more economical than a number of separate mixers
(iii) Better quality control is possible.

In some circumstances it is considered preferable only to batch the aggregates centrally and to transport these to mixers adjacent to the work being carried out.

Output depends on the type and size of plant, on the method of loading the hoppers and on the number of different sizes of aggregate to be used: 7.5 to 22 m^3 per hour is common.

A central mixing plant may be arranged in many ways. Figure 28 shows an example using ground level weighbatching controlled from an elevated position by the mixer driver, with distribution by a concrete pump.

The essential components of a central mixing plant are:

1 Adequate storage of aggregates at ground level. The layout should be such that material for two or more days work can be stockpiled within the range of the loading apparatus.

2 Overhead aggregate storage bins, to hold not less than an hour's supply at maximum output, with weighing device and some means of raising the aggregates to the storage bins such as a loading shovel, elevator, or crane and grab. Alternatively, bulkheads and star batcher may be used with each aggregate compartment opening into a weigh bin at ground level as in figure 28. This necessitates a pit to permit an aggregate skip to drop below the weigh bin. In this case the aggregate is fed to the bulkhead wall by scraper and the weighing and loading is under the control of the mixer driver.

3 Some means of storing and weighing the cement. This may be in the form of a bagged cement store, in which case batching will be based on the bag of cement, or bulk storage in a silo may be used. In the latter case the cement is weighed by a weigher integral with the discharge hopper of the silo, or alternatively, the cement may be taken to a separate weigher by screw conveyor (figure 28) or through pipes by compressed air.

4 Elevated water storage together with some means of metering the water supplied to the mixer. In some plants a compensating device is incorporated which allows for variations in the moisture content of the aggregate.

5 A mixer or mixers. Sometimes it is better to set up side by side two mixers rather than use one large capacity single mixer since mixing time is longer in a larger mixer.

6 Storage hoppers to contain the mixed concrete until fed into transport or concrete pump (figure 26, D and figure 28). In cases where tipping lorries are used for transport, the whole of the plant is raised to permit the lorries to drive under the storage hoppers for direct loading, as in figure 26 D, and in some arrangements of the components this results in a very great overall height of the plant.

To obtain the large outputs required on jobs such as nuclear power stations special equipment is used to produce outputs in the region of 60 m^3 of concrete per hour. This includes aggregate bunkers

with electronically controlled continuous weighers at the outlets. These discharge a continuous flow of weighed material on to a conveyor belt which feeds it into a continuous mixer, the whole set-up being semi-automatic and remotely controlled by push-buttons by one operator on a control platform.

Mortar mixers

Tilting drum mixers are commonly used for mixing mortar, particularly those smaller than 100T which are useful to the smaller builder (figure 26 C). Special mortar mixers are available in the form of roller pan mixers or paddle mixers; the latter are also suitable for mixing plaster.

The following pieces of contractor's plant do not fall within the preceding categories of plant. These are compressors, pumps and well points.

Compressors

These are used to provide compressed air for a wide variety of tools and are made in a variety of sizes, ranging from very small petrol or electrically driven compressors, used for paint sprayers or similar purposes, to large compressors operated by diesel engine or electric motor, which may be used to supply compressed air through a suitable main to various points on a large site (figure 29 A).

Small compressors in the range of 0.25 to 10 m^3 per minute, powered by petrol, diesel or electric motors, are used for driving a few small tools and are normally mounted on light chassis for easy movement around the site. They are usually light enough to stand on the upper floors of the building under construction. The majority of compressors used on building works fall in the 10 to 150 m^3 per minute range and are mainly powered by diesel engine, or occasionally by an electric motor, and are mounted on a two- or four-wheel chassis for movement round the site and for towing on the public highway.

Compressors may be of the rotary or reciprocating type. Although sometimes dearer in first cost and consuming more fuel, the rotary type has the advantage that they are often more reliable and lighter in weight and are simpler in design.

Pumps

Apart from concrete pumps mentioned earlier, most

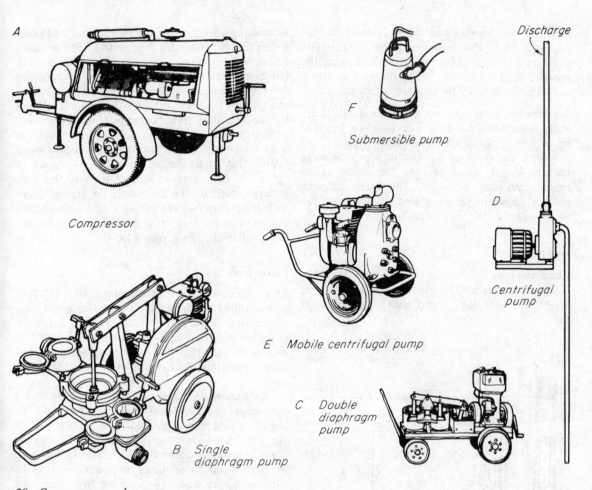

A

Discharge

F

Submersible pump

Compressor

D

Centrifugal pump

E Mobile centrifugal pump

C Double diaphragm pump

B Single diaphragm pump

29 *Compressors and pumps*

pumps used on building sites serve to keep the surface of the site and excavations free from water. They are also used sometimes to lower the water table in waterlogged ground. Pumps are operated by petrol or diesel engine or by electric motor.

Diaphragm pumps are used when the quantity of water to be handled is not great, for total heads of up to about 6 m. They are commonly used for clearing excavations as they can handle muddy water containing up to 15 per cent of solid matter. The diaphragm pump (figure 29 *B*) consists basically of a cylindrical body to the bottom of which is fixed the suction hose and the top of which is sealed by a flexible diaphragm, more or less semi-spherical in form and fixed to the cylinder at its edge. In the centre of the diaphragm is a valve which, when the diaphragm is moved up and down by the actuating

lever, opens on the downward stroke to permit water to pass on to the top of the diaphragm and closes on the upward stroke so that the water on the diaphragm is lifted to the discharge opening and, at the same time, more water is raised into the bottom of the cylinder by virtue of the suction caused. A double diaphragm pump consists of two such cylinders mounted side by side, the diaphragms being actuated by a single arm connected to the diaphragm levers at each end so that as one diaphragm is depressed the other is raised (figure 29 *C*).

Centrifugal pumps are used for moving comparatively large quantities of water free of solid matter. This type of pump (figure 29 *D* and *E*) consists of a casing, usually in volute form, which fits closely round an impeller, a disc fitted with several blades, the rotation of which draws the water in at the

53

centre of the impeller and forces it out through a discharge pipe. The maintenance of an adequate seal between the impeller and casing and round the impeller shaft is essential for satisfactory operation and various methods are adopted for ensuring this.

Submersible pumps are used for pumping 'clean' water from shafts and deep excavations for heads of up to 30 m. The submersible pump (figure 29 *F*) may have the rotor of the electric motor actually running in the water or have the motor sealed against the entry of water. The latter, although cheaper, may not remain sealed against water for very long unless the water is really clean.

Well points

These are used in lowering the ground water table. They are devices for forming small wells that can be easily sunk into the ground and withdrawn after use.

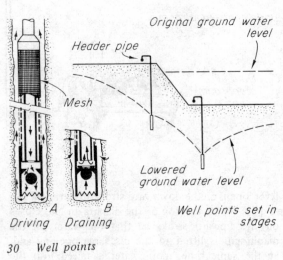

A Driving   B Draining   Well points set in stages

30   Well points

A well point consists of a 50 mm diameter steel tube enclosed at its lower end by a larger perforated tube which is covered by fine wire mesh as shown in figure 30 *A*. The driving toe at the foot encloses a simple ball valve. The points are placed about 2 to 3 m from the excavation at 0.5 to 1.5 m centres depending on the permeability of the soil. They are connected together at the top and to pumps by means of header pipes. They are driven with the aid of water pumped through the centre tube so that the jetting action of the water together with

the cutting action of the toe enables the well point to sink easily into the ground (*A*). The draining action, when pumping commences, is illustrated at (*B*). The ball valve closes the bottom of the tube and the subsoil water is drained through the perforated outer tube. The mesh prevents the perforations and the annular space from becoming blocked with soil.

Well points are effective only up to a depth of 5 m below the pump suction inlet. For greater depths they must be set in stages, as shown in figure 30. They are not suitable for very fine soils because much of the soil would be drained away with the subsoil water. For very fine soils electro-osmosis can be used, the principles of which are explained in chapter 4, page 148.

Power tools

This term covers a very large number of tools, held or controlled by hand, but operated by electricity, compressed air or petrol. They are used for many purposes and the range of variety in each type of tool is so great that it is only possible to list and illustrate the principal types and indicate their main uses.

The majority of power tools, whether operated by compressed air, electricity or petrol, have the engine or motor incorporated as an integral part of the tool. Some, however, are driven by a flexible shaft, the prime mover of which may be operated by any of the three forms of power and may be situated up to about 4 to 5 m away from the working position of the tool. Flexible shaft drive is mainly used for grinders, sanders and vibrators.

Compressed air, although an expensive form of power and comparatively inefficient mechanically, is useful where it is necessary to use a machine of the least possible weight or for heavy work such as chipping or breaking up concrete, since compressed air is the only practical means by which a hand tool can produce the really powerful blow which is necessary for these operations. Tools driven by compressed air are comparatively simple to use and maintain.

Petrol engine drive is usually found only on those tools which are comparatively large or are driven by a flexible shaft. Application of flexible shaft drive is referred to above and examples of large tools on which the petrol engine is fitted are large concrete-breaking hammers or picks, backfill tampers, and chain saws.

54

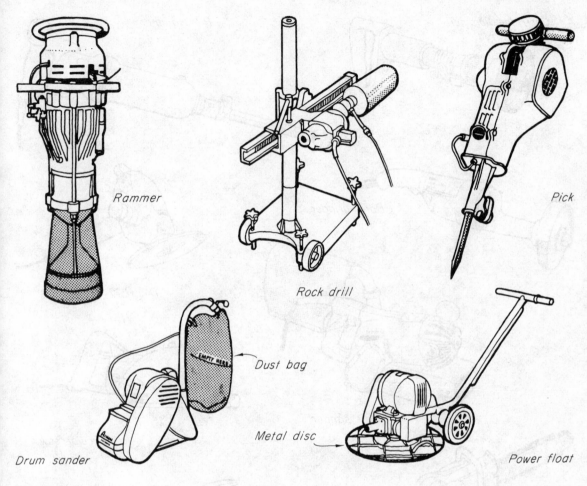

Rammer

Rock drill

Pick

Dust bag

Drum sander

Metal disc

Power float

31 *Power tools*

Electrically operated tools are also simple to maintain and have the advantage of easy starting in cold weather. The early supply of electricity, which is common on most building sites, makes possible the use of this form of power at an early stage of the contract. Alternatively small portable generators are used. The use of electrical power on a building site is also facilitated by developments such as fused distribution boards, which permit the simple plugging-in of power tools without any electrical knowledge on the part of the operator. A supply of 110 v or lower is required for power tools.

Fuller details of construction and the advantages and disadvantages of the different methods of powering are to be found in a number of publications[1].

Picks and breakers These may be operated by electricity, compressed air or petrol and may be fitted with different types of bit for working in different materials such as brick, concrete or asphalt (figure 31).

Rock drills These may be rotary or rotary-percussive in action and may be powered by electricity or compressed air, the latter usually being used for the percussive types. As well as for rock drilling, the rotary types are suitable for cutting holes up to about 65 mm in diameter in stone or marble facing and in earthenware pipes and fittings such as sinks (figure 31).

[1] See Ministry of Works Advisory Leaflets, Nos 18, 19 and 20 (HMSO)

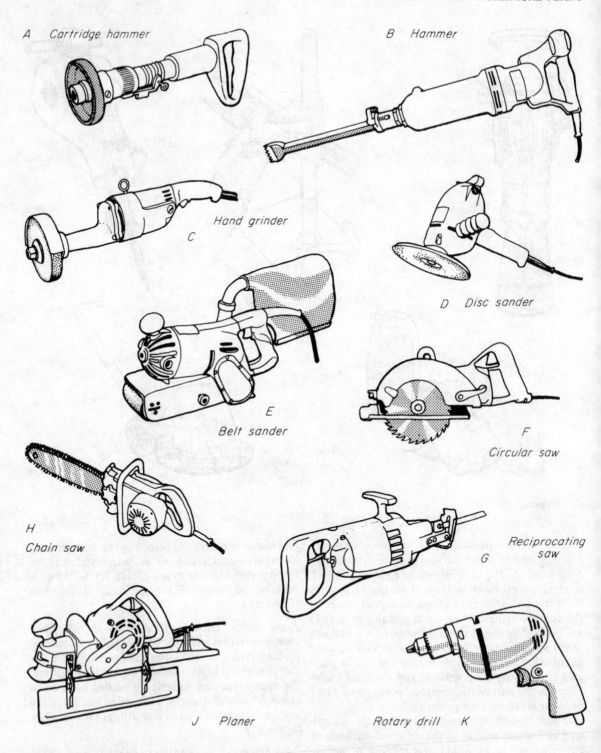

A   Cartridge hammer

B   Hammer

Hand grinder

C

D   Disc sander

E

Belt sander

F

Circular saw

H

Chain saw

G

Reciprocating saw

J   Planer

Rotary drill   K

32 Power tools

Rammers These are operated by compressed air or petrol and are used for compacting returned soil, such as the backfill in trenches and for compacting hardcore. Compacting may be by means of punning action as the rammer jumps up and down on the soil (figure 31) or by the vibrating action of a rectangular steel plate.

Hammers These may be operated by electricity or compressed air and are used for cutting holes or chases in brickwork and for bush hammering and similar operations on hard material (figure 32 *B*). They may also be used as vibrators for concrete work, being used to vibrate the formwork through a hard vulcanised rubber head so that the formwork is not damaged. They may be fitted with a wide variety of bits, including drill rotators, by means of which small holes can be drilled by rotary percussive action. As an alternative to this, the rotary action may be added to the normal percussive action of the hammer by means of a hand-operated tommy bar.

Rotary drills These may be operated by electricity or compressed air and are used for drilling holes in all types of materials. The smaller ones are commonly of pistol grip design (figure 32 *K*) and many are fitted with an extra projecting grip, by means of which the tool may be held in two hands to provide better control during drilling against the torque reaction. Attachments may be used by means of which small holes may be drilled at right angles to the spindle of the drill.

Grinders These may be driven either by compressed air, electricity or petrol engine, and are used for such operations as smoothing down the faces of *in situ* cast concrete walls and for grinding down all types of materials. They are rotary in action using abrasive wheels or discs set either at the end of the tool opposite the hand grip (figure 32 *C*) or on the underside as in the case of a disc sander. The former is called an edge wheel grinder, and is used for narrow grinding by means of the edge of the wheel; the latter, sometimes called a right angle grinder, is used for grinding broad surfaces.

Saws These are operated by electricity or petrol engine and may take the form of a circular or rotary saw, a reciprocating saw, or a chain saw. The *circular saw*, which may be fitted with different types of blade, is used for cutting timber, wall boards, plastic sheets and similar material and when fitted with an abrasive saw or disc, will cut such materials

as stone, brick and tiles (figure 32 *F*). They have cutting depths of 75 to 125 mm and a tilting base permits bevel cuts to be made. The *reciprocating saw* has a straight blade projecting at the front which operates with a backward and forward action (figure 32 *G*). Fitted with the appropriate type of blade, it may be used for cutting wood, wall boards, and metal both in pipe and sheet form. The *chain saw* consists of a guide bar about 350 mm long, projecting from the front, round which travels the cutting chain (figure 32 *H*). It is used on large baulks of timber and in tree felling.

Planers, rebaters and routers These are electrically operated. *Planers and rebaters* are fitted with two bladed, high speed rotary cutters and are used for the same purposes and in the same manner as their normal hand counterparts (figure 32 *J*). *Routers* are used with bits of varying profiles for grooving, moulding and other joinery operations.

Screwdrivers and nut runners These are usually operated by electricity but may be powered by compressed air. In general form and operation these are similar to rotary drills and as the names imply, are used to drive screws and to run and tighten nuts on bolts. They can be fitted with adjustable slipping clutches which can be preset to different values to eliminate overdriving and they are fitted with a reversible mechanism for unscrewing purposes.

Cartridge hammers These are also known as rivet or bolt guns and are used for making fixings to timber, brick, concrete and metal by means of steel pins. In essence they consist of a suitable holder to guide the pin axially as it is struck and driven home. The pins, which may be finished with a head like a nail or with a threaded end to take bolted-on attachments, are fired by means of a cartridge set off either by a built-in firing mechanism similar to a gun (figure 32 *A*) or by a firing mechanism operated by a blow from a club hammer. When a large number of similar fixings are to be made, such as skirting boards to brick or concrete backgrounds or ventilating ducting to structural steel, fixing is much quicker than by normal methods and the cost is frequently cheaper. As these tools are potentially lethal, great care must be exercised in their use. *Hand operated driving tools* which are precisely the same in principle as the cartridge hammer can be used for making fixings in brick and in some types of concrete. These are much safer than the gun since no cartridge is used, the pin being driven directly by a club hammer.

Sanders These may be operated by electricity, compressed air or petrol and are used for smoothing down wood surfaces. There are four types: (i) the *rotary or disc sander* fitted with abrasive discs up to 200 mm diameter and used for all types of joinery work (figure 32 *D*). The disc must be held at the correct angle in order to avoid pits and waves being made on the surface of the wood. It is not suitable for the final finishing of flat surfaces. (ii) the *belt sander* which uses a continuous abrasive belt 100 m wide and is fitted with two hand grips and worked over the surface in a similar manner to a plane (figure 32 *E*). It is used for smoothing flat surfaces and may be used for final finishing before painting and polishing. (iii) the *orbital sander* fitted with a sanding surface up to 115 mm by 230 mm which makes an oscillating-orbital motion at high speed. It is used for the same purposes as a belt sander, but has the advantage that the direction of grain in the wood is not important as the motion is not uni-directional. In size and form it is similar to a belt sander except that the abrasive is carried on an oscillating plate pivoted to the underside of the body. (iv) the *drum sander* which is used for sanding floors and is often called the floor surfacer. It is a larger tool than the other types of sanders, consisting basically of a rubber-covered drum, 200 to 300 mm in diameter carrying an abrasive band, mounted on a chassis which carries also the motor, the whole being pushed along by an arm or rod by the operator (figure 31). This type and also the belt and orbital sander can be fitted with dust-collecting containers or bags.

The disc or belt sanders can also be used with the appropriate types of abrasive discs or belts for metal and masonry surfaces and also for removing paints and varnishes. The disc type can also be fitted with circular wire brushes for cleaning rust and scale off ironwork and for other similar purposes.

Polishers These are almost identical to a disc sander but fitted with a flexible disc covered with various types of pads according to the nature of the polishing operation to be performed.

Vibrators and tampers These may be operated by electricity, compressed air or petrol engine. The function of these tools is to consolidate poured concrete and to release any trapped air. The general name given to this operation is tamping and for *in*

situ cast concrete in a fairly fluid state a vibrator is used. These fall into two groups, immersion or poker vibrators and clamp-on vibrators.

The *immersion vibrator* consists of a 40 to 50 mm diameter metal tube which imparts high frequency vibrations to the concrete in which it is immersed and is inserted and removed at a number of points, so that the whole of the concrete in the form is consolidated. Electrically operated immersion vibrators are the most satisfactory as they provide constant speed and the power cables are easier to handle than an air hose or a flexible shaft. The *clamp-on vibrator* is hardly a tool in the sense of those being considered here. They are heavier and more clumsy to handle than the immersion vibrator and so in fact clamp on to the formwork, which must be strong and well supported since it is vibrated as well as the concrete within it. Reference has already been made to the use of hammers on formwork for the same purpose.

The *tamper* fulfils a similar function to the vibrator, but operates by means of a succession of light taps on the surface of the concrete. It is very satisfactory for dry mix concretes and is widely used in the manufacture of precast concrete. The tamper is really a lightweight hammer which can be fitted with a variety of different sized tamping shoes and can be regulated to vary the rapidity and strength of the blows.

For large areas of surface tamping a hand-propelled beam tamper may be used. This consists of a steel-clad timber beam with handles at each end, fitted with small vibrating units, and is suitable for widths of concrete up to about 5 m.

Power floats These are also called mechanical floats and are operated by electricity or petrol engine. They are used for trowelling screeds, monolithic finishes and the surfaces of large areas of concrete and produce a very dense, smooth surface. A power float consists of a spinning metal disc, about 600 mm in diameter, which is surmounted by the engine, the whole being moved about by a projecting arm at the end of which is a crossbar handle fitted with the control (figure 31). Work is carried out much more quickly than with hand trowelling and these machines have the great advantage that they only work satisfactorily on lean and dry mixes, which produce much better results in paving work than do the rich, wet mixes. By

POWER TOOLS

this means it is possible to achieve a satisfactory
surface without a screed.

An alternative to this which also achieves a good
surface without a screed is the technique of 'early
grinding' in which the concrete surface is brought
to a finish by dry grinding with a powered grinding
machine from three to seven days after pouring.[1]

[1] Other works should be consulted for more detailed consideration of the different types of plant and tools described in this chapter and for matters of maintenance. See,
for example, *Mechanisation in Building*, H.G. Vallings (CR
Books), *Builders' Plant and Equipment*, G. Barber (Newnes-
Butterworth) and *Contractors' Plant, its organisation, operation and maintenance,* H.O. Parrack (Pitman)

3 Foundations

A foundation has been defined in Part 1 as that part of a building which is in direct contact with the ground and its function as that of transmitting to the soil all the loads from the building in such a way that settlement is limited and failure of the underlying soil is avoided. Wind loads, in this context, are assumed to cause pressure on the soil but they can, in fact, result in uplift forces at the foundations due to wind suction on the roof, particularly in extensive light single-storey buildings with flat or low-pitched roofs (see Part 1, chapter 7 for effects of wind), or due to lateral wind pressure on slender, tall structures tending to cause overturning (see Part 1, chapter 3). In such circumstances the foundations may be required to function in holding the structure down against wind uplift. Reference to this is made in chapters 5 and 9 of this volume.

Superstructure, foundations and soil act together. As indicated in Part 1 the design of the foundation cannot therefore be satisfactorily considered apart from that of the superstructure they carry. Nor should the superstructure be designed without reference to the nature of the soil on which the foundations rest. In the case of probably 75 per cent of new buildings, similar in size and character and founded on the same type of soil as the surrounding existing buildings, no problems arise in this respect. The structure and foundations of a new building can be similar to those already erected. But when a new building is much higher than those already existing or varies radically in some other way, or is built on a virgin site, its design should be developed having regard to the type of foundation likely to suit the ground. Only in this way can a satisfactory and economic solution to the design as a whole be obtained.

It is therefore necessary for the designer to have not only some knowledge of the nature and strength of the materials to be used for the foundations and superstructure, but also some knowledge of the nature, strength and likely behaviour under load of the soils on which the building will rest. An overall picture of the condition below the surface of the site is an essential factor in the selection of the type of superstructure, since it may affect fundamentally the whole planning of the building. In addition, it may affect the placing of a building or a group of buildings on a site. The science of soil mechanics, by means of site explorations and tests, provides this essential information.

SOIL MECHANICS

In the past the foundation for a building was designed on the basis of experience and although the majority were successful there were many failures due to unknown factors which experience on previous buildings had not brought to light.

Until the end of the first quarter of this century the soil had not been the subject of analytical study by engineers, although in the nineteenth century many observations were made on soils and the data recorded during the extensive construction work carried out in the development of railways and canals and in the erection of industrial buildings.

Towards the end of the nineteenth century the width of foundations was adapted to the nature of the soil by giving to the main soil types permissible bearing values and it was assumed that no settlement occurred unless the unit load on a base was greater than the bearing value of the soil on which it rested. In the first quarter of the twentieth century it was realized that every load produces settlement regardless of the permissible bearing value of the subsoil, the settlement depending on many factors other than the pressure at the foundation including the nature of the section of soil to a considerable depth and the dimensions of the loaded area. This led to the concept of allowable settlement rather than permissible bearing value and stimulated research to determine the relationship of loads to the stresses and strains set up in the soil and the behaviour of various soil types under load and to establish satisfactory methods of obtaining, classifying and testing soil samples in order to provide adequate data for foundation design. These investigations established the science of what is now known as soil mechanics.

Soil mechanics provides the engineer with information on the properties of soils which geology alone is unable to provide but which is essential in the design of any structure bearing on or against the soil and which permits a more rational approach to such problems than is possible by empirical methods based solely on experience. The practical means of providing the engineer with this necessary factual information is by thorough investigation, or exploration, of the soils on a building site and the laboratory examination of soil samples.

Site exploration

The aim of a site exploration[1], or sub-soil survey, is to provide a picture of the nature and disposition of the soil strata below the level of the ground and to obtain samples of the soils in the different strata for subsequent laboratory tests and examination.

Such an exploration should be carried out at a very early stage of a building project, commencing at the same time as the preliminary design of the structure. Because of the close connection between the soil conditions and the design of the superstructure it is inadvisable to complete the exploration before the design has been considered. This avoids unnecessary expenditure in obtaining unwanted information and the possibility of obtaining insufficient information for design purposes.

The extent of an exploration will depend on the size and type of structure, the nature of the site and the availability of local geological information. When the proposed structure is light and simple it may be cheaper to check site conditions with a few trial pits or boreholes and to use a high factor of safety rather than to carry out an extended survey (see Part 1). Should initial borings indicate considerable horizontal variations in the sub-soil a greater number would be required and in the case of clay soils an extended exploration would provide information to permit a suitable distribution of load on the foundations to be arranged so that the effects of differential settlement were minimized. Full information from an extended exploration will also permit the use of a low factor of safety with consequent economies in the foundations which can be considerable in the case of large, heavily loaded buildings.

The exploration should be taken deep enough to include all strata likely to be significantly affected by the loading from the building and this depth will depend on the type of structure, on its weight and, particularly, on the size and shape of the loaded area. Usually investigations must be made to a depth of at least one and a half times the total width of a pad foundation or three times that of a strip[2].

The cost of an exploration will, of course, vary with the type of structure and the nature of the soil but is low when compared with the total building costs: it will range from 0.1 to 1 per cent of the cost of the structure being lower for large jobs because the fixed charges for setting-up for an exploration do not vary with size of job.

As a preliminary to the sub-soil survey a surface survey will often provide much useful information of a geological nature. Topographical features give some indication of the nature of a site and of the sub-soil and should be noted.

In hilly country a low-lying flat area may indicate the silted up basin of a lake or river bed on the edges of which the thickness of silt or clay filling may vary considerably, while in flat country small mounds often indicate former glacial areas in which relatively small pockets of sand or gravel might be mistaken for a main strata or large stones in boulder clay for bed rock. Landslips form broken and terraced ground around hills and these indications would suggest instability in the subsoil.

Very few trees will grow in chalk or thin soil overlying chalk, large deciduous trees will not grow in very dry areas, conifers flourish in sandy soils and willows, rushes and reeds grow in very wet ground. These signs give some indication of the wetness of the site.

Advantage should be taken of any nearby quarry, pit or river, or any road or railway cutting, in order to examine the subsoils in the vicinity of a site, in addition to which further information may be gained from geological maps of the area and from the engineer or surveyor to the local authority.

Trial pits and boreholes

On the site itself the subsoil strata may be examined by means of trial pits or boreholes. As mentioned already the number and disposition of these will depend on the type of structure to be erected and the site conditions but they must be sufficient to provide an adequate picture of the sub-soil conditions over the site and sufficient in depth to include

---

[1] The term site exploration is used here rather than site investigation as the latter implies a broader study of a site than just the investigation of the sub-soil. For the scope of site investigations, and for fuller details of site explorations, see BS Code of Practice 2001 (1957) *Site Investigations*.

[2] See page 74 where this and other matters related to distribution of pressure in the soil are discussed.

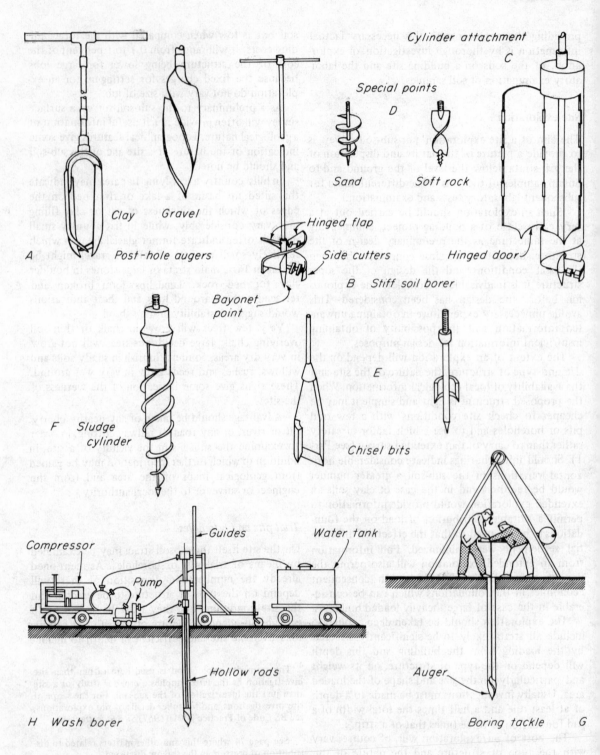

Cylinder attachment   C

Special points   D

Sand   Soft rock

Clay   Gravel

A   Post-hole augers

Hinged flap

Side cutters

B   Stiff soil borer

Hinged door

Bayonet point

E

Chisel bits

F   Sludge cylinder

Compressor

Guides   Water tank

Pump

Hollow rods

Auger

H   Wash borer

Boring tackle   G

33   Soil boring methods

all strata likely to be affected to an appreciable extent by the imposed building load. A spacing of 15 to 50 m is usual depending on circumstances.

Where practicable trial pits (see Part 1) are preferable to boreholes because the exposed rock or soil may be examined *in situ,* irregularities noted and any inclination of strata measured, but beyond a depth of about 3 m the cost of sinking a pit increases rapidly relative to that of boring so that for depths of over 6 m boreholes are invariably used.

Hand boring To depths up to 6 to 9 m in reasonably soft soils unlined boreholes may be put down by hand by means of a simple post-hole auger and extension rods. The equipment consists of a clay or gravel auger, according to the nature of the soil being penetrated, screwed to a length of gas barrel to which further lengths of barrel are added as the auger sinks (figure 33 *A*). The auger, which for this purpose is usually four to six inches in diameter, is turned by means of a tee-piece handle of gas barrel.

Other types of borers are available which facilitate boring in stiffer soils. One is shown in figure 33 *B*. The hinged upper end of the drill allows loosened soil or stones to rise upwards as boring proceeds and when the drill is raised for clearing the flap closes and retains the material on the drill. The site cutters can be adjusted to cut a hole wider than the drill to allow liners to follow the drill as it descends in loose gravel and sand where the walls of the borehole may fall in. The cutting action extends the cutters but on withdrawal a slight turn to the left draws them within the radius of the drill and allows it to be drawn upwards. In clay and other cohesive soils they admit air or water thus preventing a vacuum below the drill as it is withdrawn and facilitating its withdrawal.

A cylinder attachment can be fitted to avoid frequent clearing when boring deeper than 3 m; the hinged flap on the drill and a hinged door on the cylinder for clearing result in a form of gravel shell (figure 33 *C*). Special points can be fitted to facilitate boring in damp sand and soft rocks (*D*). Hard rocks must first be broken by chisel bits (*E*), raised and dropped to break the rock after which the drill or auger is used to extract the chips.

In free running dry sand or wet sludge the cylindrical borer shown in (*F*) can be used, the sand or sludge falling into the top of the cylinder in which it is brought to the surface.

For boring to depths greater than about 9 m it is usually necessary to use boring tackle and winch (*G*). Boring procedure generally is by hand as already described but the raising of the auger or gravel shell with its load of soil is facilitated by the use of the winch. In hard rocks or compact gravels the winch rope can be used to raise and drop the chisel bits. Lining tubes fitted with cutting shoe and driving cap can be sunk in loose sand and gravel by driving with a monkey and leader suspended from the winch, the core being removed by auger or a shell similar to those used for bored piles (see figure 61). The tubes are ultimately removed by the hand-winch or by screw-jacks. Liners are essential with soils such as loose sands and gravels but even with stiffer soils a liner for the first metre or so prevents loose top soil being knocked into the hole and acts as a guide to the auger when being inserted after clearing.

A hand rig is suitable for holes up to 200 mm in diameter and up to 24 m in depth, beyond which a power winch is usually necessary.

Mechanical boring Mechanical mobile borers speed up the boring process and reduce costs when a large number of holes is required. This type of borer is used for drilling the holes for short-bored piles and is illustrated in figure 12 *B*. For piercing very hard soil or rock various types of rotary drills may be used with hollow core-bits for the recovery of sample cores of rock.

When bedrock or other easily identified strata such as sand is to be located a quick method is by wash-boring in which the cutting bit is carried at the end of jointed tubes and as it is power rotated water is pumped down the tube and out of the bit. As it returns up the borehole, which is lined with tubes, the water carries with it particles of the soil cut by the bit from which a particular stratum may be identified (figure 33 *H*). In softer soils a bit is not required, the jet of water itself breaking up the soil to permit the liner to be sunk. The method is not suitable for obtaining a detailed section of the subsoils below a site, although driving can be stopped at intervals and samples taken below the wet, disturbed soil at the bottom of the hole by means of an extra long-reach sampling tube (see page 65).

Geophysical methods of exploration

For economic reasons boreholes, particularly if deep, frequently cannot be placed close enough to give an

63

accurate picture of the subsoil conditions. Geophysical methods may be adopted in such cases as a means of providing data between borings or of establishing the most useful positions of such borings and at the same time reducing the number necessary.

These methods can be used to obtain rapidly the depth and position of changes in strata over large areas but are only successful when the soil formations have marked differences in characteristics.

The most suitable geophysical methods for providing the information on thicknesses and depths of strata required for work on foundation studies are seismic and electrical. The seismic method of recording the reflection and time of travel of vibrations set up in the soil is useful for locating bedrock but is not suitable for use in built-up areas or on small sites. The electrical method is used for locating the position and depth of different subsoils. The most generally used of the electrical methods is the earth resistivity method which involves introducing a known electric current into the ground between two electrodes driven into the soil and then measuring the potential difference between two inner electrodes. The spacing of these electrodes controls penetration below the ground surface and calculations based on the observed currents and potentials enables the resistivity of the soil at different depths to be established. From a knowledge of the differences in resistance of various soils it is possible to determine the depth and thickness of the sub-soil strata. Some boreholes are an essential part of this method in order to prove the accuracy of the estimates and to provide data for adjusting them if necessary. They also provide the soil samples required for testing purposes.

The correct interpretation of the data obtained calls for much skill and experience and this work is invariably carried out by specialists.

In situ field tests

These are tests carried out in the field as distinct from those carried out in the laboratory and form part of the procedure for site exploration.

Vane test   This is used in soils such as soft clay and silty clay from which it is difficult to obtain good undisturbed samples (see page 65) for testing purposes. The apparatus consists of a small four-bladed vane attached to a high tensile steel rod as

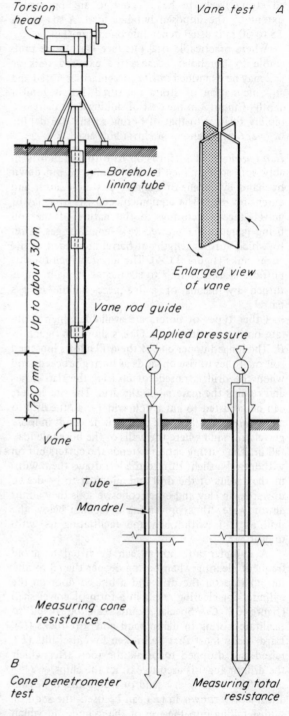

Torsion head

Vane test   A

Borehole lining tube

Up to about 30 m

Enlarged view of vane

Vane rod guide

Applied pressure

760 mm

Vane

Tube

Mandrel

Measuring cone resistance

B
Cone penetrometer test

Measuring total resistance

34   In situ soil tests

shown in figure 34 *A*. This is pushed about 750 mm into the soil at the bottom of a borehole and the force required to twist the vane is measured. This force is required to overcome the shear strength of the soil acting over the surface of the cylinder of soil turned by the vanes so that from it the shear strength of the soil can be calculated. It can be used up to a depth of at least 30 m.

Penetration test   This is used in non-cohesive sands and gravels from which undisturbed samples can be obtained only with very great difficulty. The penetrometer consists of a rod on the bottom end of which is screwed a bullet-shaped point 50 mm in diameter and on the top end a driving cap over which slides the tube of a 64 kg drop-hammer which drops through 760 mm and may be operated by hand or by a hand-winch working over a rig. Extension rods are added as driving proceeds and at intervals the number of blows of the hammer required to drive the point a given distance is recorded in order to determine the relative density of successive strata. This 'Standard Penetration Test' consists of recording the number of blows required to produce a penetration of 300 mm, the density of packing of the soil being classified as

Loose   — less than 10 blows
Medium dense   — 10 to 30 blows
Dense, or compact   — more than 30 blows

Dutch or cone penetrometer test   This is a test used to determine the bearing capacity of piles or as a rapid means of preliminary exploration. The apparatus consists of a mandrel terminating in a 60 mm diameter cone-shaped toe and outer tube. The driving load, applied by jack or increasing dead load, may be either on to the mandrel or on to the tube so that separate measurements of direct toe resistance and skin resistance may be made, or the two forces may be measured simultaneously (figure 34 *B*).

All of these tests are used only to supplement borings from which fuller evidence of the nature of the subsoil is obtained.

Loading test   This consists of applying loads to a steel plate at approximately the proposed foundation level and measuring the amount and rate of settlement which occurs under progressively increasing loads. At the beginning settlement is rapid after the application of the load but soon ceases; as the loading increases settlement will continue for a longer period between each increment of load until a point will be reached when settlement continues indefinitely. The ultimate bearing capacity of the soil is judged by the maximum load which can be applied before this occurs.

This test is of short duration and measures primarily the settlement due to direct compression and lateral displacement of the soil. It is, therefore, suitable for uniform non-cohesive soils and soft rocks in which settlements are almost entirely due to these causes but the ultimate bearing capacities obtained should be treated with reserve since the area of the test plate is smaller than the foundations and the depth of the soil affected is small; it should never be applied to deeper strata likely to be stressed by large or closely spaced foundations. The test is not suitable for c'ay soils in which consolidation settlement, taking place over a very long period of time, may be the critical factor.

Soil samples

Samples of the different soils encountered are required for testing and analysis in the laboratory. For tests such as mechanical analysis to determine particle sizes samples of all types of soil can be taken as extracted by the boring tool or from the bottom of a trial pit and placed in tins, bottles or canvas bags. For compressive and shear strength tests samples undisturbed by the boring or digging process are required because the strength properties can be altered by disturbance. For example, the shear strength of some clays may drop to one half to one quarter of their original strength when disturbed or 're-moulded'.

Samples with structure and moisture content unaltered can be taken from trial pits by a hand sampling tube when the soil is sufficiently firm and cohesive. This consists of a short length of metal tube sharpened to a cutting edge at one end and fitted to a removable handle. The tube is pushed into the soil, given half a turn to shear off the sample and extracted. The handle is removed and the hole sealed at both ends by rubber bungs.

Various sampling tools for obtaining soil specimens from boreholes have been developed, based on a thin metal tube with a cutting edge and designed to minimize disturbance as much as possible; these replace the auger as each new stratum is reached and extract an undisturbed core of the subsoil. When taken from the boring rods the cylinder is capped or sealed at both ends. The type commonly used for cohesive soils is shown in figure 35. Standard compression machines use 38 mm diameter cores but for other tests a core of not less than 75 mm diameter and about 450 mm long is required.

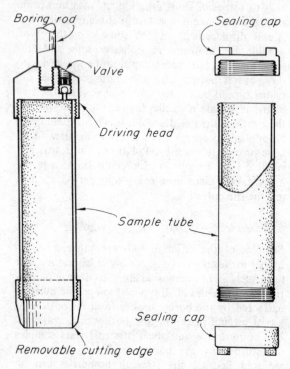

35 Soil sampling tool

It is extremely difficult to obtain undisturbed samples of most non-cohesive soils and *in situ* field tests must be made. Reasonably undisturbed cores can, however, be taken from moist sand above ground water level and for sand below ground water level undisturbed samples may be obtained by a compressed air sampler by means of which the core can be retracted into an air chamber and brought up without contact with the water in the borehole.

Sample cores of rock are obtained by the use of hollow core bits on a rotary drill. The size of core necessary to obtain a continuous sample will increase with the softness of the rock.

Undisturbed samples should be taken at each change of strata and at every 1.5 to 2 m depth of borehole and samples for classification purposes should be taken at every metre depth from the cores as they are laid out in sequence for inspection.

When ground water is encountered a sample should be taken and its level recorded.

Reports

The report of a site exploration should contain the following information:

1 Notes on all relevant topographical features
2 A complete record of the depth and thickness of the various strata in each trial pit or borehole and a note of the level of any ground water
3 A record of the point at which each undisturbed soil sample is taken
4 A full description of each stratum based on the classification of soils by grain size most commonly used in this country which recognises five broad types apart from boulders: gravel, sand, silt, clay and peat. The last, as pointed out in Part 1, is not a satisfactory bearing soil.

This classification, together with the bearing capacities of the soils and with simple tests by which these types may be identified on site, is given in table 10, Part 1.

A note should be made regarding the degree of compaction of gravels and sands and of the presence of clay or silt. Some clays break into irregular polyhedral pieces when a lump is dropped. This is most important because these 'fissured' clays can carry only about half the load they could in an unfissured state. The weight, colour and smell of all the soils may also be recorded.

Site exploration reports are based on the records of the findings in each trial pit or borehole, an example of which is given in figure 36.

Reference should be made to chapter 4, Part 1, where the subject of soils and their characteristics is introduced. Particular attention should be given to the different properties of the two extreme groups of soils, the fine-grained, cohesive soils and the coarse-grained, cohesionless soils and to the effects these have upon the behaviour of the soils under load and under different climatic conditions.

The information obtained from the exploration of a site must be extended by a laboratory examination of the soil samples collected from the various strata. This involves a number of tests to classify the soil in greater detail and to establish certain properties which govern the design of foundations.

Soil classification

In order to be able to judge the likely behaviour under load of any particular soil, it is necessary to be

66

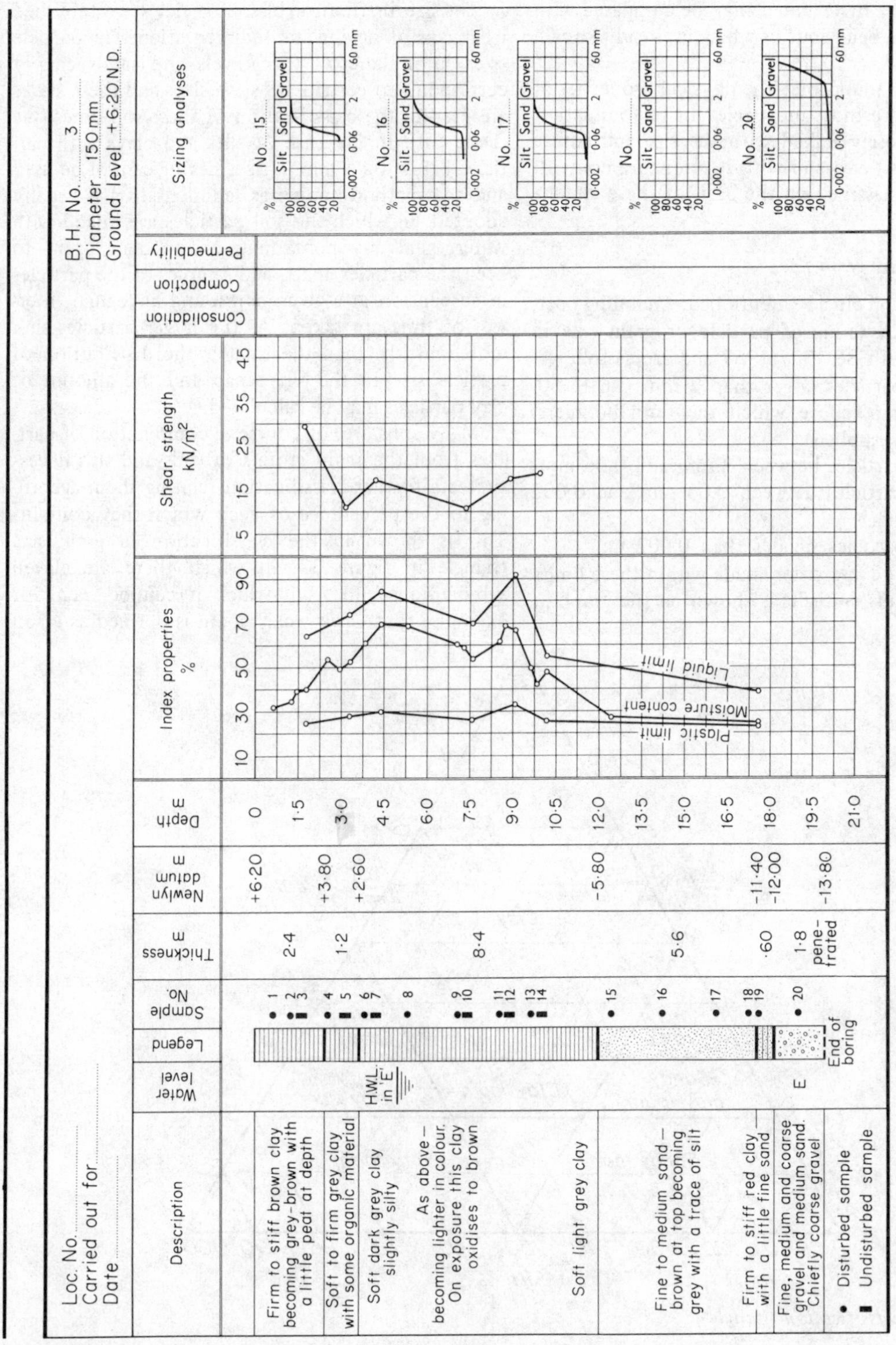

B.H. No. 3
Diameter 150 mm
Ground level +6·20 N.D.

Loc. No. ....
Carried out for ....
Date ....

Sizing analyses

No. 15 — Silt Sand Gravel — 0·002 0·06 2 60 mm — % 100 80 60 40 20

No. 16 — Silt Sand Gravel — 0·002 0·06 2 60 mm — % 100 80 60 40 20

No. 17 — Silt Sand Gravel — 0·002 0·06 2 60 mm — % 100 80 60 40 20

No. — Silt Sand Gravel — 0·002 0·06 2 60 mm — % 100 80 60 40 20

No. 20 — Silt Sand Gravel — 0·002 0·06 2 60 mm — % 100 80 60 40 20

Permeability

Compaction

Consolidation

Shear strength kN/m²: 5 15 25 35 45

Index properties %: 10 30 50 70 90 — Liquid limit — Moisture content — Plastic limit

Description	Water level	Legend	Sample No.	Thickness m	Newlyn datum m	Depth m
Firm to stiff brown clay becoming grey-brown with a little peat at depth			1, 2, 3	2·4	+6·20	0
			4, 5			1·5
Soft to firm grey clay with some organic material			6, 7	1·2	+3·80	3·0
Soft dark grey clay slightly silty	HWL in E		8		+2·60	4·5
As above — becoming lighter in colour. On exposure this clay oxidises to brown			9, 10	8·4		6·0
			11, 12			7·5
			13, 14			9·0
Soft light grey clay			15		−5·80	10·5
						12·0
Fine to medium sand — brown at top becoming grey with a trace of silt			16	5·6		13·5
			17			15·0
						16·5
Firm to stiff red clay with a little fine sand	E		18, 19	·60	−11·40	18·0
Fine, medium and coarse gravel and medium sand. Chiefly coarse gravel			20	1·8 penetrated	−12·00	19·5
End of boring					−13·80	21·0

• Disturbed sample
■ Undisturbed sample

36 *Borehole record*

able to identify it so that it may be compared with similar soils the behaviour of which is already known and recorded.

It has been found that the physical properties of a soil, which are those most relevant to foundation design, are closely linked with the size and nature of the soil particles and with its moisture content, and soils are classified on the basis of these characteristics.

Classification by grain size

The types of soil already mentioned, excluding peat, are defined in terms of particle or grain size in accordance with BS 1377:1967 and are as follows:

Gravel : particles larger than 2 mm (up to 60 mm, above which the term 'boulders' is applied)

Sand : particles between 2 mm and 0.06 mm

Silt : particles between 0.06 mm and 0.002 mm

Clay : particles smaller than 0.002 mm

The further sub-division made within these limits in gravels, sands and silts is shown on the graph in figure 38.

The size distribution of soil particles is established by means of sieving and sedimentation. The particle sizes in millimetres for gravels and sands closely correspond to certain BS sieve sizes and these sieves are used in the classification of these soil types (see figure 38). In the case of silts and clays with particles below 0.06 mm in size, sieving cannot be used and the method known as sedimentation is usually adopted, in which the soil particles are mixed with water, generally containing a dispersing agent to keep the particles apart, and samples of the particles in suspension at a given depth and at regular intervals of time are taken. As the larger particles sink more quickly than the smaller the distribution of particle size in the silt range and the amount of clay particles may be calculated.

Many soils are, of course, a combination of particles from the main groups of soils and this necessitates a further classification relating them according to the percentage of each which they contain. This is shown in the classification or grain size triangle in figure 37, in which there are eleven sub-divisions and from which it can be seen, for example, that a silty clay loam is defined as a soil

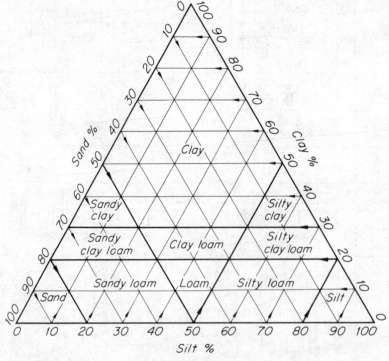

37   Soil classification triangle

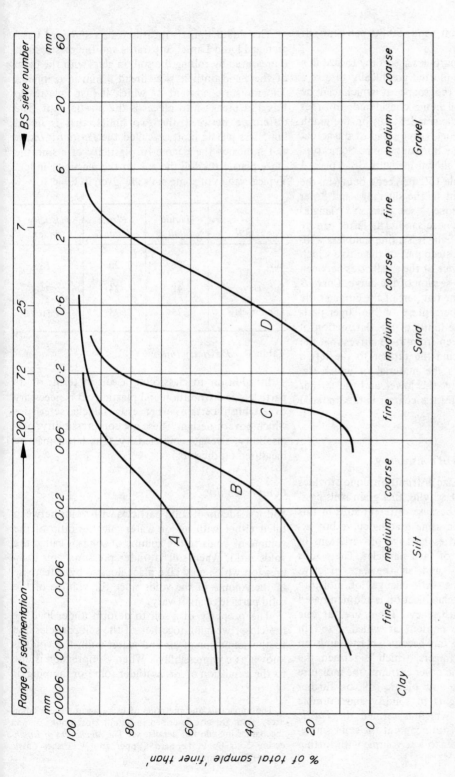

38 *Particle-size distribution curves*

containing 0-30 per cent sand, 50-80 per cent silt and 20-30 per cent clay.

The distribution of particles according to size in a sample of soil may be plotted graphically to give a 'grading curve' from the shape of which can be judged the type of soil being considered. In order adequately to spread the smaller sizes on the graph the logarithm of the particle size instead of the actual size is plotted along the horizontal axis. Some typical grading curves are shown in figure 38. Curve A indicates a clay sample (52 per cent being in the clay range, 36 per cent in the silt range and 8 per cent in the fine sand zone — see grain size triangle) and D a well-graded coarse sand. The latter can be compared with curve C which is a fine sand less well-graded as shown by the steep part of the curve which indicates a preponderance of the small sizes between the limits of the steep section of the curve. Curve B indicates a silty loam the flat part of the curve at the top indicating a low percentage of the larger particles lying between the limits of the flat section of the curve. It will be seen from these curves that the higher on the graph and the farther to the left a curve lies, the finer is the material of which the soil consists; conversely, the lower and the farther to the right a curve lies the coarser is the material represented.

Classification by limits of consistency

Classification by grain size distribution alone provides insufficient data for classifying fine grain soils passing a BS no 36 sieve because different soils in this category may have the same grading curve but exhibit different properties because of variations in the shape and nature of their particles. These soils change in consistency and, in the case of clays particularly, in volume with changes of moisture content and a further classification is adopted based on the limits of consistency. These are, at one extreme the moisture content at which a soil in liquid state on drying out becomes plastic and can be moulded in the fingers, which is termed the *Liquid Limit* and, at the other extreme, the moisture content at which the now plastic soil, on further loss of moisture, begins to solidify and crumble between the fingers, which is termed the *Plastic Limit*. At a lower moisture content the soil completely solidifies and ceases to lose volume with further loss of moisture.

The liquid limit is measured in a standard Casagrande Liquid Limit[1] apparatus and the plastic limit is measured by rolling the soil on glass with the palm of the hand until it is a thread 3 mm in diameter. The moisture content at which it just fails to do this and breaks into pieces is the plastic limit. The difference between the two limits, that is liquid limit less plastic limit, is called the *Plasticity Index* and indicates the degree of plasticity of a soil, the more plastic the soil the greater the plasticity index. Typical values of some soils are given in table 1.

Soil	Liquid Limit %	Plastic Limit %	Plasticity Index
Silt	36	20	16
Silty clay	48	24	24
London clay	75	25	50

Table 1   *Plasticity index*

In addition to classifying a soil in terms of its particle-size distribution and plasticity it is necessary to establish certain other essential characteristics which govern design. These are compressibility, permeability, strength and density. The latter may be measured on the site.

Compressibility

Soil is made up of solid particles with voids between filled either with air or water and the ratio of the volume of voids to the volume of solids is called the voids ratio. When soil is under pressure, any compression which may take place does so by a decrease in the volume of the voids since the volume of the solid particles cannot vary.

The property of a soil to deform under load by the closer wedging together of the soil particles due to the expulsion of air and water from the voids is known as compressibility. When compression is due to the expulsion of air, as under roller or tamping, it

---

[1] The apparatus and methods of carrying out these laboratory tests are not described in detail. Most books on soil mechanics include full details: see *The Mechanics of Engineering Soils* by P. Leonard Capper and W. Fisher Cassie (Spon) for example.

is called compaction and when due to the expulsion of water it is called consolidation.

Consolidation tests are carried out in an Oedometer or consolidation machine and a curve is constructed showing the relation between the voids ratio (e) and the applied pressure (p) at successive stages of consolidation, known as a p-e curve, from which values can be obtained for an equation for the calculation of settlement (see figure 45 and page 77).

Permeability

The passage of water through the voids of a soil is known as permeability and the rate of settlement under pressure depends on the ease with which any water present can flow out of the soil, that is on its degree of permeability. In sands and gravels the voids are so large that any water in the soil rapidly flows out and permits the soil particles to consolidate quickly but in the case of clay soils consolidation is a much slower process. This is because the voids in clay are minute in size compared with those in sand, for example, and are usually full of water which must force its way slowly through the fine spaces as pressure is exerted on the soil. The degree of permeability, expressed as a coefficient in metres per second, is obtained by measuring the percolation of water through a soil sample in a permeameter.

Strength of soils

The resistance to failure offered by a soil depends on its shear strength, failure occurring by surface slip. Shear resistance is ascertained in a Shear Box by means of which is measured the force required to shear a sample in half horizontally under varying degrees of vertical loading. The force applied at failure divided by the cross-sectional area of the sample gives the ultimate shearing stress.

Shear strength is considered to be made up of (i) internal friction, the resistance due to friction between the particles and (ii) cohesion, the resistance due to the tendency of the particles to hold together. Granular or coarse-grained soils derive their shear strength almost entirely from internal friction so that as vertical pressure squeezes the particles closer together the shear resistance increases. These are called cohesionless or non-cohesive soils. Fine-grained soils depend on cohesion only; however much they are loaded the particles of a soil such as clay develop no friction so that the shear resistance remains constant and equal to the cohesion of the soil. These are called cohesive soils. Intermediate types of soil exhibit both forms of resistance. The graph in figure 39 illustrates this clearly. Sample A, a sandy soil, exhibits no cohesion but shows a large angle of internal friction, that is to say there is a

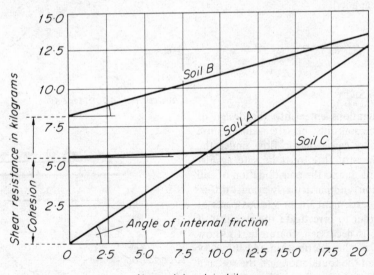

39   Shear box tests

considerable increase in shear resistance with increase in vertical load. Sample *B*, a sandy clay, because of the clay content has considerable cohesion and some frictional strength which is indicated by the relatively small angle of internal friction. The shear strength of sample *C*, a silty clay, depends almost entirely on cohesion although a slight amount of frictional strength is present. In the case of a clay the graph line would be horizontal.

This difference between the sands and clays is most important because it is in clay soils and those with small angles of internal friction that shear failure is most likely to occur. An increase in pressure on the soil due to the weight of a building will result in no increase in the strength of a clay soil whereas in the case of a sand its shear strength will rise as the weight of building increases.

Shear strength may be measured indirectly by compression tests of which there are two. One, the unconfined compression test in which a sample is compressed vertically with no lateral pressure applied, is suitable only for clays. Since the shear strength of clays is known to be half the compressive strength and as the apparatus is simple and portable the test may be used on the site. The other test is the triaxial compression test which can be used for either sand or clay, the sample being kept under lateral hydrostatic pressure while the vertical pressure is applied.

FOUNDATION DESIGN

Two main considerations enter into the design of foundations: (i) the soil must be safe against failure by shear which may cause plastic flow under the foundation, (ii) the structure must be safe against excessive settlement due to the consolidation of soil under the foundation, and particularly against differential or unequal settlement under various parts of the building. In order to investigate the stability of any foundation it is necessary, therefore, to know firstly, something of the distribution and intensity of pressure between the foundation and the soil and the intensity of pressure and shearing stresses at various points within the soil. Secondly, to know something of the mechanism of failure of the soil when over-loaded.

Pressure distribution

Distribution of contact pressure　The assumption usually made that a uniformly loaded foundation will transmit its load so that the soil is uniformly stressed is generally incorrect. The manner in which the load is distributed to the soil depends on the nature of the soil and on the rigidity or stiffness of the foundation. This is shown in table 2. It will be seen that there is a considerable variation of pressure under rigid foundations, which is illustrated in figure 40, although in practice the pressure distribution tends to become more uniform as indicated in (*A*) and (*B*) which also show the fundamentally different pattern of distribution in cohesive and cohesionless soils. The pressure will also vary with the relative density of the soil. Foundations, of course, vary between the perfectly rigid and the perfectly flexible type, in addition to which the soil may combine both cohesive and frictional properties in varying degree.

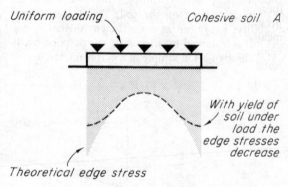

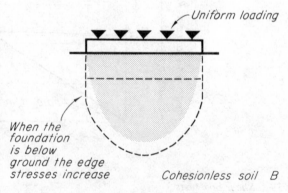

40　*Contact pressure distribution*

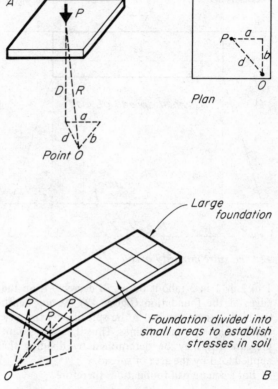

Foundation	Soil type	
	Cohesive	Cohesionless
Approaching fully flexible	Tendency to uniform distribution of pressure	Tendency to uniform distribution of pressure
Approaching fully rigid	Tendency to high stresses at edges – becoming more uniform as ultimate load is approached	Tendency to high stresses in centre – at all loads

Table 2   *Distribution of contact pressure*

This knowledge permits preliminary decisions to be made before the detailed design of a foundation is undertaken and also acts as a guide to the choice of the type of foundation by which the load will be transferred to the soil. The distribution of contact pressure has little effect on the stresses in the soil at depths greater than the width of the foundation. Its greatest significance lies in establishing the stresses which are set up in the foundation itself. In practical foundation design, based on a uniform pressure distribution, the factor of safety normally used generally covers the under-estimate of bending moments on cohesive soils. On cohesionless soils the estimate will usually be greater than the actual bending moments in the foundation.

Distribution of vertical pressure   A knowledge of the distribution of the normal vertical stress in the soil is required for the solution of settlement problems.

The intensity of vertical pressure at any point and at any depth in the soil may be established by means of Boussinesq's formula. This gives the intensity of pressure at a point 0 as

$$q = \frac{3PD^3}{2\pi R^5}$$

where q equals the intensity of pressure, P equals the applied point load, D equals the depth of the point 0 under consideration, R equals the distance between the point of application of the load and the point in

41   *Pressure intensity in soil*

the soil, O, as shown in figure 41 (A). $R = \sqrt{d^2 + D^2}$ for which d may be obtained as $\sqrt{a^2 + b^2}$. The stress intensities at any point in the soil due to a number of concentrated loads may be added. Thus a foundation transmitting a uniformly distributed load may be divided into a number of smaller areas and the load on each regarded as a concentrated load (*B*). From this the general pattern of pressure distribution under the whole of the foundation may be established. Division into 600 mm squares gives sufficient accuracy for depths below 3 m.

In practice this process involves an excessive amount of labour which may be reduced by the use of charts or tables[1]. For general preliminary purposes, however, and for establishing soil pressures in relation to foundations imposing small loads and creating small soil pressures (see chapter 4, Part 1), the average pressure at any level may be ascertained reasonably accurately by assuming a load spread of

[1] Neumark charts or Terzaghi tables

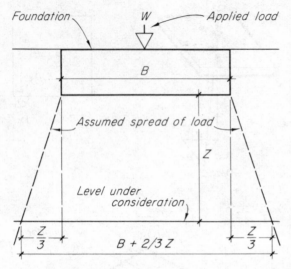

Foundation — W — Applied load

B

Assumed spread of load

Z

Level under
consideration

$\frac{Z}{3}$ B + 2/3 Z $\frac{Z}{3}$

42   Pressure intensity in soil

1 in 2 or 1 in 3 (about 60 or 70 degrees) from the edges of the foundation (figure 42). At any depth the load is then assumed to be spread over the area enclosed by the 'spread' lines. Thus the average unit pressure (q) may be determined by dividing the applied load by the area of spread.

For a square pad foundation, therefore,

$$q = \frac{W}{(B + Z)^2} \quad \text{or} \quad \frac{W}{(B + {}^2/_3 Z)^2}$$

depending on whether a 1 in 2 or 1 in 3 spread is assumed, and for a strip foundation with a load per unit run

$$q = \frac{W}{B + Z} \quad \text{or} \quad \frac{W}{B + {}^2/_3 Z}$$

At a depth equal to about the foundation breadth these results are reasonably accurate but as depth increases the pressure values obtained become greater than the maximum under the centre of the foundation as established by Boussinesq's formula. With a load spread of 1 in 3, for example, the excess is in the region of 50 per cent at a depth of twice the foundation width.

The increased pressure at any level due to the overlapping pressures from a group of closely spaced independent foundations may be determined in the same way. The spread is taken from the edges of the outermost foundations in the group and the applied load is the sum of the combined foundation loads.

A diagram may be drawn by joining points of equal pressure in the soil to produce what is known as a bulb of pressure diagram as shown in figure 43 *A*. It will be seen that the larger the loaded area, the deeper is its effect upon the soil (*B i, ii*). If a number of loaded foundations are placed near each other the effect of each is additive and if they are closely spaced one large pressure bulb is produced similar to that produced were the whole area uniformly loaded (*B iii*).

If the increase in pressure at a given depth due to a building is not greater than 10 per cent of the original over-burden pressure (see page 76) at that depth, the effect on the soil at that particular depth will usually be negligible. It has been shown in many cases that beyond the bulb of pressure bounded by the line joining points stressed to one-fifth of the applied pressure, consolidation of the soil is negligible and has little effect upon the settlement of a foundation. For practical purposes this is considered to be the significant or effective bulb of pressure. For a uniformly loaded square or circular foundation this bulb extends to a depth of approximately one and a half times the width of the foundation. For a strip foundation it extends to as much as three times the foundation width. As already noted, the pressure bulb, together with the geology of the site, indicates the depth to which investigations of the soil should be made. This is particularly important in the case of wide foundations where deep, underlying weak strata, which would not be affected by the pressures from narrow or isolated foundations, would be stressed by the wider foundations as shown in (*B ii*).

Distribution of shear stress   The distribution of shear stress and the position of maximum shear stress is important when there is a possibility of shear taking place in the soil. The maximum shear stress in the soil may be established analytically or graphically by Mohr's circle. The maximum shear stress set up by a single strip foundation carrying a uniform load of p N per m^2 is p/πN per m^2. This occurs at points lying on a semicircle having a diameter equal to the width of the foundation. Thus at the centre of the foundation the maximum shear occurs at a depth equal to half its width. The bulb of shear stresses is shown in figure 43 *C*.

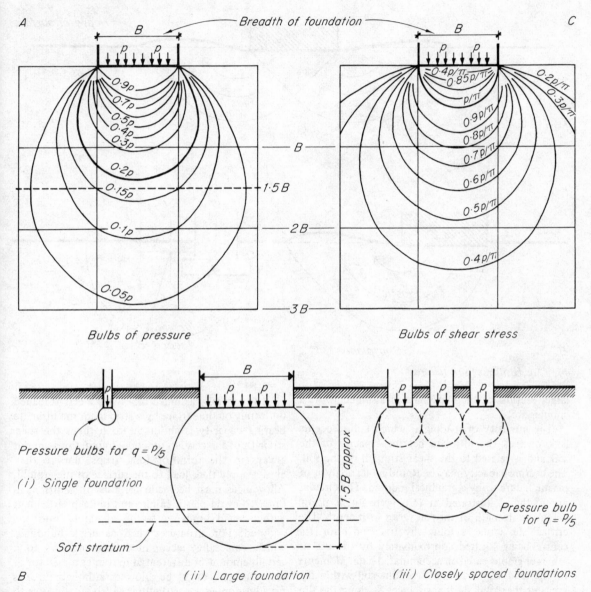

43   *Distribution of stress in soil*

Ultimate bearing capacity of soil

Should the load on a foundation be excessively high, plastic failure of the soil would occur and the foundation would sink into the ground. When a cohesive soil is surface-loaded this type of failure occurs by a wedge of soil directly under the foundation being forced downwards. This pushes the soil on each side outwards causing it to shear along a curved slip plane, so that heaving of the surface on each side takes place, as shown in figure 44 *A*. The depth of a foundation therefore must be such that the weight of soil above the base on each side together with the shear resistance of the soil is sufficient to prevent this. In practice, it is unusual for both sides of a foundation to be equally strong and failure is likely to occur on one side only, as shown at (*B*). The

75

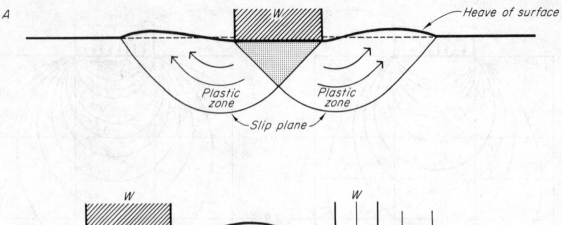

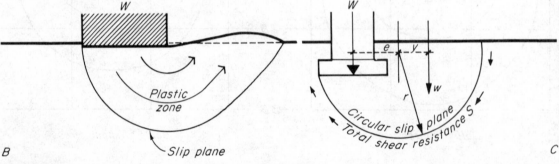

44 *Mechanism of soil failure*

larger section of the slip plane is assumed to be a circular arc.

The intensity of loading at which failure occurs is known as the ultimate bearing capacity of the soil and is related to the shear strength of the soil. The bearing capacity may be found by an analysis of plastic failure or by a graphical method. The circular arc method, illustrated at (C) figure 44, is useful when the strength of the soil varies with depth. The critical slip circle is found by trial and error, the centre being located approximately by the use of tables or graphs based on accumulated data. Moments of the load W, the weight w of the soil within the arc and the total shear resistance, S, along the slip surface are taken about the centre of the arc ($S = $ length of slip plane x shear strength of soil per unit area). For stability the sum of the moments of soil weight and shear resistance must be at least equal to We, the moment of the foundation load. The factor of safety against failure is the sum of the resisting moments divided by the moment of the foundation load:

$$Factor\ of\ Safety = \frac{(S \times r) + (w \times y)}{W \times e}$$

It is generally assumed that when a clay is loaded uniformly on the surface by a strip load, the ultimate bearing capacity is about 5½ to 6 times the shear strength of the clay. In practice, foundations are not always in the form of long strips, nor do they always apply their load to the surface of the ground. Allowances must be made for this. In addition, in order to avoid plastic failure and to keep settlement within small limits, a factor of safety must be adopted. For structures such as brick buildings, which can safely accommodate themselves to a certain amount of differential movement, a factor of safety of two might be allowed, although three is usually adopted. For structures which are sensitive to movement, such as rigid, monolithic frames, or flexible frames with light stone facings, it is necessary to allow a factor of at least three. Thus the safe uniform bearing pressure is about twice the shear strength of the soil when surface is loaded.

Normally the foundation is below the surface of the ground and a volume of soil, called the *overburden*, must be removed to form the pit. The soil below is thus relieved of the pressure due to its weight (see figure 46). This pressure may be allowed over and above that of the pressure due to the building load, so that for subsurface loading, the safe bearing pressure becomes twice the shear strength of the soil per unit area to which is added the weight per unit area of the excavated overburden.

Higher bearing pressures may be allowed on independent rectangular foundations. Although there is some difference of opinion, it is generally considered that the safe bearing pressure for a strip foundation on cohesive soils may be increased by about 30 per cent for a square foundation.

There is less risk of plastic failure with cohesionless soils than with clay, and with these the limiting conditions depend on settlement rather than on shear failure. In most cases serious settlement takes place before the ultimate bearing capactiy of the soil is reached. The ultimate bearing capacity for granular soils may be calculated from Ritter's formula, involving the depth of the foundation below the surface, the breadth of the foundation and the unit weight and angle of internal friction of the soil. In practice considerable regard is paid to the results of Standard Penetration Tests which give an indication of relative density of the soil (see page 65).

Settlement of foundations

The vertical downward movement of the base of a structure is called settlement and its effect upon the structure depends on its magnitude, its uniformity, the length of time over which it takes place, and on the nature of the structure itself. As explained in Part 1 uniform settlement over the whole area of the building, provided it is not excessive, does little damage. Unequal settlement at different points under the building, producing what is known as relative or differential settlement, may, however, set up stresses in the structure through distortion. These may be relieved in the case of a brick structure in weak mortar, for example, by the setting up of a large number of fine cracks at the joints, but in more rigid structures overstressing of some structural members might occur. The maximum amount of settlement which should be permitted, that is, the allowable settlement, in any particular circumstances is discussed on page 81.

Settlement may be caused by:

(i) the imposed weight of the structure on the soil

(ii) changes in moisture content of the soil

(iii) subsidence due to mining or similar operations,

(iv) general earth movement.

Settlement under the load of the structure may be due firstly to elastic compression, which is the lateral bulging of the soil and which takes place without a change in volume. It is usually small and occurs as construction proceeds. Secondly, it may be due to plastic flow, which has already been discussed and which must be prevented by adopting sufficient depth and bearing area for the foundations. Thirdly, settlement may be due to consolidation.

Cohesive soils Settlement in cohesive soils is dependent on the bearing pressure, the compressibility of the soil and on the depth, width and shape of the foundation. For the same bearing pressure a wide foundation will settle more than a narrow one.

Settlement due to the consolidation of cohesive soils may continue for years after the completion of the building. The compressibility of clay is appreciable, but as the reduction in volume takes place by the expulsion of some pore water, and as the permeability of clay is low, consolidation takes place very slowly. The amount of settlement in a layer of clay of thickness H equals

$$\frac{e' - e''}{1 + e'} \times H$$

where e' is the voids ratio (see page 70) from the initial consolidation due to the overburden p', and e'' is the voids ratio due to the total pressure p'', set up by the foundation load plus the overburden. The values for both e' and e'' are taken from the *p-e* curve (such as that shown in figure 45), constructed from a consolidation test carried out on a sample of

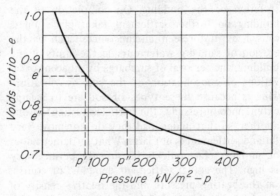

45   *Laboratory compression curve (p-e curve)*

the soil. It is possible to establish the time required for a given degree of consolidation to take place as the time is directly proportional to the square of the thickness of the soil layer, inversely proportional to the coefficient of permeability and directly proportional to the slope of the *p-e* curve. So that if the thickness of a clay stratum is known, together with the nature of the drainage conditions at its boundaries, it is possible to prepare a curve such as that shown in figure 46, showing the probable progress of settlement over a period of time.

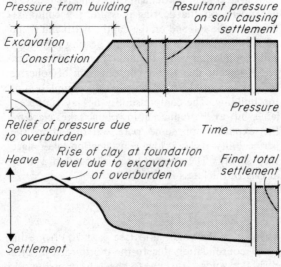

46   *Settlement of clay soils*

Cohesionless soils   In sands and gravels the amount of consolidation under load is relatively small and the rate of settlement keeps pace with the construction of the building. On completion of the building no further settlement takes place unless caused by unforeseen circumstances such as the erosion of sand by water outside the limits of the building, the removal of surcharge by adjacent excavations or consolidation of the soil by vibration. This is because these types of soil are to a large extent incompressible, the particles resting on one another, so that under load they do not move very far and settlement is not large. What settlement does occur takes place quickly because the permeability is high. The actual settlement depends, of course, on the bearing pressure and the relative density of packing of the soil particles. It also increases with the width and depth in the soil of the foundation.

Bearing pressures on soils

In the design of foundations a distinction must be made between the bearing pressure which a soil is safely capable of withstanding and the pressure which it is advisable not to exceed in practice.

The *bearing capacity* or *presumed bearing value* is the maximum pressure that a soil will bear without risk of shear failure, irrespective of any settlement which may result. It is obtained, as described above, by dividing the ultimate bearing capacity by an appropriate factor of safety, and is intended for preliminary design purposes[1].

The *allowable bearing pressure* is the maximum pressure which should be applied to the soil taking into account shear failure, settlement and the ability of the particular building to withstand settlement. The allowable bearing pressure thus depends both on the soil and the type of building concerned, and is generally less than the bearing capacity.

As explained earlier, if the soil is overloaded shear failure of the soil will result. To avoid this type of failure the soil pressure immediately beneath the foundations must not at any point exceed the bearing capacity of the soil. If the foundations rest on soil such as rock or deep beds of compact gravel (so that settlement is not a significant factor) it is sufficient to ensure that the bearing capacity is nowhere exceeded. If the soil is such that possible settlement must be considered, the soil pressure immediately beneath the foundations must not exceed the allowable bearing pressure and should, as far as possible, be equal at all points.

Uneven loading of the soil should be avoided, under both individual and combined foundations, by arranging the centre of gravity of applied loads to coincide with the centre of area of the foundation as explained in Part 1. Where some eccentricity is unavoidable the design must ensure that the allowable bearing pressure is not exceeded at any point so that relative settlement is not excessive.

Equality of applied soil pressure cannot be achieved beneath a number of independent bases if the loads on some are almost entirely dead loads but the loads on others have a high proportion of live load, since the dead loads are applied all the time but the live loads, which must be provided for, may never occur. The use of a suitably low allowable

[1] Values for bearing capacities of various types of soil are given in table 10 in Part 1.

bearing pressure may, in such cases, reduce the differential settlement to an acceptable amount. It is, however, a better arrangement, particularly on grounds of low bearing capacity, to plan the building so that the dead to live load ratio is approximately uniform in the various columns.

Settlement due to mining subsidence is confined to certain localities and this, together with general earth movement is discussed on pages 82-84.

The significance and effects of moisture changes in the soil are explained in Part 1 to which reference should be made.

Choice of foundation type

This subject is introduced in Chapter 4 of Part 1, particularly in relation to small-scale buildings, and Table 11 in that volume indicates the suitability of foundation types to the various types of soil. The underlying reasons for the recommendations made are discussed here.

The foundations and sub-structure together with the superstructure of a building are interrelated and interdependent. In order to obtain maximum strength and economy in the total structure they must be considered and designed as a whole. The larger and taller the building the more important this becomes.

The type of foundation adopted very often depends largely upon the form of construction used for the structure above. But with soils such as clay or silts, and in subsidence areas, the building itself must often be designed to react to differential settlement with no ill effect.

It is a comparatively easy matter to provide adequate foundations to small buildings, but as the size, particularly the height, of the building increases, so does the need for economy in cost of foundations increase. An increase in height increases the load on the foundations in two ways. Firstly, because the total weight of the building is greater, and secondly, because the greater horizontal forces due to wind pressure shift the resultant load on the foundations to one side, causing a local increase in soil pressure as explained in Part 1. Generally, unless conditions are exceptional, the dead weight of the building will keep the resultant within the middle third of the base. Although methods such as power boring for cylinder piles make it economically possible to build much higher on soils the nature of which previously limited the height of building, problems of settlement

are intensified. Overall settlement may be greater with the liability of greater differential settlement and the possibility of tilting of the whole structure.

In terms of overall building costs the problem of dealing with settlement must be considered from the point of view of the structure as well as of the foundation on which it rests. In extreme cases it may be essential to use a high factor of safety against ultimate failure but for economic reasons, as far as the foundations are concerned, a factor of not more than three is often desired and this may lead to considerable settlement. Although forms of structure can be designed which are flexible enough to resist safely such settlement those forms of construction which are economic in material often involve rigidity of structure, resulting in sensitivity to differential movement. The cost of structure must therefore be balanced against the cost of foundations and in some circumstances it may prove cheaper to limit settlement by using a high factor of safety so that a rigid and economic superstructure may be used.

Many factors are involved in the choice of a foundation type. The actual arrangement required in any particular case will depend upon the nature and strength of the subsoil, the type of foundation indicated by considerations of economy in the structure as a whole, the nature of the structure, the distribution of the loads from the superstructure and total weight of the building and its parts.

Foundations relative to nature of soil

A knowledge of the general manner of pressure distribution in different soils according to the type of foundation acts as a guide in making a preliminary choice of foundation. By a careful distribution of the loads from the superstructure it is possible to reduce and sometimes eliminate uneven pressure.

When it appears that settlement would be large, the bearing pressure may be reduced by adopting a greater spread of foundations, or by using a raft. It should be remembered, however, that although this procedure may be beneficial when the foundation rests directly on the compressible soil, it is not effective in preventing settlement when consolidation can take place in a deep-seated compressible stratum. This is because the wider the foundation the greater will be the depth at which the soil will be stressed. This can be seen in the comparative pressure bulb diagrams shown in figure 43 B.

When the building rests on a comparatively shallow compressible stratum, it may be cheaper to

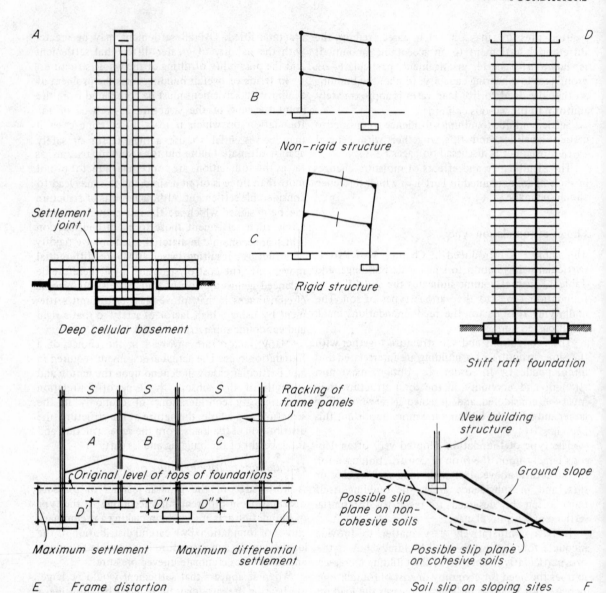

A

Settlement joint

Deep cellular basement

B

Non-rigid structure

C

Rigid structure

D

Stiff raft foundation

S S S

Racking of frame panels

A B C

Original level of tops of foundations

D' D" D"

Maximum settlement   Maximum differential settlement

E   Frame distortion

New building structure

Ground slope

Possible slip plane on non-cohesive soils

Possible slip plane on cohesive soils

Soil slip on sloping sites   F

47   Foundations, soil and superstructure

carry the foundations down to firmer soil at a lower level by means of piers or piles.

Where the compressible layers are deep-seated it is possible to limit consolidation by excavating basements so that the pressure imposed by the weight of the building is wholly or partly offset by the reduction in overburden pressure. The provision of deep basements may also be used as a means of reducing differential settlement under buildings in which some

parts are heavier than others, such as a tall tower block with lower surrounding parts of the building forming a 'podium'. A deep basement under the tower block will have the effect of reducing the pressure in the soil at that point due to the relief of overburden pressure thus reducing the difference in the pressures under the whole building (figure 47 A). The application of this method is, however, limited by the expense of the excavation which, if to be effective, may need to be very deep, and by the effect of soil movement during excavation (see page 95).

For greatest economy both in superstructure and foundations, all loads should, as far as possible, be carried directly from the point of application to the point of support. This is particularly important when the loads are great. A raft foundation may be an economic way of transferring to a weak soil the load from a large number of lightly loaded columns but when heavy load concentrations are to be transferred to the soil, deep piles passing through the weak soil to bed rock may be the cheapest solution.

Foundations relative to nature of structure

The consideration of the behaviour of soils under load shows that rigidity or flexibility of structure can influence the design of slab and raft foundations. Similarly, the behaviour of the structure under load depends not only upon the dead and imposed loads, but also upon the nature of the foundations and the manner in which the structure is supported upon them, since some of the forces acting upon the structure are those developed at its junction with the foundations (see chapter 9). Accordingly, as indicated earlier, consideration must be given on the one hand to the choice of a foundation appropriate to the type of structure likely to be most economic, and on the other hand to the possibility of using a structure suited to the soil conditions and to foundations which appear most appropriate. For example, in some circumstances on sites with weak soils the most economic building might result from the use of a few expensive types of foundation, rather than from the use of a larger number of more lightly loaded and cheaper foundations. In such circumstances the structural frame would be designed with this in mind, the columns being widely spaced in order to reduce the number of foundations required.

Design of structure to resist settlement

Since a certain amount of settlement is usually inevitable, it must be reckoned with and provision made for it in the design of the structure. The effects of differential settlement depend to a large degree upon the rigidity of the structure. For example, in structures in which beams rest simply on brick walls, unequal settlement unless excessive will not affect the internal stresses in the structural elements because these are not distorted (figure 47 B), although finishes may be affected. But in rigid continuous structures even a small amount of differential settlement of the supports will cause secondary shearing forces and bending moments in the beams and columns due to the distortion of the frame arising from the rigid joints (see figure 47 C). Buildings with rigid frames may be supported upon a stiff raft foundation (D), or the foundations of individual columns can be connected by a series of continuous beams so that the structure settles as a whole, the frame being designed to act as a very deep girder to resist any secondary stresses set up by settlement. Although secondary stresses may be set up in rigid structures by settlement of the foundations, such structures do permit forms of construction which require less resistance from the soil than non-rigid forms. This applies especially to single-storey structures where rigid frames with hinged bases may be used to prevent the transfer to the foundations of bending stresses set up by wind pressure on the superstructure above. This is discussed more fully in chapter 9.

Although the structure of a building supported on a non-rigid foundation may be of comparatively flexible construction, so that its stability is not impaired by movement, differential settlement is usually limited in order to avoid damage to claddings and finishes and excessive unevenness in the floors. The limit of this movement will depend not only upon the type of structure and the nature of the finishes, but also upon the extent to which damage and subsequent repair to finishes is acceptable. The acceptable limit may vary with different buildings. Where large movements such as occur on subsidence sites must be accommodated, special forms of construction to deal with this problem must be devised and this is discussed on page 105.

It has been suggested that the differential settlement of uniformly loaded continuous foundations and of equally loaded spread foundations of approximately the same size, is unlikely to exceed half of the maximum settlement, and that normal structures such as office buildings and flats can satisfactorily withstand differential settlements of 19 mm between adjacent columns spaced about 6 to 7.5 m apart. It is suggested that an allowable pressure should be selected such that the maximum settlement of any individual foundation is 25 mm[1]. These figures re-

[1] See *Soil Mechanics in Engineering Practice*, by K. Terzaghi and R.B. Peck, Wiley, 1967.

late only to foundations on cohesionless soils.

Since most damage arising from differential movement is due to distortion of the rectangular panels between beams and columns, and since the angle of 'racking' depends on the distance between the columns as well as on the actual amount of settlement, it is possible to formulate rules on the same principle as the deflection limitations used in beam and slab design, in which the maximum differential settlement between adjacent columns is limited to a certain fraction of the span between them. Thus the difference in the settlements of two adjacent columns, that is, the differential settlement between them, divided by their distance apart, is a measure of the severity of the settlement. There is considerable risk of damage in a normal framed building when this fraction exceeds about 1/300, so that at design stage it is advisable to so limit the estimated differential settlement that this fraction does not exceed 1/500. This is illustrated in figure 47 E. It will be seen that the distortion in frame panel A is greater than that in the centre panel B. This is obvious from inspection and also from the fact that D/S in panel A is greater than D/S in panel B. The differential settlement in panels B and C is the same, but as the distance between the columns of panel C is greater than that between those in panel B, the severity of 'racking' in panel C is less and therefore the likelihood of damage is less than in panel B. This again is clear from a comparison of the fraction D/S in each case. Having established in this way the safe estimated differential settlement, the building frame should be designed so that the strains due to the settlement are not excessive.

Although this method of relating estimated settlement to the structure is useful, the relative uncertainty and approximate nature of settlement calculations compared with structural calculations must be borne in mind. This limits their application — more refined calculations will not make the soil more uniform. Further, the flexibility of structure, generally speaking, is not continuous throughout a building due to the stiffening effect of structural walls and lift shafts. This tends to localize distortion in different parts and the resultant damage may be greater than that anticipated on the basis of a ratio of 1/1500.[1]

[1] See 'The Allowable Settlements of Buildings', by A.W. Skempton and D.H. McDonald, *Pro Inst Civ Engrs*, Part III, December 1956 and BRS Current Paper 33/75: *Settlement of Buildings*.

Foundations relative to ground movement

Reference to ground movement has been made in chapter 4, Part 1, where the effect of seasonal moisture variations and of tree roots in causing settlement on clay soil is discussed and where reference is made to the lifting effect on foundations by frost heave. These volume changes cause local movements of the soil, but mass movement of ground takes place in unstable areas: these occur as a result of subsidence in mining areas and in areas of brine pumping, landslips on unstable slopes and creep on clay slopes.

Subsidence When subsidence takes place, in addition to the total vertical movement there is a horizontal movement which exerts forces on any structure resting on or built into the subsiding ground. This can best be understood by considering the subsidence caused by the extraction of a coal seam. As the working face moves forward and support is removed from the ground behind, settlement will take place after a certain time depending on the nature of the overburden and the method of extracting the coal. This will cause a depression of the ground at the surface which will move in advance of the working face causing a tilt and curvature in the ground which is termed the 'subsidence wave'. This is illustrated in figure 48.

As the wave passes beneath a building it will subject it to changing forces according to the position of the building on the wave:
(a) At the crest of the wave a cantilever effect will be produced in the foundations and tensile forces will be set up in the ground. This tension will be transmitted to the foundations (2)
(b) On the flank of the wave the building will be subjected to eccentric loads because it is tilted out of the vertical (3)
(c) In the trough of the wave support will be removed from the centre of the foundation and a beam effect will be produced setting up tension in the foundation. Compressive forces will be set up in the ground (4)
(d) When the subsidence wave has finally passed, the building will resume its vertical position but at a lower level (5).

Although the total subsidence may be considerable the length of the wave is usually so very much greater than the maximum length of the building that day-

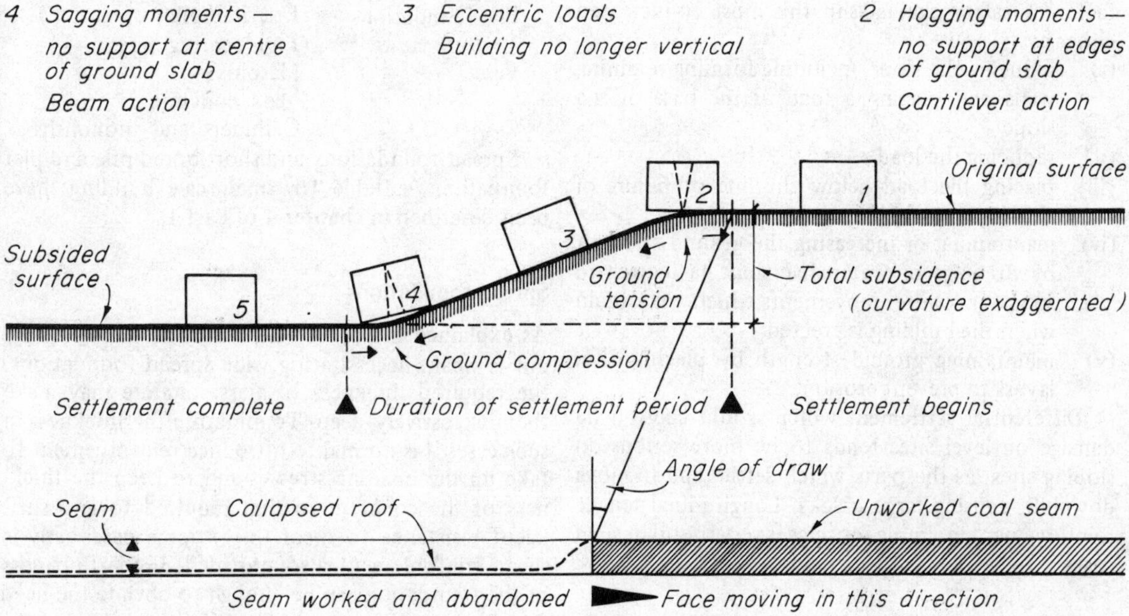

4 Sagging moments —
   no support at centre
   of ground slab
   Beam action

3 Eccentric loads
   Building no longer vertical

2 Hogging moments —
   no support at edges
   of ground slab
   Cantilever action

Original surface

Subsided
surface

Ground
tension

Total subsidence
(curvature exaggerated)

Ground compression

Settlement complete   ▲ Duration of settlement period ▲   Settlement begins

Angle of draw

Seam   Collapsed roof   Unworked coal seam

Seam worked and abandoned ► Face moving in this direction

48   Ground subsidence

to-day settlement is slight[1]. Precautions against undue damage of the structure by these movements must, however, take account of this behaviour of ground and structure during the passage of the subsidence wave. Methods used to minimize the effects of subsidence are discussed later on page 105.

Unstable slopes and creep   Unstable slopes subject to landslip may often be recognized by the characteristic uneven surface of the ground, and where possible such areas should be avoided as building sites. The tendency for the upper strata of clay soils to move downhill on sloping sites is always present and will be governed by the angle of the slope, the characteristics of the soil and other factors. Landslips on slopes previously just stable may be started by work carried out on adjoining land. Sometimes only the surface layers creep downhill and where this has occurred evidence of this will be seen in tilted fences and boundary walls and curved tree trunks. The creeping layer may vary in depth from a few millimetres to some metres, and may move on slopes as shallow as one in ten. Where buildings must be erected on such sites large scale and expensive works may be required to stabilize the ground.

Building on any sloping site, whether signs of slip or creep are evident or not, should be preceded by very careful site exploration and careful design of the foundations. Non-cohesive soils will tend to slip at different angles according to the angle of internal friction of the particular soil and this must be carefully established. The calculated shear on any inclined plane passing under the foundations, and with the upper and lower edges cutting the ground surface, should not be greater than half the total resistance available due to friction on the worst surface. (See figure 47 *F*). In cohesive clay soils slip may take place on a curved surface, as described earlier, and the calculated shear stress on any such surface passing under the proposed foundations and with its upper and lower edges cutting the ground surface should not be greater than half the shear strength likely to be available under the worst anticipated future conditions.

When building on sloping sites it is most important that water should be drained away from the uphill side of foundations.

Generally, any slope tends to be unstable and placing a load at the top increases this tendency. If site exploration and analysis indicate an inadequate

[1] For a fuller consideration of the nature of subsidence, see *National Building Studies Special Report* No. 12, 'Mining Subsidence-Effects on Small Houses'.

factor of safety against slip this must be increased either by

(i)   reducing the slope, including forming retaining walls and placing a load at the base of the slope

(ii) reducing the load

(iii) placing the load below the line of failure of the slope

(iv) maintaining or increasing the ground strength by drainage or water diversion, having regard to ground water movements which will obtain when the building is erected

(v)   maintaining ground strength by placing filter layers to prevent erosion.

Differential settlement which would cause little damage on level sites tends to be more serious on sloping sites, as the parts which settle tend to move downhill, causing large cracks. Longitudinal tensile reinforcement in simple footings is very useful in such cases.

FOUNDATION TYPES

So far in this chapter consideration has been given to the resistance of the soil to the loads imposed by building, to its behaviour under load, and to the factors which affect the choice of foundation. It now remains to continue from Part 1 the examination in detail of the actual types of foundations used in various circumstances to transfer the building loads to the soil.

The broad classification into *shallow* and *deep* foundations has been explained in Part 1 and some indication has been given of the situations in which spread, pile and pier foundations are likely to be used. This is extended here under the headings of the different foundation types now to be discussed.

The foundation types defined in Part 1 may be broken down further into the following sub-divisions each of which will be considered in detail.

Strip and slab foundations	Wide strip foundation
	Isolated column foundation
	Continuous column foundation
	Combined column foundation
	Cantilever foundation
	Balanced base foundation
Raft foundations	Solid slab raft
	Beam and slab raft
	Cellular raft
Pile foundations	Friction piles
	End bearing piles
Piers	Masonry
	Mass concrete
	Cylinders and monoliths

Spread foundations and short-bored pile and pier foundations suitable for small-scale buildings have been described in chapter 4 of Part 1.

Spread foundations

As explained in Part 1 when loads are heavy or the soil is weak, necessitating wide spread foundations, the required thickness of mass concrete may make them excessively deep. To minimise the thickness in such cases it is normal to introduce reinforcement to take up the bending stresses and to keep the thickness of the concrete to that required to give sufficient resistance to shear. In extreme cases only is shear reinforcement also provided. If the thickness of the foundation can be enough to obviate the need for shear reinforcement, the risk of cracks and the consequent corrosion of the steel by ground moisture is avoided. The detailed design of such foundations is covered in textbooks on reinforced concrete. They will be considered here in terms of their types and applications.

In spite of the greater depth of foundation considerable benefit may be derived from the use of mass concrete bases in waterlogged ground. These may often prove cheaper than the continuous pumping needed to permit the fixing of the reinforcement and the placing of concrete around it in the dry.

Wide strip foundation

This foundation may be used to carry load-bearing walls, including brick or concrete walls to multi-storey buildings. Since the loads will be heavy the necessary width of foundation will be greater than that required for a small-scale building (figure 49 *A*). The bending stresses set up by the double-cantilever action of the foundation across the base of the wall are, therefore, likely to be high and to resist these reinforcement is placed at the bottom of the slab, which will be the tensile zone, usually in the form of rods at right-angles to the length of the wall.

SPREAD FOUNDATIONS

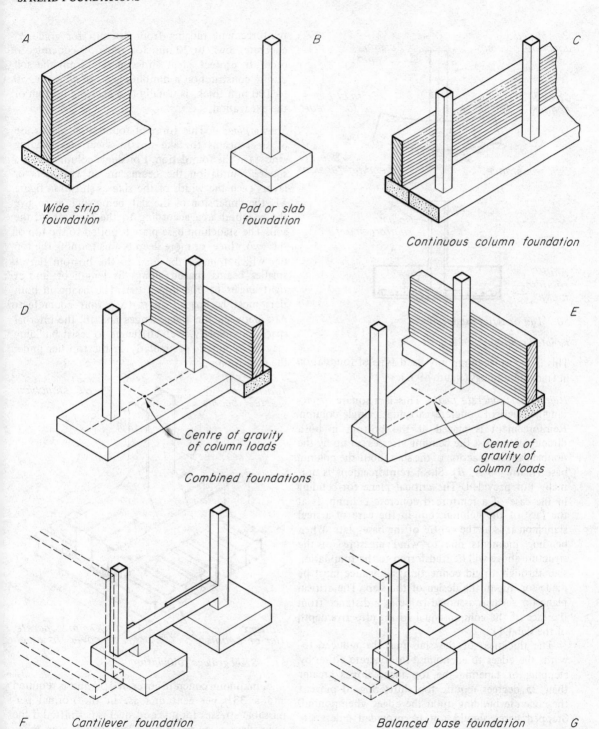

A

Wide strip
foundation

B

Pad or slab
foundation

C

Continuous column foundation

D

Centre of gravity
of column loads

Combined foundations

E

Centre of
gravity of
column loads

F   Cantilever foundation

Balanced base foundation   G

49   Foundation types

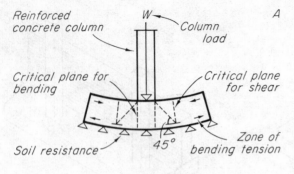

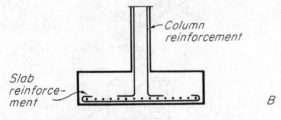

50   Pad or slab foundation

Isolated column foundations

This is the most commonly used type of foundation in framed buildings (figure 49 *B*).

Reinforced concrete pad   This is a square or rectangular slab of concrete carrying a single column. Reinforcement is placed at the bottom in both directions to resist the bending stresses set up by the double-cantilever action of the slab about the column base (figure 50 *A, B*). Shear reinforcement is normally not provided. The critical plane for bending in the case of a reinforced concrete column, is at the face of the column, but in the case of a steel stanchion it is at the centre of the base-plate. When bending moments due to wind pressure on the structure above will be transferred to the foundation slab through a rigid connection, allowance must be made for this in the design of the slab. The critical plane for shear is assumed to be at a distance from the face of the column equal to the effective depth of the slab (*A*).

The thickness of the slab may be reduced towards the edges to economise in concrete either by stepping or tapering the top face. Slopes greater than 25 degrees require top shuttering to prevent the concrete building up at the edges when poured. Stepped bases should not be provided unless site supervision is sufficient to ensure that the concrete in each base will be placed in one operation. CP 110 : Part 1 : 1972 requires a cover to the

reinforcement ranging from 40 mm for grade 25 concrete down to 20 mm for grade 50 concrete. In order to protect steel and concrete from the soil during construction a blinding layer of concrete, 50 to 100 mm thick, is usually laid over the bottom of the excavation.

Steel grillage   This form of foundation makes use of steel beams to take up the shear and bending stresses in the foundation. For single columns with a square foundation the beams are in two tiers or layers each the width of the slab as shown in figure 51, the dimension of the slab being such as to give the required area according to the strength of the soil. The stanchion base-plate is bolted to the top of the two, three, or more deep beams forming the top tier which transfer the load to the bottom tier of smaller beams spaced along the length of, and at right angles to, the upper tier. The beams in both tiers must be spaced apart to permit concrete to pass between the flange edges and fill the internal spaces completely. Web stiffeners to resist buckling may be required, particularly in the top tier under the stanchion.

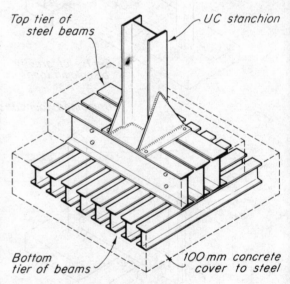

51   Steel grillage foundation

A minimum concrete cover of 100 mm is required and a $33\frac{1}{3}$ per cent increase in the normal permissable stresses for uncased steel is permitted if the beam flanges are spaced not less than 75 mm apart. The grillage foundation is expensive in steel and is not often used except when very heavy loads from

86

steel stanchions must be carried on a wide slab and where the depth of the foundation is restricted in order to keep it above the subsoil water table.

Continuous column foundation

This type of foundation supports a line of columns and is one form of combined foundation (figure 49 *C*). Although it is, in fact, a strip foundation the term 'continuous column foundation' is used here to distinguish it from the normal strip foundation carrying a wall which is subject only to transverse bending, whereas the continuous column foundation is subject to both transverse bending and longitudinal bending due to the beam action between the columns as shown in figure 52 *B*. Circumstances in which it might be used are: (i) where the spacing in one direction and loading of the columns are such that the edges of adjacent independent column foundations would be very close to each other or would overlap, and (ii) where there exists some

restriction on the spread of the foundations at right-angles to a line of columns, such as a site boundary or an existing building adjacent to a line of outer columns, which would exclude the possibility of an independent foundation of adequate size to each (figure 52 *A*). This is common in urban areas where a new building is to be erected between the existing party walls of adjoining buildings, and it is inadvisable or unnecessary to cut under the party walls to form foundations.

Assuming that sufficient width of foundation, relative to the spacing of the columns, is available to provide the necessary area of foundation, the strip is designed as a continuous beam on top of which the columns exert downward point loads and on the underside of which the soil exerts a distributed upward pressure. The stresses set up are the reverse of those in a normal continuous floor beam. The main tensile reinforcement is therefore required near the upper face between columns and near the bottom under the columns to resist the negative

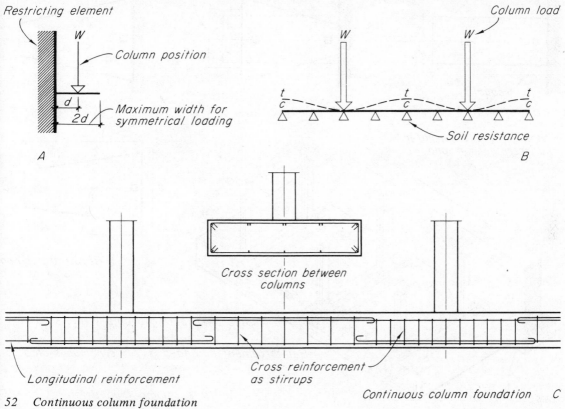

52   Continuous column foundation

bending stresses as shown in figure 52 (*C*). Transverse reinforcement must always be provided to locate the main bars during concreting, whether or not it is required to resist transverse bending.

This type of foundation is sometimes formed as an inverted tee-section in order to provide adequate longitudinal stiffness.

Combined column foundation

It has been suggested that a continuous column foundation may be used when there exists some restriction on the spread of independent foundations at right-angles to a line of columns. This assumes

that the columns can be placed far enough away from the restriction to provide sufficient area of foundation since, in order to obtain a symmetrical disposition of column relative to foundation (to ensure even stressing of the soil), the maximum width available is twice the distance from the centre line of the columns to the restriction (see figure 52 *A*). When this distance is too small to provide adequate width of foundation the columns close to the restriction may be linked to an inner line of columns on what is called a combined foundation, by means of which sufficient area and an even distribution of pressure may be obtained (figure 49 *D, E*).

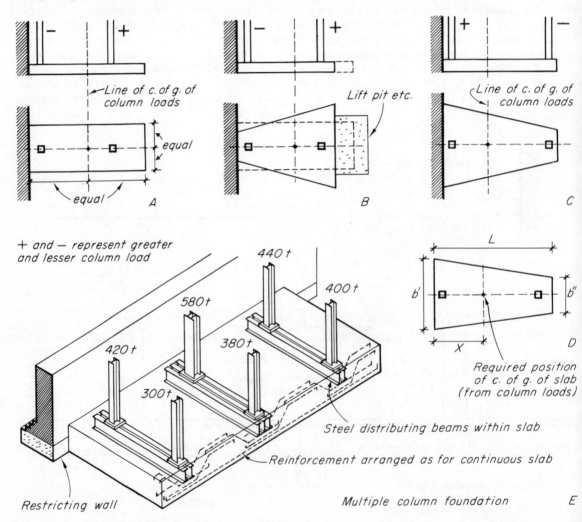

53   Combined column foundations

One or more pairs of columns may be combined in this manner, but the problem of design is the same in each case. This is the provision of a slab with an area sufficient to prevent overstressing of the soil and excessive settlement under the combined column loads, and of such a shape and proportion that its centroid lies as nearly as possible on the same vertical line as the centre of gravity of the column loads.

In the case of a pair of columns which are equally loaded or of which that next to the restriction is more lightly loaded, a simple rectangular shaped slab is used (figure 53 A). The position of the centre of gravity of the column loads is first established. This will lie mid-way between equally loaded columns or, if these are unequally loaded, nearer the more heavily loaded column. In the latter case the position is found by taking moments about a point. This will determine the length of the foundation slab since this length must be twice the distance from the centre of gravity of the loads to the line of restriction (assuming that the slab is to extend to this) in order to make the centroid of the slab coincide with the centre of gravity of the loads. The width of the slab is then determined according to the area required for the foundation.

When, in the case considered above, the necessary projection of the slab beyond the more heavily loaded column is restricted, for example, by a lift pit close to the column (figure 52 B), or when the loading is reversed and the more heavily loaded column is nearest the restriction (C), a trapezoidal base must be used. This is necessary in order to permit the centres of gravity of loads and slab to lie on the same vertical line, since the position of the centroid of a trapezium along its axis may be made to vary with changes in the proportions of the ends. The length of the slab in the first case will be limited by the restrictions at each end, and in the second case is arbitrarily fixed by extending the base just beyond the more lightly loaded column. The widths of the two parallel ends are then determined by the following equations in which the known factors of the area required, the length of the slab and the calculated position of the centre of gravity of the column loads, are used to give a trapezium the centroid of which will coincide with the centre of gravity of the loads (D):

(1) $$A = \frac{(b' + b'')}{2} \times L$$

gives $b' + b''$

Where $b' + b''$ equals the sum of the lengths of the parallel sides
A equals the required area
L equals the determined length

(2) $$X = \frac{L}{3} \times \frac{(b' + 2b'')}{(b' + b'')}$$

gives $b' + 2b''$

Where X equals the distance of the centre of gravity of the column loads from the longest parallel side b'
Insert $b' + b''$ from equation (1)

(3) $$(b' + 2b'') - (b' + b'') = b''$$ from which can be found b'.

The combined foundation slab, whether rectangular or trapezoidal in shape, will be reinforced along its length on the line of the columns, in a similar manner to the continuous column foundation, to resist the tensile stresses set up by the beam action between the columns and any cantilever action in the ends. Transverse reinforcement will also be provided.

As indicated earlier, this type of foundation may be in the form of a multiple column slab, carrying two or three pairs of columns, in which, as before, it is desirable that the centres of gravity of column loads and slab coincide (figure 53 E). This, however, is not always practicable and a certain amount of eccentricity may have to be accepted within the safe limits of soil strength and settlement. Combined foundations may be constructed as a single or double tier steel grillage in conjunction with steel stanchions (E). The loading and spacing of a single pair of stanchions may be such that sufficient width of slab will be obtained when the necessary number of steel beams in a single tier linking the stanchions has been encased with concrete to give the required 100 mm cover. If this is not the case the alternative to a bottom tier of beams at right-angles to the line of stanchions is to reinforce the concrete with transverse rods as in the case of the multiple column slab shown in (E). The use of combined foundations is not limited to the circumstances visualized above and a common or combined foundation may be provided for a number of adjacent columns where the size of independent foundations would be such that they would overlap.

Balanced foundations

These consist of the cantilever foundation and the balanced base foundation which may be used in the following circumstances instead of a combined column foundation:

(i) As an alternative to a trapezoidal combined foundation slab

(ii) When some obstruction at a column position prevents an adequate foundation being placed directly under the column. For example, when the column is placed close to the wall of an adjoining building, the foundations of which project under the column, or when a sewer passes directly under a column. In these circumstances a cantilever foundation would be used.

Cantilever foundation The foundation consists of a ground beam one end of which, cantilevering beyond a base set a short distance in from the obstructed column and acting as a fulcrum to the beam, picks up the foot of the column while the other end is tailed down by an internal column as shown in figure 49 *F*.

The loads on the beam and the two base slabs are quite simply ascertained from the loads on the two columns, and from these the beam size and slab areas may be calculated (see figure 54 *A*):

$$W_1 \times x = W_3 \times y$$

Therefore $W_3 = \dfrac{W_1 \times x}{y}$ (this counterbalancing force must be provided by the column load W_2)

$W_4 = W_1 + W_3$ (the load on the fulcrum slab)

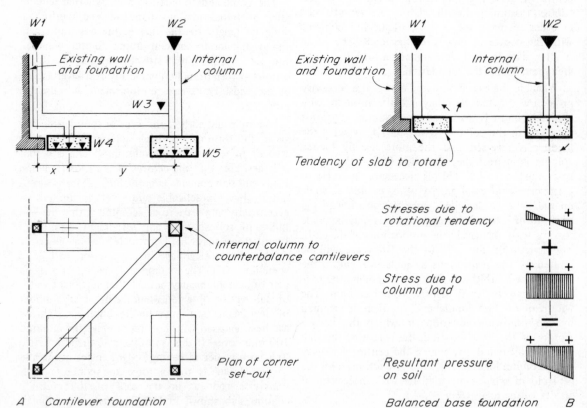

A   Cantilever foundation

Internal column to counterbalance cantilevers

Plan of corner set-out

Tendency of slab to rotate

Stresses due to rotational tendency

Stress due to column load

Resultant pressure on soil

Balanced base foundation   B

54   *Balanced foundations*

$W_5 = W_2 - W_3$ (the load on the internal slab: the 'uplift' at the end of the cantilever equal to W_3 reduces W_2 by this amount)

The counterbalancing force W_3, supplied by the internal column must be provided wholly by the dead load on this column. This must be at least 50 per cent greater than the uplift due to the combined dead and live loads on the outer column in order to provide an adequate margin of safety. Care must be taken to maintain this margin during construction. When the dead load on the internal column is insufficient for counterbalancing purposes, added weight must be provided by increasing the size of the internal base as necessary or alternatively, anchorage must be provided in the form of tension piles under the slab. Tension piles, or a mass concrete counterweight, must be used when no suitably placed column is available for counterbalancing the cantilever beam. When a corner column is involved, a diagonal cantilever beam may be used as shown in figure 54 A.

It should be noted that the short fulcrum column, shown in the illustrations is not essential. The cantilever beam may rest directly on the top of the fulcrum slab.

As indicated earlier, maximum economy is achieved if loads are transferred axially. Cantilever foundations should not, therefore, be resorted to unless it is impossible to re-align a column to permit axial support of its load.

As in a combined foundation, the cantilever foundation may be constructed in reinforced concrete or in steel. In the latter case the steel cantilever beam in conjunction with steel stanchions may bear on steel grillages or on normal reinforced concrete slabs.

Balanced base foundation In this form of balanced foundation the beam is dropped to the level of the base slabs (figure 49 G). It can be used when a base can be placed under, although eccentric to, the outer column and may be viewed as a beam balancing, or resisting the tendency to rotate on the part of the eccentrically loaded foundation slab (see figure 54 B).

In design, the foundation slab to the outer column is made large enough to take the load W_1 from the column. The tendency of this slab to rotate due to the eccentric loading is resisted by a balancing beam linking it with the inner foundation slab which, in turn will tend to rotate under the action of the balancing beam. The spread of this slab must, therefore, be sufficient not only to distribute safely to the soil the load from its own column, but also to prevent overstressing of the soil at its far edge due to the rotational tendency (B)[1]. If, for this reason, the spread became excessive, it would be necessary to use a cantilever foundation. The balancing beam must be stiff enough to fulfil its function satisfactorily. In both the cantilever and balancing beams the tensile zones would be at the top where the main reinforcement would be placed.

Raft foundations

A raft foundation is fundamentally a large combined slab foundation designed to cover the whole or a large part of the available site. Reasons for the use of a raft foundation have already been mentioned in the section on the choice of foundation type and they may now be considered in greater detail.

A raft may be used when the soil is weak and columns are so closely spaced in both directions, or carry such high loads, that isolated column foundations would overlap or would almost completely cover the site. In these circumstances it may be cheaper to use a raft when it appears likely that more than three-quarters of the site will be covered by isolated column foundations. When a raft is indicated the following considerations should be borne in mind before a final decision is made:
(1) When the depth of the weak strata down to firm soil is not much greater than about 4.5 m, it may be cheaper to use foundation piers, as these can be economically constructed to this depth
(2) When the weak soil extends to a depth greater than this, bearing piles might, in some circumstances, form a cheaper foundation than a raft
(3) When the weak soil extends to such a great depth that bearing piles would be uneconomic, a raft can be used provided that the building is reasonably compact in plan and fairly evenly loaded. When the loading is very uneven, friction piles could be used as an alternative and possibly simpler solution to the problem
(4) Treatment of the ground to improve its bearing capacity, such as vibro-compaction of sand, may be economical.

[1] See Part 1, *Eccentric loading on foundations*, pp 58, 73.

The choice of method depends on the extent of the individual loads, the size of project and the ground conditions. Each method must be compared with the others on an economic basis.

When a raft is used this reduces the ground pressure immediately beneath the foundation, but the pressures at greater depths are little less than they would have been under closely spaced isolated column foundations (see page 74). Care must be taken to ensure that the pressures at such depths are acceptable.

A raft used to distribute the building loads over a large area of weak soil may be flexible or rigid in form, as required by the characteristics of the soil, the distribution of loading and the form of structure used for the building it supports. As already explained, the rigidity of a raft has an important effect upon the distribution of pressure and upon settlement. It is possible to use a stiff raft to minimize differential settlement. For example, in the case of a building bearing on a cohesionless soil and having high load concentrations round the edges, the use of a rigid raft would result in a greater uniformity of pressure under the raft than if a flexible form were used. However, a high degree of rigidity results in a most expensive structure and a certain amount of differential settlement usually has to be accepted, the safe limits depending upon the construction of the building which is supported.

A raft may also be used as a means of bridging over weak areas on a site which otherwise is generally fairly firm. A raft should not, however, be carried directly over a local hard area in an otherwise fairly weak site. In such circumstances the hard area should be excavated to a suitable depth and backfilled with a material compacted so as to bring its bearing capacity to an amount not exceeding that of the remainder of the site.

The use of basements in foundation problems has been mentioned earlier where it is shown that a basement may be used as a means of reducing the nett intensity of pressure on the soil by a relief of overburden pressure, in order to reduce the overall settlement under the whole area of a building.

In saturated and soft subsoils, such as very wet mud or silt, it is possible to devise a foundation which will enable the building to float on the subsoil. On a liquid soil having no angle of repose, the building on its foundations would sink into the ground until a volume of soil equivalent in weight to that of itself is displaced. In other words, stability is achieved when the building is floating. For this purpose deep and stiff foundations are required. In the case of high and heavy buildings it will be necessary to sink the structure to considerable depths in order to displace sufficient soil, thus leading to considerable pressures on the walls and base. In order to resist these the basement or basements are constructed on a cellular basis as described on page 94.

A basement, as explained on page 80, may also be used under the heavier parts of a building in which large variations of loading occur, as in a building with a tall tower block surrounded closely by lower blocks, in order to produce a uniform distribution of pressure. In such circumstances, if the basement is deep, the whole of the basement must be constructed as a stiff rigid structure (figure 47 A).

A raft may also be used to provide a stiff foundation under a building the structure of which is sensitive to differential movement (see page 126).

Stiff rafts may also be used in some circumstances to overcome the difficulties of subsidence sites or, alternatively, these difficulties may be overcome by the use of an extremely flexible raft carrying a building of sufficient flexibility to deal with the movements arising during actual subsidence. This is discussed more fully on page 105. The flexibility of a raft foundation varies, as in the case of normal beams, with the depth and the degree of stiffness between the parts and, where great stiffness is required, a deep beam or cellular structure is essential.

Design of rafts generally   In order to distribute evenly the pressure on the soil, the centre of gravity of the building loads as a whole should lie on the same vertical line as that of the raft. This is particularly important since the raft form is used on weak yielding soils where a slight unevenness of pressure will cause considerable differential settlement. Although, as seen in combined slab foundations, equal loading of the columns is not essential in order to make the centre of gravity of the loads coincide with the centroid of the raft, this is facilitated if the plan of the building is symmetrical and without projecting elements. The ideal is a simple, regular shaped raft carrying symmetrically arranged, equally loaded columns and on which any heavier parts of the structure are grouped symmetrically about the axis. If practical limitations on the extent of the raft at the sides make it impossible to bring about the coincidence of the two centres of gravity, it may be necessary to make the raft irregular in shape in order to reduce the eccentricity to a minimum. In circumstances where some part of the raft must be unevenly loaded to an excessive degree, or where a considerable projection or arm in the plan form is unavoidable, the unevenly loaded or the projecting section should be on a separate raft with the building structure above also separated. This will avoid adverse effects on the structure arising from differential settlement due to the uneven loading on the subsoil.

The layout of drain and service pipes should be considered at design stage so that the effects of holes and ducts passing through the raft may be taken into account. When founding on very soft clay or silt, in which long-term settlement may occur, provision should be made for relative movement between the building and any services entering the building, if necessary by means of flexible connections as suggested in Part 1.

The weight of a building structure is not uniformly distributed over the raft, but is concentrated at the wall or column points and, since a completely rigid raft is not practicable in most cases, there is the tendency for greater pressure to occur under these points, giving a non-uniform distribution of pressure under the raft. Nevertheless, to simplify design procedure, other than in the case of an extremely flexible raft, the assumption is frequently made that the pressure distribution is uniform or varies uniformly under the raft, and that the upward reaction of the soil on the raft is uniform. To find the reaction of the soil per m^2, the total load from the structure is divided by the area of the raft and from this may be calculated the thickness of the slab and the reinforcement required for the raft, the design process being similar to that for an inverted floor.

Problems due to subsoil water arise during the construction of basements on account of the upward pressure of the water. The total dead weight of the basement and building above must, when completed, exceed the maximum upward pressure of the water. When the basement is under construction, however, it is without the superstructure load and must be prevented from floating. If the dead weight of the basement alone is insufficient for this, one of

a number of methods can be used to do so.

(a) The subsoil water level may be lowered by pumping or other means. This is usually essential in the initial stages of construction at least

(b) The basement may be temporarily filled with water to a height such that its weight, together with the dead load of the basement, resists the upward pressure of the subsoil water

(c) Holes may be formed in the floor through which the subsoil water may enter the basement and prevent the build up of external pressure. When construction has advanced enough to provide sufficient dead load, the holes are sealed

(d) Ground anchors may be used to prevent flotation (see page 145).

Pumping to keep the water level down is most commonly used, but when construction is sufficiently far advanced the use of the other methods, particularly (b) which is the simpler of the two, enables pumping to be stopped. If a basement was to be used only to reduce pressure on a weak soil, it might prove cheaper to use piles in order to avoid the construction of a basement in such conditions.

Rafts may be divided into three types according to their design and construction (i) solid slab, (ii) beam and slab, (iii) cellular (figure 55).

All are basically the same, in consisting of a large, generally unbroken area of slab covering the whole or a large part of the site. The thickness of slab and size of any beams will be governed by the spacing and loading of the columns and the degree of rigidity required in the raft.

Solid slab raft   This type of raft consists of a solid slab of concrete reinforced in both directions. Light solid slab rafts are used for the small load-bearing wall type of building, such as houses, or for light framed structures where the bearing pressures are relatively low and these are described in Part 1. For larger and more heavily loaded buildings, this type of raft is often economic only up to a thickness of about 300 mm. Unless the column and wall loads are very heavy, the slab is reinforced top and bottom with two-way reinforcement. When loads are heavy, more reinforcement is required on the lines of the columns to form column bands similar to a plate floor. The reinforcement to the panels between would then be two-way reinforcement mainly at the top. If the slab is situated at ground level, it is generally desirable to thicken the edge of the slab or to form a downstand beam (figure 55) of suff-icient depth to prevent weathering away of the soil under the perimeter of the raft (see Part 1). In some cases deep edge beams are required to contain the soil under the slab in order to transfer the load to a lower level. This is illustrated in the three-storey building shown in figure 55. The ground floor slab, 300 mm above ground level, is stiffened by cross and edge beams to act as a raft, the pressure from which is transferred to the ground below through soil filled within and contained by the deep edge beams before the floor slab is cast.

A thin solid slab raft divided into small sections, held together by steel mesh at mid-thickness, may be used for light framed buildings on sites liable to subsidence. For design reasons, discussed later, the raft rests on the surface of the ground, acting as the ground floor slab and no downstand edge beams are provided, soil erosion being avoided by a few feet of paving round the building.

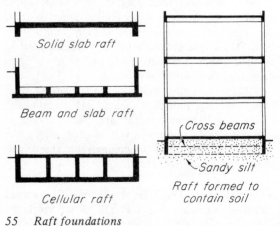

Solid slab raft

Beam and slab raft

Cross beams

Cellular raft

Sandy silt

Raft formed to contain soil

55   *Raft foundations*

Beam and slab raft   When loading or requirements of rigidity necessitate a slab thickness greater than 300 mm it may be cheaper to use beam and slab construction, which may be visualized as an inverted floor, the slab bearing directly on the ground and the beams projecting above it (figure 55). The rein-forcement is arranged accordingly. On sites where the weak strata is overlaid by a thin layer of com-paratively stiff soil at building level, economy in construction could be effected by placing the beams below rather than above the slab. The beams could then be cast in excavated trenches of the appro-priate size without the use of shuttering, and the

necessity of a sub-floor would be avoided. The slab at the bottom, however, has the advantage of prov-iding stiffness to the bottoms of the beams, which will be in compression in the zones of negative bending moment between columns. Where possible, the raft should be cantilevered beyond the outside lines of columns so that the bending moments in the inner slabs are reduced.

When there is some unevenness of loading on the columns of a frame, but where loading and strength of soil are such that a full raft is not required, beams may be used to link isolated column found-ations to give some degree of rigidity in order to limit differential settlement.

Cellular raft   When stresses in the raft are high, and particularly when great rigidity is required, the beams must be deep and when the overall depth is likely to exceed 900 mm a cellular form of construction is adopted. This consists of top and bottom slabs with edge and intermediate beams in both directions forming a hollow cellular raft (figure 55). When such a raft is extensive in area and great rigidity is required to reduce differential settlement, the depth may need to be as much as a full basement storey or more using reinforced concrete cross walls monolithic with the floors. The cellular basement is also necessary when very deep basements are used to reduce overall settlement by a relief of over-burden pressure (figure 47 *A* and *B*). As already ex-plained cellular basement construction is one means of dealing with these problems in the case of very tall buildings, as both the need to limit overall settlement and to reduce differential settlement often arise.

A normal cellular raft is completely cellular, but in the case of a basement reinforced concrete walls, monolithic with the floors and dividing the basement area into small compartments, make it unsuitable for many purposes. An alternative method may be used in which heavy reinforced concrete columns and floor beams are designed to form Vierendeel girders[1] in both directions. To obtain sufficient rigidity, the junctions of top and bottom of the columns with the beams will normally need large haunches. This restricts the floor area of the base-ment and in some circumstances it may be preferable to use pile foundations instead of a cellular basement as an alternative, although possibly more expensive method.

[1] See page 192

In soft soils a cellular raft is constructed and sunk as a monolith as described on page 104. The soil is supported by the sides as excavation proceeds and on completion a base slab is laid at the bottom. In the case of large rafts it may be necessary to divide the raft area into a number of separate monoliths which, after sinking, are joined together by *in situ* cellular construction to form a single unit.

In deep excavations there is an upward movement of the excavated surface called 'heave' which is caused by the pressure of the overburden at the sides of the excavation forcing up the base of the excavation (see figure 46). If a very deep basement is needed in soft clay the heave and, subsequently, the ultimate settlement of this, may be the determining factor in deciding the depth to which the foundation should be taken. If sufficient relief of overburden pressure cannot be obtained the excess pressure may be taken by piles and if the piles are are driven prior to the excavation, the amount of heave can be reduced.

Pile foundations

These may be defined as a form of foundation in which the loads are taken to a low level by means of columns in the soil on which the building rests.

It is a method of support adopted (a) on sites where no firm bearing strata exists at a reasonable depth and the applied loading is uneven, making the use of a raft inadvisable, (b) when a firm bearing strata does exist but at a depth such as to make strip, slab or pier foundations uneconomical, that is at depths over 3 to 4.5 m, but not so deep as to make use of a raft essential. Piles to depths of 18 m are common and in exceptional circumstances they are used to depths of 30 m or more, but piles over this length are considered long. (c) in shrinkable clay soils as a means of founding below the zone of seasonal moisture movement and of dealing with the problems of moisture change in the soil caused by vegetation, especially trees and shrubs.

The short bored piles used to found below the zone of seasonal moisture movement in clay soils are described in Part 1 where the effects of trees on this type of soil are also described. The methods of protecting buildings from the consequences of these effects are indicated in table 11 of that volume. Drying of the soil below large trees can extend to 4.5 m or more and buildings close to standing trees should be supported on bored piles deep enough to pass

below this drying zone. When mature trees are felled the clay below will swell over many years as it takes up water and buildings erected before the swelling is complete will be subject to uplift. To avoid adverse effects on the structure reinforced bored piles should be used to anchor the building, with the top 3 m sleeved from the surrounding ground. The ground floor must be suspended with the supporting beams bearing on the piles well clear of the ground.

Piles are also used when pumping of subsoil water would be too costly or timbering to excavations too difficult to permit the construction of spread foundations.

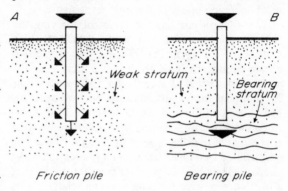

Friction pile *Bearing pile*

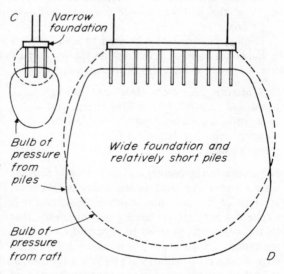

56 *Pile foundations*

Irrespective of the type piles may be divided into two categories according to the manner in which they lower the level of the applied pressure (i) friction piles, (ii) end bearing piles (figure 56 *A, B*). Most piles do, in fact, carry the load by a combination of friction and end bearing.

Friction piles which transfer their load to the surrounding soil by means of the friction between their surfaces and the soil, and to a slight extent by end bearing, are used in deep beds of clay and silt as an alternative to a raft. They may also be used in conjunction with a raft, as already described, when the latter cannot be taken deep enough to obtain sufficient relief of overburden pressure to keep settlement within acceptable limits. Such foundations should be designed with great care as they may result in unacceptable differential settlement.

The friction pile has the effect of carrying the bulb of pressure to a low level so that the high stresses are set up in the soil at a level where it is strong enough to resist them rather than near the surface where it is weaker. It may be assumed that friction piles form a raft imposing the load upon the soil about two-thirds down their depth. In order to obtain an effective lowering of the bulb of pressure it is essential that the ratio of pile length to building width should be high. The wider a foundation is the deeper will its bulb of pressure penetrate and unless the piles in a wide building are long relative to its width they will make little difference to the actual depth of the bulb of pressure (compare (*C*), (*D*), figure 56).

Friction piles may also be used to act in tension in holding a building down against the wind uplift referred to at the beginning of this chapter.[1]

End bearing piles carry their load through weak strata and transfer it to a firm stratum on which their ends rest. They may be viewed as simple columns receiving lateral restraint from the soil through which they pass.

Pile groups The bearing capacity of a pile depends upon its length, its cross-sectional area and the shear strength of the soil into which it bears and it is frequently necessary to use a group of piles, rather than a single pile, in order to obtain adequate bearing capacity. Such a group is termed a 'pile cluster' (figure 57 *A*). For reasons of stability a group of at least three piles is often used under any heavy load rather than a single pile. This, in addition, also provides a margin of safety should one be defective in some way. It is, however, not always necessary to use three piles if displacement and rotation are

limited by beams connecting the pile caps (see below).

End bearing piles in a cluster should be placed not less than twice the least width of the piles centre to centre, and friction piles not less than the perimeter of the piles centre to centre.

In the case of non-cohesive soils the soil tends to

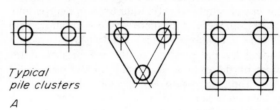

Typical pile clusters

A

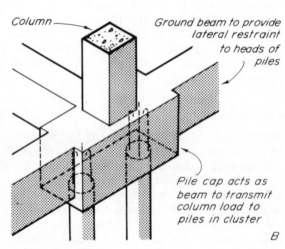

Column — Ground beam to provide lateral restraint to heads of piles

Pile cap acts as beam to transmit column load to piles in cluster

B

57   Pile clusters

consolidate during driving but in cohesive soils it tends to be remoulded by displacement upwards and outwards, with a loss of strength. If this remoulding is not excessive the soil partly regains its strength after rest and by adequately spacing the piles the remoulding process is minimized.

The tops of the piles in a cluster are connected by a block of concrete termed the 'pile cap', reinforced to transmit safely the column load to the heads of the piles in the cluster (see figure 57 *B*). Adjacent caps are usually linked to each other by ground beams of reinforced concrete, this always being done when there are less than three piles under a load.

[1] Alternatively, ground anchors may be used for this purpose. See page 145

The beams provide lateral rigidity to the tops of the piles and must be able to resist any bending due to eccentric loading on the piles. This is important when the piles are in soft ground giving little lateral support. The beams will normally also carry the lower floor slab and walls of the structure above.

The safe bearing capacity of a pile or cluster of piles may be established in a number of ways:

(1) By a loading test on a trial pile or cluster, by means of kentledge bearing directly on the cap. The test load is usually up to 1½ times the working load

(2) By measuring the 'set' of the pile, which is the distance it sinks into the ground under a given number of blows of a hammer of given weight dropped through a given distance. With this data the bearing capacity is calculated by means of dynamic formula. Dynamic formulae are reliable in non-cohesive soils and hard cohesive soils but in cohesive soils other than 'hard' a check by calculation should be made as in (3)

(3) In cohesive soils, by calculation based on the measured shear strength of the ground at various levels

(4) By means of a small diameter penetration test, making use of the cone penetrometer developed to measure the soil resistance at its foot and the frictional resistance along its outside surface (page 65). These measurements can be taken at varying depths over a building site and the lengths and bearing capacities of the piles predetermined before work begins

(5) By previous experience with similar piles nearby.

As will be seen from the earlier part of this chapter, tests will not give a complete picture of the final settlement of the completed structure as a whole, since it will produce a deeper bulb of pressure than the single pile or cluster of piles (figures 43 and 56), and this must be considered separately having regard to the factors on which this depends.

Soil compaction Piles may be used in non-cohesive soils such as silt, loose sand or gravel, as a means of compacting it by the vibration of driving in order to increase its bearing capacity. This necessitates the use of a large number of piles over the area of a site and alternative methods of compacting have been developed. One is a mechanical technique known as vibro-flotation, in which compaction is effected by a large, heavy poker vibrator about 1.80 m long supported by a mobile crane. Water jets incorpor-

ated at the lower end produce a saturated mass of soil into which the vibrator sinks under its own weight until it has reached the depth required for compaction. A column of compacted soil up to 3 m in diameter is produced and by arranging the compaction points so that these columns overlap full compaction over any area is achieved. Another technique for loose, wet sands compacts by means of an electric spark across electrodes at the end of a pipe sunk by water jetting, which produces spheres of compaction about 3 m in diameter. These can be made to overlap vertically and horizontally across a given area.

Types of piles

Piles are considered as *driven* or formed *in situ* according to whether they are preformed and driven into the ground by blows of a hammer or are formed on site by placing concrete and reinforcement in a borehole[1].

Some types of pile are, in fact, basically formed *in situ* but a shell or former into which the concrete is poured is driven into the soil and from the point of view of soil mechanics are considered as driven piles. To distinguish them from driven precast concrete, timber of steel piles they may be termed 'driven tube' piles.

Driven piles These may be of timber, steel or precast concrete. The latter, either normally reinforced or prestressed, is most commonly used in building work. Timber piles are not much used in building work except for minor purposes, such as supports to manholes in running sand for example. The use of timber piles is restricted by the limitations in cross sections and lengths available. Splices are unsatisfactory. In addition the supply of large numbers of timber piles is difficult at short notice. Timber tends to rot if not always submerged and difficulties are encountered in the use of timber piles if there is hard driving in the early stages. Steel piles in the form of open ended box or H-sections, or pipes, are used when the length is very great, especially when driven to rock to carry heavy loads. These conditions would require large section heavy concrete piles, difficult to handle and liable to damage, and in such circumstances they would be used instead of steel only in situations where the

[1] These are also referred to as *displacement* and *non-displacement* (or *replacement*) piles respectively.

latter would be subject to heavy corrosion.

Precast concrete piles may be square, hexagonal or round in section and may be solid or hollow. 18 m long precast piles are normal and lengths up to from 15 to 18 m can be obtained from stock. Large section piles over 30 m long have been used; they are usually cast on or near the site. The toe of a precast pile is usually fitted with a cast iron or steel protective shoe and the top is protected during driving by a steel cap or 'helmet' in order to prevent spalling. Shoes are not essential in soft ground but special shoes are provided for entering rock.

Precast piles may be obtained cast in units of varying but standard lengths up to 10 m, which can be joined by special steel caps cast in each end to permit any length of pile to be formed. They may also be obtained cast in segments 1 m long, assembled by spigot and socket joint, for light loadings, These have a central hole for a reinforcing or post-tensioning bar.

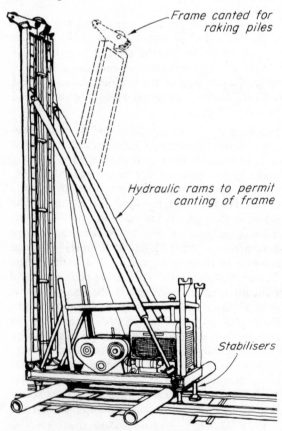

Frame canted for raking piles

Hydraulic rams to permit canting of frame

Stabilisers

58   Piling frame

The reinforcement in precast concrete piles is designed to resist the stresses set up in transport and slinging and in driving, as well as those set up by the working load. Prestressing may be used, in which case the whole pile may be prestressed as cast, or long piles can be formed of factory made precast sections transported to the site where they are made up into full pile lengths and post-tensioned prior to driving.

The pile is supported between guides or 'leaders' in a vertical steel piling rig or frame (figure 58) sufficiently high to take the pile and to allow room above the cap for a hammer, or for the drop of a weight, by means of which the pile is driven into the ground. Pile driving frames are normally mobile and arranged to rake over to drive piles at an angle when required. For piles less than 12 m in length an excavator may be equipped with hanging 'leaders' to act as a pile driver (see figure 10 *G*). Driving by a weight or drop hammer is performed by allowing a weight or 'monkey', attached by cable to a steam or diesel operated winch, to fall on to the head of the pile in successive blows. The winch raises the monkey each time to the top of the drop and is then released to allow the monkey to fall. The distance of drop is usually about 1.50 m although drops up to 3 m are sometimes used.

Pile driving hammers rest directly on the pile and travel down with it. Driving is more rapid than by a drop hammer because of the large number of blows delivered per minute. They are operated by steam or compressed air as single- or double-acting hammers. Petrol operated hammers are also available, acting on the principle of earth rammers. With a single-acting hammer the motive power simply lifts the hammer after which it falls by its own wieght. With a double-acting hammer the steam or air assists in forcing it down on to the pile after having raised it. The weight of the striking parts of a double-acting hammer is much less than that in a single-acting hammer but the former gives a more rapid succession of blows.

Driven piles are most suitable for open sites where the length required will be constant and where there is ample headroom for the driving frames. Considerable vibration and noise is set up during driving and on some sites this would preclude the use of this type of pile.

In certain types of soil, such as fine, clean sands and fine gravels, driving is facilitated by the use of a jet of water discharged under pressure at the toe of

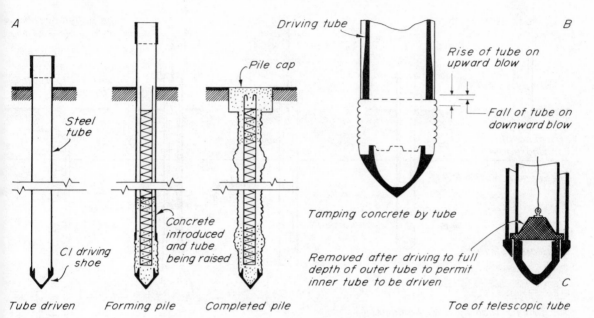

A

Steel tube

CI driving shoe

Tube driven

Concrete introduced and tube being raised

Forming pile

Pile cap

Completed pile

Driving tube

B

Rise of tube on upward blow

Fall of tube on downward blow

Tamping concrete by tube

Removed after driving to full depth of outer tube to permit inner tube to be driven

Toe of telescopic tube

C

59   *Top driven tube piles*

the pile to displace the soil underneath. Various arrangements of jets are used and the water is taken to the toe through a pipe or pipes running through the pile from the top. Towards the end of driving the water is cut off so that the pile is driven to rest for the last 1.20 to 1.40 m through undisturbed soil.

Driven tube piles   These are more suitable than driven precast concrete piles on sites where there is likely to be a considerable variation in the lengths of piles. They may be formed in various ways using steel or concrete tubes. In some forms driving is from the top, in others from the toe by means of a mandrel or by a drop hammer falling through the length of the tube. In some forms the tube or shell is left permanently in position, in others it is withdrawn as concreting proceeds. In common with driven piles vibration occurs under the blows of the hammer, the degree of which depends upon the particular system used.

Where driving is from the top a steel tube of required diameter is fitted with a cast-iron conical shoe, placed in a driving frame and driven down by a drop or a steam hammer in the same way as a precast concrete pile (figure 59 *A*). Reinforcement, in the form of a cage made up of longitudinal rods linked by helical or horizontal binding with spacers

at intervals to ensure adequate concrete cover, is lowered into the tube. Concrete is then poured in through a hopper or skip as the tube is gradually withdrawn leaving the driving shoe behind.

Tamping of the concrete may be done by a drop hammer falling on to each charge of concrete as the tube is slowly withdrawn, consolidating it and forcing it into close contact with the surrounding soil. In other systems, using a tube with a thickened bottom rim, the concrete is tamped by the tube itself which, as it is withdrawn, is subjected to rapid up and down blows from a steam hammer, causing it to vibrate and consolidate the concrete by its thickened edge (figure 59 *B*).

The normal tube length is from 12 to 15 m, but this may be extended when necessary by screwing on a further length when the first has been driven or, alternatively, the tube may be used as a leader for a precast concrete extension pile driven through the tube and taking the conical shoe with it. This has the advantage of reducing frictional resistance since the driving force has to overcome only the resistance of the extension pile as it is being driven, the upper tube remaining fixed in position. The same result may be obtained by the use of telescopic tubes in which the upper length after it has been driven to its full extent acts as a leader to an inner

99

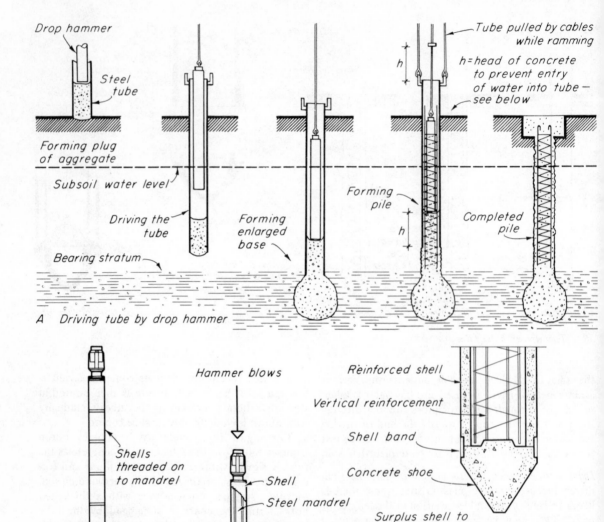

Drop hammer

Steel tube

Forming plug of aggregate

Subsoil water level

Driving the tube

Bearing stratum

Forming enlarged base

Forming pile

h

Tube pulled by cables while ramming

h=head of concrete to prevent entry of water into tube — see below

Completed pile

A   Driving tube by drop hammer

Hammer blows

Reinforced shell

Vertical reinforcement

Shell band

Concrete shoe

Shells threaded on to mandrel

Shoe

Pile ready for driving

Shell

Steel mandrel

Surplus shell to be removed

Blows from hammer transferred directly to shoe by mandrel

Pile ready for reinforcement and concrete

Driving the pile

Completed pile

B   Driving tube by mandrel

60   Toe driven tube piles

100

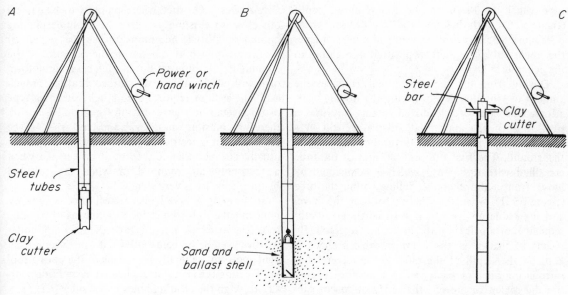

A

*Power or
hand winch*

*Steel
tubes*

*Clay
cutter*

Sinking tubes through clay

B

*Sand and
ballast shell*

Sinking tubes through ballast

C

*Steel
bar*

*Clay
cutter*

Tube driving (when
necessary)

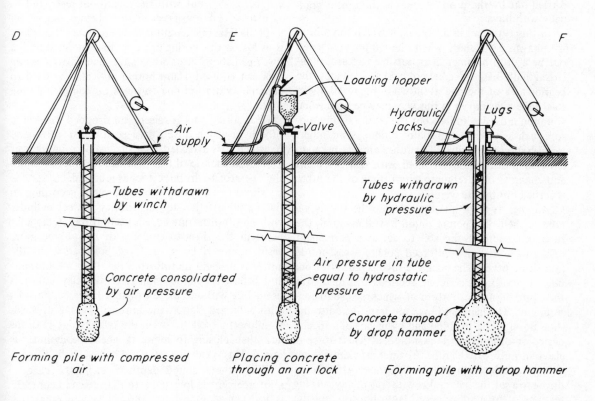

D

*Air
supply*

*Tubes withdrawn
by winch*

*Concrete consolidated
by air pressure*

Forming pile with compressed
air

E

Loading hopper

Valve

*Air pressure in tube
equal to hydrostatic
pressure*

Placing concrete
through an air lock

F

*Hydraulic
jacks*

Lugs

*Tubes withdrawn
by hydraulic
pressure*

*Concrete tamped
by drop hammer*

Forming pile with a drop hammer

61   Bored piles

tube which is driven to the full required depth (figure 59 C). Both these methods can be used where piles must pass through the water of a river or lake, the upper tube being left in position as a casing to the concrete filling.

Driving from the toe substantially reduces the amount of vibration set up during the driving operations. One type of pile driven in this way uses a steel tube of the appropriate diameter held in a driving frame with the open bottom end resting on the ground. The first 600 or 900 mm of the tube are filled with dry gravel which is compacted by blows from a drop hammer falling within the tube (figure 60 A). When the friction between the gravel and the inside of the tube is sufficiently great subsequent blows pull the tube down into the soil. By means of marks on the hammer cable a check is kept on the depth of the plug and more gravel is introduced when necessary so that the tube is continually sealed to prevent the entry of soil or subsoil water. Reinforcement and a semi-dry concrete mix is placed as described earlier and tamping is carried out by the drop hammer as the tube is gradually withdrawn.

In this type and in all types in which the tube is withdrawn, particularly where the tamping is carried out by a drop hammer, it is possible to form an enlarged bulb base of concrete which increases the bearing area of the pile as shown in figure 60. Light steel tube piles driven in this way may be used when vibration must be limited but the nature of the subsoil, such as waterlogged soil or running sand, might make boring impossible. The tubes are left permanently in position and are filled with reinforcement and concrete. Any required lengthening of the tubes is carried out by site welding.

Another type of pile driven from the toe uses a tube or shell of concrete round a steel mandrel, the tube being left in the soil to act as a permanent casing to a core of reinforced concrete cast inside (see figure 60 B). The concrete tube is built up from precast reinforced concrete sections threaded over the mandrel, at the bottom of which is a solid concrete shoe. The hammer blows are transmitted to the shoe through the mandrel and by means of a special device at the top of the mandrel sufficient force is also transmitted to the tube to overcome skin friction and to enable it to follow the shoe. After being driven to a set the mandrel is withdrawn, any surplus sections of tube are removed from the top and the reinforcement and concrete is placed inside.

Bored piles   On sites where piling is to be carried out close to existing premises, or for underpinning buildings with the minimum of disturbance, driven or driven tube piles are unsuitable because of the amount of vibration set up. In such cases some form of bored pile is used in which the vibration is much less. The system requires only a light shear-leg type of rig instead of a large driving frame and is useful, therefore, on sites where levels vary and where space or headroom is restricted. The use of bored piles is usually cheaper than any form of driven pile when a few only are required or when the required bearing capacity is very small.

In forming a bored pile a steel tube is sunk by removing the soil from inside it by means of a coring tool (figure 61 A, B), the tube then sinking under its own weight or being driven down by relatively light pressure, generally by means of the coring tool itself acting through a steel channel passed horizontally through the clearing holes in the sides (C).

The tube is made up from screw-coupled sections varying from 1 to 1.35 m in length, so that work can be carried out with the headroom restricted to as little as 1.80 m, fresh sections being coupled on until the final pile length is reached. When headroom is as low as this boring must be done by hand with a low rig. Labour costs are greater than when a coring tool can be used. Hand boring may also be used to eliminate vibration due to the dropping of a coring tool.

The coring tool is raised and dropped inside the tube by a winch driven by a petrol or diesel engine to which it is attached by a steel cable running over a pulley at the top of the shear-legs. A clutch enables the operator to drop the tool as required.

The type of coring tool used varies according to the nature of the subsoil. In clay a steel cylinder with a bottom cutting edge is used. It has rectangular holes in the sides to enable stiff clay to be more easily removed by a spade or crowbar when the cutter is brought up full from the tube (A). In sands and ballast a cylinder similar to the clay cutter is used, but without the extracting holes in the side, with a hinged flap at the bottom opening upwards to allow the soil to enter the cylinder as it drops and then closing to retain it when the cylinder is raised. This is called a sand and ballast shell (B).

When the required depth is reached, a cage of reinforcement is lowered into the tube and concrete is introduced in batches through a hopper as the tube is gradually withdrawn either by block tackle operated by the winch or by hydraulic jacks operating against lugs on the top of the tube (*F*). Tamping can be carried out by a drop hammer or by compressed air.

In the latter method, which eliminates vibration during the tamping process, after each batch of concrete is introduced a pressure cap is screwed on the top of the tube and compressed air is admitted which forces the concrete down and out against the surrounding ground (*D*). With this method a bulb foot, similar to that formed by a drop hammer, may be formed in ballast by forcing cement grout into the ballast surrounding the foot of the pile.

When ground water is present it can be kept out of the tube during concreting by compressed air, in which case the concrete is placed through an air-lock attached to the top of the tube in place of the normal pressure cap as shown in (*E*). This consists of a cylindrical steel hopper with a feed hole at the top and a large valve at the bottom. Admission of air first blows out the water through a pipe lowered down the tube. This pipe is then closed, and by maintaining the air at sufficient pressure to balance the head of subsoil water, the tube is kept free of water. Concrete is then placed in batches through the air-lock, each batch being forced down and out by a temporary increase in the air pressure as the tube is being raised.

In using bored piles or driven tube piles where the tube is withdrawn, the possibility of 'necking' must be borne in mind. This is liable to occur when the external water pressure is more than the pressure of the concrete where it flows out of the bottom of the tube. It may also occur when the pile is in very soft clay which will try to press back into the pile when the tube is withdrawn if the concrete pressure is insufficient. Compressed air on the concrete helps to overcome this if it is maintained at the critical time. To minimise this difficulty it is possible to introduce into the tube, instead of poured concrete, a core pile made up of short precast concrete units assembled initially on a central steel tube, with reinforcing bars passing through holes round the units. As the casing tube is withdrawn grout is introduced into the central hole under compressed air, filling all voids in the pile and passing through holes to fill the space between the core pile and the sides of the bore hole.

Cylinder piles These are large diameter bored piles which may be from 600 mm to 2 m or more in diameter and possess the advantages of the traditional cylinder foundation which has been used for many years by engineers to carry very large loads.

One system, developed essentially for use in cohesive soils, uses an auger for drilling the hole. The auger is fixed to the end of a long square vertical kelly bar which passes through a turntable by means of which it is rotated. A standard excavator may be rigged specially for this purpose as shown in figure 10 *H,* or specially designed gear consisting of a high retractable vertical jib to carry the rotating kelly bar may be fixed to a heavy mobile wagon. These piles are generally unreinforced except where they pass through a soft stratum near the ground level and may be sunk to a depth of 30 m or more in suitable conditions. A lining is not normally used except in cases where a granular stratum overlays a clay stratum, when the hole may be lined to a depth of from 7.50 to 9 m. Drilling is rapid and shafts can be sunk into firm clay at a speed of up to 3 m per hour. The base of the piles may be expanded up to two or three times the diameter of the shaft by means of an under-reaming tool consisting of a cylinder from which cutting wings are made to extend when at the base of the shaft.

Another system uses steel linings from 900 mm to 1.20 m in diameter for the full length of the boring, which is carried out by a specially designed turn grab carried on a square kelly bar in the same way as the augers described above. The grab is capable of boring through most types of soil up to the hardness of soft rock so that the system is not limited to firm cohesive soils. The steel lining is made up of inter-locking sections each about 3 m long, the first of which has a cutting edge. The lining is forced down, and ultimately drawn up, by means of hydraulic jacks working on a steel collar clamped tightly to the top section by a horizontal jack. The rotating grab is fitted with three hinged blades which are opened for drilling and then closed to retain the excavated material for withdrawal. The blades are also used to form an under-reamed base when this is required.

Instead of a lining it is possible to support the sides of the boring by a bentonite suspension, as described under diaphragm walls, page 143, the boring being carried out through the fluid.

Cylinder piles have a number of advantages over normal types of piles in terms of cost and construction time, particularly when heavy column loads have to be carried. A group of normal piles can be

103

replaced by a single large diameter cylinder pile which reduces drilling costs and construction time. In many cases a group of friction piles can be replaced by a single under-reamed cylinder pile which carries its load mainly in end bearing. For example, loads of 2000 tonnes can be carried on a single cylinder in suitable soil, such as the lower levels of the London clay. In most cases, because a column can be carried by a single cylinder, pile caps can be greatly reduced or eliminated, thus shortening the construction time and reducing the cost. In addition, the soil strata in the boring can be inspected as the work proceeds and concreting can be carried out more satisfactorily than in the case of normal *in situ* piles, and can be properly inspected. These piles, because of their capacity to carry very heavy loads, are valuable as foundations to very tall buildings founded on soft soils where the alternative to cylinder piles would be a large number of long friction piles.

In practice little is known about the relation between base and shaft resistances, but tests indicate that if the base resistance is to be fully developed large settlements are to be expected. This must be carefully considered when under-reamed piles are used. The under-reamed pile can resist large tensile forces and is therefore useful where tension piles are required.

Jacked piles Where complete freedom from vibration and noise is essential a jacked-in pile can be used in which precast sections of reinforced concrete are successively jacked down one after the other. The jack works either directly on solid precast sections joined together by lengths of tubular steel grouted in a central hole as they sink into the ground (figure 65 *D*), or on the top of a built-up steel mandrel which forces down a concrete shoe and tubular precast sections, the completed tube being filled with reinforcement and concrete.

When these piles are used for underpinning work as shown in figure 65 *D* the jack operates against the underside of the foundation, otherwise it works against a travelling kentledge. The expense of jacked-in piles limits their use.

Pier foundations

Brick and concrete piers

When a good bearing stratum exists up to 4.5 m below ground level, brick, masonry or mass concrete

foundation piers in excavated pits may be used. At this depth, unless the nature of the ground necessitates considerable timbering or continuous pumping, piles are not always economical. This type of pier foundation is described in Part 1, page 57.

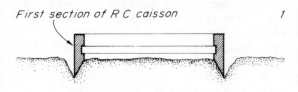
First section of R C caisson 1

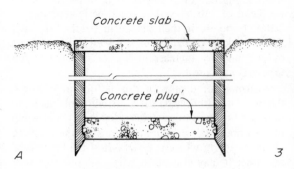
RC caisson extended as excavation proceeds 2

Concrete slab

Concrete 'plug'

A 3

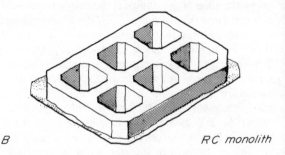
Cylinder foundation

B RC monolith

62 *Cylinder and monolith foundations*

Cylinder and monolith foundations

These are a form of pier foundation used when very heavy loads must be carried through waterlogged or unstable soil down to bed rock or to a firm stratum.

A *cylinder foundation* is constructed by means of a caisson, a cylinder of steel or concrete which provides support to the sides of the excavation and by means of which soil and water can be excluded. The caisson is from 1.80 m upwards in diameter and is built up as high as is practicable before excavation begins (see figure 62 *A1*). The soil is excavated from within by grabbing, hand clearing being used usually only at the cutting edge (*A2*), and the caisson sinks to the required depth under its own weight or with the aid of kentledge. The bottom edge of a concrete caisson is formed into a cutting edge usually protected by a steel shoe. After sinking, the caisson is left permanently in position. It may either be filled with concrete or sealed at the bottom with a concrete 'plug' and at the top with a concrete slab (*A3*). The interior may be left empty or filled with sand or water.

Cylinders which are open at top and bottom, known as *open caissons*, are used where ground conditions permit. When it is necessary to keep the excavation free of water and silt, a *pneumatic caisson* is used. This has a working chamber with a closed top fitted with an air-lock through which men and materials must pass in order to permit the air within the caisson to be maintained at sufficient pressure to prevent the entry of water. Open caissons can be taken to depths up to 45 m or more; theoretically there is no limit to the depth. In stiff clay soils considerable weight is required to overcome the skin friction as the caisson sinks and in these conditions cylinder piles, already described, form a suitable alternative. The use of bentonite slurry, however, pumped between caisson and soil can now be used as a lubricant. This is especially necessary when very large diameter caissons, 30 to 60 m in diameter, forming the total base of a building structure, are being sunk. Pneumatic caissons may be used for any types of soil, but the depth to which they may be sunk is limited to about 34 to 37 m below the water surface level. This corresponds to an air pressure of about 345 kN/m^2 which is the greatest pressure in which men can work. In practice pneumatic caissons are generally not sunk much beyond 30 m.

A *monolith* is a rectangular open caisson of steel or concrete with a number of wells for excavation, used when loading necessitates a very large area of of foundation (see figure 62 *B*). Heavy mass or reinforced concrete monoliths have the advantage of greater weight. These can be very large depending upon the nature of the job. The initial lift of a monolith might be as high as 3 m or more constructed on a temporary concrete slab which would be smashed and removed prior to sinking. Further lifts about 1.80 m high would be constructed until the required depth had been reached. As with cylinders monoliths may be sealed at the top and bottom or be filled in solid according to the requirements of a particular job.

The advantages of cylinder and monolith foundations over normal pile foundations are that the soil may be inspected at the base and during the course of excavation and the concrete can reliably be placed in position. The area of a caisson should be so related to that of the superstructure above that it is great enough to allow for some deviation from its correct position during sinking.

Foundations on subsidence sites

The behaviour of soil and buildings under the action of subsidence has been described on page 82. A number of methods are used to minimize the effects of mining subsidence upon a building; of these the following are commonly used.

(i) The foundation of the building may be designed as a strong reinforced concrete raft, generally in cellular form to give adequate stiffness, or as a stiff system of beams with three-point support on the soil through spherical bearings on piers or pads. The foundation structure, together with the building supported by it, will tilt as the ground moves, but the building will not distort and will, therefore, be free from damage. These methods are expensive, particularly the construction of stiff cellular rafts, and are generally used only for large structures, since the cost of the foundations is likely to be much in excess of the value of any damage caused in a small building.

(ii) The building itself may be designed as a rigid frame structure or be strengthened by reinforcement to act as a single structural element, capable of resisting without cracking all the stresses set up as it tilts, and as the support from the ground reduces in area and varies in position with the movement of the ground. With both these methods it is desirable to break up long buildings into short blocks, connected only by some form of flexible joint.

(iii) In contrast to the above methods, the building and its foundation can both be designed as flexible elements, the construction of the structure and its cladding being detailed to accommodate the differential movements of the soil without damage either to structure or finishes.

(iv) Hydraulic jacks may be used to raise or lower the building at different points to keep pace with the subsidence movement and thus avoid tilting. With this method the building structure would rest on reinforced concrete or prestressed concrete foundation beams bearing on concrete bases, provision being made for the jacks to be inserted at the points of support where adjustment is required[1]. The jacks can be individually and automatically adjusted as the settlement takes place by connecting them to a central control system.

When the method adopted requires a long building to be broken down into short independent units, these should not exceed about 18 m in either direction in order to keep differential settlement within reasonable limits. If the independent units must be connected this is accomplished by movement or slip joints in walls, roof and floor slab. These gaps must be sealed in such a manner that freedom of movement is permitted — thin corrugated metal can be used at the walls and some form of sliding element at roof and floor.

As an alternative to corrugated metal, rendering on expanded metal has been used. When notice is given by the National Coal Board that movement will take place the rendering is cut down to permit 'concertinaing' to take place, and when the movement is complete and the gaps have again opened up they are made good permanently.

Protection within the limits of these small units should provide resistance against the forces set up in the ground and against those set up by the cantilever and beam actions of the foundation (see page 82). This is done by providing steel reinforcement disposed according to the stresses set up in the foundation slab by the different external ground forces. The amount of the ground forces transmitted to the slab may be limited by reducing as much as possible the friction between the slab and the ground. This is accomplished by laying the ground or foundation slab on building paper on a bed of sand, shale or other granular material 150 to 300 mm thick.

Resistance to the forces due to the flexure of the foundation at the crest or trough of the wave can be provided only at considerable expense by means of a stiff and rigid foundation, often in heavy and deep cellular form. In small and light structures the cost of such foundations, sometimes as much as 10 per cent of the cost of the building, is likely to be greater than the cost of repairing any damage caused by unrestricted flexural movements, and foundations of this type would be economical only in large, multi-storey structures in which the high cost of the foundations would be distributed over a total floor area much greater than that of the foundation itself.

Damage to the structure arising from the flexural movements of the foundation when this is not rigid are minimized by using, as far as possible, flexible forms of construction such as timber framing and brickwork bedded in weak mortar to prevent serious cracking. Light reinforcement introduced in brickwork at top and bottom and around openings also assists in this respect, especially when the differential movements are likely to be severe. In addition to these precautions flexural movement can be reduced by siting the building with the shorter axis in the direction of the likely maximum curvature when this can be ascertained.

These methods accept as inevitable a considerable amount of damage to the structure, requiring repair after subsidence is complete, or they make use of heavy, rigid and costly foundations designed to resist the flexural stresses. Even so in many cases experience has shown these to be inadequate to prevent extensive damage to the structure. This is probably due to the fact that the deeper the structure is in the ground the more likely it is to be affected adversely by the horizontal ground forces.

An alternative method suitable for buildings of light construction, in which the foundation and superstructure are integrated into a wholly flexible structure, uses a foundation designed as a thin, flexible slab which accommodates itself to the curves and tilt of the subsidence wave. The building structure is formed as a light, completely pin-jointed frame capable of adjusting itself to the differential vertical movements transferred to it through the foundation slab (see figure 63). Claddings designed

[1] See 'Underpinning and Jacking Buildings Affected by Mining Subsidence', by J.F.S. Pryke, MA (Cantab), AMIStructE, *The Chartered Surveyor*, April 1960.

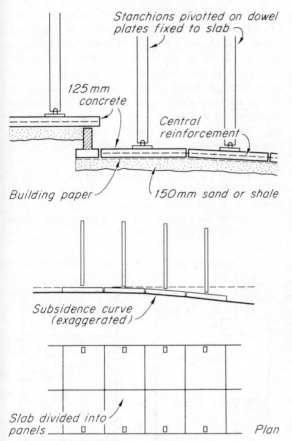

125 mm concrete

Stanchions pivotted on dowel plates fixed to slab

Central reinforcement

Building paper

150mm sand or shale

Subsidence curve (exaggerated)

Slab divided into panels

Plan

63 *Flexible structure and foundation*

to permit free movement, such as tile hanging, weatherboarding or hung concrete slabs, are used and windows are designed in timber with sufficient clearances to accommodate the maximum anticipated distortion.

The foundation slab, laid on sand or shale, consists of a 125 mm thickness of concrete reinforced at the centre of the thickness and divided into panels not greater in area than 18.5 m². The panel edges are painted with bitumen so that the slab will bend fairly freely with the ground. The primary function of the reinforcement is to hold these slabs together. The sand or shale bed, together with the lightness of structure, reduces friction and minimizes the ground forces transmitted to the slab.

Where the loading on columns is such as to require spreading over a relatively large area local top and bottom reinforcement is placed in the slab

but no thickening of the slab occurs. Alternatively, reinforced precast concrete foundation pads are used to which the slab reinforcement is tied when the slab is cast up to them.

Since no flexural stresses are set up in the slab as a whole the need to limit its length in order to keep differential settlements within small limits does not arise. The length of the building is not, therefore, dictated by this consideration but rather by that of the maximum length which can economically be held together by tensional reinforcement. This appears to be about 55 m for a single-storey building reducing to 43 m for one of three or four storeys, up to which height it is considered that this form of foundation is applicable.

Rigidity is provided in the pin-jointed frame structure by specially designed diagonal steel braces incorporating coiled springs at the ends. These do not move under normal wind and dead loads but react to the greater stresses due to subsidence. One of the springs always acts in compression, the other remaining inoperative. When the subsidence wave has passed the springs bring the frame back to its normal position and continue to control it under normal loads[1].

This type of flexible foundation and building structure is, of course, suitable for soft or made up ground, on which differential settlements are likely to develop after the building is completed, as well as for normal sites. There is no reason why its use for lightly loaded buildings, for which it was designed, should be limited to subsidence sites.

A form of roller bearing has been used for heavy single storey structures, in which two sets of rollers at right angles to each other in a tray, are situated at the foot of portal frames to permit free horizontal movement of the soil and foundations in any direction without transfer to the structure[2].

[1] For full details of this system and of problems connected with it, such as the passing of services through the floor slab, see *Architects' Journal*, May 8, 1974 and October 10 and 24, 1957. This system of construction is marketed under the name of *Clasp* by Brockhouse Steel Structures Limited.

[2] See also Institution of Civil Engineers *Report on Mining Subsidence*, 1959.

UNDERPINNING

The term underpinning is applied to the process of excavating under an existing foundation and building up a new supporting structure from a lower level to the underside of the existing foundation, the object being to transfer the load from the foundation to a new bearing at a lower level. This may be necessary for any of the following reasons:

(i) when excessive settlement of a foundation has occured

(ii) to permit the level of adjacent ground to be lowered, for example where a new basement at a lower level is to be formed

(iii) to increase the load bearing capacity of a foundation.

Before underpinning operations commence, the structure of the building should first be examined for weaknesses such as poor brickwork or masonry and for effects of settlement which may be accentuated during the course of the work. In very old or badly damaged buildings, the structure may require strengthening by grouting up cracks and loose rubble masonry, for example, or by tying-in walls by tie rods or prestressing cables. Temporary support should be provided by adequate shoring and by strutting up of openings (see chapter 11). In some circumstances a wall or column must be relieved of all loads bearing on it by strutting floors and beams down to a solid bearing clear of the underpinning. During the course of underpinning high structures it may be advisable to keep a check on any

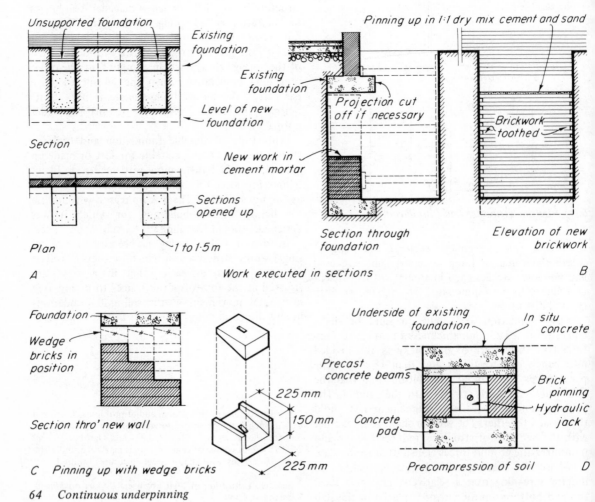

Unsupported foundation
Existing foundation
Level of new foundation
Section
Sections opened up
Plan
1 to 1·5 m

A

Pinning up in 1·1 dry mix cement and sand
Existing foundation
Projection cut off if necessary
New work in cement mortar
Brickwork toothed
Section through foundation
Elevation of new brickwork

Work executed in sections

B

Foundation
Wedge bricks in position
Section thro' new wall

C   *Pinning up with wedge bricks*

225 mm
150 mm
225 mm

Underside of existing foundation
In situ concrete
Precast concrete beams
Brick pinning
Hydraulic jack
Concrete pad
Precompression of soil

D

64   Continuous underpinning

possible movement by taking readings of plumb bobs suspended from high points on the structure and by making checks on levels from time to time. 'Tell-tales' should be fixed to check any further movement at points where it had previously occurred.

Continuous underpinning

Where possible walls should be underpinned with the minimum of disturbance to the structure and to the occupants of the building, and in most cases the work can be carried out without the support of needles passing through the wall to the interior. The work should be carried out in sections or 'legs' in such a sequence that the unsupported lengths of existing foundation over excavated sections are equally distributed along the length of the wall being underpinned (figure 64 A). In buildings other than small lightly loaded structures, the sum of these unsupported lengths should not at any one time exceed one-quarter of the total length of the wall, or in the case of a very weak wall, or one carrying heavy loads, one-fifth to one-sixth of the total length. Unless unavoidable, a section should not be excavated immediately adjacent to one which has just been completed. On a low, lightly-loaded building such as a two-storey house with the walls in good condition, it is common practice to work to a sequence in which not more than one-third of the length is unsupported at any one time. The lengths of wall which can be left temporarily without support over the excavated sections will depend upon the thickness and general state of the wall and its foundation and upon the load which it will be carrying during the course of the work. Advantage is taken of the 'arching action' of normally bonded brick and stone walls over openings and the length of sections can usually be from 1 to 1.50 m.

As each section is excavated, timbering will generally be required, the depth of each stage of excavation depending upon the firmness of the soil. Poling boards or, in loose soils permitting only small depths to be excavated at a time, horizontal sheeting will be used to support the soil faces under the wall. When the nature of the soil or other circumstances precludes the withdrawal of poling boards or sheeting as underpinning proceeds, these are usually formed of precast concrete about 1 m by 300 mm by 50 mm thick. Holes through the thickness permit grout to be pumped through to fill solidly any voids between the soil face and the back of the boards. This is essential in order to prevent the voids closing up afterwards and causing settlement at the back of the wall.

When the full depth of the excavation is reached, the new concrete foundation will be placed if brick underpinning is being used or, if concrete is being used either for a normal wall or a retaining wall, the first lift of the concrete. (For details of retaining wall construction see 'Timbering for excavations', chapter 11.) The ends of the brickwork in each section should be left toothed ready for bonding with the next section (figure 64 B); all concrete work should have grooves formed to provide a key for the next section of concrete. Any horizontal reinforcement must project, being turned up against the sides until the adjacent section is excavated. It is then turned down to be spliced with that in the next section. All brickwork should be in cement mortar.

After cleaning off the underside of the existing foundation, brickwork is built up to within about 25 to 38 mm of the foundation. When time has been allowed for the mortar to set and shrink, the brickwork is pinned up with half-dry 1:1 cement and sand rammed in hard with a 25 mm board to make solid contact with the foundation above. In thick walls it is necessary to step down the brickwork from back to front in order to pin up satisfactorily, each step being pinned up in turn, commencing at the back. It is possible to pin up with specially made two-piece wedged-shaped engineering bricks, bedded in stiff mortar at the top of each step which, as they are driven home, reduce the mortar thickness to a minimum. The two pieces are keyed together by mortar squeezed into a vertical slot in each as the wedge is tightened (see figure 64 C).

Concrete underpinning, which must be placed over front shutters, can generally be taken only to within about 75 mm of the existing foundation. This, after the concrete has had time to set and shrink, is filled with half-dry 1:1 cement and sand or a dry mix fine aggregate concrete, rammed home with a mechanical tamper. Many Local Authority surveyors insist on a 75 mm gap being left over brickwork so that it may be pinned up in this way, particularly in the case of thick walls. Alternatively, pressure grouting can be used to pin up. Daywork joints in concrete walls which are to be water resistant must be freed of laitance. This can be washed away from freshly-placed concrete by high-pressure water spray within 1½-2 hours of placing, leaving a good, clean key. When the adjacent sections are opened up the exposed concrete faces may need to be hacked to form a good key and wire brushed to remove any remaining laitance and loose material. The keyed surface should be thoroughly wetted and brushed over with a thin coat of 1:2 cement and sand mortar immediately before fresh concrete is placed against it.

Underpinning in each section should commence as soon as possible after the bottom has been exposed, and be carried out as quickly as possible. No earth face should be exposed overnight. If delay is likely the bottom should be protected by a blinding layer of concrete or the last few inches of excavation should be held over.

When the walling is weak or when sections must be opened up in lengths greater than those suggested above or where the safe bearing pressure of the soil is likely to be exceeded during the underpinning operations, it may be necessary to provide temporary support in the form of needles passing through the wall or under the foundation, or vertical struts placed directly under the foundation. Needles of timber or steel should be of ample size to avoid deflection and should be kept in close contact with the structure by means of folding wedges or jacks. Supports must be well clear of the area of underpinning work and taken on to a solid bearing. When the needles pass through the wall above foundation level, subsidiary needles or springing pieces hung from the main needles may be required to carry the section of wall below the level of the main needles, particularly if the walling is weak (see figure 66 A).

In order to minimize the number of needles penetrating within the building it is possible to use widely spaced main needles which pass through the wall and carry a beam parallel to the outside face of the wall. This acts as a fulcrum to intermediate counterweighted cantilever needles which penetrate only the thickness of the wall (figure 66 B). When support by needles is impracticable, temporary steel struts bearing on concrete pads may be inserted beneath the existing foundation. The underpinning work may then be built up or cast between them before they are removed or, alternatively, they may be left in position and eventually cast in with new-concrete work.

The method of underpinning and final pinning up described above, if carefully executed, will result in negligible movement due to settlement of the new work, but in large and heavy buildings, a system of precompressing the soil on which the new work will bear is adopted. By this means consolidation of the soil is effected before the load from the underpinning above is applied and subsequent settlement is avoided. In this method a pad of reinforced concrete is cast on the bottom of the excavation and when it has matured a hydraulic jack of the normal type or a Freyssinet flat jack is fixed to it at the centre of the underpinning section. Precast concrete beams are then placed on top of the jack and concrete is placed on them up to the underside of the existing foundation (figure 64 D). Before this has set it is jacked up so that all voids in the underside of the foundation are completely filled with the concrete. When the concrete on the beams has matured a predetermined load is applied by the jacks so that the soil is compressed and its bearing capacity is 'pre-tested' in advance. When all the work has been carried out in this way, the spaces between the jacks in adjacent sections are pinned up with engineering bricks in cement mortar which, when set, permits the jacks to be removed and the remaining spaces to be filled up tightly.

Hydraulic jacking may also be used to ensure a positive contact between the face of the excavated soil at the back of each section and any supporting poling boards or sheeting. Each stage of excavation is supported at the back by 100 to 150 mm of *in situ* reinforced concrete acting as a permanent poling board. Temporary struts back to the opposite excavated face incorporate hydraulic jacks by means of which a predetermined pressure is applied to the concrete poling board about 24 hours after casting, while it is still green: at this stage the concrete is still flexible and this ensures with greater certainty close contact with the soil face and the elimination of any voids behind the concrete.

Prestressing can be applied to continuous *in situ* concrete underpinning carried out in the normal way to form a continuous beam under the existing foundations. This is done by sinking some sections to the full required depth to act as supporting piers to a beam formed between them by the intermediate sections which are not sunk so deep. These are formed into a continuous beam by post-tensioning with cables, passed through holes formed in the sections as they are cast and anchored at the ends.

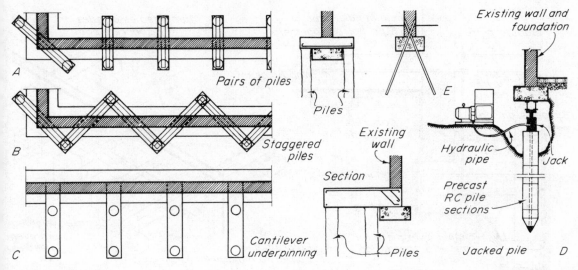

65   Piled underpinning

In addition to providing continuous support under the foundations, which may be essential when the superstructure is in a very poor condition, this method keeps to a minimum the amount of deep excavation which must be carried out.

Piled underpinning

Continuous underpinning as described above may in some circumstances be impracticable or uneconomic. The presence of subsoil water with certain types of soil might make the work extremely difficult, or when the load has to be transferred to a great depth the cost of continuous underpinning might be excessive. In such circumstances piles may be used with the load from the wall transferred to them by beams or needles. In order to avoid vibration which would be undesirable in many cases of underpinning, bored or jacked-in piles are used.

The arrangement of beams and piles depends on many things, including the state of the structure and the necessity or otherwise of avoiding disturbance to the use or contents of the building. The most straightforward method is probably to sink pairs of piles at intervals along the wall, one on either side, connected by a needle or beam passing through the wall just above or immediately below the foundation, whichever best suits the circumstances (see figure 65 A). Reinforced concrete or steel beams can be used for the needles. The latter occupy less space and

are generally simpler to fix. An alternative method which may sometimes be more suitable is to stagger the piles on each side of the wall (B). This has the effect of halving the number of piles required, although each will take a greater load, and of lengthening the needles. If the positions most convenient for sinking the piles do not coincide with the most suitable positions for the needles, two beams, one on each side of the wall and parallel to it may be carried by the piles and will support the needles which will pass through the wall at the most suitable points.

These methods, because of the internal piles, necessitate work within the building being underpinned. Where this is not possible, groups of two or more piles along the outside of the building may be used to support the wall by means of cantilever capping beams projecting under the wall (C).

As an alternative to cantilevered supports, jacked-in piles can be used immediately under the existing foundations in order to avoid internal work. These piles have been described on page 104 and the method is illustrated in figure 65 D. It is, of course, essential that the weight of the structure be sufficient to provide adequate dead weight as a reaction to the jacks as the piles are being forced down.

Quite a different method of piling[1] makes use of

[1] A patented system known as *Pali Radice* (Fondedile Foundations Limited).

111

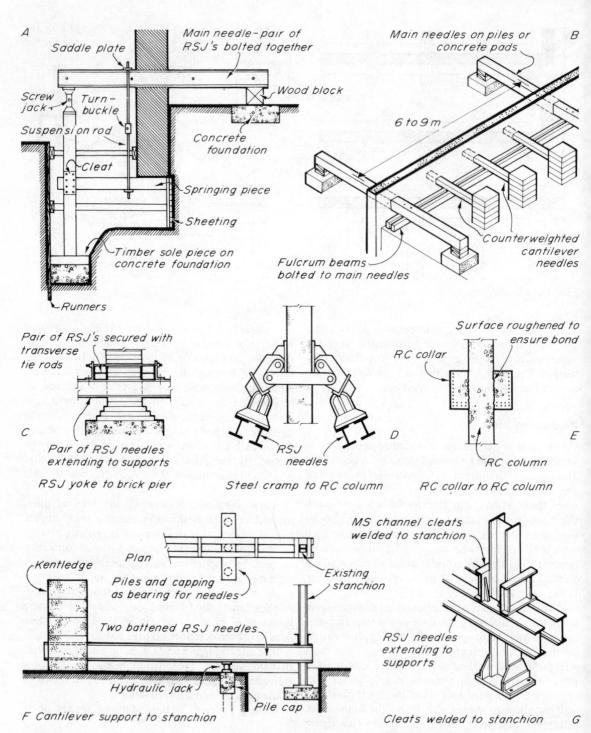

A
Saddle plate
Main needle–pair of RSJ's bolted together
Screw jack
Turn-buckle
Suspension rod
Wood block
Cleat
Concrete foundation
Springing piece
Sheeting
Timber sole piece on concrete foundation
Runners

Main needles on piles or concrete pads
6 to 9 m
B
Fulcrum beams bolted to main needles
Counterweighted cantilever needles

Pair of RSJ's secured with transverse tie rods
C
Pair of RSJ needles extending to supports
RSJ yoke to brick pier

RSJ needles
D
Steel cramp to RC column

Surface roughened to ensure bond
RC collar
E
RC column
RC collar to RC column

Kentledge
Plan
Piles and capping as bearing for needles
Two battened RSJ needles
Hydraulic jack
Pile cap
Existing stanchion

MS channel cleats welded to stanchion
RSJ needles extending to supports
Cleats welded to stanchion   G

F Cantilever support to stanchion

66 Underpinning

small diameter (120 to 280 mm) drilled friction piles passing through the actual foundation (figure 65 E), thus eliminating the necessity for needles to transfer the load to the piles. The piles are bored at an angle by means of a rotary drilling rig, either directly through the foundation or from a point up the wall as shown in figure 65. When the hole has been drilled reinforcement is introduced, a cement and sand grout is pumped in under pressure to form the pile and the lining tube is extracted. Formation of the piles is quiet and free of vibration and can be carried out with a headroom of only 1.80 m.

When the superstructure and its foundation is in a weak condition, continuous beam support to the wall may be essential. In small, lightly-loaded buildings such a beam, projecting beyond the wall face, can sometimes by supported adequately on a line of single piles only outside the building. In other cases the beam must be carried by needles bearing on piles on each side of the wall or on cantilever capping beams as described above. The supports may, however, be spaced at greater intervals and the beam designed to span between them. When support to the wall may be given above the level of the foundation the beam can be formed in a number of ways which simplify construction. First by the use of 'stools'. These are concrete blocks with holes in them through which reinforcing bars can be passed. When very deep beams are required the stools are formed of a number of vertical steel bars welded to top and bottom plates. Holes are cut in the wall at approximately 1 m centres at the required level and the stools are inserted (figure 242 C). These act as props to the wall above when the intervening sections of the wall between them are cut away. When this has been done for the full length of the underpinning, reinforcement may be threaded through the stools and tied in position. Shuttering is then erected and the concrete placed to form a continuous beam incorporating the stools as part of it. The gap at the top of the beam is pinned up as described earlier.

Another similar method makes use of prestressed concrete beams and involves no *in situ* casting of concrete. Precast blocks or segments, with holes cast through them for a prestressing cable, are inserted in turn next to each other in holes cut in the wall, each being pinned-up on top and wedged underneath until the whole of the wall is supported by a line of these blocks (figure 242 D). The joints between them are filled with dry mix mortar rammed tight and

openings are left at each end for anchorage and stressing of the cable. A post-tensioning cable is then threaded through the full length, tensioned and anchored at each end to form a single, homogeneous beam. Alternatively, stressed steel underpinning beams may be used made up of 600 mm to 1.50 m lengths of steel beams with steel diaphragm plates welded to each end and drilled for high tensile torque bolts. Each length is inserted in turn and pinned up in the same way as the segments of a post-tensioned concrete beam, so that the wall above is continuously supported (figure 242 E). The beam is inserted in the wall with the joints between the lengths of beam left slightly open on the tension side, and after packing solid at the top, the bolts are tightened by a torque wrench. In closing the joints, the lower flange of the beam is stressed so that the beam takes up its load without deflecting.

The methods just described are also useful when new openings must be formed in existing walls and permit the work to be carried out without the dead shoring and strutting which is otherwise necessary as described in the section on shoring in chapter 11.

Underpinning to column foundations

In underpinning framed structures, the main problem is to provide satisfactory support to the columns while they are being underpinned. Before excavation is begun these must be relieved of load by dead shores under all beams bearing on them.

Reinforced concrete columns and brick piers can be supported by means of a horizontal yoke formed of two pairs of rolled steel beams set in 25 to 50 mm deep chases in the sides of the columns (figure 66 C). The bottom pair are large enough and long enough to act as needles to transfer the load to temporary support and the upper pair are at right-angles to the needles and bear on them. The pairs of beams are tied together by transverse tie rods or angles. An alternative method, which avoids chasing into the sides of the column and which can deal with greater loads, is to grip the column or pier in a heavy steel cramp designed to grip more tightly as it takes up the weight of the column (D). The base of the cramp bears on needles on opposite sides of the column which transfer the load to supports well away from the column base. Reinforced concrete columns may be supported by a reinforced concrete collar (E). Steel stanchions can be supported on steel needles by steel angle or channel cleats welded to the stanchion flanges and bearing on the needles (*G*). Lightly loaded stanchions can sometimes be carried by a large diameter steel pin, passed through a hole drilled in the web and bearing on the needles.

In all cases the supports to the needles must be far enough away from the column or pier to avoid collapse by dispersion of the pressure on to the sides of the excavation, and also to allow sufficient working space. The needles must have good solid bearings, strong enough to take the loads transferred to them from the columns and this may sometimes necessitate the use of piles at the bearings. If underpinning is due to the excavation of a new basement under the columns, the temporary pile supports may have to penetrate below the level of the new basement to a depth sufficient to permit them to act as free standing columns when the main excavation is carried out and while the column is being underpinned.

Circumstances may make it impossible to provide support to the needles on two sides of a column or stanchion and a counterweighted cantilever support must then be used. This would be formed of a pair of long, deep steel needles bearing on a fulcrum support some distance from the column in the form of a base slab or piled foundation (see figure 66 *F*). One end of the needles would pick up the yoke or cleats attached to the column and the other would be tailed down by kentledge or tension piles. A hydraulic jack placed between the needles and the fulcrum support enables the latter to be 'pre-compressed' and any settlement taken up to prevent movement of the column.

Ground treatment

Suitable soil such as gravel and sand may be strengthened, or its permeability decreased, by pressure grouting, and water in sand and silts can be frozen by refrigeration. The task of underpinning (and of normal foundation construction) can often be simplified by the application of these processes to the soil and in some circumstances they can be used as an alternative to underpinning. In the latter case, where permanent effects are required, grouting would be used, since freezing is only practicable as a temporary measure. Grouts may be solutions of various chemicals which gel in the soil or suspensions of cement, fly-ash or bituminous emulsion[1].

[1] See *Code of Practice* 2004:1972, 'Foundations', Part 6, for a description of these and other geotechnical processes.

4 Walls and piers

Types of walls and the functional requirements of walls are discussed in chapter 5 of Part 1 where the construction of masonry and timber frame walls is described in detail. The behaviour of the wall under various conditions of loading is discussed in chapter 3 of that volume. Some further aspects of the strength and stability of walls are examined here as a preliminary to the determination of wall thickness which is described later on page 120.

Eccentric loading

Floor and roof loads on walls are frequently applied on one side, rather than on the axis of the wall or bearing. In the case of concrete floor slabs with a span of less than thirty times the wall thickness it may be assumed that the load is applied through the centre of the bearing. However, the deflection of a floor tends to concentrate the load on the bearing edge of the wall and this tendency will be greater with flexible floors, such as normal timber construction or longer span concrete floors, and for these an eccentricity of one-sixth of the bearing width should be assumed. Even if the bearing of such a floor covers the full width of the wall it is desirable to assume this eccentricity[1].

As explained in chapter 3 of Part 1 such eccentric loading tends to cause bending in the wall, having the effect of increasing the compressive stress in the wall on the loaded side and of decreasing it on the opposite side (figure 67 A where p represents the stress due to the load applied axially). The simple formula by means of which the intensity of these stresses is established is given on page 56 of Part 1. This variation in stress must be borne in mind for the increased stress on the loaded side could become greater than the safe compressive strength of the wall and, if the eccentricity is too great, tensile stresses will be set up in the side opposite that on which the load is applied (figure 67 B).

This has a marked effect upon the strength of the wall, particularly so in masonry walls in which the tensile strength is small and which, in practice, is usually ignored[2]. It is assumed that cracking occurs when tensile stresses are set up. Thus when the eccentricity is greater than one-sixth of the thickness and tension occurs, part of the wall will cease to function structurally. It is also assumed that on cracking a redistribution of the compressive stress takes place such that the resultant passes through the centroid of the stress triangle (figures 67 C). The working portion of the wall is thus reduced to three times the distance of the point of application of the

---

[1] See BRS Digest 61, *Strength of brickwork, blockwork and concrete walls*

[2] The tensile strength of the wall is governed not by the tensile strength of the mortar used but by the adhesional strength or bond between the units and the mortar. Failure invariably occurs as a result of a breakdown of this bond rather than from a failure of the mortar or units in tension. Bond is extremely variable and average ultimate values range from 0.28 to 0.55 MN/m^2 depending on the mortar mix, workmanship and weather conditions at the time of building. CP 111:Part 2:1970 *Structural Recommendations for Load Bearing Walls*, recommends that no reliance should normally be placed on the tensile strength of masonry. Where the masonry units are prepared before laying in accordance with CP-121:Part 1:1973 *Walling* the Code does, however, with certain provisos allow discretion to the designer to take into account tensile stresses in bending up to a maximum of 0.14 MN/m^2. (Table 12.6, page 166, of *Principles of Modern Building*, Vol 1, 3rd Ed (HMSO), shows the effect of different mortar mixes and walling units upon the transverse strength of walls)

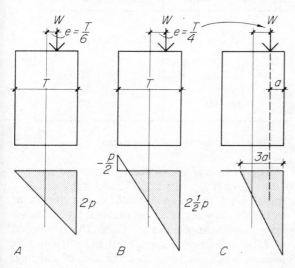

67   *Effect of eccentric loading*

load from the bearing face of the wall and the maximum stress on this reduced working area is established from $\dfrac{2 \times W}{3 \times a}$ and for the eccentricity shown would be $2\frac{2}{3}\,p$. In the calculation of walls this is allowed for by the application of reduction factors to the safe compressive strength of the walling material (see page 122).

Overturning

Lateral loads, such as the pressure of wind or of stored materials on normal walls and of earth on retaining walls, tend to bend and overturn the wall as shown in chapter 3 of Part 1. This causes the line of resultant pressure to become eccentric and sets up a stress distribution similar to that caused by eccentric loading, with a reduction in working area when cracking occurs. In free-standing walls this may overcome by making the wall thick enough to bring the resultant well within the base or by buttressing. In buildings account is taken of horizontal and vertical lateral supports, which may be present in the form of floors and cross walls or a structural frame, to provide added resistance against overturning to that of the wall itself (see figure 24, Part 1).

Buckling

As explained in chapter 3 of Part 1 short walls or piers ultimately fail by crushing, but as the height increases they tend to fail under decreasing loads by buckling. For purposes of calculation the height of a wall is generally based on the distance between floors, assuming these give adequate lateral support to the wall. The actual proportion of the floor to floor dimension to be considered for this purpose and the methods of providing adequate lateral restraint by means of floors are discussed later.

The manner in which these possible causes of failure are accounted for in the regulations governing the minimum thickness of walls is discussed under the section on 'Calculation of wall thickness'.

Loading

This is considered in terms of (i) the dead load, (ii) the live load, (iii) wind load.

The *dead load* is the weight of all parts of the structure bearing upon the wall together with its own self-weight. Weights of building materials used as a basis for estimating such dead loads are given in BS 648.

The *live load* is the weight of superimposed floor and roof loads including snow. Minimum loadings on floors and roofs are given in CP3: Chapter V: Part 1: 1967; reference is made to these in chapters 6 and 9.

In most types of building it is unlikely that all floors would be fully loaded at the same time. In the design of walls, piers and columns the Code allows for this by permitting varying percentage reductions of the total superimposed floor loads according to the number of floors carried (see table 3). No reduction is permitted for plant or machinery specifically allowed for or for warehouses, storage buildings, garages and office areas used specifically for filing and storage or for factories and workshops unless designed for loadings of $5\ \text{kN/m}^2$ or more and provided the loading assumed for the wall or pier is not less than it would have been if all floors had been designed for $5\ \text{kN/m}^2$ with no reductions. When a wall or pier supports a beam, the reduction permitted in respect of a single span of beam supporting not less than $46\ \text{m}^2$ of floor (see 'Floor loading', chapter 6, page 238) may be compared with that given by table 3 and the greater of the two taken into account in the design of the wall or pier.

Number of floors including roof carried by member under consideration	Per cent reduction in total imposed load on all floors carried by member under consideration
1	0
2	10
3	20
4	30
5 to 10	40
over 10	50

Table 3   *Reduction of floor loads*

The *wind load* is assessed by a method given in CP3 : Chapter V : Part 2 : 1972 which is based on a 'basic' wind speed appropriate to the district in which the building is to be situated, taken from a map of isopleths given in the Code. This is adjusted to take account of topographical and other factors such as height above ground and size of building, and is then converted by formula to a dynamic pressure. The actual pressure or suction exerted on a particular surface of the building (eg wall or roof) is then established by the application of an appropriate coefficient to the dynamic pressure.

The distribution of wind pressure on walls is not uniform. It is greater near the eaves, and near the

mm external cavity walls and 175 mm solid internal loadbearing walls, and five storeys using only 102.5 mm internal walls. This necessitates careful consideration of such design aspects as (i) the choice of a suitable cellular plan form, (ii) maintaining a suit-

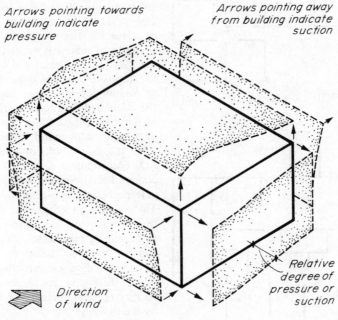

Arrows pointing towards building indicate pressure

Arrows pointing away from building indicate suction

Relative degree of pressure or suction

Direction of wind

68   *Pressure and suction due to wind*

ends of a wall high suctions are possible as shown in figure 68 where the distance of the curved 'planes' from the faces of the building indicates the degree of pressure or suction on surfaces of the building. The Code gives coefficients to take account of this.

SOLID MASONRY WALLS

For many years loadbearing brickwork has provided the cheapest structure for houses and blocks of flats from three to five storeys in height, and, broadly speaking, loadbearing brick or block wall construction will still produce the cheapest structure for small-scale buildings of all types where planning requirements are not limited by its use[1]. Until masonry walls were calculated on a scientific basis great heights required great thickness of wall, but with the development of adequate design methods as exemplified in CP 111, it is now possible to build fourteen to fifteen storeys high with only 280

able proportion of height to width of building to keep wind stresses to a minimum, (iii) running concrete floor slabs through to the outer face of external walls to reduce the eccentricity of floor load (see page 115).

A suitable plan form is one in which the floor area is divided into rooms of small to medium size and in which the floor plans repeat on each storey. This results in the walls running through in the same plane from foundations to roof and the floors, being of moderate span, transferring moderate and distributed loading to the walls. Maisonettes, flats and hostels are building types in which these requirements may be fulfilled.

For high-rise buildings the disposition of the walls must be such that they provide structural rigidity in the direction of both axes of the building so that some form of cellular plan (figure 69 A) or one with

[1] See figure 74 and also page 180

117

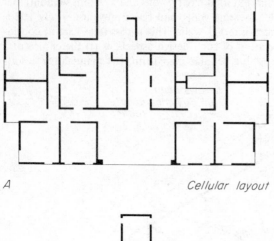

A   *Cellular layout*

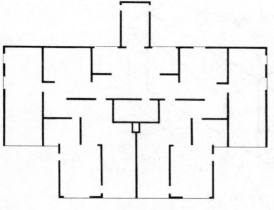

B   *Crosswall layout*

69   *Plan forms for masonry construction*

cross walls in both directions (*B*) is essential.

Reference is made in chapter 5 of Part 1 to the empirical method of determining wall thicknesses by reference to the height and length of wall and to other conditions laid down for this purpose in building by-laws and regulations[1], and it is explained why this method is generally used for small traditional types of building.

In the case of buildings over three storeys, however, the wall thicknesses are invariably determined by means of calculations. This is common practice also with some smaller types of loadbearing wall structures other than two-storey domestic buildings, especially when a large proportion of openings is required which would reduce the effective wall area to a minimum. The reasons for this are that in using the empirical method it is not possible to

take into account stronger materials than those laid down as a minimum, thus to gain the advantage of reduced thicknesses, nor the actual loads likely to bear upon the walls in various parts. In addition the limitation on the width of openings necessitated by an assumed maximum loading and a specified minimum strength of material tends to be restricting in design. The determination of wall thickness by calculation, having regard to actual loads and varying strengths of materials, leads to more economic construction in most cases and to a greater freedom in the disposition of openings. The thickness of a wall or column determined in this manner is based upon the calculated load to be carried, the actual strength of the materials used and the slenderness of the wall or column under consideration.

In the County of London, the London Building Acts (Amendment Act), 1939, Part III, Section 26, confers discretionary powers on the District Surveyor in respect of the construction of all public buildings. These come directly under the jurisdiction of the District Surveyor as regards all structural matters. In examining proposals for a public building he may consent to their being in accordance with the London Building (Constructional) By-laws in respect of wall thicknesses for loadbearing or panel walls, or he may require a higher standard than that required by the By-laws.

Strength of masonry walls

The strength of a masonry wall depends primarily upon the strength of the units and of the mortar. In addition the quality of workmanship is important. Variations of 25 to 35 per cent in the strength of brickwork have been found due to bad workmanship, mainly on account of badly mixed mortar and imperfect bedding of the bricks. This is an important consideration in high-strength masonry walls where all units should be properly bedded and all vertical joints flushed up solid.

The effect of the strength of the units and of the mortar is illustrated by the graphs in figure 70 which are based on tests on brickwork. It will be seen that the strength of the wall increases with the strength of the individual bricks, but not in direct proportion to the unit strength. This is due to the significant effect of the mortar, which is shown in the adjoining graph, where the effect of varying mortar

[1] London (Constructional) By-laws Part VII and Building Regulations Schedule 7

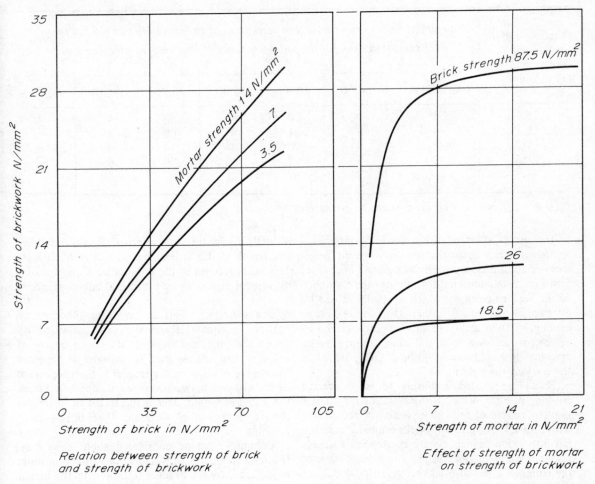

Relation between strength of brick
and strength of brickwork

Effect of strength of mortar
on strength of brickwork

70   Strength of brickwork

strengths is related to the strength of the wall. It can be seen that little advantage is gained, from the point of view of ultimate wall strength, in increasing the strength of the mortar beyond a certain point for any particular grade of brick. In terms of actual mortar mixes this is illustrated in table 4 which shows that although a cement mortar will give greatest strength a cement-lime mortar with 50 or 60 per cent of the cement replaced by lime gives only a slightly lower strength of brickwork. A straight non-hydraulic lime mortar will generally give only a brickwork strength of less than half that attained with a cement or cement-lime mortar. It should be noted that the optimum strength of brickwork is obtained with a 1:3 basic proportion for the mortar.

The mortars suggested to give maximum strength with various grades of brick are:

Low strength bricks (10.35 N/mm^2)
1 cement : 2 lime : 9 sand
Medium strength bricks (20.7 to 34.5 N/mm^2)
1 cement : 1 lime : 6 sand
High strength bricks (48.3 N/mm^2 or more)
1 cement : 3 sand : (lime may be added up to ¼ volume of cement)[1]

[1] For a full discussion on the strength of masonry walls see *Principles of Modern Building*, Volume 1, 3rd Ed (HMSO), from which much of the above has been extracted

119

Proportion of cement and lime to sand (by volume)	Strength of brickwork expressed as percentage of strength of brickwork built in 1:3 cement mortar, for the following ratios of lime:cement by volume						
	All cement	50:50	60:40	70:30	80:20	90:10	All lime
1:1	–	72	70	66	58	47	–
1:1½	–	87	84	77	68	56	–
1:2	96	94	90	84	74	60	–
1:3	100	96	92	87	79	65	48
1:4	–	92	87	81	71	59	–

Table 4   *Effect of mortar proportions on strength of brickwork*

The use of these relatively weak mortars has a significant effect upon the weather resisting properties of solid masonry walls as explained in chapter 5 of Part 1. In addition, they improve the resistance of the wall to cracking, which is usually caused by differential movement rather than by excessive loading, by their ability to 'give' slightly to take up movements of the wall and thus often prevent cracking or, should movement be so great, distribute the cracking throughout the joints.

Bricks are graded according to their 'average crushing strength' which is the average strength of a random sample of ten bricks tested in accordance with BS 3921. The prices of bricks normally available fall into groups which closely correspond to these grades of strength. The crushing strengths of some of the common types of bricks are given in table 39 of *MBC: Materials*. These range from 3.5 to 172 N/mm^2 for machine moulded bricks and from 7 to 59 N/mm^2 for hand moulded bricks.

Calculation of wall thickness

The London By-laws and the Building Regulations both accept the recommendations of CP 111 : Part 2 : 1970 as a basis for determining the thickness of walls and piers. The Code defines certain terms in relation to this.

Slenderness ratio   Reference has already been made to the failure of walls and columns[1] under decreasing loads with increasing height. The ratio of height to thickness as a measure of this tendency is termed the slenderness ratio. This is defined as the ratio of effective height to effective thickness, but in the case of walls it may be based alternatively on the effective length if this is less than the effective height. This takes account of the stability which is provided by vertical as well as by horizontal lateral supports.

Effective height   This is based on the distance between adequate lateral supports[2] provided by floors and roof and depends upon the degree of support they are assumed to provide. The greater the degree of support the smaller is the proportion of the distance between centres of support taken as the effective height. This is illustrated in figure 71, A to C in respect of walls and D, E in respect of columns.

Columns must be considered about both axes. If lateral support is provided in one direction only, as indicated by the beam in (D), the effective height relative to that direction will be as shown, but in the other direction it must be twice its height above the lower support. In the absence of any top support (E) the latter value must be taken relative to both directions.

Where the wall between two openings constitutes a column as in (F) its effective height is based upon the height of the taller of the openings, Z. Where Z does not exceed H/2 the effective height is 1½ Z. Where Z exceeds H/2 it must be taken as 1½ Z or H, whichever is the less.

[1] A column is defined as an isolated member of which the width does not exceed four times its thickness. The term 'pier' is defined as a thickened section of a wall placed at intervals along its length (see figure 71).

[2] Defined as 'support which will restrict movement in the direction of the thickness of the wall, or, in relation to a column, movement in the direction of its thickness or width'

CALCULATION OF WALL THICKNESS

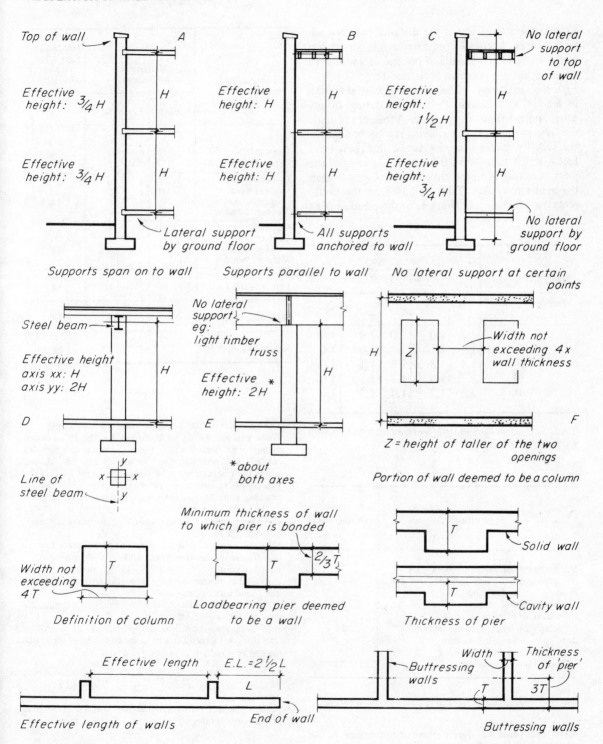

Top of wall ⟶   A   B   C   No lateral
                                                                               support
                                                                               to top
                                                                               of wall

Effective   H   Effective   H   Effective   H
height: 3/4 H   height: H   height:
                                                                 1 1/2 H

Effective   H   Effective   H   Effective   H
height: 3/4 H   height: H   height:
                                                                 3/4 H

                                                                               No lateral
                    ⟵ Lateral support   All supports   support by
                      by ground floor   anchored to wall   ground floor

Supports span on to wall   Supports parallel to wall   No lateral support at certain
                                                                            points

Steel beam ⟶   No lateral   Width not
                            support   Z   exceeding 4 x
Effective height   eg: H   wall thickness
axis xx: H   light timber
axis yy: 2H   truss

D   Effective   *
                            height: 2H

Line of   y   *about   Z = height of taller of the two
steel beam   x x   both axes   openings
              y   Portion of wall deemed to be a column

                            Minimum thickness of wall
                            to which pier is bonded   T

Width not   T   T   2/3 T   ⟵ Solid wall
exceeding
4 T
                            Loadbearing pier deemed   T
Definition of column   to be a wall   ⟵ Cavity wall

                                                            Thickness of pier

            Effective length   E.L.= 2 1/2 L   Width   Thickness
                                                        ⟵ Buttressing   of 'pier'
                                         L   walls
                                                                      T   3T
Effective length of walls   End of wall
                                                            Buttressing walls

71   Calculated walls

121

Effective length   This is the distance between adjacent piers, buttresses or intersecting or return walls. The effective length of the end of a wall with no end stiffening is shown in figure 71.

Effective thickness   This is the actual thickness of a solid wall excluding plaster, rendering, or any other applied finish or covering. Allowance is made for any stiffening piers which may be bonded to the wall by multiplying the actual thickness by a factor which varies with the size and spacing of the piers, resulting in an effective thickness greater than the actual thickness. Table 5, based on the Code, gives these factors[1]. Buttressing or intersecting walls

Thickness of pier	Pier spacing				
	Width of pier				
Thickness of wall	6	8	10	15	20
1.0	1.0	1.0	1.0	1.0	1.0
1.5	1.2	1.15	1.1	1.05	1.0
2.0	1.4	1.3	1.2	1.1	1.0
2.5	1.7	1.5	1.3	1.15	1.0
3.0	2.0	1.7	1.4	1.2	1.0

Table 5. *Determination of effective thickness of wall stiffened by piers*

Type of wall	Maximum slenderness ratio
Unreinforced brickwork or blockwork set in hydraulic lime mortar	13
ditto in buildings not exceeding two storeys	20
Brickwork or blockwork set in other than hydraulic lime mortar	27
ditto in walls less than 90 mm thick in buildings of more than two storeys	20

Table 6   *Maximum permitted slenderness ratios*

[1] The code does not permit these modifications to the effective thickness when the slenderness ratio is based on the effective length of the wall

122

Description of mortar	Mix (parts by volume)			Hardening time after completion of work†
	Cement	Lime	Sand	
Cement	1	0-¼*	3	days
	1	½	4½	7 / 14
Cement-lime Cement with plasticizer§ Masonry cement‖	1 / 1 / –	1 / – / –	6 / 6 / –	14
Cement-lime Cement with plasticizer§ Masonry cement‖	1 / 1 / –	2 / – / –	9 / 8 / –	14
Cement-lime	1	3	12	14
Hydraulic lime	–	1	2	14
Non-hydraulic	–	1	3	28

*   The inclusion of lime in cement mortars is optional

†   These periods should be increased by the full amount of any time during which the air temperature remains below $4.4°C$ plus half the amount of any time during which the temperature is between $4.4°C$ and $10°C$

‡   Linear interpolation is permissible for units whose crushing strengths are intermediate between those given in the table

§   Plasticizers must be used according to manufacturers' instructions

‖   Masonry cement mortars must be used according to manufacturers' instructions, and mix proportions of masonry cement to sand should be such as to give comparable mortar crushing strengths with the cement: lime: sand mix of the grade

¶   A longer period should ensue where hardening conditions are not very favourable

Note: Where the cross-sectional plan area of a wall or column does not exceed 0.3 m^2, the basic stress should be multiplied by a reduction factor equal to

$$0.75 \times \frac{A}{1.2}$$

where A is the area (in m^2) of the horizontal cross-section of the wall or column

Table 7   *Basic stresses for masonry walls*

Basic stress in MN/m² corresponding to units whose crushing strength (in MN/m²) ‡ is:

Brickwork members									Blockwork members								
2.8	7.0	10.5	20.5	27.5	34.5	52.0	69.0	96.5 or greater	2.8	3.5	7.0	10.5	14.0	21.0	28.0	35.0	52.0
0.28	0.70	1.05	1.65	2.05	2.50	3.50	4.55	5.85	0.28	0.35	0.70	1.05	1.25	1.70	2.10	2.50	3.50
0.28	0.70	0.95	1.45	1.70	2.05	2.80	3.60	4.50	0.28	0.35	0.70	0.95	1.15	1.45	1.75	2.10	2.80
0.28	0.70	0.95	1.30	1.60	1.85	2.50	3.10	3.80	0.28	0.35	0.70	0.95	1.10	1.35	1.60	1.90	2.50
0.28	0.55	0.85	1.15	1.45	1.65	2.05	2.50	3.10	0.28	0.35	0.55	0.85	1.00	1.20	1.45	1.70	2.05
0.21	0.49	0.70	0.95	1.15	1.40	1.70	2.05	2.40	0.21	0.23	0.49	0.70	0.80	1.00	1.20	1.40	1.70
0.21	0.49	0.70	0.95	1.15	1.40	1.70	2.05	2.40	0.21	0.23	0.49	0.70	0.80	1.00	1.20	1.40	1.70
0.21	0.42	0.55	0.70	0.75	0.85	1.05	1.15	1.40	0.21	0.23	0.42	0.55	0.60	0.70	0.75	0.85	1.05

Table 7   *continued*

may be considered as piers of width equal to the thickness of the intersecting wall and of a thickness equal to three times the thickness of the stiffened wall (figure 71).

If a column has no lateral support or has support in both directions the effective thickness will be based on the least dimension. When the column has support in one direction only slenderness ratio values must be established using effective thicknesses based on both dimensions, and the larger be adopted.

CP 111 lays down maximum values for the slenderness ratio of masonry walls which are shown in table 6. In many buildings where loading is light and the necessary wall thickness is small the slenderness ratio becomes the controlling factor, limiting as it does the height for any given thickness of wall.

Permissible stress The stresses permitted in a wall or column are regulated according to the strength of the bricks or blocks, the type of mortar to be used and the slenderness ratio of the wall or column. Basic stresses, arising from combined uniformly distributed dead and superimposed loads and related to the strength of the units and the type of mortar used, are given in CP 111 and are shown in table 7.

For members with slenderness ratios up to six the basic stresses may be used. Above this ratio the permissible stresses in the wall or column must be established by the application of a reduction factor to the basic stress. Values for this factor for varying slenderness ratios and eccentricities of loading are given in table 8[1].

[1] A reduction factor need not be applied for sections within one-eighth of the height of the member above or below the lateral support

Slenderness ratio	Stress reduction factor*			
	Axially loaded	Eccentricity of vertical loading as a proportion of the thickness of the member		
		1/6	1/4	1/3†
6	1.00	1.00	1.00	1.00
8	0.95	0.93	0.92	0.91
10	0.89	0.85	0.83	0.81
12	0.84	0.78	0.75	0.72
14	0.78	0.70	0.66	0.62
16	0.73	0.63	0.58	0.53
18	0.67	0.55	0.49	0.43
20	0.62	0.48	0.41	0.34
22	0.56	0.40	0.32	0.24
24	0.51	0.33	0.24	–
26	0.45	0.25	–	–
27	0.43	0.22	–	–

* Linear interpolation between values is permitted
† Where in special cases the eccentricity of loading lies between $\frac{1}{3}$ and $\frac{1}{2}$ of the thickness of the member, the stress reduction should vary linearly between unity and 0.20 for slenderness ratios of 6 and 20 respectively

Table 8   *Reduction factors for slenderness ratios*

Where additional stresses are caused by eccentricity of loading and/or lateral forces (see page 115 and Part 1, page 57) the maximum stress resulting from the sum of these and that due to axial loading may exceed the permissible stress referred to in the previous paragraph by 25 per cent, provided the excess is due solely to eccentricity of loading and/or lateral forces. Where concentrated loads, as at girder or lintel bearings, cause further additional stresses of a local nature, the maximum total stress may exceed the permissible stress by 50 per cent.

The basic stresses given in table 7 have been based on certain ratios of strength of wall to strength of unit (see page 118) where the proportion of the height to thickness of the unit is about that of the normal brick. It has been shown[1] that these stresses are unnecessarily low when deeper blocks are used and the Code allows for this by permitting the basic stresses to be increased by a modification factor with increasing ratio of height to thickness (table 9). The Code includes clauses limiting the application of these factors in respect of solid concrete blocks with crushing strengths greater than 35 MN/m^2 and all other units with crushing strengths greater than 5.5 MN/m^2.

Ratio of height to thickness of brick or block	0.75	1.0	1.5	2.0 - 3.0
Factor	1.0	1.2	1.6	2.0

Linear interpolation between values for the reduction factors is permissible

Table 9   *Modification factors for unit shape*

In order to keep the thickness of walls within reasonable limits and, preferably, of the same thickness for the full height of the building, particularly in the case of cross wall construction, variations in the types of bricks and in the mortar mixes are made according to the stresses at different heights. Excessive variation is uneconomic and leads to difficulties in supervision on the site. Sufficient flexibility in strength can, however, be obtained in most buildings by the use of three to four grades of bricks with one or two mixes of mortar. Ordinary fletton bricks, which are cheap, when used in 215 mm thick imperforate cross walls are adequate for flat blocks up to four or five storeys in height when the floor spans are short. Concrete blocks may be used instead of bricks. When these are specified advantage may be taken of the permitted increase in the basic stresses on blocks having a height to thickness ratio greater than 0.75 (see table 9). The London By-laws lay down that a non-calculated brick or block wall shall have a thickness at any level not less than one-sixtieth of the height measured from that level to the top of the wall. A minimum thickness of 190 mm at any point is required in the case of an external wall, whether calculated or not.

Walls built of materials of differing strengths bonded together are less important now as load-bearing structures since the general practice is to use a thin 'veneer' of non-structural facing material attached to a structural backing, but provision is made for dealing with such a combination in two ways. The weaker material may be considered to be used throughout the full thickness and the permissible stress established on that basis. Alternatively, the area of that portion of the wall built of the strongest material only may be consid-

[1] See Part 1, page 113

ered as carrying the load, in which case the permissible stress is established using a slenderness ratio calculated on the thickness of that material alone.

Random rubble walling should be based on permissible stresses of 75 per cent of the corresponding stresses for coursed walling of similar materials.

The *design process* may be summarized briefly as follows:

1 Calculate total load (W) per metre run of wall or on column at level under consideration ('Loading' – pages 116, 238, 305)
2 Assume wall or column thickness and establish slenderness ratio (page 120)
3 Establish any eccentricity of loading (page 115)
4 Ascertain appropriate stress reduction factor (RF) (table 8)
5 Establish bearing area per metre run of wall or of column (A)
6 Establish 'equivalent basic stress' = $\dfrac{W}{A \times RF}$.

(Allow for any increase in stress for eccentric or local loading – page 124)

7 Select grade of brick and mortar with strengths appropriate to the equivalent basic stress (table 7)

PLAIN MONOLITHIC CONCRETE WALLS

A plain monolithic concrete wall means a wall of cast *in situ* concrete containing no reinforcement other than that which may be provided to reduce shrinkage cracking, together with a certain amount round openings. The monolithic reinforced concrete wall is considered in chapter 5 and in this chapter reference will be made only to the plain concrete wall of either normal, no-fines or lightweight concrete.

As with reinforced concrete walls they are most economic when used both to support and to enclose or divide, provided they are at reasonably close spacing. That is to say, up to about 5.5 m apart. They are, therefore, used mainly for housing of all types, both as external and internal loadbearing walls, when low building costs can be attained.

Dense concrete is generally used for high buildings although no-fines concrete has been used for heights up to ten storeys in this country. In Europe blocks as high as 20 storeys have been constructed with no-fines loadbearing walls more cheaply than with a frame.

Plain monolithic concrete walls suffer certain defects which, in some respects, makes them less suitable as external walls than other types. With normal dense aggregates the thermal insulation is low[1] and the appearance of the wall surface may be unsatisfactory, requiring some form of finishing or facing. In addition the unreinforced concrete wall, and particularly the no-fines wall, is unable to accommodate itself to unequal settlement as does a reinforced wall by virtue of the reinforcement or a brick or block wall to a certain extent by the setting up of fine cracks in the joints. Thus, as a result, large cracks tend to form in the wall. Nevertheless, where foundations are designed to reduce unequal settlement to a minimum such walls can successfully be used.

Aggregates used for dense plain concrete are natural aggregates conforming to the requirements of BS 882, air-cooled blast furnace slag and crushed clay brick. Aggregates for light-weight concretes are foamed slag, clinker, pumice and any artificial aggregate suitable for the purpose. No-fines concrete may be composed of heavy or lightweight aggregate.

Dense concrete walls are constructed from concrete made with a well-graded aggregate giving a concrete of high density. The London By-laws require the thickness of any concrete external or party wall to be not less than 150 mm thick and CP 123. 101, 'Dense Concrete Walls', recommends a similar thickness for external walls.

In most buildings the thickness of any type of plain concrete wall must, by reason of other functional requirements, be thicker than the minimum dictated by loadbearing requirements. An example of this is the dense concrete separating wall, which must be 175 mm thick in order to provide an adequate degree of sound insulation between houses and flats (see *MBC: Environment and Services*).

Lightweight aggregate concrete walls will give better thermal insulation than dense concrete when used for external walls but care must be taken in the choice of aggregate for external use because of the danger of excessive shrinkage and moisture movement occurring with certain types. Clinker has a corrosive action on steel and should not be used if

---

[1] See *MBC: Materials* chapter 8. See also tables 1 and 2, CP 123.101, *Dense Concrete Walls,* for thermal transmittance coefficients for solid and cavity dense concrete walls.

shrinkage reinforcement is to be incorporated.

All types of lightweight aggregate concrete are more permeable than dense concrete, and where the wall is exposed to the weather a greater thickness of cover to the steel is required, with possibly the further protection of rendering. Concrete with a wide range of density and compressive strength can be obtained by the selection of appropriate aggregate and mix (see table 66, page 169, *MBC: Materials*).

No-fines concrete walls are constructed with a concrete composed of cement and coarse aggregate alone, the omission of the fine aggregate giving rise to a large number of evenly distributed spaces throughout the concrete. These are of particular value in terms of rain exclusion as explained in Part 1, chapter 5. No-fines concrete is suitable for external and internal loadbearing walls or for panel wall infilling to structural frames.

The weight of no-fines concrete is about two-thirds that of dense concrete made with a similar aggregate. Aggregates graded from 19 mm down to 9.5 mm are used with mixes of 1 to 8 or 10 for gravel aggregate and 1 to 6 for light-weight aggregates. The aggregate should be round or cubical in shape and no more water should be used than that required to ensure that each particle of aggregate is thoroughly coated with cement grout without the voids being filled. The hydrostatic pressure on formwork is only about one-third of that of normal concrete. This is an advantage since horizontal construction joints should be minimized and formwork one or two storeys high can be employed without it being excessively heavy. Any normal type of shuttering can be used or open braced timber frames faced with small-mesh expanded metal are suitable. These permit inspection of the work during pouring. The economies which can be effected in the formwork are important in view of the high proportion of the total cost of concrete work accounted for by the formwork alone.

No-fines concrete walls should not be subjected to bending stresses nor to excessive eccentric or concentrated loads. Slender piers and wide openings are, therefore, unsuited to no-fines construction. Isolated piers should not be less than 450 mm in width or one-third the height of adjacent openings.

The bond strength of no-fines concrete is low but for openings up to about 1.5 m wide the walling itself may be reinforced to act as a lintel, provided there is a depth of wall not less than 230 to 300 mm

above the opening. As a precaution against corrosion the steel should be galvanized or coated with cement wash and bedded in cement mortar. For wider openings an *in situ* or precast reinforced lintel of dense concrete is generally necessary. Even when the wall above openings is not required to act as a lintel to carry floor or roof loads, horizontal reinforcement equivalent to a 13 mm steel bar should be placed above and below all openings. In buildings with timber floors the steel above the openings in external walls is usually made continuous.

For small scale domestic buildings a wall thickness of 200 mm is usually structurally sufficient. For external walls, unless a lightweight aggregate or internal insulating lining is used, a 300 mm thickness will be necessary. Wall thicknesses of 300 mm are used for multi-storey loadbearing wall construction. Lateral support to the walls must be provided by positively anchoring the floors to the walls by some means such as vertical steel dowels set in the wall and secured to the floor reinforcement or to timber joists (see figure 72 *B*);

Because of its weakness in tension, walls of no-fines concrete are sensitive to differential settlement. Particular attention must, therefore, be paid to the design of the foundations. For small buildings the lower part of the walls and the strip foundation should be of dense concrete, reinforced if necessary. For high buildings adequate stiffness is usually obtained by the use of rigid reinforced dense concrete cellular foundations[1].

Note on table 10. Intermediate values for other mixes may be found by interpolation

* These requirements may be deemed to be satisfied if two-thirds of the value is obtained at 7 days.

† The average cube strength for mix design purposes should be 2.10 MN/m^2 in excess of the cube strength specified unless more definite information is available

‡ The attainment of this strength increases the density of the concrete to an extent where the thermal insulation properties may be impaired

[1] For detailed recommendations for the use of no-fines concrete see *Post-War Building Studies* No.1, 'House Construction': Appendix (HMSO). See also *MBC: Materials*

Cement	Aggregate	Nominal mix	Volume of aggregate per 50 kg of cement		Cube strength* within 28 days after mixing		Maximum permissible stresses
			Fine	coarse	Preliminary test	Works test	
			m³	m³	MN/m²	MN/m²	MN/m²
Portland cement, Portland blast-furnace cement and other cements included in CP 110	Concrete with: (i) Natural aggregates to BS882 (ii) Air-cooled blast-furnace slag (coarse aggregate) to BS1047	1:1:2 1:1½:3 1:2:4 1:3:6 1:4:8	0.03 0.05 0.07 0.10 0.14	0.07 0.10 0.14 0.20 0.28	40 34 28 15 12	30 25.5 21 11.5 8.6	7.6 6.5 5.3 2.4 1.7
Portland cement, Portland blast-furnace cement and other cements included in CP 110	Concrete with: (i) Foamed blast-furnace slag to BS 877 (ii) Clinker aggregate to BS 1165 (iii) Such other artificial aggregates as may be suitable having regard to strength durability and freedom from harmful material	Proportions to be selected to give the required cube strength				14.0 11.0 8.3 5.5 2.8	2.6 2.1 1.6 1.1 0.5
	No-fines concrete with: (i) Natural aggregates to BS 882	Special mixes		0.28		7.0†‡ 3.5	1.3 0.6
	(ii) Air-cooled blast-furnace slag to BS 1047	1:8		0.28		2.8	0.5

Table 10   *Basic stresses for plain concrete walls*

Thickness of plain concrete walls

The procedure in calculating is the same as for masonry walls but using different permissible stresses based on varying types and grades of concrete (table 10)[1] and a different set of reduction factors (table 11) to apply to these stresses for slenderness ratios over fifteen, up to a maximum of twenty-four. An increase in the permissible stresses in a plain concrete wall may be made when the ratio of its storey height to length is less than 1½. This varies linearly from zero at a ratio of 1½ to 20 per cent at a ratio of ½ or less. The length of the wall in this case is either the overall length or the length between adjacent openings.

The same increases may be made in the permissible stress in respect of eccentric and concentrated loads and lateral forces as for masonry walls.

It should be noted that notwithstanding the thickness established by calculation the London By-laws require the thickness of an external or party wall of concrete to be not less than 150 mm.

Slenderness ratio	15	18	21	24
Reduction factor	1.00	0.90	0.80	0.70

Linear interpolation between values for the reduction factors is permissible

Table 11   *Plain concrete walls: reduction factor for slenderness ratios*

Shrinkage reinforcement may be required in *in situ* cast concrete walls, other than those of clinker aggregate or no-fines concrete, particularly in external walls, in order to distribute the cracking due to setting shrinkage and thermal movement, and thus minimise the width of the cracks. Where this reinforcement is considered to be necessary, the Code recommends that it should be not less in volume than 0.4 per cent of the volume of the concrete in an external wall. It also makes recommendations in respect of internal walls, the positioning and distribution of the reinforcement and the provision of extra reinforcement round openings where shrinkage effects are greatest.

As the drying shrinkage of no-fines concrete is low, reinforcement for this purpose is not usually necessary except, perhaps, with some lightweight aggregates, because the stresses set up by the slight shrinkage are relieved by the formation of fine cracks round the individual particles of aggregate. Shrinkage reinforcement may also be omitted from dense concrete walls where the mix is lean and of low shrinkage and where end restraints on the walls are small and work can be carried out continuously[2].

Lateral support

Adequate lateral support or restraint to a wall as defined in the footnote on page 120 is provided by a concrete floor or roof slab with a 90 mm bearing on a masonry wall or 75 mm on a concrete wall, irrespective of the direction of span. Where the bearings are less than this or where the slab only butts against the wall metal anchors must be used built into the slab at least 380 mm, at spacings of 1.8 m.

Restraint is less certain with a timber floor, the joists of which bear on the wall. It is non-existent when the joists run parallel to the wall. Adequate restraint may be obtained, however, with timber floors and roofs by means of positive metal anchors. These should be at least 30 mm by 5 mm cross section and 400 mm long with split and upset or other forms of ends for building into solid walls (figure 72 *A, B, C*). 50 mm anchors, split and bent up or down at least 150 mm, should be used for cavity walls (*E, G*) or twisted anchors set in concrete blocks should be used (*F*). They should be securely fastened to the joist ends at 1.2 m intervals in buildings over two storeys high or up to 1.8 m in buildings of one or two storeys. To obtain the stiffness necessary to provide lateral support where the joists run parallel to the wall similar anchors engaging at least three joists are used, with any joist strutting being adjacent to the anchors (*B*). The strutting between these joists may with advantage be solid. Experience has shown that the stiffening effect of partitions in small houses usually makes unnecessary the special anchoring of floors to the walls.

It is not usually necessary to provide such anchorage at loadbearing partitions but the joists should be securely spiked together where they bear on the wall in order to ensure an effective tie. In the case of cross wall construction where the joists must not pass through the wall, metal ties

[1] The Code also gives a table of stresses for special mixes of structural grade concrete for use in plain concrete walls

[2] See page 221

128

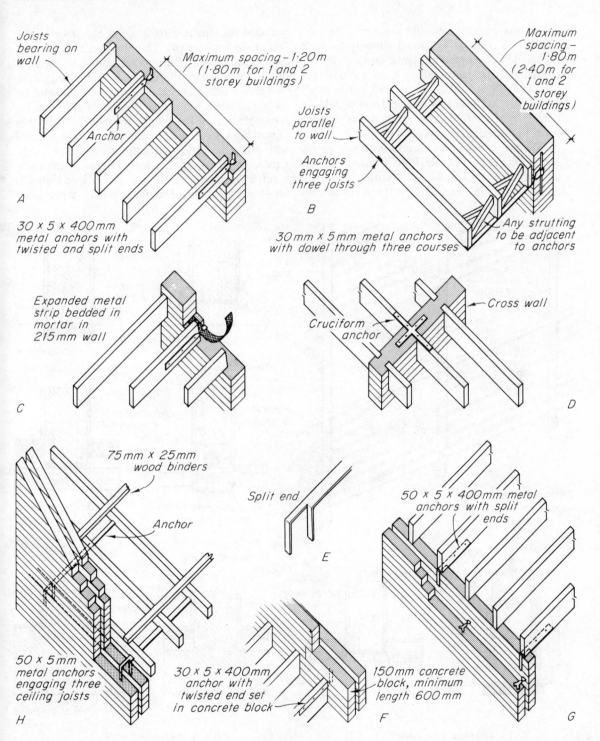

Joists bearing on wall

Maximum spacing – 1·20 m (1·80 m for 1 and 2 storey buildings)

Anchor

A

30 x 5 x 400mm metal anchors with twisted and split ends

Maximum spacing – 1·80 m (2·40 m for 1 and 2 storey buildings)

Joists parallel to wall

Anchors engaging three joists

B

30 mm x 5 mm metal anchors with dowel through three courses

Any strutting to be adjacent to anchors

Expanded metal strip bedded in mortar in 215 mm wall

C

Cruciform anchor

Cross wall

D

75 mm x 25 mm wood binders

Anchor

Split end

E

50 x 5 x 400mm metal anchors with split ends

50 x 5 mm metal anchors engaging three ceiling joists

H

30 x 5 x 400mm anchor with twisted end set in concrete block

150 mm concrete block, minimum length 600 mm

F

G

72   Lateral restraint to walls

fixed to the ends of opposite joists may be used at similar intervals (*D*). Lateral support to gable walls should be provided by similar anchors fixed to the ceiling joists. Since ceiling joists are normally lighter than floor joists, and are not stiffened by floor boards, sufficient lateral stiffness should be attained by fixing binders across joists running parallel to the wall as shown at (*H*).

Stiffening of floor joists is also necessary should a sound insulating floating floor be used. In this case the floor boards will not stiffen the floor structure and the necessary stiffness should be provided by strutting right across the floor on the line of the anchors (*B*). 150 mm by 25 mm boards should also be notched into and spiked to the ends of the joists where they bear on loadbearing walls.

Timber wall plates should not be built into the walls and when a spreader is required a 50 mm by 8 mm MS bearing bar should be used. The reduction in wall thickness caused by the built-in ends of joists will result in higher stresses at the floor levels. Care must be taken, therefore, to ensure that there is sufficient strength and brick- or blockwork should be of a high standard at these points. When joists

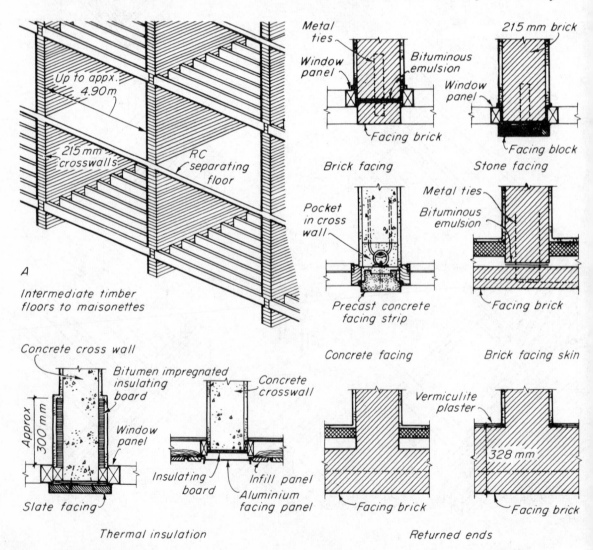

A

Intermediate timber floors to maisonettes

Brick facing · Stone facing

Concrete facing · Brick facing skin

Thermal insulation · Returned ends

73   Cross wall construction

bear on both sides of the wall, as in cross-wall construction (*D*), and good bonding may be difficult, the infilling between the joists may be of *in situ* concrete.

Piers and walls providing lateral support to a loadbearing wall must be of sufficient height and thickness to provide efficient support and must be effectively bonded at the intersection. Block bonding should not normally be used but always a fully bonded junction. Where large blocks are used or where block bonding is necessary, satisfactory bond at the intersections may be obtained by the use of 38 mm by 5 mm metal ties, about 600 mm long with the ends bent up 50 mm. These should be spaced not more than 1.2 m apart.

CROSS WALL CONSTRUCTION

This is a particular from of loadbearing wall construction in which all loads are carried by internal walls running at right angles to the length of the building (see figure 73 *A*).

The majority of buildings require dividing up by internal partitions or separating walls. In certain types in which these occur at sufficiently close and regular intervals and where a high degree of fire resistance and sound insulation is required, necessitating relatively thick and heavy walls, an economic structure results when the walls are made to carry the loads from floors and roof. Typical examples are terraces of houses, maisonette, flat and hotel blocks and schools.

The advantages of this form of construction are:
(i) Simplicity of construction — the walls consist of simple unbroken runs of brick- or block-work or *in situ* concrete

(ii) Projecting beams and columns are eliminated
(iii) It lends itself to repetition and standardization of both structural and non-structural elements and thus to the prefabrication of the latter
(iv) The external walls, being free from load, may be designed with greater freedom in the choice of materials and finish
(v) As a result of these factors, construction costs are low.

Concrete cross wall construction in multi-storey buildings over four to five storeys in height is normally reinforced and is called box-frame construction. It is discussed later in chapter 5. At this point smaller buildings only will be considered, built of bricks, blocks or plain concrete.

Although not a disadvantage in the type of building to which cross wall construction is most suited, planning is restricted by the fact that the walls must run up in the same plane on all floors, from foundations to roof. It is desirable for economic reasons that the walls should also run in an unbroken line on plan from front to back of the building although this is not structurally essential. For maximum economy the walls should also be free of openings as far as possible. A further restriction on planning is the limitation which must be placed on the spacing of the walls in order to obtain an economic combination of walls and other elements of the structure. Of these, the floors and external walling are the most important.

Spacing of walls

Cross walls should be spaced at regular or regularly repeating intervals along the building in order that a limited number of floor spans can be standardized in terms of thickness, reinforcement and formwork.

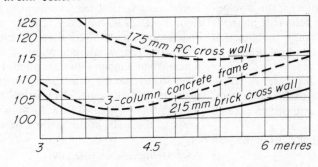

Based on five storey construction with 2.7 m storey height and 9m depth of building

Centres of cross walls or frames

74   *Cross walls: relative cost of structure*

Thus also external wall elements, other than those of brick or blocks, may be standardized in a limited number of units.

The optimum spacing is about 4 to 4.25 m but spacing may range from 3.4 to 5.5 m without varying the cost of the structure itself by more than a few per cent. Within this range any increase or decrease in the cost of the floor structure is balanced by the respective decrease or increase in the cost of the walls. This can be seen from the graph in figure 74. Although within the limits of 3.4 to 5.5 m the cost of the structure remains reasonably constant, the cost of external cladding increases directly with the increase of wall spacing. Analysis shows that while there is usually little to be gained from the point of view of cost and structure in using intermediate cross walls within plan units of up to 5.5 m in width, they should be introduced for greater widths.

Design generally

The design of the walls is carried out on the basis of CP 111, the loading and the stresses in the walls being determined in the manner described in the previous pages.

A cross wall structure will normally be stable against the pressure of wind parallel to the walls but will, in most cases, require some longitudinal bracing against wind pressure at right angles to the walls. This is because there is little rigidity at wall and floor junctions and the materials of which the walls are constructed have little tensile strength. The floor loads and the self-weight of the walls set up 'pre-compression' in the walls which enables the latter and the floor-to-wall joints to resist tensile stresses up to the limit of this 'precompression'[1]. In many cases this may be insufficient to prevent the development of tensile stresses and the following methods may be adopted, either singly or in conjunction with each other, to provide longitudinal bracing to ensure overall stability:

(i) The use of a staircase or lift tower of which the walls at right angles to the cross walls act as buttresses and are assumed to take all lateral forces
(ii) The provision of longitudinal walls in the plan to take the lateral forces
(iii) The return of the ends of the cross walls as an 'L' or 'T' for the same purpose

(iv) The introduction of certain piers and walls in reinforced concrete, or the forming of the joints between walls and floors in reinforced concrete
(v) The introduction of longitudinal reinforced concrete beams.

External cavity panel walls of brick and clinker block alone can provide sufficient rigidity to the structure of buildings up to five storeys high and of reasonable length, provided the area of window is not excessive and the workmanship is sound, especially at the bonding of panels to cross walls.

Bricks, blocks or plain *in situ* concrete are all suitable materials for this form of construction. The advantage of blocks in terms of stress allowances has already been mentioned. They have the added advantage of being quickly laid, especially in areas such as Scotland, for example, where they are widely used. Plain concrete walls have in some cases proved as cheap as comparable block walls, particularly where the opportunity has been taken to use a crane for rationalizing the building operations.

The graph in figure 74 shows the relative costs of brick and concrete cross walls together with those of a reinforced concrete frame for five-storey buildings. At this height the unreinforced wall, typified by brick, and the frame become competitive. But the reinforced concrete wall[2], even at its most economic spacing of about 5.5 m is not economic for such low buildings.

In cross wall construction the *in situ* solid reinforced concrete floor gives greatest rigidity. Hollow block floors and precast floors may be used if the junction with the wall can be cast *in situ*, their effectiveness depending upon the degree of rigidity which can be obtained at this point. This rigidity can be increased, whatever the type of floor, by introducing vertical reinforcement in the wall at the floor level. This, however, is more practicable for concrete walls than for those of bricks or blocks where the rods would complicate the laying of the units. When timber floors are used, for example in three-storey houses or as the intermediate floors in maisonettes (see figure 73A), lateral support to the walls must be obtained from these by anchoring the joists to the walls in the manner described on page

[1] See 'Overturning' on page 116

[2] See page 221 for reference to the omission of reinforcement from concrete cross walls

128. Where the joists run parallel to the walls and bear on precast concrete cross beams the latter, if spaced not too far apart, may provide adequate restraint. No longitudinal rigidity to the structure as a whole is provided when timber floors are used.

In Part 1[1] reference is made to the problem of waterproofing the exposed ends of cross walls. When the walls terminate in returned ends for reason of stability or are completely covered by the outer skin of a cavity panel wall the ends have adequate protection as illustrated in figure 73. When, however, the wall ends are exposed on elevation between thin infilling panels as shown in the three examples in figure 73, for reasons given in Part 1 some positive weatherproofing is essential in order to prevent moisture penetration across the short distance between the exposed and internal faces of the wall. The methods adopted involve the use of a damp-proof barrier behind a brick facing or the use of other forms of facing with a minimum number of joints, as shown.

In addition to the possibility of water penetration at the ends of cross walls that of heat loss also arises. This is particularly so in the case of dense concrete walls. It can cause condensation on the inner wall faces near the external walls and is avoided by the use of thermal insulation. Depending upon the relationship of the infilling panels and the wall ends, the insulation may be applied to the inner faces or to the external face of the cross wall as shown in figure 73.

REINFORCED MASONRY WALLS

Reinforced masonry, usually in the form of reinforced brickwork, has been used to a considerable extent in the United States of America and in earthquake countries such as India and Japan. It has been used to a limited extent in Great Britain since the nineteenth century. The reinforcement enables the brickwork to withstand tensile and shear stresses in addition to the compressive stresses which it is capable of bearing alone.

Horizontal reinforcement may be bedded in the bed joints of the masonry as shown in the brick lintels described and illustrated in chapter 5 of Part 1, but vertical reinforcement necessitates the use of slotted, recessed or perforated bricks or the use of special bonds, such as rat-trap bond (see Part 1), within the cavities of which the reinforcement may be grouted, as it may be within the cavities of hollow concrete blockwork as shown in Part 1.

Brick walls, with the appropriate reinforcement within them, can be made to act compositely with reinforced concrete floors above and below them to form deep box girder cantilever beams, the floor slabs acting as flanges and the walls as webs[2]. This is a reasonable technique when the component parts are already being used in a system as in, say, masonry cross wall construction, but it should be emphasised that care should be taken not to use brickwork for work which clearly calls for reinforced concrete.

CP 111 : Part 2 : 1970 recommends that the design of reinforced masonry should follow the general principles adopted for reinforced concrete but provides design factors such as permissible stresses, modular ratios and required cover to reinforcement. The Code lays down minimum amounts of cover to reinforcement in external walls and in walls in contact with the soil, but the London By-laws prohibit the use of reinforced masonry where exposed to weather or moisture from soil.

Prestressing in the form of post-tensioned high-strength brickwork can be used for tall columns to overcome the tensile stresses caused by eccentric loading and lateral wind pressure. This permits the overall size to be kept to a minimum. It can be done by building a hollow column, the central space accommodating post-tensioning cables or bars (see chapter 5, page 229) which are anchored to the foundation and to a steel plate at the top. This is simpler than building up a combination of solid brickwork and normal reinforcing bars.

CAVITY LOADBEARING WALLS

The advantages of the cavity wall in terms of weather resistance and thermal insulation have been described in Part 1, chapter 5, and the function of the wall ties in linking the two relatively thin leaves to provide mutual stiffening is explained. This is particularly important where one leaf only is loaded, the other then acting as a stiffener. Details of the necessary spacing of the ties in the wall are given and reference is made to the need to avoid the use

[1] See Part 1, page 85

[2] See 'Reinforced Brickwork Box Beams' – *Technical Report SCP 3:* Structural Clay Products Ltd

of lime or the weaker cement-lime mortar mixes in order to ensure adequate bond of ties to wall.

Differential movement between the two leaves is likely to occur in large areas of external cavity walling, which over a period of time may weaken the bond of the ties in the joints. This, together with slight deformations of light wire ties, will reduce their effectiveness as a stiff link between the leaves. In the case of buildings not exceeding four storeys or 12 m in height, whichever is less, it is considered that the outer leaf may be uninterrupted for its full height. In taller buildings these movements can be sufficiently limited by supporting the outer leaf at vertical intervals of not more than three storeys or 9 m whichever is less and by dividing the wall horizontally into separate panels not exceeding 12 to 15 m wide. In such circumstances strip metal ties are preferable to the lighter wire ties. The vertical sub-division may conveniently be formed by running floor slabs through to the outer face at appropriate levels, since in most buildings over three storeys in height some of the floors, even in maisonette blocks, will be of reinforced concrete. The detailing of the vertical joints between the wall panels will depend on the type of structure (see figure 140).

Wall thickness

CP 111 provides for calculation on the same basis as for solid walls, provided that where both leaves share the load the effective thickness of the wall is taken as two-thirds of the sum of the thicknesses of the two leaves. This takes into account the fact that a cavity wall has less lateral stiffness than a solid wall equal in thickness to the sum of the thicknesses of the two leaves.

Where one leaf only supports the load the permissible stress for that leaf may be based on the slenderness ratio calculated from either (i) the effective thickness given above, or (ii) the effective thickness taken as the actual thickness of the loadbearing leaf. Except where the supporting leaf is more than twice the thickness of the other leaf, a lower slenderness ratio and a higher permissible working stress result by using (i). When the load is carried by both leaves, one of which is of weaker material, as is the case when a lightweight inner leaf is used with a brick outer leaf, the permissible stress to be used for both leaves should be based on the weaker of the two materials, and determined by using an effective thickness equal to two-thirds of

134

the sum of the two leaves. This is in accordance with the provision made for solid walls built of two materials in Clause 314 of the Code.

The London Building (Constructional) By-laws and the Building Regulations both provide for establishing the thickness of loadbearing cavity walls either by the empirical method or by calculation in accordance with the provisions of CP 111.

Under the empirical method the London By-laws permit loadbearing external or party walls to be constructed as cavity walls in one-storey buildings with a maximum height of wall of 3 m but no restriction on the length and, with the exclusion of buildings of the warehouse class, in the following (i) the topmost storey of a building, with the same height limitation but no restriction on the length of the cavity wall, (ii) in two-storey buildings if the wall does not exceed either 7.5 m in height or 9 m in length and (iii) in the topmost two storeys of buildings of greater height with the same limitations on the dimensions of the cavity wall portion as in (ii). These limitations assume a leaf thickness no greater than that laid down in the By-laws and referred to below.

The Building Regulations lay down no dimensional limitations such as these for a cavity wall but relate its required overall thickness to the height and length of the wall by requiring it to be not less than either the sum of the cavity width and leaf thicknesses given below or the thickness which would be required for a solid wall of the same height and length plus the width of the cavity, whichever is greater.

Generally, the Building Regulations require each leaf to be at least 100 mm thick and the London By-laws, 90 mm. In both, however, an exception is made in the case of walls to a single storey house or to the top storey of a house, where the inner leaf only may be 75 mm thick provided the length of the wall does not exceed 7.5 m (Regulations – 8 m) and its height 3 m, or 5 m if a gable wall. As pointed out in chapter 5, Part 1, the Building Regulations require double the number of wall ties, and the roof load to be carried by both leaves, when the inner leaf is reduced to 75 mm. Both sets of regulations require a cavity width of not less than 50 mm or more than 75 mm.

CP 111 lays down a general minimum thickness for both leaves of 75 mm, but it should be noted that the London By-laws still require 90 mm as a minimum even where the wall thickness is calculated on the basis of the Code.

PANEL WALLS

Masonry and monolithic walls may be used as non-loadbearing filling to the panels formed by the structural columns and beams of a framed building or to areas of calculated masonry walling. In the latter case the London By-laws refer to such a part of the wall as a 'panel wall'. The term is used here to cover both applications.

In a framed structure most of the wall elements are assumed to sustain no loads from floors and roof. Compressive strength is therefore of less importance than transverse strength to withstand pressure from wind or stacked materials. Although heavy construction can provide good fire resistance, durability and sound insulation, the compressive strength of materials such as brickwork and concrete is not, therefore, generally fully utilized when they are used for panel walls, and considerable weight is added to the structural frame. To overcome this disadvantage, in addition to the use of lightweight concrete for masonry and monolithic panel walls, other means of providing the infilling or enclosing element have been devised to reduce the deadweight while still fulfilling the functional requirements of a non-loadbearing external wall. These are dealt with later under 'External facings and claddings'.

Investigations by the Building Research Station have shown that panel walls and the supporting frame act together giving considerable rigidity and strength to the frame. In some tests the supporting beam was wholly in tension, the wall acting with it as the compression element. If this composite action is taken into account in order to economize on the actual beam elements or to provide lateral rigidity to the frame against wind pressure on the building as a whole, then the compressive strength of a panel wall does become a significant factor.

The strength and stability of a panel wall depends upon its stiffness, which in an unreinforced wall will vary with its thickness, upon the nature of the edge fixing and, in the case of brick or block walls, on satisfactory bond between mortar and unit. Good edge fixing on all four sides is essential in order to develop maximum strength against lateral forces. Although greatest pressure can be sustained when the edges are actually built into the frame to give a rigid fixing, the normal mortar joint is usually adequate for this purpose. This can be improved when necessary by the use of metal ties built in wall and frame. In masonry walls the lateral strength is related to the bond existing between the mortar and the walling units. The table referred to in the footnote on page 115 shows clearly that an increase in the tensile strength of the mortar by the use of a high cement content does not result in a proportional increase in the strength of the wall. But the use of walling units which provide a better bond with the mortar, such as hollow clay blocks, especially those with undercut grooves on the bed faces, does result in a significant increase in the lateral strength of the wall.

The Building Regulations make no specific reference to panel walls. These would be designed in accordance with the appropriate sections of CP 111 having regard to the dead weight of the wall, the lateral pressure to be resisted and the high pressures and suctions caused by wind near the eaves and corners of a building (see page 117). For normal conditions and panel size, a 215 mm solid or 280 mm cavity wall is usually sufficient.

The London By-laws impose a minimum thickness of 190 mm for a solid masonry panel wall and a maximum height of 7.5 m. The actual thickness must be at least one-eighteenth of either the height or the length, whichever is the less (figure 75 A). Cavity panel walls are permitted and these must comply with the requirements for cavity loadbearing walls except that the inner leaf may be 75 mm instead of 90 mm thick and may be of hollow blocks. The height must not exceed 7.5 m and the area must be not more than 19 m² . The height or length (whichever is the less) must not exceed 4 m (B).

These provisions, in the case of a 190 mm solid wall, permit a panel 3.42 m high of unlimited length, or of the maximum height of 7.5 m and 3.42 m long. In the case of a cavity wall, working to the maximum figures given for either the height or length, the limits of the panel are 4 m by 4.75 m and 7.5 m by approximately 2.5 m.

The By-laws limit the extent to which a panel wall may project beyond its supporting beam (see figure 75). A solid panel must not overhang by more than one-third of its thickness and a cavity panel by not more than one-third of the thickness of the overhanging leaf, unless the base of the wall is built solid for a height equal to the full thickness of the wall, in which case it may overhang by one-third of the full thickness.

This leads to some practical difficulties when it is desired to carry a brick panel over the structural frame to give a continuous brick facing to the

135

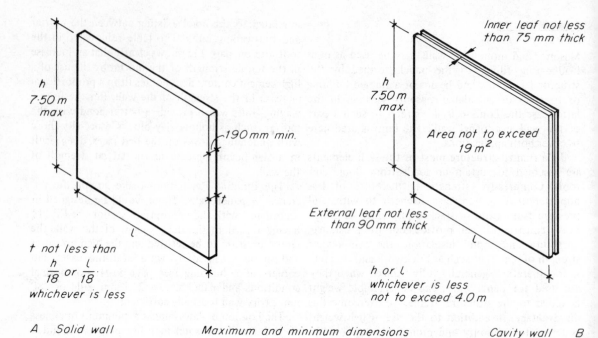

h 7·50 m max

190 mm min.

t

l

t not less than

$\frac{h}{18}$ or $\frac{l}{18}$

whichever is less

A   Solid wall

Inner leaf not less than 75 mm thick

h 7.50 m max.

Area not to exceed 19 m²

External leaf not less than 90 mm thick

l

h or *l* whichever is less not to exceed 4.0 m

Maximum and minimum dimensions   Cavity wall   B

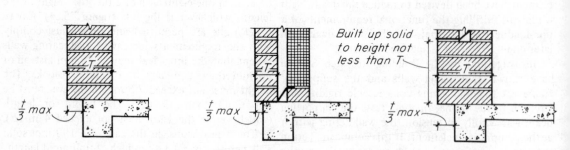

T

$\frac{t}{3}$ max

T

$\frac{t}{3}$ max

Built up solid to height not less than *T*

T

$\frac{t}{3}$ max

Maximum permissible overhang beyond supporting member

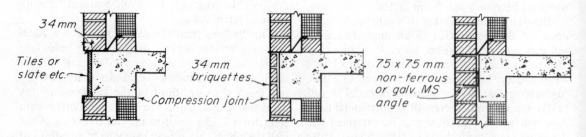

34 mm

Tiles or slate etc.

Compression joint

34 mm briquettes

75 x 75 mm non-ferrous or galv. MS angle

Methods of facing over supporting member of a panel wall

75   Panel walls

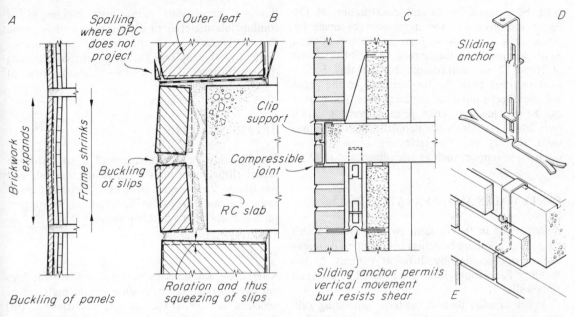

A Brickwork expands / Frame shrinks

Buckling of panels

Spalling where DPC does not project

B Outer leaf

Buckling of slips

Clip support

Compressible joint

RC slab

Rotation and thus squeezing of slips

C

Sliding anchor permits vertical movement but resists shear

D Sliding anchor

E

76   *Movement joints to panel walls*

building. With a 328 mm thick panel no problem arises since one-third of the thickness will allow a half-brick cover to the structural members. But such a thick wall is excessively heavy and, in any case, would rarely be required for most normal frame spacings. The usual 215 mm wall, or cavity wall with a 102.5 mm outer leaf have permissible overhangs of only 72 mm and 34 mm respectively, both less than a half-brick. Cut bricks have been avoided by the use of special briquettes, 62 mm or 24 mm on bed, to face the supporting beam, but some difference in colour is to be expected at times, particularly in the case of the thinner briquettes. These will dry out after rain more quickly than the thicker brickwork above and below and will show up as bands of lighter tone.

An alternative is to use metal supporting angles fixed to the structural beam and accommodated in the thickness of a brick bed-joint. This permits a 102.5 mm brick cover to the beam and at the same time keeps the projection over the support down to one-third of the leaf thickness. The angle support must be of non-ferrous metal to comply with the provisions of the By-laws. These methods are illustrated in figure 75.

Monolithic concrete panel walls may be used and if constructed of lightweight concrete they may provide a high degree of thermal insulation in external walls. Satisfactory edge fixing is given by the normal bond with structural members either of concrete or encased steel, and by tying in the shrinkage reinforcement to the structural members.

When brick panel walls are used in a reinforced concrete frame it is essential to make provision for vertical differential movement between panel and frame caused by the irreversible expansion of the brickwork (see note to table 13 and *MBC: Materials* page 125) and the shrinkage and creep of the concrete frame. As a result of these opposing movements stresses may be set up in the panels which may cause distortion and failure of the brickwork, such as displacement and spalling of the bricks next to the supporting beams and, occasionally, buckling of the panels. (figure 76 *A*).

The possibility of damage is most likely when cavity panels are used which overhang the support and the latter is faced with briquettes or stone slabs. In this case the pressure on the panel tends to make the outer leaf rotate at the bearing, particularly if, through inaccuracies in the frame, this leaf has a support less than two-thirds of its thickness, and the briquettes are crushed and made to break away from the frame (figure 76 *B*).

In practice adequate support for the outer leaf

137

must be ensured by careful construction of the frame and provision for movement be made by means of a compression joint at the top of each panel. In most cases the normal bed joint thickness of 10 to 12 mm will be sufficient for this. Such a joint removes both the top restraint to the panel and the direct support to the briquettes. The former can be provided by some form of anchor which gives lateral restraint while permitting vertical movement (C, D) and the latter by some form of mechanical support such as that illustrated in (E)[1].

PARTY WALLS AND SEPARATING WALLS

A party wall in the London By-laws means a wall separating adjoining buildings belonging to different owners or occupied by different persons. In the Building Regulations these are referred to as separating walls.

In the London By-laws the term 'separating wall' means a wall separating different tenancies within the same building, such as the walls between flats and maisonettes, and such walls are not necessarily required to comply with the provisions for party walls.

A party wall may, or may not, be loadbearing. Its primary function is to prevent the spread of fire between adjoining buildings and to provide an adequate degree of sound insulation. In the London area the thickness of a party wall must be not less than 170 mm if constructed of masonry or 150 mm if of concrete. The only limitations in the Building Regulations are that such a wall shall be of a stated degree of fire-resistance and shall be, for certain building types, non-combustible throughout[2]. As mentioned under 'Cavity walls', within certain limits of height and length, and for certain classes of building, a party wall may be constructed as a cavity wall. But the cavity wall, once considered desirable for reasons of sound insulation, is for all practical purposes now known to be no better in this respect than a 215 mm solid wall (see *MBC: Environment and Services*).

The London By-laws prohibit the building-in of timber or other combustible material to the required thickness of a party wall, except the ends of beams, joists, purlins and rafters in houses not exceeding three storeys provided they do not extend beyond the centre line of the wall and are encased in not less than 90 mm of non-combustible material. The

Building Regulations permit the building-in of combustible material to a separating wall provided the fire resistance of the wall is not rendered ineffective thereby (see Part 1, page 208, regarding the significance of this relative to the bearings of joists on the wall).

RETAINING WALLS

The function of a retaining wall is to resist the lateral thrust of a mass of earth on one side and sometimes the pressure of subsoil water. In many cases the wall may also be required to support vertical loads from a building above.

Stability

A retaining wall must be designed so that (i) it does not overturn and does not slide, (ii) the materials of which it is constructed are not overstressed, (iii) the soil on which it rests is not overstressed and circular slip is avoided on clay soil.

The pressure on the back of the wall is called 'active' pressure and tends to overturn the wall and push it forward. The force exerted by the earth on the front of the wall in resisting movement of the wall under the active pressure is called the 'passive' earth resistance (figure 77 A). The actual thrust of retained earth on a wall and its direction and point of application can only be determined approximately. There are several theories for assessing the active pressure and the passive resistance which are explained in detail in most textbooks on the theory of structures. Active water pressure is avoided by the provision of a vertical rubble drainage layer, or drainage counterforts, discharging through weep holes in the wall or, if the wall encloses a basement, through lateral drainpipes at the back of the wall as shown in figure 77 B, C, which conduct the moisture to a main drain. In the case of clay soils,

[1] For a more detailed consideration of this problem see *Brick Development Association Technical Note Vol 1, No.4:* 'Some observations on the design of brickwork cladding to multi-storey rc framed structures', Donald Foster, September 1971, and *DOE Housing Development Notes, III Construction:* '1 Measures to avoid or cure the crushing of brick panels in concrete frame buildings', April 1973

[2] For the general requirements regarding fire resistance, see chapter 10

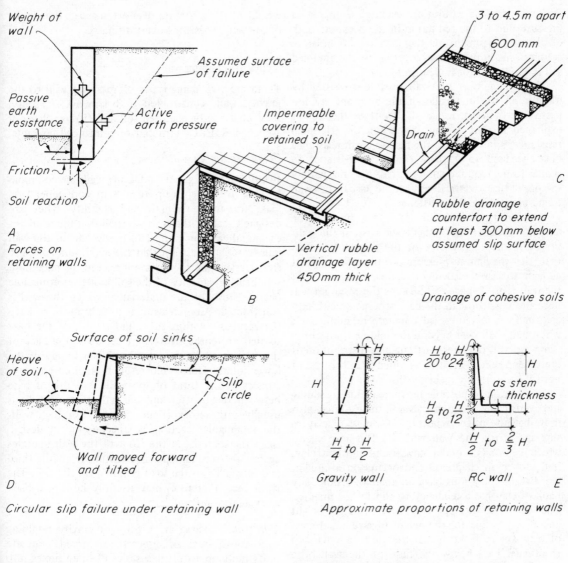

Weight of wall

Assumed surface of failure

Passive earth resistance

Active earth pressure

Friction

Soil reaction

A

Forces on retaining walls

Impermeable covering to retained soil

Vertical rubble drainage layer 450 mm thick

B

3 to 4.5 m apart

600 mm

Drain

C

Rubble drainage counterfort to extend at least 300 mm below assumed slip surface

Drainage of cohesive soils

Surface of soil sinks

Heave of soil

Slip circle

Wall moved forward and tilted

D

Circular slip failure under retaining wall

H

$\frac{H}{7}$

$\frac{H}{4}$ to $\frac{H}{2}$

Gravity wall

$\frac{H}{20}$ to $\frac{H}{24}$

H

as stem thickness

$\frac{H}{8}$ to $\frac{H}{12}$

$\frac{H}{2}$ to $\frac{2}{3}H$

RC wall

E

Approximate proportions of retaining walls

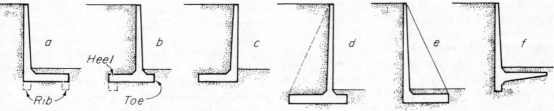

a   b   c   d   e   f

Heel

Rib   Toe

Typical reinforced concrete retaining walls

F

77   Retaining walls

this will prevent saturation leading to increased pressure due to the reduction in the shear strength of the soil arising from the increase in moisture content. An impermeable covering to the retained soil is also useful for this purpose.

The tendency of the wall to slide is resisted by friction on the underside of the base and by the passive earth resistance at the front of the wall as explained in Part 1, page 57. When the frictional resistance is insufficient, the passive earth resistance must be used to increase the total resistance to sliding to the required amount. This may necessitate the provision of vertical ribs on the base of the wall to increase the passive resistance.

Overturning will occur (i) if the line of the resultant pressure falls outside the base of the wall, or (ii) if the eccentricity of the resultant is such that the maximum pressure at the toe is great enough to cause settlement leading to rotation of the wall (Part 1, pages 57-58). As the resultant will normally pass through the base at some eccentricity and, in fact, in reinforced concrete retaining walls will usually fall well beyond the middle third of the 'stem' or wall thickness, a triangular distribution of pressure on the soil must be assumed as in the case of normal eccentrically loaded walls.

In addition to possible movement due to sliding and overturning, when on clay soil the wall may also move because of the tendency of a mass of clay to slip and carry the wall with it. This movement, which can occur under any foundation on clay, is described in chapter 3 and occurs on a circular arc. Because clay soils have no angle of repose, any bank of clay has a tendency to slip in this manner. The safe angle of slope is a function of the height and it is possible for the height of a retained mass of clay soil to be such that the arc of circular slip is situated well below the base of the wall. The strength and stability of a retaining wall has no bearing upon such soil movement beneath it and no variations in the detailed design of the wall would affect its overall stability in this respect (see figure 77 D). When this type of failure appears likely, sheet piling may be used, taken to a depth below the slip circle sufficient to prevent movement taking place.

In basement retaining walls, sliding or rotation can be overcome when necessary by making the active pressures on each side of the basement counteract each other through the floors. The

weight of the structure over often assists the weight of the wall in resisting overturning.

Types of retaining wall

There are two main types of retaining wall (i) the gravity wall, constructed of brickwork, masonry, mass concrete or enclosed soil and (ii) cantilever or L-shaped walls constructed of reinforced concrete.

Gravity retaining walls

In building work gravity retaining walls are commonly used for heights up to 1.8 m and depend on mass for their strength and stability. They are designed so that the width is such that the resultant of lateral earth pressure and the weight of the wall falls in such a position that the maximum compressive stress at the toe does not exceed the maximum safe bearing capacity of the soil, nor the permissible bearing stress of the material of which the wall is constructed. An endeavour is usually made to keep the resultant within the middle third of the base so that no tensile stresses are set up at the back of the wall at its bearing on the soil or in any of the lower joints. This would result in high compressive stresses at the front of the wall. A width of base between one-quarter and one-half of the height is usually satisfactory (figure 77 E). For high walls the rectangular section is uneconomic in design, since the material at the front of the wall operates against the resultant passing within the middle third and also adds to the load imposed on the soil. The front face, therefore, may usefully be sloped back as explained in Part 1[1].

Crib walls

Crib walls These are a form of gravity retaining wall constructed of precast reinforced concrete units built up to form a series of open boxes into which suitable free-draining soil is placed to act as an integral part of the retaining wall (figure 78).

The wall consists of a concrete foundation strip on which the first layer of units is accurately levelled and into which they key. The units consist of stretchers laid parallel with the face of the wall and headers, with ends formed to interlock with the stretchers, laid at right-angles to them. The face of

[1] It should be noted that the London By-laws limit the difference in height of the ground levels on either side of a normal wall to four times the thickness of the wall at the higher level unless the wall is adequately buttressed

wall should be battered to a slope not greater than 1 to 6 or less than 1 to 8, although it may be vertical where the height is less than the length of the headers. The filling should be of soil able to develop high friction such as coarse sand, gravel or broken rock and should be extended about 900 mm behind the crib to form a drainage layer.

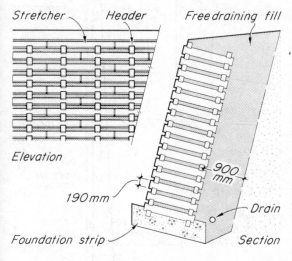

78   Retaining walls – crib walling

The crib wall is suitable for terraces and embankments and in single width as shown in figure 78 for heights up to 5 m. For greater heights up to 9 or 10 m, and when poor quality common infill is used, double or triple width cribs are required in order to obtain adequate mass. These use longer headers and three or four rows of stretchers.

Some forms are designed with deep stretchers to give a solid face to the wall, but with open joints between to provide free drainage to the infill.

This type of wall uses less concrete and is quicker to construct than a mass concrete gravity wall in circumstances where it is suitable.

Cantilever retaining walls

In the cantilever retaining wall the practice of sloping the front face of a gravity retaining wall is carried further so that full benefit may be derived from the advantages which arise in doing so. This is particularly necessary in high retaining walls where the size of a gravity wall would be excessive. As the resultant of the lateral pressure and weight of wall falls well outside the thickness of the wall, or vertical stem, high tensile stresses are induced and it is necessary to use reinforced concrete so that the stem can act as a vertical cantilever. These walls are more economical in the use of materials, occupy less space and weigh less than gravity walls.

Different forms of reinforced concrete walls are illustrated at F, figure 77. That shown in (a) is most commonly used in building structures where it is not possible to excavate behind the stem of the wall. When some excavation can be carried out behind the wall, advantage should be taken of this to form a base projecting partly in front and partly behind the stem as shown in (b) so that the weight of the soil on the heel, or back portion of the base, can assist in counterbalancing the overturning tendency. The most economical arrangement is usually that in which the length of the heel is approximately twice that of the toe. The form shown in (c) has the whole of the base under the retained soil. Because of the increased stability given by the weight of the retained soil, the base may be shorter than in type (a). The inherent economy of types (b) and (c) may, however, be counterbalanced by the cost of excavation when the soil must be removed in order to construct the base slab, and type (c) may not be a suitable form when the stem carries superimposed loads from a structure above, since these would be concentrated on the toe.

When the height of the wall is over 7.5 m, and the thickness of the wall might be excessive, it is usually cheaper to use a counterfort retaining wall as shown in (d) and (e), in which vertical ribs called 'counterforts' act as vertical cantilevers. Type (e) with the counterforts at the front of the stem is also called a 'buttressed retaining wall'. The pressure of the soil on the wall is transferred to the counterforts by the wall slab which spans horizontally between them. Similarly, the base slab is designed to span between the counter-forts. In very tall counterfort walls it is cheaper to use horizontal secondary beams spanning between the counterforts with the slab spanning vertically between them. If the spacing of these beams is varied down the height of the wall, the bending moments in each span of the slab can be kept the same so that the same thickness of slab can be maintained throughout the full height of the wall.

141

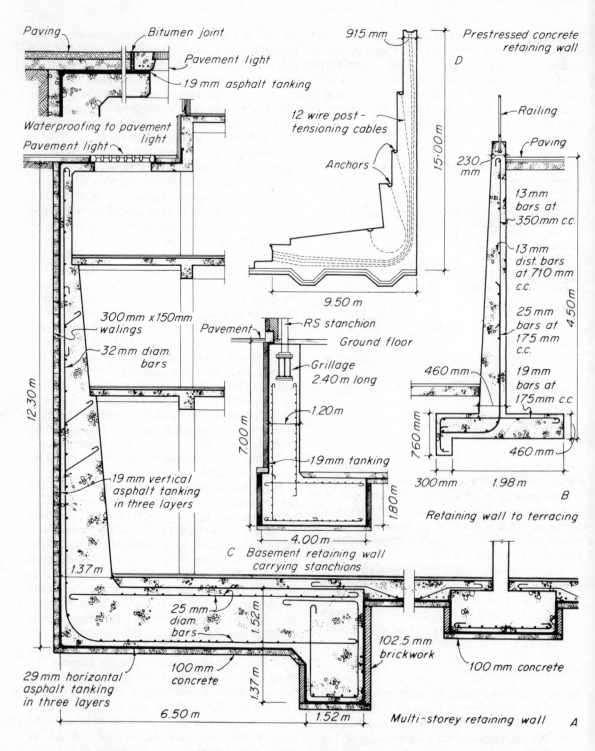

Paving

Bitumen joint

Pavement light

19 mm asphalt tanking

Waterproofing to pavement light

Pavement light

915 mm

12 wire post-tensioning cables

Anchors

Prestressed concrete retaining wall

D

15.00 m

9.50 m

300 mm x 150mm walings

32 mm diam. bars

12.30 m

19 mm vertical asphalt tanking in three layers

1.37 m

25 mm diam. bars

29 mm horizontal asphalt tanking in three layers

100 mm concrete

6.50 m

Pavement

RS stanchion

Ground floor

Grillage 2.40 m long

1.20 m

19 mm tanking

7.00 m

1.80 m

4.00 m

C Basement retaining wall carrying stanchions

1.52 m

1.37 m

1.52 m

102.5 mm brickwork

Railing

Paving

230 mm

13 mm bars at 350 mm c.c.

13 mm dist. bars at 710 mm c.c.

25 mm bars at 175 mm c.c.

460 mm

19 mm bars at 175 mm c.c.

760 mm

460 mm

4.50 m

300 mm

1.98 m

B

Retaining wall to terracing

100 mm concrete

Multi-storey retaining wall

A

79 Retaining walls

142

In the case of deep basements formed as cellular rafts, the cross walls or cross frames can act as the buttress counterforts, and in normal single-storey basements columns from the superstructure above can sometimes be used in a similar way, if suitably spaced and running down on the plane of the wall. Alternatively, in some circumstances the wall may span between the basement and ground floor.

When frictional resistance to sliding is insufficient, it may be necessary to form a projecting rib on the underside of the base in order to increase the depth of earth providing passive resistance. This may be in any convenient position from the extremity of the toe to the heel as shown dotted on type (a). The best position is at the heel so that the great bearing pressure under the toe of the base can prevent the spewing of the soil in front of the rib. A rib is usually essential when the vertical load is small compared with the lateral thrust of the earth as in a freestanding wall such as that at (B), figure 79. This sometimes occurs also in a building with a high basement retaining wall such as the multi-storey wall illustrated at (A). The construction of the base at an angle assists in increasing the resistance to sliding and at the same time will result in a more even distribution of pressure on the soil. The sloping up of the base in this way is most useful when incorporated with a projecting rib under the stem of the wall as at (F-f), figure 77.

The approximate proportions of the type of wall commonly used in basements are shown in figure 77 E.

When the base of a cantilever retaining wall projects entirely in front of the stem as a toe, the main reinforcement is arranged vertically on the tension side of the stem, that is, nearest the retained earth, and is carried round on the under side of the base slab (figure 79 A, B, D). Shear stresses are not often so great as to require the provision of shear reinforcement. If any part of the base projects under the retained soil, the downward pressure of the soil will set up a reverse bending moment and the tension side of the slab under the soil will be at the top where the horizontal reinforcement must be placed (B).

When a retaining wall carries a substantial superstructure load (C) the stem may need to be thicker and, due to the concentrated load at the heel, the base can sometimes be shorter. Since the stem acts as a column reinforcement is required at both faces.

In the counterfort wall the counterforts are designed as vertical cantilevers fixed at the base. When they are on the same side as the soil retained, the compressive edges are stiffened by the stem slab but there is no such stiffening when they act as buttresses on the front of the stem, so the edges may need thickening to resist the tendency to buckle. The stem and base slabs are designed as slabs continuous over the counterfort or buttress supports, the base slab distributing the pressure on to the soil. The disposition of the steel reinforcement in the base will depend on whether counterforts or buttresses are used. In the first case there will be tension at the bottom of the slab between the counterforts due to the downward pressure of the soil it carries. In the second case there will be tension at the top due to the upward pressure of the subsoil below.

Prestressing can be applied to high retaining walls and permits the thickness to be kept to a minimum. Set backs, at suitable intervals on the base and stem, provide positions for the anchorage of the post-tensioning cables as indicated at (D).

The stem and base of the cantilever retaining wall may be tapered as the bending moments reduce, but it is often cheaper not to do so in order to simplify shuttering and the placing of the concrete.

Diaphragm walls

This type of wall was used originally for purposes of water cut-off, especially to intercept seepage below dams. In construction work it is now taken to mean a reinforced concrete wall cast in a deep trench for use as a retaining wall (figure 80 A).

The trench is excavated to the required depth in alternate sections up to about 6 m long, depending on the soil, either by grab or by rotary or percussive methods (B). As excavation proceeds the soil is replaced by a bentonite mud slurry which is produced from a clay having thixotropic properties which cause the slurry to gel when at rest, thus giving support to the sides of the excavation and keeping out subsoil water. During excavation the mixture of excavated material and slurry is pumped out and sieved, leaving most of the slurry to be returned to the trench.

On completion of each panel prefabricated cages of reinforcement are lowered through the slurry and concrete is placed through a tremie pipe. As the concrete rises the bentonite slurry is displaced and can be re-used until its gelling properties reduce to the point when it requires re-conditioning or re-

143

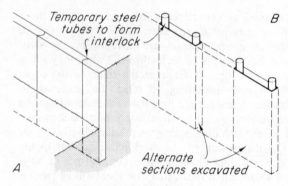

B

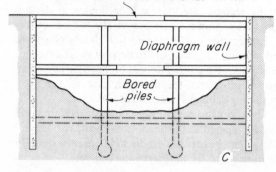

80   Diaphragm walls

placing. The intermediate panels are then excavated and the wall completed following the same procedure.

Wall thicknesses from 450 mm to 1 m by grabbing and to 1.5 m by rotary or percussive methods are possible with depths from 30 to 45 m.

Diaphragm walls may also be constructed by contiguous bored piling in which bored piles are formed in line touching each other or slightly overlapping to provide continuity. Pile diameters of 375 mm and upwards are used.

The surface exposed on excavation will, with both these methods, require some facing up and any weaknesses in the joints may need to be sealed by injecting grouts, unless a lining to the wall with a cavity and a hollow floor is provided as described on page 148.

Since these walls have no toe as in the cantilever retaining wall stability is provided either by struts in the form of the floor structures (figure 80 C) or by ties in the form of ground anchors (figure 81 B).

In the former method initial stability may be provided at the top by temporary girders while the

excavation is carried out between the walls and the basement structure is constructed upwards from the lowest level. The diaphragm walls must be sufficiently deep beyond the lowest basement floor to achieve adequate passive resistance from the soil. Alternatively, construction may follow the excavation downwards, the floor structures being constructed at each stage of excavation as shown at (C), figure 80. If internal columns are part of the structure these must be formed as piles rising the full depth of the basement before excavation

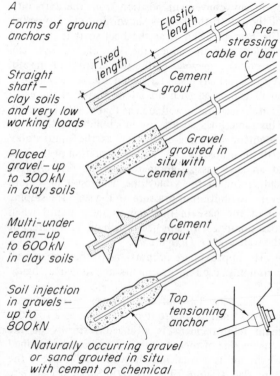

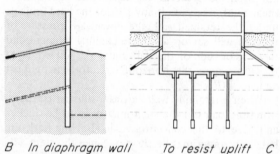

81   Ground anchors

commences.

In the latter method no temporary top strutting nor permanent strutting by floors is required. Ground anchors consist of a prestressing cable or bar with a fixed anchorage which holds the cable in the ground at one end, and a tensioning anchor at the other (figure 81 *A*). The cable is inserted in a hole formed by rotary or rotary percussive drilling or by vibro driving depending upon the nature of the soil, the hole being lined where necessary. Anchors are formed in different ways according to the nature of the soil and the loads to be sustained as shown in figure 81. In their application to diaphragm retaining walls they are inserted as excavation proceeds thus providing stability as the wall is exposed (*B*).

Ground anchors may also be used in construction work to overcome uplift due to flotation of relatively light structures in high groundwater (*C*). either as a permanent measure or as a temporary measure during construction (see page 93), or to overcome uplift of foundations due to wind forces (see page 60).

WATERPROOFING

Methods adopted to prevent the lateral penetration of rain and the penetration of ground moisture through walls are described in chapter 5, Part 1. Methods of providing protection against entry of subsoil water under pressure and of dealing with the problem of rising damp in existing walls are discussed here.

Waterproofing of basements

When walls and floor form a basement below ground level, the penetration of moisture through the sides of the wall as well as through the floor is commonly prevented by the application of an unbroken membrane or coating of suitable impermeable material over the whole of the walls and floor. When the basement is below the level of subsoil water, the latter will be forced through the walls and floor under pressure and special precautions must be taken, either by completely enclosing the basement in a waterproof 'tank' of impermeable material or by the use of high-grade dense concrete, with or without an integral waterproofer, for walls and floor. Prestressed concrete can also be used for this purpose.

Waterproof tanking

This is most commonly carried out in asphalt although alternatives to this are asphaltic bitumen on a fabric base applied in three layers with all joints lapped and sealed, or tough plastic sheeting similarly sealed at all joints. This has the advantages of flexibility and rapid installation. Asphalt in tanking work is laid in three coats to a total thickness of not less than 29 mm on horizontal surfaces and not less than 19 mm on vertical faces. All internal angles are reinforced by means of a fillet 50 mm on the face, formed in two coats. All joints in the coats must be broken by at least 150 mm in horizontal work and 75 mm in vertical work. The asphalt membrane may be placed either on the outside or the inside face of the structure. When placed on the inside it may be necessary to provide 'loading' walls and floor of sufficient strength to prevent the asphalt being forced off the structure by the pressure of water.

External tanking   The advantages of placing the asphalt on the outside face as shown in figure 79 *A* and figure 82 *A* are: (i) the structure itself provides the necessary resistance against the pressure of water on the asphalt (ii) the asphalt keeps the water out of the structure. Faults due to poor workmanship or to settlement are, however, difficult to locate and remedy, because the point at which the water enters the internal face may be a considerable distance away from the fault in the asphalt. In spite of this the advantages of the external membrane are such that it is usual to adopt this method in all new buildings.

The asphalt to the floor is laid on a 75 to 100 mm blinding layer of concrete and is usually protected immediately by a 50 mm fine concrete screed. The concrete floor slab which is then laid must either be thick and heavy enough to resist the upward pressure of any subsoil water or be suitably reinforced to fulfil the same function. The vertical asphalt to the walls should be protected externally by a protective skin. This is usually 102.5 mm of brickwork, but the requirements of the job may necessitate the use of thin *in situ* reinforced concrete or precast concrete walings (figure 79 *A*), for reasons given on page 109.

The asphalt may be applied either direct to the retaining wall or to the protective skin, according to the circumstances of the job. When working space is available behind the wall the wall is first built and

145

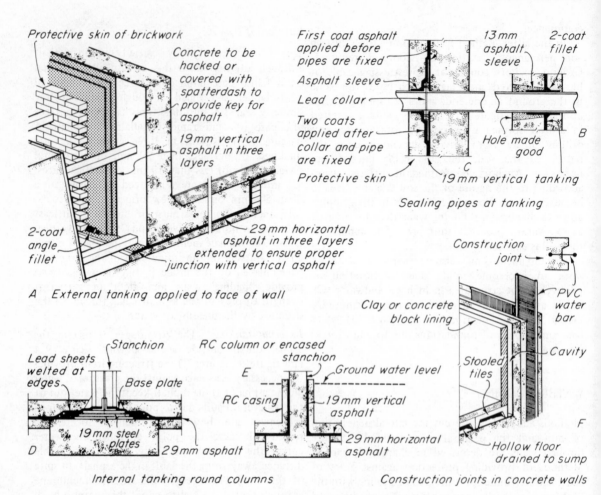

Protective skin of brickwork

Concrete to be hacked or covered with spatterdash to provide key for asphalt

19 mm vertical asphalt in three layers

2-coat angle fillet

29 mm horizontal asphalt in three layers extended to ensure proper junction with vertical asphalt

A   External tanking applied to face of wall

First coat asphalt applied before pipes are fixed

Asphalt sleeve

Lead collar

Two coats applied after collar and pipe are fixed

Protective skin

13 mm asphalt sleeve

2-coat fillet

Hole made good

B

19 mm vertical tanking

C

Sealing pipes at tanking

Construction joint

Clay or concrete block lining

PVC water bar

Cavity

Stooled tiles

Hollow floor drained to sump

F

Construction joints in concrete walls

Lead sheets welted at edges

Stanchion

Base plate

19 mm steel plates

29 mm asphalt

D

RC column or encased stanchion

Ground water level

E

RC casing

19 mm vertical asphalt

29 mm horizontal asphalt

Internal tanking round columns

82   Waterproofing of basements

the asphalt then applied direct to the face of the wall which, if of brickwork, should have the joints raked out to a depth of 13 mm to form a key. If of concrete, it should be treated in some way to give a rough surface for the same purpose (figure 82 A). Where possible the excavation on the outside of the wall should be sloped back to avoid the use of timbering, but if strutting is essential this must be arranged so that the position of the struts can be. changed to permit the asphalt to be applied at the strut points. The protective skin would most suitably be of 102.5 mm brickwork built up as the asphalting is carried out, so that the struts may be repositioned to bear on the brick skin and thus avoid possible damage to the asphalt (A). When no space at the back of the wall is available, for example in

underpinning work or on a confined site where the basement extends to the boundary of the site, the asphalt is applied to the protective skin. In these circumstances temporary support to the soil face behind the protective skin must be maintained until the new retaining wall has been constructed, and the asphalt work must be carried out in short lifts as each section of the wall rises as described on page 404.

In the case of a reinforced concrete wall which is cast direct against the asphalt, care must be taken that the asphalt is not damaged or pierced by reinforcing bars or tamping rods. In some circumstances when the basement wall is of concrete, even when working space is available behind the wall, it may be cheaper to build up the brick skin as a thin,

146

self-supporting wall, stiffened by piers at intervals, to which the asphalt can be applied. The concrete wall can then be cast between the asphalted skin and inside formwork.

The pits for column foundations and retaining wall bases are covered with a blinding layer of concrete and the sides lined with 102.5 mm brick-work built up off the edge of the blinding layer. The whole is then tanked with asphalt, the upper edges being joined to the asphalt layer in the floor (figure 79 *A*).

Internal tanking Application of the asphalt to the inside face of the structure is largely confined to existing buildings where it would normally be required as a damp-proofing measure rather than as a tanking against water under pressure. When used as tanking, apart from the disadvantages of not protecting the structure from water and of requiring loading walls and floor to protect and hold the asphalt in position against the pressure of water, there are problems involved in waterproofing round columns.

If the internal tanking is carried over the top of the column foundation slab, the column will pierce the asphalt skin. It is thus necessary to form a 'pipe' of asphalt round the column rising from the horizontal layer to some distance above the highest surface level of the subsoil water as indicated in figure 82 *E*. A reinforced concrete casing will then be required round the sleeve to prevent it being forced off the column face. As in the case of the walls and floor, the column or steel stanchion is permanently, or at least periodically saturated. An alternative method which may be used for steel stanchions in a new building is to sandwich two or three lead sheets between steel base plates, large enough to project at least 229 mm beyond the plates all round (*D*). The edges of the lead sheets are welted all the way round and the exposed faces painted with bitumen, after which the asphalt is worked between them up to the edges of the base plates and to the base of the stanchion. This can make a reasonably watertight junction provided the water pressure is not excessive. However, neither of these methods is really satisfactory and in new buildings it is better to drop the asphalt skin completely under the foundation slab.

Although asphalt is not likely to be over-stressed under a foundation slab, the pressure on the asphalt due to heavy loads over confined areas should be considered and be limited to 65 to 86 kN/m^2.[1]

Service pipes and drainpipes must often pass through asphalt tanking and some provision must be made to prevent water entering at these points. The usual method when the water pressure is not high is to form an asphalt sleeve round the pipe about 300 mm long and extending an equal distance on both sides of the line of the asphalt tanking (*B*). This sleeve is applied before passing the pipe through the wall after it has been thoroughly cleaned, scored and painted with a coat of bitumen. The asphalt tanking is then worked up to the sleeve and the joint is reinforced with a fillet. When the water pressure is high, and is likely to force water between the sleeve and the surface of the pipe, a metal collar is incorporated at the junction of the pipe and the tanking to form a seal (*C*). The collar may be formed by a 3 to 6 mm plate welded on the pipe or by a lead sheet sandwiched between a flanged pipe joint as shown. The collar should project a minimum distance of 150 mm all round the pipe and both faces should be painted with bitumen before the asphalt is worked round it. Water-proofing round pavement lights is shown in figure 79 *A*.

Waterproof structure

High-grade thoroughly consolidated concrete can be highly impervious to moisture, but in practice it is not easy to obtain an impervious structure owing to the difficulty of working round strutting and forming proper construction joints, which are usually weak points. A number of precautions are generally taken to overcome this weakness. PVC or copper water-bars may be incorporated at the joints (figure 82 *F*). When long lengths are being cast it is preferable to leave a gap of 450 to 610 mm between adjacent sections and to fill these later after the edge faces have been prepared in the manner described on pages 109-10. This minimizes the extent of shrinkage. Vibration should be used in order to obtain maximum density of concrete and an integral waterproofer may be incorporated with the mix, although it is inadvisable to rely on this in the absence of first quality concrete[2].

[1] Natural Rock Asphalt and asphalt complying with BS 1097 with a 50/50 mixture of Trinidad Epure and Residual Bitumen have been tested satisfactorily to 130 kN/m^2

[2] See Ministry of Works Advisory Leaflet No. 51, *Watertight Basements, Part 1*

When dependence is on the impermeability of the concrete rather than on waterproof tanking, the possibility of damage and inconvenience through leakage may be overcome by constructing a hollow floor consisting of a concrete topping over special half-round or stooled flat tiles, and building up a lining of clay or concrete blocks 50 to 75 mm thick and 50 mm in front of the face of the basement walls (F). Should any leaks occur the water will drain into the hollow floor from which it will run into a sump constructed for the purpose and fitted with a float controlled electric pump which will come into operation when the sump fills. In some circumstances, this method, even allowing for the cost of periodic pumping, can be cheaper than full waterproof tanking.

The prestressing of concrete, because it maintains the whole of the concrete in compression, prevents cracks occurring under load and also has the effect of closing up shrinkage cracks at working joints. Since high-quality concrete must be used for prestressed concrete work and prestressing overcomes the weakness at construction joints, it is possible to obtain a waterproof construction with greater certainty than with ordinary reinforced concrete. However, prestressed concrete is not likely to be used solely on this account but where for structural reasons it appears appropriate, advantage can be taken of its impermeable qualities.

Rising damp in existing walls

The commonest cause of rising damp in existing walls is the absence of a damp-proof course. In some cases this can be cured by forming a narrow external trench or dry area against the base of the wall. This will permit the damp to evaporate outwards before it rises to a level where it can penetrate to the inside of the building. When the wall is thick, evaporation from the inside of the wall may be assisted by inserting porous high capillary tubes along the base of the wall. These are tubes about 50 mm in diameter made of high-capillary earthenware inserted in holes formed in the wall and sloping upwards slightly from the outside face. They should penetrate about two-thirds the thickness of the wall and must be bedded in a weak mortar in order to provide sufficient capillarity. This is used very dry to minimize shrinkage and breaking away from the surrounding wall. The spacing of the tubes depends upon the degree of dampness in the wall but will usually be in the region of 450 to 900 mm apart. The functioning of the system depends on the fact that evaporation of the moisture through the tube causes a fall in temperature and an increase in weight of the air in the tube. This damp air slips out from the bottom of the tube and is replaced by fresh air: thus a continuous circulation of air is set up and evaporation is continuous[1]. This method can also be used without cutting a trench as an alternative to inserting a normal damp-proof course above ground line although it is now generally used with water-repellent barriers as a means of drying out the dampness already in the wall.

The traditional method of inserting a damp-proof course in an existing wall is to cut out at intervals short sections of the wall at the appropriate level on the lines of normal underpinning work and rebuild them to incorporate a damp-proof course of engineering bricks or other suitable material. This is a lengthy and expensive operation, and methods have been developed which involve sawing either by hand or power-driven saw a narrow slot in a mortar bed joint into which is driven a metal damp-proof membrane. By this method the work can be carried out much faster and at less than half the cost of the traditional method. It is however, only suitable for walls in which the courses are reasonably straight and in which the walling is sound. Unsound brickwork and loose rubble in the core of thick stone walls will fall and block the slot[2].

Silicone water-repellent may also be used to form a water barrier at the base of a wall. Holes penetrating well into the thickness of the wall are drilled about 150 mm apart and a silicone-rubber latex solution is introduced through them until the full thickness of the wall is saturated and a damp-proof barrier formed. This method is particularly suitable for walls the nature of which would make difficult the insertion of a damp-proof course.

An electrical method of drying out damp walls and maintaining them in a dry state is based on the phenemenon of electro-osmosis. The principle of this is that if an electric current is passed between two electrodes buried in a capillary material any free water in the material will flow towards the cathode or negative electrode. In practice, a number

[1] See *Stones of Britain,* by B.C.G. Shore, FAMS, LRIBA (Leonard Hill Ltd), pages 235-237

[2] See *Building Research Station Digest* 27 for a full description of this method

of positive electrodes, consisting of spirals or loops of copper strip, are embedded in holes drilled to a depth of about two-thirds of the wall thickness from 200 to 500 mm apart. These are connected by a copper wire embedded in a chase in the wall. Negative electrodes connected by insulated cable are buried in the soil at or below foundation level. The leads connecting the wall and soil electrodes are connected to accessible junction boxes so that an electric current can be applied if required. This is seldom necessary because by short circuiting the electrodes sufficient current to operate the system is obtained due to the potential difference naturally existing between a damp wall and the soil on which it rests and from which derives its dampness. One particular advantage of the system is that it is not necessary to fix the wall electrodes below the floor because the wall below the electrodes will be dry down to the soil electrodes at the foundation level.

Waterproof rendering, or corrugated bituminized fibre-based lathing for plaster may be applied to the internal face of damp walls to protect finishes. These methods will not, however, dry out the wall.

EXTERNAL FACINGS AND CLADDINGS

The traditional masonry technique of bonding stone with a brick backing produces a structure which combines strength and satisfactory appearance and in which the facing material acts structurally with the backing material. The advent of the framed structure with solid infilling panels led to the use of thin applied slabs of stone or precast concrete as an external facing rather than ashlar work, and the increasing use of framed structures naturally resulted in attempts to reduce the dead load of the non-structural enclosing construction. This factor, to-gether with the development of lifting devices for use on building sites, led to the technique of constructing walls with large, light elements or 'claddings' attached to the structural frame. This separation of the finishing process from that of the basic structure helps to reduce the number of operatives on the site at any given time and reduces the risk of finishes being damaged by following trades.

Fixing techniques will vary according to the nature of the 'background' structure, the type of facing or cladding system used and whether it is applied as the work proceeds or after the completion of the structure. In discussing the various methods by which building structures may be finished externally it is convenient to subdivide them under the following headings:

Facings   Methods of finishing which require a continuous background structure to give the necessary support and fixing facilities for the materials forming the external face of the building. In no case, except in ashlar work, will the facing materials take loads other than their own weight.

Claddings   The term 'cladding' is here taken to mean a method of enclosing a building structure by the attachment of elements capable of spanning between given points of support on the face of a building, thus eliminating the necessity for a continuous background structure. A cladding element will generally be large enough to take a large part of the wind force acting on the building and must be strong enough to transfer this load to the basic structure. Claddings may be heavy elements such as precast concrete slabs, or lightweight elements such as metal, plastic or asbestos cement profiled sheetings, or glass curtain walling.

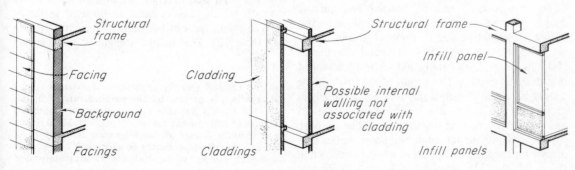

83   External finishes

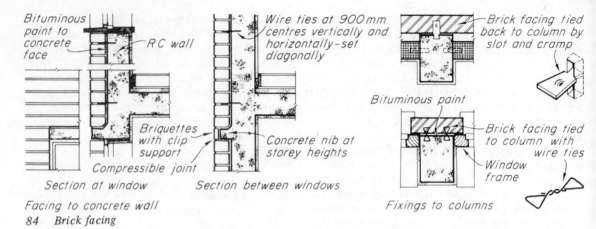

Bituminous paint to concrete face

R C wall

Briquettes with clip support

Compressible joint

Section at window

Wire ties at 900 mm centres vertically and horizontally-set diagonally

Concrete nib at storey heights

Section between windows

Facing to concrete wall

84 Brick facing

Brick facing tied back to column by slot and cramp

Bituminous paint

Brick facing tied to column with wire ties

Window frame

Fixings to columns

Infilling panels A method of providing enclosure by large, fairly light elements which are generally based on some light framing of timber or metal. These elements differ essentially from claddings in that they are fixed between the members of the structural frame of the building, rather than applied to the face of the frame to form a skin. In addition to supporting their own weight, infilling panels must be strong enough to support wind loads and transfer them to the main structure through properly designed fixings.

These categories of external finishes are shown diagrammatically in figure 83.

Facings

Brickwork

Although brickwork is naturally associated with load-bearing structures its weathering properties and excellent appearance and range of colour make it a very suitable facing to other materials such as structural concrete. When used as a non-structural facing, bricks need not be bonded and various straight jointed patterns may be employed. The brickwork should be tied to the background structure with twisted wire ties or metal cramps at 900 mm centres horizontally and vertically arranged in a diagonal pattern. The weight of the brick facing should be taken at each storey level by metal angles or by projections from the structure (see figures 75 and 84).The facing over openings should be supported on nibs or metal angle supports unless the brickwork is reinforced to act as a lintel as shown in figure 72, Part 1[1]. Where brick facing is applied to

a concrete background, the latter should be painted with bituminous paint to prevent staining of the exterior by salts in the concrete. A compressible joint should be provided at each storey height (figure 84) for reasons given on page 137.

Natural stone

The choice of stone for facing will be influenced by such factors as the design of the building, its situation and the aesthetic and technical considerations relevant to each particular case. Natural stone may be broadly classified as follows in order of hardness and durability and approximate weight:

Igneous rocks, such as granite
$$2560 - 3200 \text{ kg/m}^3$$
Metamorphic rocks, which include marbles, slates, quarzite $2630 - 3040 \text{ kg/m}^3$
Sedimentary rocks, such as limestone and sandstones $1950 - 2750 \text{ kg/m}^3$

Igneous and metamorphic rocks are extremely hard, durable and water resistant and are capable of taking a high polish. They are of high density and strength which permits them to be used as facing slabs in thicknesses of between 13 mm and 50 mm (see table 12). Due to the strength of these stones

[1] In the London area the application of external facings and claddings is covered by the requirements of By-law 6.15. The District Surveyor must approve the nature and thickness of material and method of support. Metal fixings to all external facings or claddings must normally be of stainless steel, phosphor bronze or aluminium bronze alloy unless their position in the work ensures adequate protection from corrosion

Stone	Slab size (mm)	Minimum thickness (mm)	Mortar for bedding and jointing
Limestones			
Ancaster Freestone	610 x 450	25	Lime and stone dust
	1220 x 610	38	
	1525 x 760	50	
Ancaster Weatherbed	900 x 450	25	Lime and stone dust
	1525 x 610	38	
	1830 x 900	50	
Doulting Freestone	760 x 610	50	2:5:7 cement, lime, stone dust
Hornton	1070 x 450	38	Spot bedding: with 1:3 cement/sand; jointing:lime with 10% cement
Painswick	760 x 450	75	2:5:7 cement, lime, stone dust
Portland	760 x 760	50	2:5:7 cement, lime, stone dust
	1525 x 900	75	
	1830 x 1220	100	
St Adhelm (Box Ground)	760 x 450	50	ditto
Sandstones			
Auchinlea Freestone	610 x 450	100	1:1:6 cement, lime and sand
Berristall	1220 x 610	50	
Bolton Wood	900 x 610	50	
Crosland Hill (York Stone)	1830 x 300	50	1:3 cement and sand
	2440 x 685	75	
Darley Dale	1525 x 900	75	1:1:4 cement, lime, stone dust
Dunhouse	760 x 610	75	
Pennant	900 x 900	50	1:5 cement and stone dust
Woodkirk (York Stone)	2440 x 900	75	Hydraulic lime and stone dust
	3050 x 1525	100	
White Mansfield	610 x 450	25	1:3 cement and sand or stone dust
	610 x 610	38	
	900 x 760	50	
Granites			
Corrennie	1525 x 760	38	Lime or cement and sand
Kemnay	2440 x 1220	38	ditto
Rubislaw	900 x 610	25	
Shap	1220 x 610	50	
Slates			
Sawn	1525 x 760	38	Up to 1830 x 760 mm available in some slates
	1220 x 610	25	but 1220 x 610 mm recommended as average size slab
	1220 x 610	19	Only in string or apron courses up to 530-610 mm high
	900 x 450	13	Only for small areas of small slabs bedded solid to wall
Natural Riven	610 x 610	25	
	450 x 300	13	ditto
Marbles	Irrespective of	19	For facing only up to first floor level
	slab size	25	For facing rising above first floor level
	ditto	38	Many Local Authority surveyors require this as a minimum

Note: The maximum size against each stone indicates that recommended by the quarry for facing slabs, or the maximum size available from the quarry.

The mortar, where indicated, is that recommended by the quarry.

Table 12   *Thickness of natural stone external facing slabs*

comparatively thin slabs may be fixed with light cramps and wires without risk of the edges splitting. Sedimentary rocks, formed by the redisposition of older rocks by the action of air or water, are generally softer, less durable and more absorbent and exhibit a highly laminar structure. Such stones should be laid with their natural bed at right-angles to the face of the wall[1]. This principle applies whether the stone is bonded into a brick or block background or is used as a non-bonded slab facing. Consistent with this, certain minimum thicknesses are desirable for facing slabs according to the nature of the stone and table 12 indicates the thicknesses recommended by the quarries for a number of typical building stones suitable for external facings. This list, of course, is not comprehensive, and does not imply that other good building stones are not suitable for this purpose. It will be seen that some sedimentary rocks may be used as thin as 38 to 50 mm for slab facings, but this thickness does not afford much material at the edges to give a secure anchorage for cramps. A minimum thickness of 75 mm or even 100 mm is recommended for many stones. Very thin slabs are not always the most economic, depending upon the fixing problems on the job. Slabs should be about 0.28 to 0.37 m^2 face area and limited to 55 to 68 kg in weight if they are to be man-handled.

Care should be taken that limestone and sandstone are not used in juxtaposition in such a way that the soluble salts formed by the decomposition of the limestone are washed on to the sandstone, since this may produce rapid decay of the sandstone. Similar decay may occur where cast stone and sandstone are placed in juxtaposition and brick can be damaged in the same way. Stone facing on a background of brick or concrete should always be coated on the back with bitumen, or the background itself should be similarly coated. This is necessary in order to provide a barrier against the movement of salts in the bricks, mortar or concrete. These, unless isolated in this way, are taken up in solution by rain passing through the joints or the slabs, are deposited in the stone when the rainwater evaporates through the facing and cause staining and decay of the stone.

Metal anchorages　Natural stone facings are fixed to the 'background' structure by means of metal fixings which are designed primarily to hold the slabs back to the wall and keep the faces in correct alignment[2]. The fixings can assist in relieving stones below of the weight of those above, but the weight of the facing should usually be brought back to the structure at 2.4 to 3.0 m intervals by angles bolted to the structure or by suitable projections (figure 85). All metal anchorages should be of non-ferrous metal such as stainless steel or bronze, since galvanizing and bitumen painting of iron or steel give temporary protection only and may result in subsequent staining and spalling of the stone (see footnote page 150). All metal ties should be non-ferrous and not inferior to the specification given in BS 1243.

Figure 85 shows various types of metal anchorages and supports for sedimentary stones. These are usually in strip form and reasonably substantial when supporting the weight of a thick slab. The ends are accommodated in mortices or grooves cut in the edges of the slabs. The thinner igneous and metamorphic slabs are invariably secured with wire cramps and dowels as shown in figure 86 which necessitates less labour on the hard dense stone, drilling and surface sinking only being required. Dowels are sometimes used with cramps in the vertical joints of large sedimentary slabs to hold back and align the adjacent slabs.

The ends of all forms of anchors and corbels must be fixed firmly to the background and this is usually done by bedding in mortar in mortices cut, or in the case of concrete cast, in the background material. Forming of individual mortices in concrete can be avoided by casting in dovetail section pressed metal channels to form horizontal or vertical slots which accommodate the dovetail-shaped ends of various types of cramps as shown in figures 85 and 86.

Setting, jointing and pointing　Sedimentary stones are usually bedded and backed up solidly in mortar, and suitable mortars must be used[3]. The width of the joints should not be less than 5 mm, metal strips being used as screeds. The finish to the face of the mortar joints may be carried out as the work proceeds in the same mortar used for bedding. Alternatively, joints may be raked back 19 mm and pointed with specially prepared mortar of selected colour and equal strength to the bedding mortar. Marble, granite and slate facings are fixed with a 13

[1] See page 102, *MBC: Materials*

[2] Non-bonded slab facings only are covered here. See Part 1, chapter 5 for ashlar facing

[3] CP 121.201, *Masonry, Walls ashlared with natural or cast stone*, gives suitable mortars for various stones. See also table 12

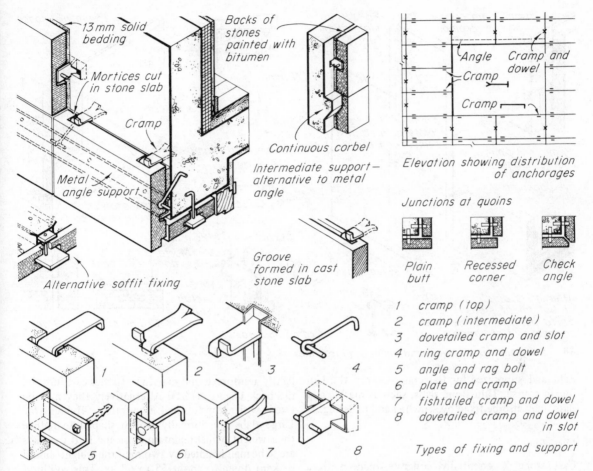

Types of fixing and support

1 cramp (top)
2 cramp (intermediate)
3 dovetailed cramp and slot
4 ring cramp and dowel
5 angle and rag bolt
6 plate and cramp
7 fishtailed cramp and dowel
8 dovetailed cramp and dowel in slot

85   Stone facings – sedimentary stones

mm air space behind the slab formed by setting the stone against mortar dabs (figure 86). This technique allows for differential movement of the facing and structural background and also prevents staining and 'blooming' of the faces of the slabs. In addition the backs of marble slabs are painted with shellac or bitumen paint.

In recent years there have been cases of the loosening of facing slabs at the fastenings, particularly on the gable walls of very high reinforced concrete buildings. This is due to the accumulative effect of 'creep' in the great height of concrete wall and to overcome this, mastic joints at every floor, or not more than 6 m apart, are now generally required.

Fixing An example of sedimentary stone slab facing is shown in figure 85. In this instance the stones are secured by metal fixings, distributed as indicated, which give each stone some support and keep the faces in alignment. Support over openings is given by an angle rag-bolted to the structure. As already mentioned, the weight of the facing should be brought down to a positive bearing on the structure at 2.4 to 3.0 m intervals by similar angles or by concrete corbels.

Fixing methods for igneous and metamorphic stones are also shown in figure 86. 'S'-hooks are used to align the edges of adjacent slabs when fixing is by wire cramps into one edge only of two adjacent edges. The mortar dabs position the 'unfixed' edge relative to the background and the 'S'-hook prevents it falling outwards. Alternatively, double toed cramps are used. Positive bearing is provided by metal corbels set into grooves in the backs of adjacent slabs, each corbel thus sharing the weight of a slab.

Small slabs of slate and quartzite are bedded

153

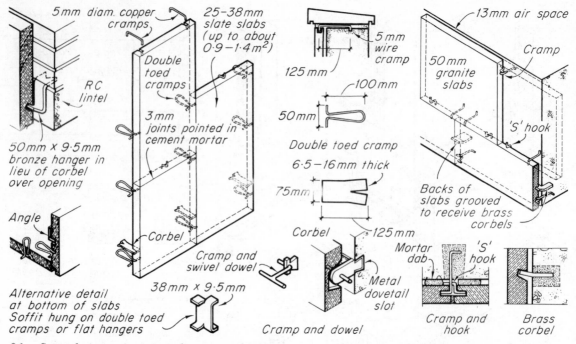

86 *Stone facings – igneous and metamorphic stones*

solid and supported by flat cramps every third or fourth course. Thin stone slabs, such as marble and slate, may also be fixed by screwing and pelleting.

Cast stone

Cast stone is essentially concrete made with a crushed stone aggregate and white cement. The manufacture of this material is the job of specialist firms, and it is most important that the stone be fully matured before building it in since the material, like any other concrete product, shrinks on setting and hardening.

Cast stone may be used as a facing in a similar manner to natural stone but as the material is cast it is possible to lightly reinforce the slabs which can be thinner than their natural stone equivalent, although it has been shown that by careful manufacture slabs of 38 to 64 mm thickness may be unreinforced. This avoids the possibility of surface staining which results from insufficient cover to steel reinforcement.

Concrete

Facing slabs may be cast in concrete in a similar way to cast stone slabs. They are about 50 mm thick

lightly reinforced and may have fixings cast into the the back (figure 87 *A*). Alternatively they may be fixed with conventional non-ferrous fixings (*B*). Large slabs are normally trough shaped to reduce their weight whilst maintaining strength. If the slabs are to be man-handled by two men the weight should be kept down to about 55 to 68 kg. This will limit the area of 50 mm slabs to about 0.45 m^2. The slabs should be carried at about 3 m vertical intervals by corbels or by bonding-in blocks cast on the backs of the slabs (*A*). Since the dimensions of concrete slabs will vary, sufficient tolerance must be provided in the joints. Mortar joints should be not less than 6 mm wide and mastic or dry jointing not less than 3 mm wide.

The panels may be finished in a variety of colours and textures by careful choice of aggregate which may be exposed by washing or spraying the surface before hardening or by scrubbing or wire brushing. Polishing or acid spraying may be carried out after manufacture. Patterned timber, plastic or rubber moulds may be used to provide surface texture (see page 425 and footnote and also *MBC: Components and Finishes,* chapter 16). Reinforcement should have at least 32 mm outer cover. Concrete slabs used as permanent shuttering are shown at (*C*). In this

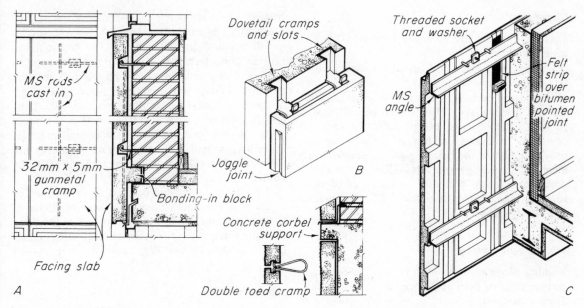

87   *Concrete facings*

example the slabs are held by adjustable bolts and square washers on steel angle framing to ensure alignment whilst the concrete is poured. The joints are rebated and allow plenty of tolerance.

Terrazzo slabs   These are composed of concrete with an 8 mm facing of terrazzo. The slabs, which should be painted on the backs and rear edges with a sealing compound, are fixed in similar ways to stone and concrete slabs.

Terracotta and faience

These materials are described briefly in *MBC: Materials,* chapter 5. For external work they are produced in the form of slabs 300 mm x 200 mm, 450 mm x 300 mm and 610 mm x 450 mm by 25-32 mm thick (figure 88 *A, B*). They are scored or dovetailed on the back to give a good key and may be fixed with non-ferrous metal cramps in a manner similar to thin stone slabs. Hollow blocks are also produced which are filled with fine concrete and bonded into brickwork. The blocks are limited to .084 m^3 volume and are usually used for such elements as sills and copings (C). The protective glaze on faience only covers the face and a small strip of the return surfaces. Arrises are rounded. Joints should be 6 mm wide.

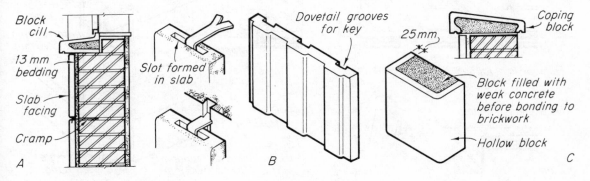

88   *Terracotta and faience facings*

Glazed tiles

Tiles for external use are made from a plastic clay which produces an 'open' body capable of resisting the action of frost. Tiles should be specified as 'exterior frost resisting glazed tiles', and are bedded on a rendered backing in mastic or mortar. Concrete surfaces, particularly soffits, should be well hacked for key. When covering a large surface tiles are best contained in panels, or broken up into separate panels to reduce the risk of cracking due to movement of the background. The edges of the panels should have open weathered or mastic filled joints to allow for movement (see also *MBC: Components and Finishes,* chapter 15).

Tile and slate hanging

Roofing tiles and slates may be used as facings to vertical surfaces both as decoration and protection. They should not be hung in positions where they are vulnerable to damage. Tiles are hung on 38 mm by 19 mm battens, preferably fixed over counter battens at about 400 mm centres. A lap of 38 mm is adequate requiring a gauge of 114 mm for the battens, the tiles being double nailed and laid to break joint, as in roofing practice. Some examples are shown in figure 89 *A* to *D*. The bed joints in rat-trap bond (see Part 1, page 93) are at approximately 114 mm centres and nibless tiles may be used nailed direct to the joints as shown in (*B*). Slate hanging is similar to roofing practice, the same gauges relative to each size of slate being used. Although not recommended as a general practice, slates may be nailed directly into brick jointing and a suitable size can be chosen to work in with the courses[1].

Tile and slate hanging is often used in relation to infilling panels and (*E* and *H*) show methods of treating junctions with walls. At openings the tiling or slating may be butted against projecting frames (*F*) or, in the case of tiling, angle tiles may be used (*G*).

Asbestos cement

This material may be used in the following ways:
Asbestos slate and tile hanging, siding Asbestos

[1] For the application of tiles and slates to roofing see *MBC: Components and finishes,* chapter 18

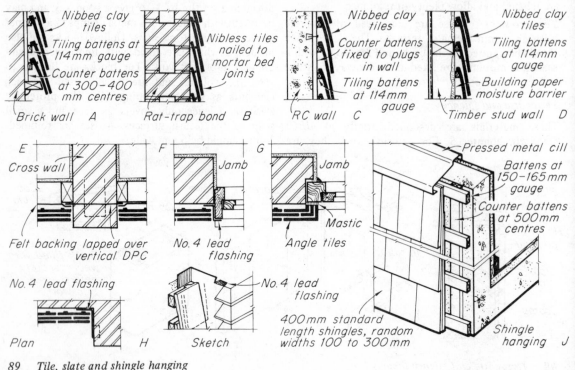

89　*Tile, slate and shingle hanging*

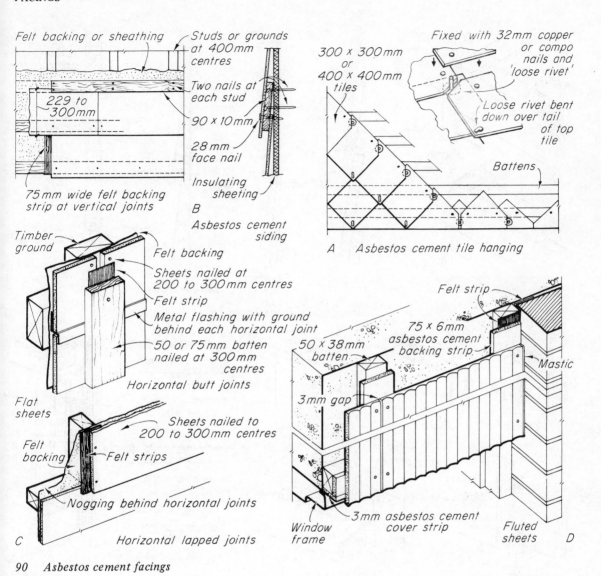

Felt backing or sheathing

Studs or grounds at 400mm centres

Two nails at each stud

90 x 10mm

28 mm face nail

Insulating sheeting

229 to 300mm

75 mm wide felt backing strip at vertical joints

B   Asbestos cement siding

Fixed with 32mm copper or compo nails and 'loose rivet'

300 x 300mm or 400 x 400mm tiles

Loose rivet bent down over tail of top tile

Battens

A   Asbestos cement tile hanging

Timber ground

Felt backing

Sheets nailed at 200 to 300mm centres

Felt strip

Metal flashing with ground behind each horizontal joint

50 or 75mm batten nailed at 300mm centres

Horizontal butt joints

Flat sheets

Felt backing

Felt strips

Sheets nailed to 200 to 300mm centres

Nogging behind horizontal joints

C   Horizontal lapped joints

Felt strip

75 x 6mm asbestos cement backing strip

50 x 38mm batten

Mastic

3mm gap

Window frame

3mm asbestos cement cover strip

Fluted sheets   D

90   Asbestos cement facings

slates are rectangular and are centre nailed. Asbestos tiles are square, laid diamond pattern, and are double nailed at the centre. In addition a loose rivet is used to hold the slates or tiles together in threes (figure 90 A). Asbestos siding is basically flat sheeting used in strip sizes 600 mm to 1.2 m long. Fixing is to battens with a 25 mm or 38 mm lap at the head and vertical joints are protected by felt backing strips (B).

Asbestos cement sheeting   This may be fixed to grounds either with lapped horizontal joints with a felt strip behind the vertical joints, or with horizontal butt joints, Z-flashings and vertical cover strips (C). Fluted sheeting is usually fixed with countersunk screws (D).

Permanent shuttering   For limited areas of facing such as to spandrels, it is possible to use flat or profiled sheeting as a shutter lining to be retained as permanent facing, bonded to the concrete. This avoids evidence of fixings and provides a solid backing to the sheeting, which is brittle and otherwise easily cracked in vulnerable positions.

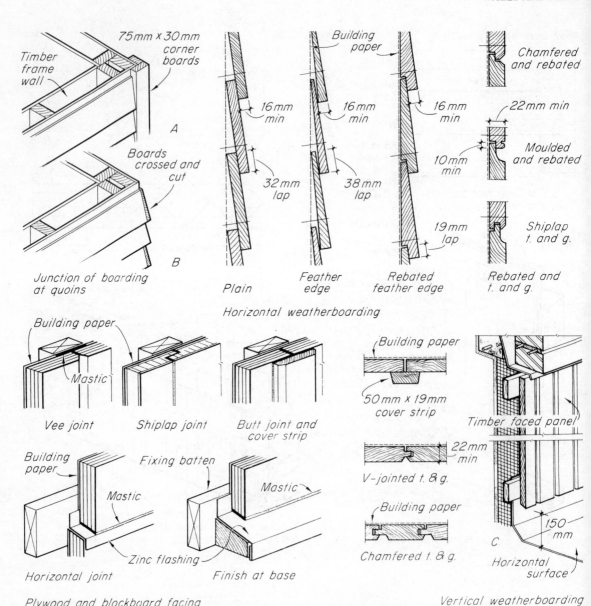

Horizontal weatherboarding

Plain | Feather edge | Rebated feather edge | Chamfered and rebated | Moulded and rebated | Shiplap t. and g. | Rebated and t. and g.

75mm × 30mm corner boards

Timber frame wall

A

Boards crossed and cut

B

Junction of boarding at quoins

Building paper

16mm min | 32mm lap | 16mm min | 38mm lap | 16mm min | 10mm min | 19mm lap | 22mm min

Building paper

Mastic

Vee joint | Shiplap joint | Butt joint and cover strip

Building paper

Mastic | Fixing batten | Mastic

Zinc flashing

Horizontal joint | Finish at base

Plywood and blockboard facing

Building paper

50mm × 19mm cover strip

V-jointed t. & g. | 22mm min

Building paper

Chamfered t. & g.

Timber faced panel

150 mm

C

Horizontal surface

Vertical weatherboarding

91   Timber facings

Timber

Where by-laws and regulations permit[1], timber may be used to face walls in the following ways:

Weatherboarding   This is nailed directly to the sheeting of a timber frame wall (figure 91 *A, B*) or to grounds plugged to concrete or brickwork at about 1200 mm centres for 25 mm boards fixed hori-zontally (400 - 450 mm for 16 mm boards) and 600 mm centres for 25 mm boards fixed vertically. Diagonal boarding 25 mm thick should be fixed to vertical grounds at about 760 mm centres. Nails used to fix the boarding should be composition nails, since hammering may break the coating of sherad-

[1] The Building Regulations do not permit softwoods (except Western Red Cedar and Sequoia) as external facing unless preservative treated

158

ized or galvanized nails. Copper nails will usually produce characteristic staining. Secret nailing, where appropriate, helps to prevent corrosion staining of unpainted boarding. Hessian based bituminous felt or building paper should be placed immediately behind the boarding[1].

Boarding should be free to move on at least one side and the lower edge should be kept at least 150 mm above any horizontal surface to prevent splash staining. Vertical boarding should be fixed so that the lower edges are free to allow water to drain off easily (C) but where this is not possible, for example when the boards are in grooves or rebates, they should be treated with preservative or be sealed.

Types of horizontal weatherboarding and methods of treating the quoins are shown in figure 91, together with types of edge-joints to vertical boarding which provide for movement. Frames to openings are usually made to project to form a finish to the boarding as shown for tile hanging.

All timber used as ground work behind boarding should be treated with preservative before fixing.

Panelling   This may be fixed to suitable groundwork plugged to concrete, brickwork or similar background to provide larger unbroken surfaces than weatherboarding. Plywood and blockboard are used since they have greater dimensional stability than natural timbers. The plywood or blockboard must be resin bonded with a WBP (weather and boil-proof) adhesive complying with BS 1204 to prevent delamination. The Building Regulations require a minimum thickness of 8 mm for external facings. Unless a special plywood is used which has all laminates pressure impregnated before being bonded together, timbers liable to fungal attack should be pressure impregnated before being fixed. This should be done after the boards have been cut to size. Mastic bedding or pointing in contact with the timber should be of the non-oiling variety unless the edges are sealed with polythene tape. The sheets should be screwed with brass screws at all edges and the joints caulked with a non-staining caulking compound. Suitable spacings for grounds are 10 mm sheet – 610 mm, 13 mm sheet – 800 mm, 16 and 19 mm sheet – 1070 mm.

Treatments of vertical and horizontal joints for this type of facing are·shown in figure 91.

Shingle hanging   Edge grain sawn cedar shingles hung vertically are twice nailed at the centre to battens at 150 mm or 165 mm gauge. Each shingle is held by two other nails from the course above it and the resulting covering is extremely weather tight. Composition, copper or best hot spelter galvanized nails should be used (figure 89 J).

Metal

Metals such as steel, stainless steel, copper, bronze, aluminium and lead may be used for facings to masonry and concrete structural backings in the following ways:

Fully supported sheeting   This is similar to roofing technique, the sheets being joined together with welts and standing seams and secured by cleats nailed to horizontal timber insets in concrete or battens in screed as shown in figure 92 A. Lead sheet is too heavy to rely on cleat fixing and the sheets are held by a number of brass screws driven into lead plugs in the backing and finished-off with lead burned dots.

Profiled metal sheeting and panelling   This application of metal sheeting consists of sheets pressed or extruded into various profiles to give rigidity, thus enabling fewer fixings to be used. Alternatively, thin panels may be stiffened by welding ribs or angle stiffeners to the backs of the sheets or the panels may be cast. Metals used in this way include:

(a) *Aluminium alloys*   These are available in different types and various finishes[2]. Typical profiled sections and applications of aluminium panel facings to masonry and concrete walls are shown in figures 92 B and 93 A, B. Aluminium should be protected from damp mortars and plasters since it is liable to alkaline attack. A protective paint should be used to dissociate it from steel or other metals. Plastic or fibre washers may be used to keep sheeting clear of steel fixing bolts and similar fixings.

(b) *Steel*   Carbon steel panels, apart from weathering steels[3] require protection against rusting. This can be achieved in various ways[4]. Some of these methods give colour and variation in texture. The most effective protective finish is vitreous enamelling, which can be applied in any colour and finished

[1] See Part 1, page 130 regarding the need for a 'breather' type material in this position on timber frame walls

[2] See chapter 9, *MBC : Materials*

[3] eg *Cor-Ten* steel

[4] See chapter 9, *MBC : Materials*

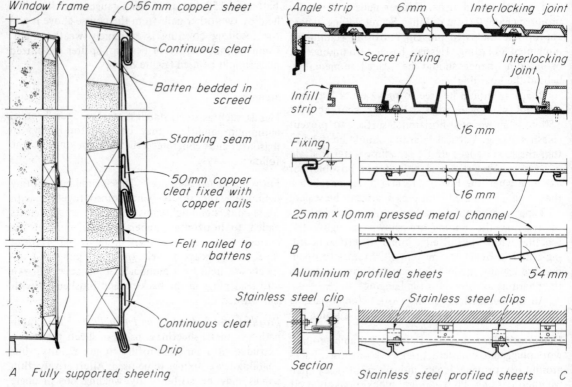

Window frame — 0·56mm copper sheet

— Continuous cleat

— Batten bedded in screed

— Standing seam

— 50mm copper cleat fixed with copper nails

— Felt nailed to battens

Stainless steel clip

Continuous cleat

Drip

A   Fully supported sheeting

Angle strip   6 mm   Interlocking joint

— Secret fixing   Interlocking joint

Infill strip

16 mm

Fixing

16 mm

25mm × 10mm pressed metal channel

B

Aluminium profiled sheets   54 mm

Stainless steel clip   Stainless steel clips

Section

Stainless steel profiled sheets   C

92   Metal facings

in full gloss, matt, semi-matt and eggshell surface textures. Other finishes include the application of stone dust by a refractory process and asbestos, applied as a felt or by spraying on a zinc coating or primer. Chemical colouring of plated surfaces is possible but needs to be given a lacquered finish; weathering will depend on the efficacy of the lacquer. An example of steel panels fixed with clips and lugs screwed to timber groundwork is shown in figure 94. The panels are butt-jointed and caulked with a mastic compound.

(c) *Stainless steel*   This is a general name for a number of steel alloys.[1] The material is pressed into panels using 1 mm to 0.8 mm metal and is more economic when used in narrow strips, say 250 mm to 300 mm wide, than in large sheets. The apparent waviness of flat sheets can be reduced by embossing or by profiling, keeping the flat surfaces down to 100 or 125 mm in width. Fully corrosion resistant welds are possible which are invisible when polished. All fixings should be of stainless steel or be heavily protected carbon steel. Figure 92 C shows an example

of pressed stainless steel applied as a facing on metal fixings bolted to the structural background.

(d) *Bronze*   This copper alloy has greater mechanical strength than sheet copper. It is expensive, can be obtained in plain and profiled sheets and weathers to a dark brown/black with a slight green patina[1]. Bronze facing to a concrete wall, screwed to a supporting framework of mild steel flats, angles and channels, is shown in figure 93.

Plastics

Of the vast number of plastics at present in existance, only certain types are suitable for external use as facings and reference should be made to *MBC: Materials*, chapter 13, where these are described.

Glass

Opaque glass forms colourful and easily cleaned surfaces. The different types available and the methods of fixing sheets as external facings are

[1] See chapter 9, *MBC : Materials*

160

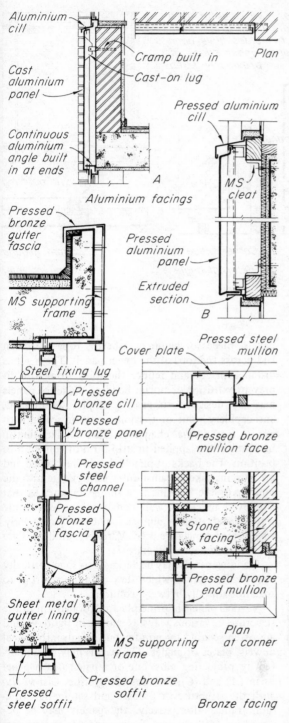

Aluminium
cill

Cramp built in   Plan

Cast-on lug

Cast
aluminium
panel

Pressed aluminium
cill

Continuous
aluminium
angle built
in at ends

A

MS
cleat

Aluminium facings

Pressed
bronze
gutter
fascia

Pressed
aluminium
panel

Extruded
section

B

MS supporting
frame

Steel fixing lug

Cover plate

Pressed steel
mullion

Pressed
bronze cill

Pressed
bronze panel

Pressed bronze
mullion face

Pressed
steel
channel

Pressed
bronze
fascia

Stone
facing

Sheet metal
gutter lining

Pressed bronze
end mullion

Plan
at corner

MS supporting
frame

Pressed bronze
soffit

Pressed
steel soffit

Bronze facing

93   Metal facings

discussed in chapter 12, *MBC: Materials.* Typical fixing details are shown in figure 95 for facings up to 2.4 m high. Above this fixing clips to project over the face of the glass or cover strips must be used.

Mosaic and flint

Glass or ceramic mosaic facings are available in a variety of colours. The mosaic is usually gummed to a paper backing, the paper being soaked off when slabbing up is completed (see chapter 15, *MBC: Components and Finishes).*

Flints may be natural or 'knapped'[1]. They require a sand and cement bed some 125 mm thick to enable the stones to be set in well over half their length.

Renderings

External renderings are discussed in *MBC: Components and Finishes* chapter 14. Edge protection is important and methods used to provide this are shown there.

Claddings

Concrete

In addition to their use as facings and permanent shuttering, precast concrete panels may be designed to act as cladding to a structural frame, independent of·any infill walling.

Weight is usually reduced to a minimum by casting the body of the panel as thin as possible, rigidity being achieved by casting ribs at the edges and at intermediate positions on larger panels as shown in figures 96 and 97. The main reinforcement is concentrated in the ribs, the thinner body of the panels being lightly reinforced with galvanized wire or left unreinforced in many cases. Thin slabs (say 50 mm thick) of constant thickness without ribs should be reinforced with welded galvanized mesh to reduce the risk of rust staining on the faces. In all cases reinforcement should be kept back 25 mm to 32 mm from the weather face of any concrete panel or slab.

Panels may be cast in factories or on site, and in either case joints must provide adequate tolerance since panels can rarely be produced to finer dimen-

[1] See page 118, Part 1

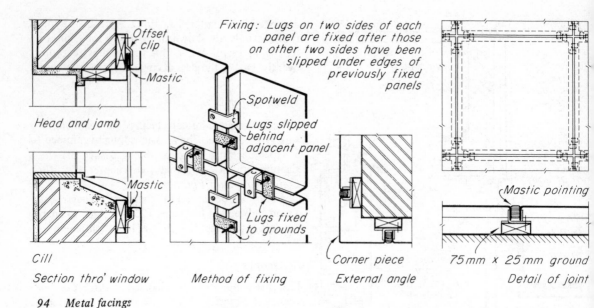

Fixing: Lugs on two sides of each panel are fixed after those on other two sides have been slipped under edges of previously fixed panels

Offset clip

Mastic

Head and jamb

Mastic

Cill

Section thro' window

Spotweld

Lugs slipped behind adjacent panel

Lugs fixed to grounds

Method of fixing

Corner piece

External angle

Mastic pointing

75mm x 25mm ground

Detail of joint

94   Metal facings

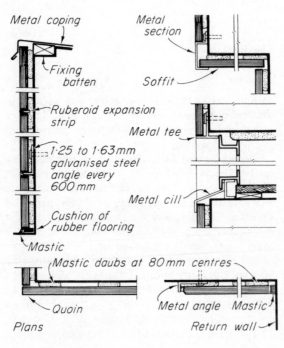

Metal coping

Fixing batten

Ruberoid expansion strip

1·25 to 1·63mm galvanised steel angle every 600 mm

Cushion of rubber flooring

Mastic

Mastic daubs at 80mm centres

Quoin

Plans

Metal section

Soffit

Metal tee

Metal cill

Metal angle   Mastic

Return wall

95   Glass facing

sional tolerances than plus or minus 5 mm. In most cases greater allowances should be made. To assist in obtaining the correct width bed joint between

panels, pieces of hemp or asbestos rope are often laid across the tops of the lower panels. These act as distance pieces and relieve the joint from load to prevent extrusion of mortar bedding. Edges of panels may be designed to give a simple, self-drained, anti-capillary joint or the joint may be sealed with mastic (see *Joints*, chapter 2, Part 1). Pointing in mastic is either applied in strip form or gunned into position. The backs of panels should be drained and precautions taken against entry of moisture into the interior of the building in a similar manner to that employed in cavity wall construction, since some water penetration is almost inevitable, particularly where large panels are used. Various joint details are illustrated in figures 96 and 97.

Cladding panels may be designed to span vertically between floors from which they obtain support or to span horizontally between columns.

Vertical panels in principle are 'hooked' on to the structure in various ways as shown in figure 96 and are often secured by non-ferrous or galvanized steel dowels, plates or angles. The panels may be supported by projecting nibs, one or a pair to each panel as in (A) and (C), the nibs being either cast *in situ* with the structure or precast and set in the structural frame. Alternatively, the panels may be designed to gain support from continuous string or stringer units (B) or from beams (D).

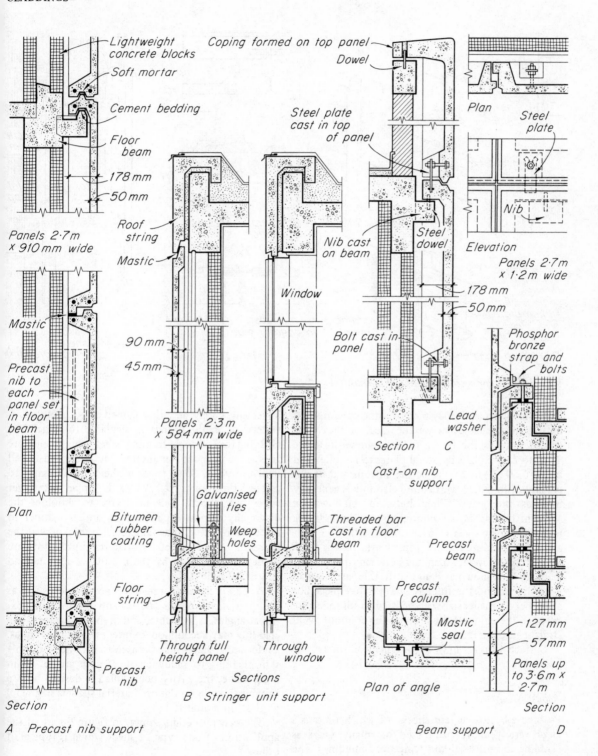

Lightweight concrete blocks

Soft mortar

Cement bedding

Floor beam

178 mm

50 mm

Panels 2·7m × 910 mm wide

Mastic

Precast nib to each panel set in floor beam

Plan

Bitumen rubber coating

Floor string

Precast nib

Section

A Precast nib support

Roof string

Mastic

90 mm

45 mm

Panels 2·3 m × 584 mm wide

Window

Galvanised ties

Weep holes

Threaded bar cast in floor beam

Through full height panel

Through window

Sections

B Stringer unit support

Coping formed on top panel

Dowel

Steel plate cast in top of panel

Nib cast on beam

Steel dowel

Bolt cast in panel

Section   C

Cast-on nib support

Precast column

Mastic seal

Plan of angle

Beam support

Plan

Steel plate

Nib

Elevation

Panels 2·7m × 1·2m wide

178 mm

50 mm

Phosphor bronze strap and bolts

Lead washer

Precast beam

127 mm

57 mm

Panels up to 3·6 m × 2·7m

Section

D

96   Precast concrete claddings – vertical

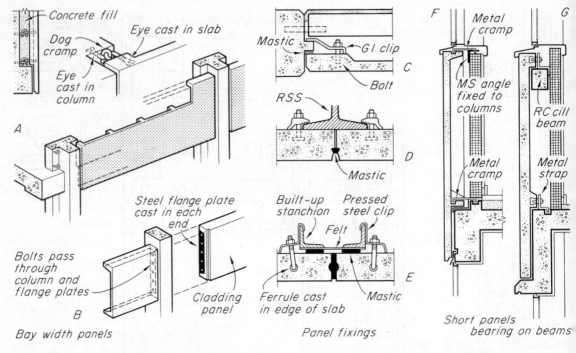

97   Precast concrete claddings – horizontal

Horizontal panels may pass over the columns and butt against each other on the face, or they may extend only to the return faces as shown in figure 97 *A* to *E*. Fixing is by metal plates (*B*) or clips and bolts (*C, D, E*) or by projecting stirrups and *in situ* concrete filling (*A*). Horizontal cladding is generally used as the panel filling between windows on adjacent floors, and an alternative to column fixing is to design the slabs to bear on the structural floor and to be stabilized at the top by attachment to an intermediate support spanning between the columns at cill level, as shown at F and G. In this case the slab need not extend the full bay width in one piece but a number of smaller slabs can be used to fill the space between the columns. Rebates on the columns avoid straight through joints.

Surface finishes may be obtained as for facings (page 154).

Asbestos cement

Profiled asbestos cement sheets, of which there is a large variety, have been used for many years as cladding to industrial and temporary buildings. The sheets are lap jointed at the sides and the upper and lower edges may be either lapped or butt jointed. A variety of metal fixings is available to attach the cladding to steel angle framing, steel tubing, concrete rails, or timber rails or inserts. These are illustrated in figure 98 *A*. Some types of sheets are designed to conceal the fixings (*B*). Where the upper and lower edges are lapped, it is necessary to shoulder the diagonally opposite panels in a similar manner to interlocking tiling to avoid excessive thickness at the junction of four sheets. Horizontal butt joints are rendered waterproof by the use of a Z-flashing and additional fixings (*C*).

A comprehensive range of accessories such as flashings, filler pieces (*D, E*) and internal and external angles, is available, and in many instances the cladding may accommodate integral window frames. Underlining sheets may be used to produce a double skin cladding or sandwich sheet with greater insulating value, the cavity being filled if desired with a suitable flexible insulation material such as slag wool or glass fibre quilt (*C*).

The general details shown in figure 98 are of the application of one type of sheet which is typical of others.

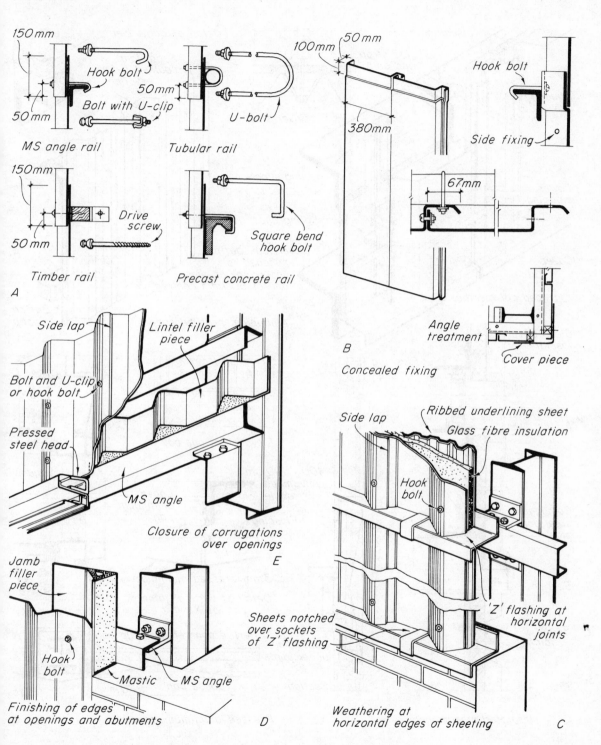

150mm

Hook bolt

50mm

Bolt with U-clip

50mm

MS angle rail

U-bolt

Tubular rail

150mm

Drive screw

50mm

Timber rail

Square bend hook bolt

Precast concrete rail

A

100mm 50mm

380mm

Hook bolt

Side fixing

67mm

Angle treatment

B

Concealed fixing

Cover piece

Side lap

Lintel filler piece

Bolt and U-clip or hook bolt

Pressed steel head

MS angle

Closure of corrugations over openings

E

Side lap

Ribbed underlining sheet

Glass fibre insulation

Hook bolt

'Z' flashing at horizontal joints

Jamb filler piece

Hook bolt

Mastic MS angle

Finishing of edges at openings and abutments

D

Sheets notched over sockets of 'Z' flashing

Weathering at horizontal edges of sheeting

C

98 *Asbestos cement claddings*

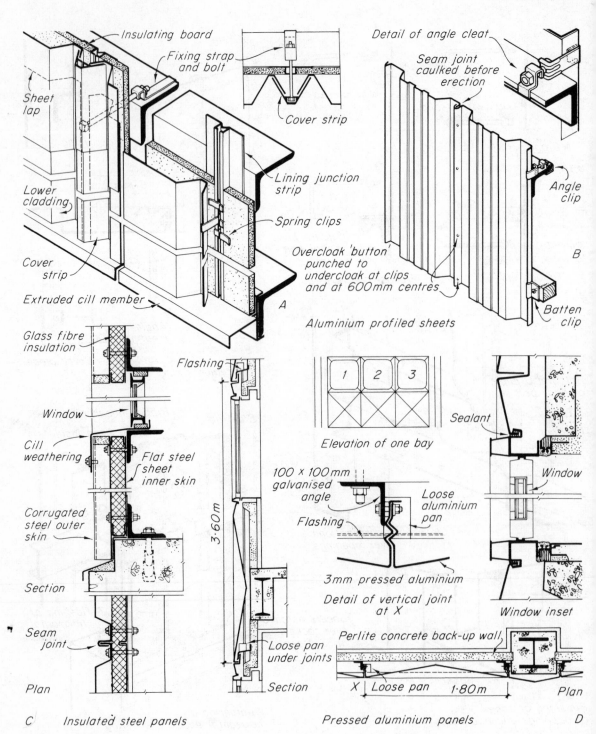

Insulating board

Fixing strap and bolt

Sheet lap

Lower cladding

Cover strip

Extruded cill member

Cover strip

Lining junction strip

Spring clips

A

Detail of angle cleat

Seam joint caulked before erection

Angle clip

Overcloak 'button' punched to undercloak at clips and at 600mm centres

Batten clip

Aluminium profiled sheets

B

Glass fibre insulation

Window

Cill weathering

Corrugated steel outer skin

Flashing

Flat steel sheet inner skin

Section

Seam joint

Plan

C   Insulated steel panels

3·60 m

Loose pan under joints

Section

Elevation of one bay

100 × 100 mm galvanised angle

Flashing

Loose aluminium pan

3mm pressed aluminium

Detail of vertical joint at X

Perlite concrete back-up wall

X   Loose pan   1·80m

Pressed aluminium panels

Sealant

Window

Window inset

Plan

D

Pressed aluminium panels

99   Metal claddings

Metal

The metals used for facings may be employed in the form of sheet claddings. Generally, aluminium and steel are used for this purpose. Metal cladding sheets are profiled to give the necessary stiffness to span between fixing rails and may be fixed directly to these members or be lined with an insulating material as shown in figure 99 *B, A*. Alternatively, panels may be fabricated with internal and external metal surfaces and an insulating material sandwiched between them (*C*). Methods of attachment vary from simple fixings such as hook bolts and screws for single thicknesses of sheeting to secret fixings based on concealed clips (*B*) or specially designed cover strips which accommodate square bolt heads in concealed slots (*A*).

Large pressed aluminium cladding panels can be made in storey height units, suitably profiled to obtain rigidity and prevent vibration, with integral window units (*D*). Interlocking vertical joints allow for thermal movement and resist water penetration. Any water penetrating the vertical jointing is drained downwards in the cavities formed within the joint, trapped in catch pans at the base and conducted to the face of the cladding through the horizontal jointing.

Plastics

Cladding panels may be fabricated from GRP, that is glass fibre reinforced plastics, the resin used generally being polyester (see chapter 13, *MBC: Materials*). They can be made in large areas and one or two-storeys in height. GRP is light in weight and easily mouldable into simple or complicated shapes by 'laying up' by hand or, for mass production, by mechanical spray. If made without filler or pigment the material is translucent. Fillers produce an opaque material and by the use of pigments a wide colour range is possible. In view of the low elastic modulus of GRP and the thinness of the material used units must be suitably profiled to curved or bent shapes or be made of sandwich or stressed skin construction with a core of expanded or foamed plastic in order to obtain adequate rigidity.

Panels of single skin construction are curved or bent and generally stiffened round the edges by return flanges (figure 100). Additional bracing in the form of steel or timber cross members, or ribs formed in the plastic, is sometimes required to stiffen the panel as a whole, especially if it is wide and shallow in profile. Long narrow units are relatively stiffer than wide ones because of the reduction in effective span, but they result in a greater number of joints to be formed during assembly and greater risk of failure at these points. Where necessary for structural reasons flanges to units and fixing points may have metal stiffening members incorporated during laying-up (figure 100).

When sandwich or stressed skin construction is used curved or bent shapes are not essential because of the inherent stiffness of this form. The foamed plastic forming the core provides good thermal insulating qualities to the panels.

Where window units are incorporated in the panels they are commonly retained in position by filler strip type structural gaskets (figure 100. See also Part 1, figure 5 *A*).

Panels may be attached directly to the structure by bolting through metal lugs bonded into the panel or by bolting through the flanges to lugs or brackets fixed to the structure as in figure 100. Fixing may be at top and bottom only as shown in which case the adjacent flanges are bolted together, the joint being sealed with mastic or with a gasket of foam strip or of neoprene. An alternative method of fixing is by bolting the flanges to a framework of timber or of light steel angles or tees.

When panesl are wide stress build-up due to thermal movement can be avoided by fixing one edge rigidly, the other being only retained in position by fixings.[1]

Curtain walling

The term is here taken to mean a system of cladding comprising a frame or grid of members fixed to the face of a structure, usually at each floor level, and an infilling of panels, glazed or solid, as may be required to perform both the functions of window and wall.

The curtain wall must fulfil the same functional requirements as any other system of external walling and it will be considered here under these headings. The main problem in the design of curtain walls lies in the framework which holds the panels

[1] See also Building Research Establishment Digest 161: *Reinforced plastics cladding panels* and a series of papers on glass fibre reinforced plastics in buildings in the *Architects Journal* March 21, April 4 and May 2, 1973 for further information on plastic claddings generally

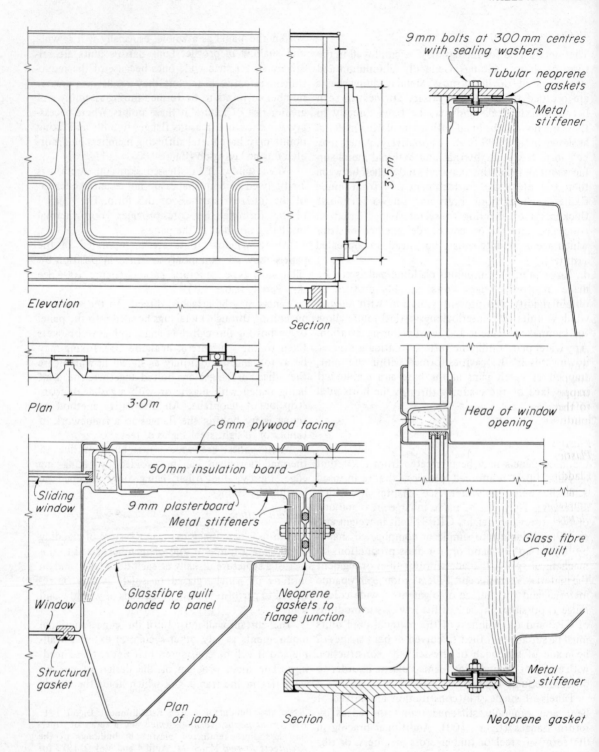

Elevation

Section

3·5 m

Plan

3·0 m

9mm bolts at 300mm centres
with sealing washers

Tubular neoprene
gaskets

Metal
stiffener

8mm plywood facing

50mm insulation board

9mm plasterboard
Metal stiffeners

Head of window
opening

Sliding
window

Glass fibre
quilt

Glassfibre quilt
bonded to panel

Neoprene
gaskets to
flange junction

Window

Structural
gasket

Plan
of jamb

Section

Metal
stiffener

Neoprene gasket

100   *Plastic claddings*

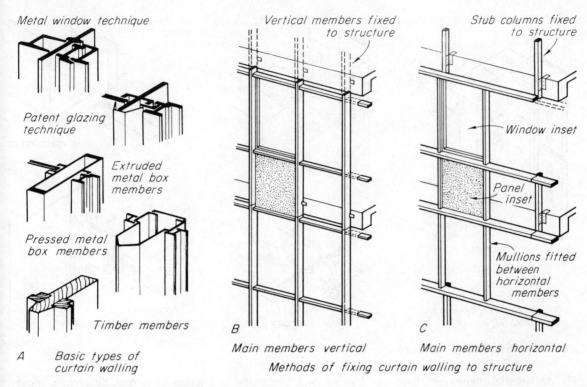

Metal window technique

Patent glazing technique

Extruded metal box members

Pressed metal box members

Timber members

A   Basic types of curtain walling

Vertical members fixed to structure

B   Main members vertical

Stub columns fixed to structure

Window inset

Panel inset

Mullions fitted between horizontal members

C   Main members horizontal

Methods of fixing curtain walling to structure

101   Curtain walling

and this is normally of metal or timber. Three methods are adopted in the construction of metal curtain walling based on (i) the patent glazing principle, (ii) metal window fabrication technique, (iii) extruded or pressed box mullions and transoms (figure 101 *A*). Patent glazing, with its carefully arranged drainage channels and weep holes, and the metal window technique of 'factory cladding' were both used long before the term 'curtain wall' was introduced.

In the majority of proprietary systems of curtain walling, the main members are vertical and are fixed to and span between the floor slabs or beams. Horizontal members, heads, cills and transoms, are fitted between these verticals (*B*). A method of fixing to the structure in which the principal members are horizontal, thus permitting the employment of normal weathered jointing technique, uses 'stub' columns to support the framing (*C*). These form a projection on the inside of the building unless absorbed in the cavity between the outer panel and a back-up wall, should this be required by fire regulations.

Strength and stability   Curtain walling, like other claddings, carries only its own weight between supports, but it must be capable of resisting wind forces and of transmitting them to the structure. Wind loads increase in severity as the height above ground and the degree of exposure increase and the members of the framing and their fixings must be designed accordingly. Fixings should be of stainless steel or non-ferrous metal (see footnote on page 150 regarding the requirements of the London By-laws) and so designed that two-thirds of the fixings employed are capable of resisting the wind forces on the walling. This provides a margin of safety in the event of the failure of one fixing and prevents progressive failure of a number of fixings.

Thermal movement of all parts will occur. In order to maintain dimensional stability this movement must be limited or be allowed to take place freely. Differential movement is likely to occur between (i) the framing and the structure, (ii) the vertical and horizontal members of the framing, (iii) the framing and the infilling panels, (iv) the inner and outer surfaces of composite panels.

169

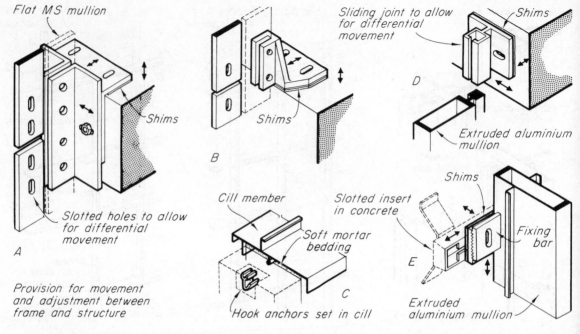

Flat MS mullion

Shims

Slotted holes to allow for differential movement

A

Provision for movement and adjustment between frame and structure

B

Shims

Cill member

Soft mortar bedding

Hook anchors set in cill

C

Sliding joint to allow for differential movement

Shims

D

Extruded aluminium mullion

Slotted insert in concrete

Shims

Fixing bar

E

Extruded aluminium mullion

102   *Curtain walling – movement*

Differential movement between framing and structure is due to the fact that the curtain walling is more exposed to varying external influences than the structure which it protects. The latter, in addition', may achieve a fairly constant temperature due to its higher thermal capacity and to internal heating and air conditioning. Movement between the framing and the structure and between the members of the framing system is provided for in the detailed design of joints and fixing devices.

Methods of attachment designed to allow for differential movement of the structure and framing are shown in figure 102 *A* to *D*. To prevent the accumulation of thermal movement[1] in mullions extending over several storeys, each section is jointed to the next in a manner which allows each to move without influencing the other (figure 102 *A* and figure 103 *A,C*).

Fixing devices must, in addition, be capable of adjustment in any direction to provide for inaccuracies in the structural surfaces to which the framing is attached. Bolt holes should be slotted and packing pieces or shims used to provide for movement and adjustment. Plastic washers should be interposed between adjacent surfaces to allow adequate tension in the bolts combined with sufficient

reduction in friction to permit differential movement (figure 102 *A, B, D, E*).

Movement between members is allowed for by sufficient tolerance in the joints or by the use of split members. Details of typical spigotted joints with space for expansion movement between members are shown in figure 103 *A* and *C*. Details of typical split member construction are shown at (*B*) and (*C*). In (*B*) the mullions only are split to allow for horizontal movement. (*C*) shows two examples of the split mullion and transom technique which permits erection of the walling in large pre-assembled units.

Differential movement between infilling panels and framing is accommodated by sufficient tolerance to permit free movement at the edges of the panels, which necessitates either flexible weatherproofing at the joints or drainage of the joints. Glazed panels must be detailed with care. Contraction of the frame rather than expansion of the glass is likely to be the cause of breakage. Clear glass should have an edge clearance of 3 mm for widths up to 760 mm and 5 mm for widths over 760 mm. Opaque and heat absorbing glass should have 6 mm clearance, irrespective of its dimensions. Only two setting blocks

[1] See table 14, page 235

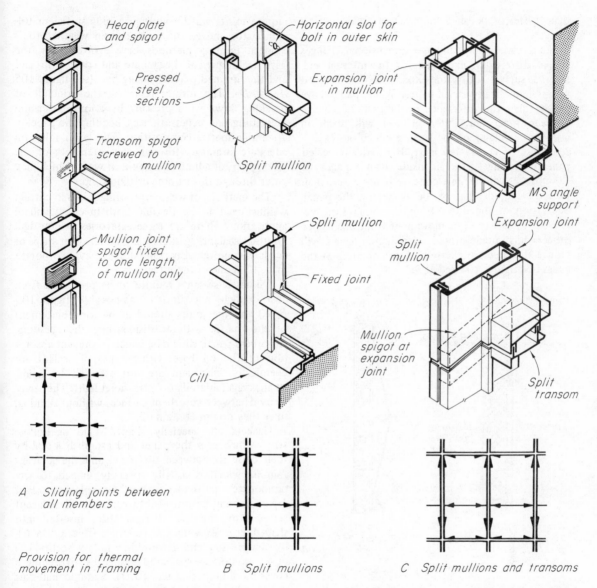

Head plate and spigot

Pressed steel sections

Transom spigot screwed to mullion

Split mullion

Horizontal slot for bolt in outer skin

Expansion joint in mullion

Mullion joint spigot fixed to one length of mullion only

Split mullion

Fixed joint

MS angle support

Expansion joint

Split mullion

Split mullion

Cill

Mullion spigot at expansion joint

Split transom

A Sliding joints between all members

Provision for thermal movement in framing

B Split mullions

C Split mullions and transoms

103 Curtain walling – movement

(lead or hardwood) should be used to obtain the correct bottom clearance and up to 3 mm face clearance should be allowed between glass and frame and glass and bead.

Internal stresses may be set up by differential expansion due to differences in temperature at the edges and centre of a glass panel (particularly in coloured, opaque or heat-resisting glass) caused by shading of the edges by frame or beads. To prevent this it is recommended that the maximum edge cover should be not more than 10 mm. In this connection it should be noted that edges of the glass must be smooth and free from shelling, chipping or grazing. This is a particular problem with wired glass, which should not be used in coloured form and only in transparent or translucent forms if well ventilated at the rear to assist in cooling (see figure 106 *E*). The temperature difference between the edges and centre of a glass panel will be less if the frame is dark in colour since this will attain higher temperatures

171

than lighter or polished materials of greater reflectivity.

Solid panels may undergo dimensional changes due to differential expansion of the internal and external surfaces or by the expansion of air trapped within hermetically sealed panels. Where panels have sealed edges the difference in dimensional change between inner and outer surfaces will result in warping or bulging of the surface which has expanded most or contracted least. Hermetically sealed panels will bulge if the air inside them expands, or the surfaces will become concave if the external air pressure exceeds that of the air within the panel. Separation of the outer from the inner skin will prevent transmission of movement from one to the other whilst predeformation of metal panels will reduce the pressure exerted·on the framing at the edges (see figures 104 and 105*F*).

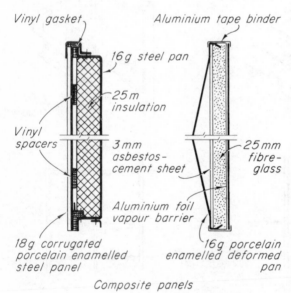

Vinyl gasket

Aluminium tape binder

16 g steel pan

25 m insulation

Vinyl spacers

3 mm asbestos-cement sheet

25 mm fibre-glass

Aluminium foil vapour barrier

18 g corrugated porcelain enamelled steel panel

16 g porcelain enamelled deformed pan

Composite panels

104   Curtain walling – composite panels

Weather resistance The problem of providing weather tightness in curtain walling is aggravated by the use of impervious infilling materials such as glass, metal or plastic. Unlike natural materials such as stone or brick, which are partially absorbent, these materials are unable to take up moisture. Under wind pressure a large volume of water running down the face of the wall will, therefore, attempt to enter through the joints, since these are the only potential points of entry. Such impervious facings, therefore,

require joints which, while providing sufficient tolerance for movement, still remain weatherproof. Modern jointing methods have evolved from the basic techniques of the 'rebate and cover strip' and and the 'drained patent glazing bar' (see figure 105 *A, B, D*). The former relies on the principle of breaking down wind pressure by a joint cover and preventing rain penetration by bedding the panel in a suitable mastic composition. The latter relies on adequate drainage, through the interstices of the glazing bar, of a limited amount of water which may enter through the capping or fixing clip.

The methods of weatherproofing joints in curtain walling need to be flexible both in effect and in application. There are three categories of non-rigid jointing: sealants, gaskets[1] and metal cover strips or beads. A comprehensive joint design may incorporate more than one of these features.

Oil-based sealants require to be protected from dust, sunlight and air as far as possible (figure 105 *B, C*) and the joints should allow for replacement of the mastic without dismantling the cladding. Absorbent panels should be sealed to prevent absorption of the oil base. Other types of sealant are chemically inert and are not affected by acids, alkalis, vegetable oils or ultra-violet light. They may be used between absorbent surfaces without staining, since they do not contain oil.

Gaskets are specially shaped solid or hollow strips whcih grip the panel and establish a seal by their close fit between the panel and the adjacent framing which is usually specially shaped to accommodate a particular section. As explained in Part 1 gaskets may be forced into contact with adjacent surfaces by the use of filler strips inserted into shaped grooves with special tools (figure 105 *E*), by compressing and deforming a hollow section (*F*) or by compressing a solid gasket by tightening a cover moulding against it (*G,H*). Suitable materials are synthetic rubbers such as neoprene and butyl (cured). Plastics such as PVC should only be used where they are protected from direct exposure and weathering or are easily renewable. Where neoprene gaskets are to surround a panel the corners may need to be preformed or shop welded. PVC can be fused together by cutting with a hot knife across the mitre.

In using metal cover strips for jointing the panel

[1] See 'Joints', chapter 2 Part 1 and *MBC: Materials,* chapter 16

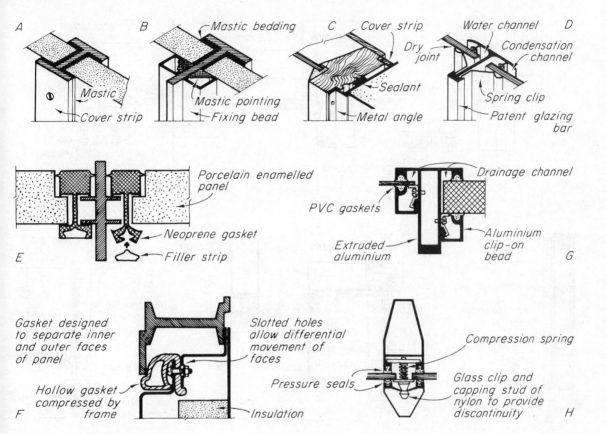

105 *Curtain walling – weather resistance*

is held against a flange by a sprung clip bead or capping (figures 105 *G* and 106 *E*). Any water entering through the joint between the panel and retaining clip is drained downwards through the mullion.

Thermal insulation and condensation The framing system and the infilling panels must achieve a satisfactory level of overall thermal insulation. The thermal insulation may be provided separately from the curtain wall in which case it will be supported on the structure, or it may be incorporated in the curtain wall itself in the panel construction. A number of proprietary panels are available, none of which are much more than 50 mm thick, which provide adequate insulation. The framing system itself, however, permits heat loss, since the members are in direct contact with inside and outside air. This leads to a flow of heat across the members which act as 'cold bridges' and results in condensation on the inner surfaces. This cold bridge effect can be red-

uced by discontinuity (figures 105 *H* and *106 A*), insulation of the inside surfaces (figure 106 *B*) or insulation of the member by external protection. Few proprietary systems of curtain walling in Great Britain incorporate any such refinements.

The problem of interstitial condensation needs particular attention in curtain walling systems which are infilled with well insulated panels having impervious external facing materials. In well heated buildings the internal relative humidity is generally higher than that of the external atmosphere. This results in a vapour pressure differential between inside and outside. Since most insulating materials are porous, it is possible for moisture-laden internal air to penetrate the interstices, where condensation will occur if the temperature gradient through the thickness of the panel is steep enough to fall below the dewpoint temperature of the air in the panel. In homogenous porous materials such as brick, the condensate will evaporate to the atmosphere, but

173

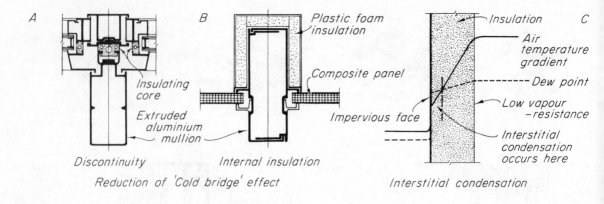

A Insulating core / Extruded aluminium mullion

Discontinuity

B Plastic foam insulation / Composite panel / Impervious face

Internal insulation

Reduction of 'Cold bridge' effect

C Insulation / Air temperature gradient / Dew point / Low vapour –resistance / Interstitial condensation occurs here

Interstitial condensation

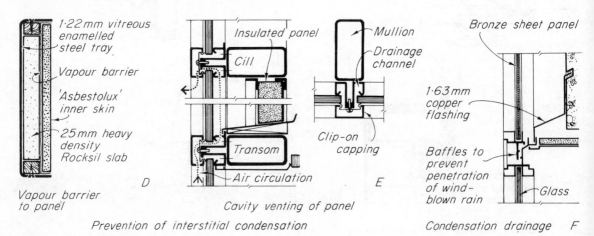

D 1·22 mm vitreous enamelled steel tray / Vapour barrier / 'Asbestolux' inner skin / 25 mm heavy density Rocksil slab

Vapour barrier to panel

E Insulated panel / Cill / Transom / Air circulation / Mullion / Drainage channel / Clip-on capping

Cavity venting of panel

F Bronze sheet panel / 1·63 mm copper flashing / Baffles to prevent penetration of wind-blown rain / Glass

Condensation drainage

Prevention of interstitial condensation

106   *Curtain walling – thermal insulation*

where an impervious external skin prevents evaporation the trapped moisture can do damage to the back of such materials as stove enamelled steel, or painted glass (see figure 106 C). Furthermore, insulating materials become less effective when their conductivity is increased by dampness.

A vapour barrier placed on the room side of an insulated panel will help to reduce the passage of moisture-laden air from the room into the insulation. Such a vapour barrier may be formed of well lapped and sealed aluminium foil, bitumen-coated paper or two coats of aluminium primer finished with two coats of oil paint on plaster. It is, however, difficult to maintain a vapour seal at the joints between panel and framing elements, and some residual vapour will penetrate into the panel unless careful end sealing is achieved or hermetically sealed panels are used (figures 104 and 106 D).

Where insulation is separated from the outer facing, either by use of double skin construction (figure 106 E) or where a back-up wall of insulating material forms part of the construction (figure 237 A), condensation on the inner face of an impervious external panel can be reduced if the cavity is vented as indicated in figure 106 E. Proper drainage should be provided to carry away any condensate which may form. In high buildings particularly, on the faces of which upward wind currents can develop, any drainage openings should be detailed to prevent rain being blown up into the cavity as shown at (F).

Sound insulation   Sound insulation against airborne noise can only be achieved by providing walls of adequate mass or by discontinuity in the construction. The panels generally available for use in curtain wall systems are usually very light and discontinuity is difficult to achieve. The back-up wall required by fire regulations can contribute mass to a panel wall and, if discontinuity can be achieved between this and the external facing element, reasonable sound reduction can be expected.

The problem of structure-borne sounds is increased by the use of a framing system which has elements common to a number of rooms. In theory this can be alleviated by introducing discontinuity at the junctions of framing members and by cushioning the anchorages, although few proprietary systems incorporate such devices.

Fire resistance   Reference is made in chapter 10 to the problem of fire resistance in curtain walling.

Materials of construction   Aluminium is favoured by many manufacturers because of its lightness, moderate cost and ease of forming. Due to the high coefficient of thermal expansion (see table 14) large thermal movements will occur which must be allowed for in detailing. Steel anchorages should be given a coat of protective paint or be isolated by plastic or fibre washers.

Mild steel, either hot rolled or cold formed, is used for framing elements and in sheet form for panels. Hollow sections are usually built up. Despite the disadvantages of potential corrosion, mild steel is cheap and strong, and is widely used for curtain walling construction. Unlike carbon steel, stainless steel is corrosion resistant, although having a higher coefficient of expansion. The material is expensive and narrow strips are cheaper than wider and squarer sheets, so that it is cheaper to build up facings to panels from a number of interlocking strips. The material may be used for framing, or more usually for pressed cover sections to aluminium or mild steel cores.

Bronze is expensive but has excellent weathering qualities. It is used for panels and pressed cover sections to steel cores.

Where building regulations permit, timber may be used for framing and panelling. The material is cheap and has the advantages of being non-corrosive, easily shaped and of low thermal transmittance. Careful detailing is necessary to allow for moisture movement by including sufficient tolerances between panels and framing. Flexibility is achieved by loose tongueing and interlocking sections (figure 107 *D*).

If correctly used glass is an excellent panel material since it is comparatively cheap, weather resistant and corrosion resistant. Coloured wired glass should not be used since it is susceptible to thermal breakage. Transparent or translucent glass placed in front of a coloured surface with a cavity between should be removable to permit dust and condensaiton staining to be cleaned off the rear surface.

Asbestos silica, an asbestos composite is frequently used as an insulating core in composite panel construction. It may be stove enamelled. Asbestos cement has low thermal movement, is non-combustible and is comparatively light and it is cheap. It has, however, a high moisture movement and gets more brittle with age. Specially densified sheets are stronger and may be integrally coloured and polished. A wide range of colours is obtained by applying coloured skins during manufacture, some of which are suitable for external use. Chlorinated rubber paints and silicone paints reduce water absorption and repel surface water.

As stated on page 160, only certain groups of plastics are suitable for external use. Plastic infill panels may be cellular or the material may be used as an external or internal lining bonded to other materials in composite panel construction[1].

Infilling panels

Most materials used in the forms of cladding described in the previous section may be employed in the construction of infilling panels. The technical problems of weather exclusion, thermal and sound insulation, condensation and fire resistance associated with infilling panels are identical with those encountered in curtain walling.

Brick and block infilling panels are considered under 'Panel walls'. Lightweight infilling panels present the same problems of edge jointing as heavy panels but are even more liable to thermal or moisture movement and need careful design at the junction with the structure, particularly at head and jambs.

Metal-framed infilling panels

Complete storey height infillings, sometimes referred to as 'window walls', are constructed in a similar manner to curtain walling systems, the vertical framing elements being so fixed at the base and head as to allow for thermal movement (figure 103 *A*). Methods of providing suitable fixing at jambs to

[1] For a fuller discussion of all aspects of curtain wall design see *Light cladding of buildings* by Michael Rostron 1964 (Architectural Press) from which much information in this section has been drawn

allow for movement are illustrated in figure 107 *A*). Spandrel height panels may be screwed or clipped to metal angle framing fixed to the floor slabs or between columns.

Concrete infilling panels

This term is usually applied to the forms of horizontal cladding which span between columns and which are described on page 164.

Timber-framed infilling panels

It has been common practice for many years to infill between brick walls with timber-framed walling based on traditional stud walling technique, the timber being cut ond assembled on site.

Trends towards prefabrication have brought about the introduction of pre-assembled frames which can be erected as units and finished on site or are already finished before erection. It is common practice to construct the panel with a frame of rebated sections, similar to a large timber window frame with all necessray mullions and transomes, and to infill this with glazing, openable sashes and fixed panels in the manner of curtain wall technique. The junctions of head, sill and jamb with the structure are usually designed to allow sufficient erection tolerances. Typical details of this type of infilling panel are shown in figure 107 *B* and *C.*

Proprietary curtain wall systems provide a further method of timber infill panel walling. These consist of various sections forming the main frame into which are fitted various types of panels. A typical example is shown at D.

In all methods of installing panels or frames (including window frames between structural elements) it is an advantage to form a rebate to reduce the risk of wind blown rain entering the joint directly. This may be provided by shaping the basic structural material such as brick or concrete, or by forming the rebate with a facing element as shown in some of the illustrations in figure 73.

MOVEMENT CONTROL

The nature of movements in buildings generally is referred to in the next chapter on page 234, where references are given to more detailed discussions of the subject elsewhere. Only the practical steps necessary to prevent damage to walls through movement will be discussed at this point.

Settlement movement

The methods used to minimize differential settlement have been described in the previous chapter. In masonry walls, provided the mortar is not too strong, slight movements can be taken up in the joints without damage to the wall and in normal circumstances no special provision is made. Monolithic concrete walls are more sensitive to movement, and when these are not reinforced, particularly with no-fines concrete, reinforced foundations may be required. Tall buildings, may require a cellular raft to provide a foundation of sufficient stiffness.

Moisture movement

Most walling materials absorb water and in doing so they expand; on drying out they contract. This is termed moisture movement, and it is usually reversible except in concrete, mortars and plasters. These have an initial drying shrinkage on first drying out after setting which exceeds any future reversible movement. The magnitude of moisture movement varies with the material and in some it is so great as to necessitate special measures in construction. Table 13 taken, from *Principles of Modern Building,* Volume 1, gives a broad grouping of materials relative to their moisture movement[1].

Special precautions[2] should be taken with walls built of sand-lime or concrete bricks or concrete blocks and these are primarily the use of weak mortars and the provision of vertical movement joints to control cracking. These are referred to under *Blockwork* in chapter 5 of Part 1, where the advantage of a weak mortar is described and the need to break long walls by control joints into panels with lengths not exceeding one-and-a-half times to twice the height is stressed (see figure 108 *A*).

Control joints are essential in long imperforate walls and should be provided at intervals of 6 to 9 m. Such joints may be vertical or may follow the bonding lines in toothed fashion and are formed by raking the mortar in the joint to a depth of 13-19 mm and caulking it with a mastic sealant. Where stability of the panels requires them non-ferrous

[1] For percentage movements in individual materials see Table 5, *MBC: Materials*

[2] See also *Principles of Modern Building,* Volume 1, 3rd Ed (HMSO)

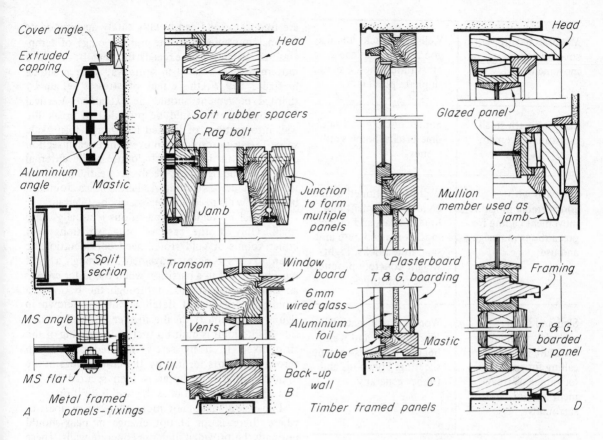

Cover angle
Extruded capping
Aluminium angle
Mastic
Split section
MS angle
MS flat
Metal framed panels – fixings
A

Head
Soft rubber spacers
Rag bolt
Jamb
Junction to form multiple panels
Transom
Vents
Cill
Window board
6 mm wired glass
Aluminium foil
Tube
Back-up wall
B

Plasterboard
T. & G. boarding
Mastic
C

Head
Glazed panel
Mullion member used as jamb
Framing
T. & G. boarded panel
D

Timber framed panels

107   Infilling panels

metal dowels are placed across the joint in every other course with provision made to ensure that sliding can take place as the panels on each side move, such as covering them with greased paper. Alternatively, grooved blocks with a neoprene extrusion may be used (figure 108 *B*).

The stress concentration which occurs in the narrow sections of wall above and below window openings can be distributed through the wall by steel reinforcement in bond beams in the lintel and cill courses (see figure 84, Part 1) or by masonry reinforcement in the bed joints immediately above and below these courses respectively (figure 108 *C*). Such reinforcement must not be carried across any control joints in the walling, nor should any applied finishes on the wall faces. As suggested in Part 1 these narrow sections of wall can be avoided by designing doors and windows within storey-height infill panels as shown in (*D*), the vertical joints on each side constituting control joints between the infill panel

and the imperforate blockwork panels on each side.

Shrinkage in structural concrete is normally restrained by reinforcement (see page 128). When shrinkage or contraction joints are considered desirable in a concrete wall to control random cracking and the joint must be water resistant, as in the case of retaining walls, a PVC water bar or stop is incorporated in a similar manner to that shown in expansion joint (*B*) in figure 140. At a contraction joint, however, the reinforcement would carry across the joint line and no gap would be formed in construction although precautions would be taken to prevent bonding of the concrete on each side of the joint. Those materials with large moisture movements are generally most suited to internal use and some slab and sheet materials in this category are commonly used in the construction of partitions. Methods of providing for movement in these are suggested in table 15, Part 1.

Materials having very small moisture movement	Well-fired bricks and clay goods*. Igneous rocks. Most limestones. Calcium sulphate plasters
Materials with small moisture movement	Some concrete and sand-lime bricks. Some sand-stones
Materials with considerable moisture movement calling for precautions in design and use	Well-proportioned ballast concretes. Cement and lime mortars and render-ings. Some concrete and sand-lime bricks. Light-weight concrete products. Some sandstones
Slab and sheet materials with large moisture movement calling for special technique of treat-ment at joints and surrounds	Wood-cement materials. Fibrous slabs and wall-boards. Asbestos-cement sheeting. Plywoods and timber generally

*Note: Fired clay products may show an initial expansion on wetting that is irreversible. It has been shown in Aust-ralia that walls built with 'kiln-fresh' bricks may expand twice as much as similar walls built with bricks that had been exposed to the weather for a fortnight. A slow moisture expansion, greater than that shown by the standard tests, may also take place over a period of years. Some striking failures due to this effect have been reported from the United States of America (see also page 137)

Table 13   *Degrees of moisture movement*

Thermal movement

The cause and magnitude of thermal movement is described in the next chapter (page 234).

Brick walls up to about 30 m in length which are free to expand can usually safely accommodate thermal movements in the normal range of temp-eratures. When a wall exceeds this length, or where movement is restrained in some way such as shown in figure 139 *F*, the effect of stresses set up by thermal movement should be examined. Vertical expansion joints should be provided to break the wall into sections of limited length and to provide space for an unrestrained movement in each section of 9.5 mm in 12 m. Such joints in the external walls of buildings must be made weatherproof by the use of non-extruding resilient filling. Joints in boundary walls need not necessarily be filled in.

Thermal movement in partitions is usually quite small. Only in the case of long partitions near boiler rooms or cold stores need consideration be given to this type of movement, although a joint must be formed in a partition when the latter crosses an expansion joint carried through the floors of a building. Methods of detailing this are shown in figure 140 *F* and *G*. In the former the wall on one side of the joint is built up first and one wing of the strip fixed to it by screws inserted and driven through the slots formed by the cut-out lugs in the opposite wing. The latter is then secured by the lugs which are bedded in as the wall is built up.

Expansion joints not more than 15 m apart or where there is an abrupt change in plan should normally be provided in dense concrete walls. These may be formed as shown in figure 140 *A*, *B*, or with crimped copper strips (*K*) as in (*H*) and (*J*). Applied finishes may require special treatment as at (*C*, *D*).

Reference is made on page 134 to the use of vertical expansion joints to limit differential thermal movement between the inner and outer leaves of large areas of cavity walling, and methods of dealing with thermal movement in curtain walling are des-cribed on page 169.

Unless the roof slab is arranged to slide on the head of a wall a joint should be provided in the wall where one is provided in the roof slab.

In positions where a sliding joint is required as at (*F*) in figure 139, this can be detailed as shown in figure 140, *E*.

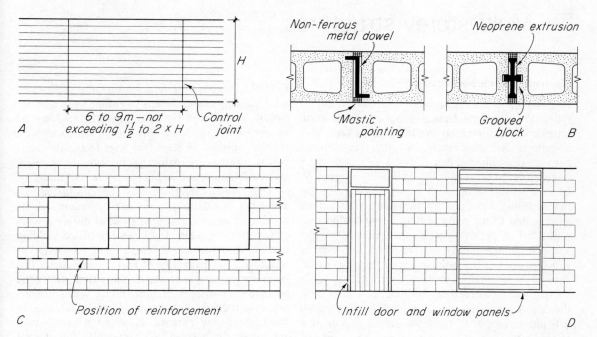

A

6 to 9m – not
exceeding 1½ to 2 × H

Control joint

H

Non-ferrous metal dowel

Mastic pointing

Neoprene extrusion

Grooved block

B

C

Position of reinforcement

Infill door and window panels

D

108   Movement control in concrete masonry

5 Multi-storey structures

The frame and the loadbearing wall

High-rise construction is used for many building types. For some the framed structure is the most suitable form, for others the loadbearing wall. The advantages and disadvantages of these two forms have been described in Part 1.

In a framed structure, the wall being relieved of its load bearing function, it is possible to fulfil the the enclosing functions by forms of construction more suited to the purpose than heavy loadbearing walls, and to provide a structure lighter in weight and often more quickly erected. Such a system is widely used but the point made in Part 1 must be stressed here: that in certain circumstances loadbearing wall construction, fulfilling as it does both the loadbearing and enclosing/dividing functions, will prove cheaper in terms of overall cost than a framed structure. This is particularly so in case of individual small-scale and domestic buildings, although in the case of large programmes involving one basic type of small-scale building, permitting bulk ordering of materials and the use of a prefabricated system to reduce site labour, a framed structure may prove more economical than one based on the loadbearing wall. In the case of taller buildings with suitable plan forms, such as some types of flat blocks, the wall used as a loadbearing element can again produce a cheaper structure than a frame. This has been more fully discussed in terms of brick and plain concrete walls on pages 117 and 132. See also page 222.

The cost of the structural frame of a building is generally a small proportion only of the total building cost and varies on average from 10 to 40 per cent of the total cost.[1] Small economies in the structure will, therefore, usually have a slight effect only on the total cost. Nevertheless, since the cost of the structural frame is often the largest single item in the total cost, it should be kept to a minimum, although it must always be borne in mind that the cheapest structure may not necessarily produce the cheapest building. For example, reduction in the cost of the structure at the expense of increased depth of beams will lead to an increase in the cost of finishings, particularly external claddings, due to the increased overall height of the building. As a further

example, in some proprietary forms of prefabricated steel building frames beams of rather wide span are closely spaced and the resulting light load carried by the wide span beams produces beams which are slightly wasteful in steel (see page 182). The extra cost is, however, more than balanced by the economies of standardisation and economy in roof cladding.

The frame, therefore, must be considered not alone but in relation to the other elements of the building. The comparative analysis of different ways of constructing a simple two-storey office building made by Stillman and Eastwick-Field, referred to in footnote 2 on page 139 of Part 1, emphasizes this point very clearly and at the same time shows the economic advantages of loadbearing wall construction over framed construction for small buildings. In the first scheme external loadbearing cavity walls are used, 'punctured' at intervals for normal windows; in an alternative scheme a structural frame is used with solid and glazed infilling panels. Although the cost of the frame in the alternative scheme is shown as little more than half that of the loadbearing walls, since the frame cannot enclose space the cost of the infilling panels must be added to this to make it comparable in functional terms with the wall construction, so that the total cost of the frame and infilling panels is shown to be rather more than one-and-a-half times the total cost of the walls together with the windows.

The high cost of the glazed infilling panels in this particular example could be reduced by replacing them with cavity wall panel construction, but in a small building such as this the final result would be quite illogical since the cavity wall itself is capable of carrying the floor and roof loads without the frame. The justification of a framed structure in circumstances such as these rests on the advantages given by a frame in other directions, such as the large areas of glazing which are considered necessary in some building types.

[1] The higher proportions would be for buildings in which the structure predominates, in which services are few, little internal sub-division is required and the area of enclosing element to floor area is small, as in warehouses, or may not exist, as in car parks.

CHOICE OF APPROPRIATE STRUCTURE

Multi-storey wall construction may be carried out in masonry or in plain or no-fines concrete and the use of these is described in chapter 4. The use of reinforced concrete for this type of construction is described later in this chapter. For multi-storey framed structures both steel and reinforced concrete are used. Although a structural frame must be considered in relation to the other elements of the building it is, nevertheless, necessary to consider it separately in broad terms at the outset of a scheme, as in this chapter, in order to establish a basis from which to begin.

There is no simple formula by means of which the most appropriate structure for a particular building may be selected and because there are so many variable factors affecting choice and economy these can be discussed only in general terms.

Site

The available site, its cost, the nature of the subsoil and planning restrictions on the height of buildings in the particular area must all be considered. When the cost of the land is low and a large area is available, the automatic choice of a multi-storey structure is illogical. Apart from increased construction costs, such things as the extra cost of external walling due to the increased ratio of external wall to floor area in a tall building, with its repercussions on the heating of the building in terms of greater heat loss[1], make the serious consideration of a low structure essential. But where the cost of land is high, as in central city areas, a multi-storey structure must invariably be used in order to reduce to a minimum the required area of site to economize on its cost.

The height of a tall structure may be limited by economics or by height regulations. Construction costs increase with building height and additional storeys lead to increased column, foundation and wind-bracing costs, so that for each particular job there is an economical height limit at which the structure may be stopped, even though it may be lower than that permitted by regulations. The economical height may be determined by assessing the return on investment, this being calculated with regard to the area of the building, land costs, the number of floors and the cost per unit volume or area, as well as many other factors. This, of course, necessitates the use of some form of cost analysis

and cost planning. Above the economical limit there are diminishing returns on the money invested in the building. If the maximum height should be limited by height regulations rather than by the cost of construction, it may be necessary to place the maximum number of storeys within the permitted height and this will have a profound influence upon the choice of a type of floor to give the thinnest possible construction. In commercial buildings the rental value of any extra floors gained in this way must be capitalized before comparing the cost of a selected floor with that of alternative floor systems.

The type of soil on which the structure is to rest will influence its design, above as well as below ground. For example, in the case of soil with a low bearing capacity and high compressibility, on which differential settlement might occur, a low structure imposing minimum point loads on the soil through simple foundations would be most suitable, but if the location of the site made it necessary to build high over a small area, expensive foundations such as piles or caissons taken to a firm stratum might be required. In this case it might prove cheaper to increase the span of beams in the frame in order to restrict the number of columns and thus the number of expensive foundations. The relationship between soil, foundations and superstructure is considered in more detail in chapter 3.

Type and use of building

An analysis of the problem on the basis of the site considerations – a balancing of costs and use requirements as they arise from site conditions – will usually give in broad terms an indication of the most suitable kind of structure, and this will lead to a final choice from a few alternative systems within the broad lines suggested. This choice will be influenced by the nature and use of the building and the architect's conception of the finished building.

Plan forms requiring a repeating unit both horizontally and vertically, and in which room widths may be multiples of the structural unit, lead logically to the use of a simple skeleton frame. The intensity and nature of the superimposed loading must be considered so that a suitable type of floor structure can be selected and the most economic span chosen for the particular floor type. When a square structural grid can be adopted, or at least one in which the

[1] See Part 1, page 46.

larger side of each bay does not exceed one-and-a-third times the length of the shorter side, the use of a two-way reinforced floor slab with four-edge support becomes economically possible, as it can result in a thinner and lighter slab when load and span conditions are suitable. Certain building types, such as flats, may be more cheaply constructed in load-bearing wall construction than as a frame. The standard of natural lighting required from outer walls will influence the height from floor to ceiling according to the depth of the rooms on plan.

These factors will involve the consideration of column centres, wall spacings, storey heights and infillings and claddings and of the implications of the provision of mechanical services, especially heating and air-conditioning with the ducts involved. Should the solution suggested by these considerations be at variance with that suggested by site considerations, then a review of both must be made in order to arrive at a final satisfactory solution.

Span and spacing of beams

Span　The column spacing in a framed structure, which determines the span of the beams, is perhaps the most important factor influencing the cost of the structural framework. Generally, the cheapest frame will be that with the closest column spacing and, therefore, the lowest dead weight[1]. Apart from frames with exceptionally tall columns, that is in those of normal storey height, a column costs less than a beam. Although in practice the selection of the most economic structure may not always be possible, because of the space requirements of the building, it is desirable to bear in mind this economic factor as a 'yardstick' by means of which can be assessed the cost consequences of adopting a particular solution.

Spacing　The spacing of the beams is dependent upon a number of factors:

    (i)　The economic span of floor slabs
    (ii)　The economic loading of beams
    (iii)　The economic loading of columns

In the case of a building not requiring wide span beams the economic span of the floor slab will be a decisive factor in fixing the spacing of the main floor beams. For multi-storey buildings such as flats, offices and other types with light and medium loading this will be from 3 to 4.5 m, to give a floor slab in the region of 125 to 150 mm thick. Greater spacing will necessitate the use of a deeper floor of

tee, hollow beam or other type of construction. A detailed consideration of floor construction will be found in chapter 6.

In establishing the spacing of beams in a structural frame another important influence, apart from the requirements of planning, is the relationship between load and span of beams. Since both the bending moment and, especially, the deflection of a uniformly loaded beam are more rapidly increased by an increase in span than by an increase in the load per metre run, it is generally uneconomical to carry light loads over long spans. This can be seen from the expressions for bending moment and deflection:

$$\text{BM} = \frac{wl^2}{8} \quad \text{and} \quad d = \text{constant } (c) \times \frac{wl^4}{\text{EI}}$$

The bending moment increases with the square of the span l, and the deflection with fourth power, but both only directly with the load per metre run, w. If the span of a beam is doubled the deflection will be sixteen times as great but if the load per metre run is doubled the deflection will be only twice as great. In other words, an increase in span necessitates a much greater increase in the 'I' value of the beam in order to keep deflection within acceptable limits, compared with that necessitated by a proportionate increase in the load per metre run. This being so, when beam spans must be large it will usually be cheaper to space the beams further apart, particularly when floor loading is light or medium, even though the load per metre run on the beams is thereby increased. The savings in cost due the smaller number of beams required will be greater than the increase in cost due to the wider floor slabs and the increase in size of the remaining beams and columns due to the greater load they carry.

Heavy loading on wide span beams also has economic repercussions on the columns of a frame. Variations in loading cause only relatively small variations in column weights since the decisive factor in most columns is stiffness against buckling, so that where beam spans are great an increase of the unit load on the beams due to spacing them at wider intervals has the effect of imposing greater loads on fewer columns without necessitating a great increase in their weight. This is particularly beneficial where the column heights are great and the greater tendency to buckle results in large

---

[1]　Although weight is not the only factor in minimising cost: see page 187.

column sizes.

It will be appreciated from what has already been said that for a given span of floor slab, which will establish the load per metre run on the supporting beams, there will be an economic span for the beams. For types of buildings with light to medium floor loads and normal concrete floors used at their economic span this will be from 4.25 to 6 m.

When the column centres have been tentatively established, studies of typical bays are made for comparative purposes in terms of construction and costs. An analysis will be made of the various possible framings in structural steel or reinforced concrete or in both, some of which will prove to be more economic than the others. Wherever possible the frame layout should be based on a regular 'grid', a term derived from the pattern of beams at each floor level, this pattern being determined by the set out of the columns of the frame. The advantages of a regular grid are given in Part 1, page 141.

The imposed loads on building frames are transmitted through successive units or stages to the foundations, that is, for example, from floor slab to beam and from beam to column, and the total cost of the structure is made up of the cost of each successive stage. Generally, to achieve maximum economy the number of stages through which the load is transferred should be kept to a minimum, although since they are interdependent and modifications to one will affect the others, the introduction of an extra stage may in some circumstances be justified. For example, the introduction of secondary beams may result in a saving of so much dead weight in the floor slabs as to make it economically advantageous to do so.

As a general principle loads should not be transferred horizontally if it is possible to carry them axially downwards from the point of application. Thus setbacks on section resulting in columns bearing on beams should be avoided where possible, as this necessitates the use of heavy or deep beams and increases the cost of the structural frame, as does the introduction of wide span beams on lower floors, carrying a number of columns in a closely spaced grid above (see figure 111 *B, D*).

Wind pressure

Another factor to be considered in the selection of the most appropriate structure is the need for the structure to be able to resist wind pressure. This is a most important consideration in the case of very tall buildings as far as both foundations and superstructure are concerned. The recommendations of CP 3 (chapter V) concerning wind pressure on buildings as a whole and upon claddings are described in chapter 4. They apply to framed structures as well as to loadbearing wall structures.

The means adopted in wall structures to provide stiffness against wind forces are described in the section on reinforced concrete wall construction on page 222 and in the case of masonry structures, on pages 117 and 132.

A structural frame may be considered as acting under pressure of wind either as a cantilever in which the floors remain plane or as a portal frame in which the floors bend. In analysing the structure in terms of wind resistance the height-to-width ratio is extremely important in deciding in which of these ways the structure is likely to react. A very tall, slender building will act more nearly as a cantilever while a short, squat structure will act as a portal frame. In practice most buildings act partially in both ways.

In whichever way a particular structure may behave it will move under wind pressure, in the same way that a cantilever deflects under load, and in the case of tall buildings particularly, where the movement could be large, this must be limited if it is not to result in cracks in finishes and discomfort to occupants of upper floors. In heavy structures with heavy, stiff claddings, wind forces are not likely to produce much lateral deflection, but in framed structures with light curtain walls the basic structure required by dead and imposed loads may have to be stiffened in order to limit lateral movement, or side-sway. A maximum permissible deflection at the top may be fixed and the bending moments produced by this may be calculated and the structure designed accordingly. In the United States a maximum deflection of 0.002 times the height of the building is used as a limit for buildings having a rigid outer skin, and 0.001 times the height for buildings with curtain walls which provide little rigidity to assist in reducing the deflection. The necessary stiffness or wind-bracing can be obtained in a number of ways, used either separately or in combination:

(a) by the use of deep beams or girders producing very stiff joints with the columns (steel and reinforced concrete)

(b) by the use of brackets or gussets to produce stiff beam to column joints (steel)

(c) by the use of diagonal bracing in vertical panels of the frame (steel and precast concrete)

(d) by constructing solid walls called shear walls in suitable positions or by using stair wells or lift shafts, if these are constructed in monolithic reinforced concrete, running the full height of the building so that they act as stiff, vertical members to which wind loading is transmitted by the floors. Shear walls may be introduced in panels of the frame to fulfil the same stiffening function as diagonal braces (steel and reinforced concrete).

The method of windbracing to be adopted must be considered at the beginning of the design stage because it will have a bearing upon floor thicknesses and floor types, beam depths and column sizes and the introduction or not of stiffening walls.

In multi-storey buildings any tendency to uplift at the foundations will usually be more than counterbalanced by the total weight of the building, which will keep the resultant pressure within the middle third of the base. In extreme cases, however, when buildings are very tall and slender, and especially if they are of considerable length, it may be necessary to resist uplift by means of tension piles or ground anchors incorporated in the foundations.

FRAMED STRUCTURES

Choice of material

The materials most suitable for the construction of multi-storey structural frames are steel and reinforced concrete. Timber, apart from its use in domestic type buildings up to about three storeys high[1], and the aluminium alloys, by reason of their particular characteristics, are not suitable for this purpose. A comparison of the characteristics of these four structural materials is given in chapter 9.

The principal factors influencing the choice between steel and concrete are

 (i) the availability of materials and labour
 (ii) cost
 (iii) speed of erection
 (iv) possibility or otherwise of standardizing the sizes of the structural members
 (v) size and nature of site
 (vi) fire resistance required

With these in mind the two materials may be compared in order better to understand the consequences of the choice of one or the other.

Site considerations

Both steel and cement are made under factory conditions and both are subject to British Standards. The strength of steel is controlled and established during manufacture, but that of concrete is often dependent upon strict supervision on a building site. On a large contract effective supervision and the use of carefully controlled batching plant is likely to lead to better quality concrete than on a small contract. Doubt regarding the possibility of obtaining first class supervision for a particular job might lead to the choice of steel.

The fabrication of the individual members of a steel frame is carried out 'off-site' in the steel contractor's shop and erection only is necessary on the site itself. The erection of the steel is carried out by skilled labour, quickly and accurately within small tolerances. Completion at an early stage in the building programme permits the laying of floors and roofs and the rapid sealing-in of the whole structure to permit other trades to follow on continuously without interruption through adverse weather. The fixing of lower floors can, of course, be commenced before the frame has been completed up to roof level and while the upper members are being fixed in position. The obstruction of floor space by the vertical props required for *in situ* cast floors may be avoided by the use of telescopic shutter supports off the beams or by the use of precast concrete or steel deck floors.

In *in situ* cast concrete structures all work, except possibly the cutting and bending of reinforcement, is carried out on the site, the erection of formwork and placing of reinforcement by skilled and unskilled labour and usually the making and placing of concrete by semi-skilled or unskilled labour. The proportion of site work to prefabricated work is greater in the case of *in situ* concrete and building operations are complicated by formwork and centering, which has to be in position for a considerable time to allow the concrete to strengthen before the frame can take up loads and subsequent lifts be commenced. Much site space is needed for the storage and mixing of materials, although in

[1] See Part 1, chapter 6.

184

some districts this problem can be overcome when the site is restricted in area by the use of 'truck-mixed' concrete. The type of formwork used and the nature and size of the building will, of course, have an effect upon the actual method of work and the speed at which it progresses.

Cost and speed of erection

With a repetitive skeleton frame and the employment of a suitable contractor, reinforced concrete is likely to be cheaper in cost[1] but, if *in situ,* generally longer in construction than steel, although construction can be rapid with careful and detailed planning of the site work, good site organization and properly trained operatives and good supervision.

To some extent cost and speed of erection are dependent upon the availability of materials and labour. When materials are scarce prices rise and delivery periods become extended. When they are abundant prices and delivery periods fall.

On expensive sites speed of erection to permit early occupation of the completed building is a most important consideration and is sometimes the decisive factor in the choice of the framing material.

The use of precast concrete framing members for multi-storey structures helps to overcome some of the disadvantages of *in situ* work and reference should be made to page 207 where this is discussed.

Design considerations

In situ reinforced concrete often allows greater flexibility in design because of its monolithic nature and because it is not confined to standard sections. There is, however, an economic limit set to this flexibility by the increased costs of formwork for excessive changes of section and, generally, reinforced concrete can be used most economically when the structure is designed to permit the maximum re-use of a minimum amount of formwork. This fact is not only of vital importance to the structural engineer but also to the architect designing in reinforced concrete, for it imposes on him a strict discipline in terms of form and detail of which he must be aware during the design period of any project.

When the plan for some reason or other must be irregular, reinforced concrete may prove to be cheaper in cost and quicker in construction since a large number of complicated connections might be necessary in a steel frame.

Sometimes it is desirable to extend a framed building. When the frame is of steel the problem is relatively simple, the new members being bolted to the existing after they have been exposed by the removal of any fire-resisting casing. With reinforced concrete it is necessary to expose sufficient of the reinforcement in the existing members to obtain a satisfactory bond with the new work or, alternatively, sufficient to permit the welding on of extension bars to provide the necessary bond.

In choosing between steel and concrete it must be borne in mind that each material should be used in the way which best exploits its particular potentialities. Solutions which are favourable to steel are not necessarily so to concrete, and if comparisons are to be realistic it is important that they should be made on the basis of the cost of a building as a whole based on the structural solution most favourable to each material.

Fire resistance

Reinforced concrete itself, with normal cover to the reinforcement, provides a substantial degree of fire resistance which can be varied by variations in cover thickness and other means. Steel, because of its behaviour under the action of fire (see page 371), requires a protective cover varying in nature and thickness according to the standard of fire resistance required. The material commonly used for this purpose in the past and still very widely used at present is concrete, poured round the steel members within a timber formwork. Although support for this formwork can be obtained from the structural frame itself the process slows down the progress of work. Alternatives are sprayed asbestos and hollow protection in the form of various combinations of renderings or plaster on metal lathing or in the form of asbestos board, by means of which fire protection periods of up to four hours can be attained. These methods have the advantage of reducing the weight of the casing and speeding up the progress of work, but are not all necessarily cheaper than, nor even as cheap as the concrete casing. An alternative to protective casings is to cool the steel by water in the event of

---

[1] Although the introduction of high yield-point steel, hollow fire-protection techniques and automatic machines in fabricating shops have made it possible for steel, in favourable circumstances, to provide the cheaper method.

fire by techniques which have now been developed (see chapter 10).

THE STEEL FRAME

The majority of multi-storey steel frames erected up to the present time have been, and still are, designed and constructed on the same general principles as the first multi-storey frames erected over sixty years ago. They are designed on the assumption that the beams are simply supported at their junctions with stanchions. Certain definite advances in many directions have been made as a result of research carried out in the period between the two world wars and subsequent to the last, which lead to a more rational design approach and to economies in the use of steel. Before discussing these it is proposed as a basis to examine in some detail what might be called the 'classic' steel frame still largely adopted.

As shown in the small-scale examples in chapter 6, Part 1, the frame consists of horizontal beams in both directions and vertical columns called stanchions, usually all of standard rolled sections of various sizes, joined together by welding and bolts with cleats, to form a stiff, stable skeleton capable

of transmitting its own dead weight and all other loads to the foundations, to which the feet of the stanchions are bolted. All the steel members are commonly encased with concrete or other suitable material to which reference has already been made, the thickness depending on the degree of fire resistance required. The beams carry one of the many types of *in situ,* precast concrete or steel floors described in chapter 6.

Basic elements of construction

The multi-storey steel frame is built up of hot-rolled mild steel sections[1] standardized in shape and dimensions. These are described and illustrated in Part 1, page 142. There are beam, channel, angle and tee sections and of these the ·beam section is the basis of the frame for both beams and stanchions. Where necessary these can be strengthened by the addition of plates welded to the flanges although this is rarely necessary since both Universal beam and column sections are produced in related 'families' of different weights, the use of which eliminates a great deal of plating and compounding for heavily loaded beams, and

[1] See page 203 for a description of cold-formed steel sections and their uses.

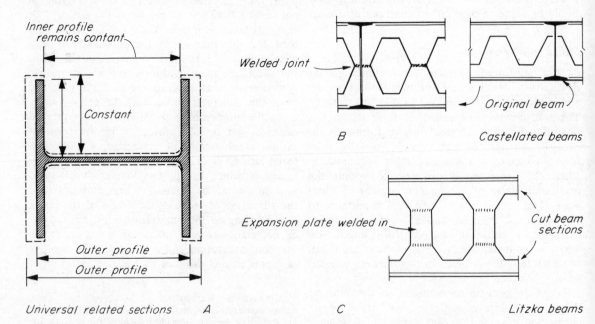

Inner profile remains contant

Constant

Outer profile
Outer profile

Universal related sections   A

Welded joint

Original beam

B   Castellated beams

Expansion plate welded in

Cut beam sections

C   Litzka beams

109 Structural steel sections and built-up beams

the need for flange and seating plates to stanchions (see figure 109 A).

A form of deep beam cut from a standard beam section is known as the 'castellated' beam and this also has the advantage of eliminating plating. This is shown in figure 109 B. The web of a Universal beam is cut along its length on a castellated line by a flame cutter and the two halves are then placed 'point' to 'point' and welded at the junctions to produce an open web beam of the same weight per metre run as the original section but with a depth one-and-a-half times that of the original and, therefore, having a greater moment of inertia. The insertion of plates between the 'points' as shown in (C), to produce what is known as a Litzka beam, extends the depth of the beam further at the expense of only a small increase in weight. These beams, or girders, are particularly suited to long span use where the loading is light, where stiffness rather than loadbearing capacity is the critical consideration.

A range of rectangular hollow steel sections is available which, because of their efficient shape are particularly useful for columns and other compression members.

High-tensile steel can be used for the sections instead of mild steel. It is more costly but its high yield point permits greater working stresses, usually 50 per cent more than for mild steel, resulting in a saving in weight which can offset the extra cost of the material. Deflection criteria may limit the development of the maximum stresses since its elastic modulus is approximately the same as that of mild steel, although methods have been devised to overcome this (see page 199). Even so an ultimate saving in weight of 8 to 10 per cent is possible. Although not widely used for complete frames, it is useful where a local reduction in size of beam or stanchion is essential and this in turn leads to a reduction in the amount of casing and to a saving in height and floor space.

The number of standard and non-standard sections now available to the designer is large and necessarily so if design is not to be limited. Nevertheless, the cheapest frame is not necessarily that which contains least material, although weight can be a useful broad guide, and it is usually more economical to standardize the size of sections to some extent on any particular job rather than to relate the size of every member exactly to the calculations. Any saving effected by using the minimum size in every case may be outweighed by the cost of excessive 'firring' and packing out in order ot obtain some degree of uniformity in the finished overall beam and stanchion sizes, and by the cost of making more complicated connections. It must also be borne in mind that the material is cheaper than labour, that sections held in stock are cheaper and more readily available than others, and that it is cheaper to buy a large number of members of the same size than to buy a few lengths of a number of different size sections. A regular grid layout for the frame, as indicated in Part 1, assists such a standardization by equalizing the loading on different parts.

Frame layout

Planning requirements permitting, economy in the steel frame is obtained by the adoption of a regular and reasonably close spacing of stanchions. Short span beams are more economical per unit of length than long span beams (see chart, figure 110).

For normal conditions of loading the cheapest beam will be the deepest available section giving the required strength. Although this may be in excess of the strength required it can still be cheaper than a shallower but heavier section. Take as an example a beam spanning 6 m carrying a total distributed load of 150 kN. The shallowest beam suitable is a 254 x 146 weighing 43 kg per metre run which will carry 155 kN, but a 356 x 127 beam will carry 175 kN over the same span yet weighs only 39 kg per metre run, resulting in a saving of 24 kg of steel on the beam. As indicated earlier such savings must be balanced against increased costs which may arise in other parts of the building due to increased overall floor heights. The savings on the frame are likely to be greater than the increases in other directions in the case of large buildings of the warehouse class, where the area of external walls and internal walls and partitions is small relative to the total floor area.

As a preliminary rough guide an economic depth of from one-fifteenth to one-twentieth of the span should be allowed for beams. A deflection not exceeding 1/360 of the span is normally considered satisfactory. Stanchions are usually Universal beam or broad flange sections.

The simplest form of frame is a skeleton 'cage' made up of Universal beams and stanchions with the members standardized as far as possible and

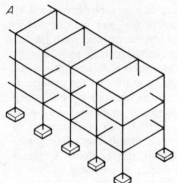

Simple 'cage' frame

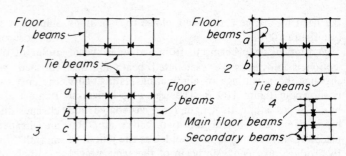

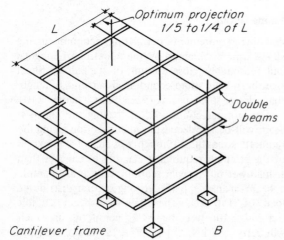

L

Optimum projection
1/5 to 1/4 of L

Double beams

Cantilever frame   B

Layout	Member	SPANS	
		Practicable range (m)	Economic range (m)
1	Floor beams	3.65 to 15.25	4.25 to 6.00
	Slab	2.40 to 7.30	3.00 to 4.25
2	Floor beams	3.00 to 15.25 for economy 'a' should not be more than 1⅕ x 'b'	4.25 to 5.50
	Slab	As for 1. above	
3	Floor beams	a.c. 3 to 15.25 b. 1.8 to 15.25 for economy 'b' should be from 1/8 to 2/3 of (a+c)	4.25 to 5.50 2.40 to 3.00
	Slab	As for 1. above	
4	If spacing of floor beams is greater than 5.50 m secondary beams may be used to keep slab span within the economic limits		

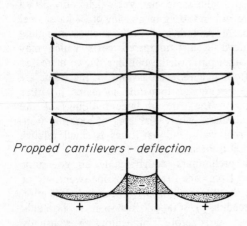

Propped cantilevers – deflection

Propped cantilevers – bending moments

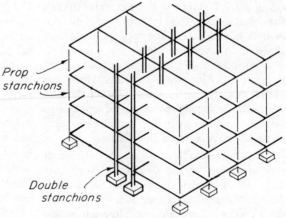

Prop stanchions

Double stanchions

C   Propped cantilever frame

110   Steel frames

the weight kept to a minimum, having regard to all other relevant factors (figure 110 *A*). Of the beams at each floor level those in one direction will be 'floor' beams carrying the floor slabs and those at right-angles to them will be 'tie' beams which are necessary to provide lateral stability to the skeleton frame as explained in Part 1. As a steel frame is made up of one-way spanning stages a rectangular rather than a square grid layout is generally the most economic (see 1 to 4, figure 110).

By projecting the ends of the main beams beyond the outer stanchions as cantilevers the negative moments set up at the supports serve to reduce the positive moment in the centre spans, which can then be longer without an increase in the size of the beams or, alternatively, a reduction in the size and weight of the beams can be made[1]. The optimum projection for the cantilever bays, producing approximately equal positive and negative bending moments, is one-fifth to one-quarter of the overall length. The external walls are freed of stanchions and services can be freely run round the building, but stanchions some distance inside the walls are required and this may be inconvenient in respect of the planning of some types of building. The double beams or double stanchions which must be used in order to achieve the continuity of the cantilever beams at the supports, result in more complicated construction (see figure 110 *B, C*).

If the cantilever projections are made long in proportion to the centre span the floor space is uninterrupted except for the central stanchions which may possibly be in the line of corridor walls, but span for span this is more extravagant in material because of the very high cantilever moments. In addition, wind stresses in the stanchions are high because of the small distance between the stanchions, and this method is usually justified only when some form of good lateral bracing is incorporated such as lift shafts, gable walls or suitably placed cross walls which can resist the wind load.

The outer end of long cantilevers may be simply supported by props or struts in the plane of the walls to give what is known as propped cantilever construction as shown in figure 110 *C*. In this, although the centre span is small the negative bending moments over the supporting stanchions and the deflection of the cantilevers are reduced as shown on the diagrams in (*C*), even though

'hogging' of the centre span may still occur[1]. The props are designed with free or hinged ends at the floor levels. Their function is to provide end support to the cantilevers, to transmit to the main structure the wind load at each floor level and to transmit their proportion of the vertical load to the foundations. They can be comparatively small in cross section as they provide no rigidity to the frame against lateral wind pressure and, as a consequence of this, they will transmit no turning force to the foundations due to wind pressure.

If it is not desired to cantilever out at all floors, it may still sometimes be necessary to set in the outer stanchions at the lower floors either for architectural reasons or in order to simplify the problems of foundations adjacent to a building line. This may necessitate heavy compound or plate cantilever girders, particularly if headroom requirements prevent the use of the most economic depth (figure 111 *A*).

Certain types of buildings require wide span beams throughout; others, based generally on a close, regular grid, require a variation of that grid to permit the provision of large, unobstructed floors at certain levels, such as ballrooms and restaurant areas in hotels, or the provision of wide shopfronts. These will require stronger beams in the form of some type of girder (figure 111 *B*).

Girders

Spans greater than the economic limits suggested earlier require the use of the deeper, heavier beam sections. In the past, when the loads were heavy, it was the practice to increase the flange area by the addition of plates to the top and bottom flanges of a standard beam section, forming a 'compound girder'. The wide range of Universal sections now available and the ease with which sections may now be automatically welded up has made the practice rare.

For heavier loads and for spans in the region of 15 to 36 m, where a section larger than that obtainable from standard beam sections would be required, it is necessary to use plate girders or trussed girders built up of plates or smaller standard sections. The weight of steel in relation to the cube of the building is increased in such cases,

[1] See Part 1, page 61.

189

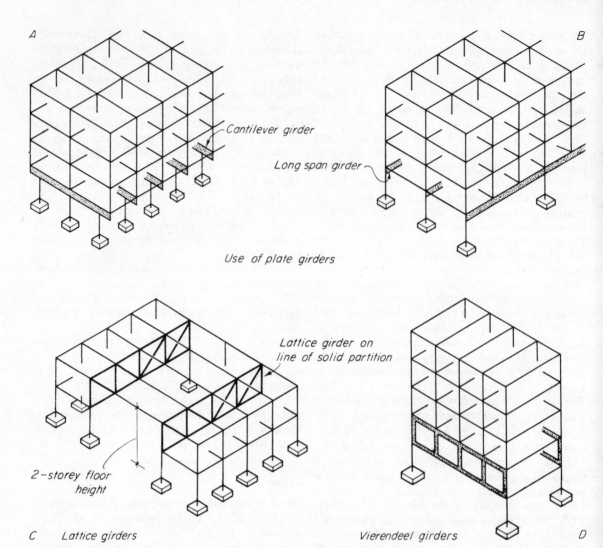

A

B

Cantilever girder

Long span girder

Use of plate girders

Lattice girder on
line of solid partition

2-storey floor
height

C   Lattice girders

Vierendeel girders   D

111 Steel frames

resulting in an increase in cost. This may be minimized if ample depth can be allowed for the girders. The increased depth of lever arm results in a decrease in flange stresses, and this in a considerable reduction in the weight of the flanges, at the expense of a slight increase only in the weight of the thin web of a plate girder or the slender braces of a trussed girder.

Plate girders

A plate girder consists of a web plate welded to flange plates to form a large I-section, the thickness of the flanges being varied according to variations in stress by curtailment of the flange plates as the stresses reduce. When necessary the webs are strengthened against buckling by stiffeners cut to fit between the flanges and welded to them and to the web (figure 112 A). Stiffeners are usually placed at the ends and under any point loads and then at equal distances between at a spacing not greater than one-and-a-half times the depth of the web. For very large spans and heavy loads it may be necessary to increase the thickness of the web plates towards the bearings where the shear increases, in the same way that the flange plates are increased

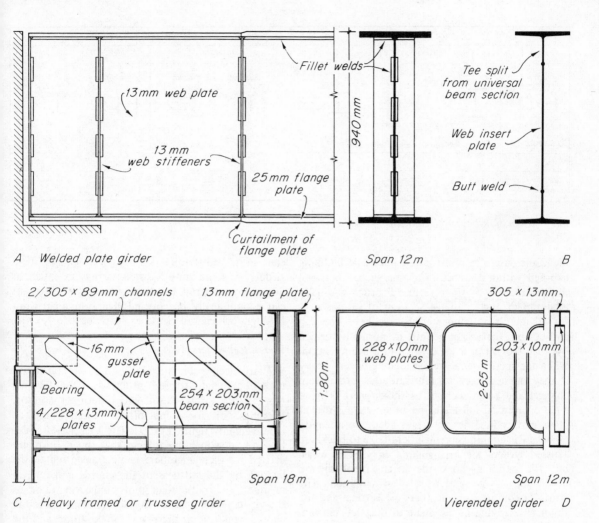

A   Welded plate girder

Fillet welds

13mm web plate

13 mm web stiffeners

25mm flange plate

Curtailment of flange plate

Span 12 m

Tee split from universal beam section

Web insert plate

Butt weld

B

2/305 × 89mm channels

13mm flange plate

16 mm gusset plate

Bearing

4/228 × 13mm plates

254 × 203mm beam section

1·80 m

Span 18 m

C   Heavy framed or trussed girder

305 × 13mm

228 ×10mm web plates

203 ×10mm

2·65 m

Span 12 m

Vierendeel girder   D

112   Steel girders

where the bending stresses increase.

An alternative to a flat flange plate is a tee split from a Universal beam section between two of which the web plate may be welded (B).

When extra wide flanges are necessary, or to give lateral stiffness in long beams, the girder is is made with two webs to form a 'box girder'. These are not so liable to twist or bend laterally as a single-web girder.

The economic depth for plate and box girders may be taken as one-twentieth of the span. When a depth shallower than the economic limit is used, careful consideration must be given to deflection to ensure that this is not excessive. The width of the flanges is normally from one-fortieth to one-fiftieth of the span. If wider than this adequate flange stiffening is necessary. For spans much greater than 18 to 21 m plate girders are very heavy and become uneconomic.

Trussed or lattice girders

Where wide spans must be covered and sufficient depth is available, the use of a deep, triangulated beam is likely to be cheaper than a plate girder. These are known as trussed, lattice or framed girders and usually, in a multi-storey frame where height is likely to be limited, can most easily be accommodated in a storey height, for example, on the line of an internal dividing wall (figure 111 C)

191

Bolted connections				Direct welded connections			
305 × 165 × 46 kg UB		203 × 203 × 46 kg UC		305 × 165 × 46 kg UB		203 × 203 × 46 kg UC	
e = 252·5mm	e = 100 mm	e = 201·5mm	e = 100 mm	e = 152·5mm	e = 3mm	e = 101·5mm	e = 3·5mm
100		100					
Z = 646·4 cm³	Z = 99·54 cm³	Z = 449·2 cm³	Z = 151·5 cm³	Z = 646·4 cm³	Z = 99·54 cm³	Z = 449·2 cm³	Z = 151·5 cm³
Relative stress in stanchion f = M/Z = We/Z							
$\frac{W \times 25\cdot25}{646\cdot4} = \cdot04$	$\frac{W \times 10}{99\cdot54} = \cdot10$	$\frac{W \times 20\cdot15}{449\cdot2} = \cdot05$	$\frac{W \times 10}{151\cdot5} = \cdot07$	$\frac{W \times 15\cdot25}{646\cdot4} = \cdot024$	$\frac{W \times \cdot3}{99\cdot54} = \cdot003$	$\frac{W \times 10\cdot15}{449\cdot2} = \cdot02$	$\frac{W \times \cdot35}{151\cdot5} = \cdot002$

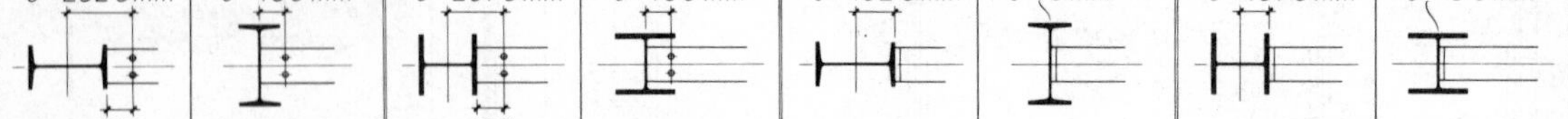

113   Stanchion/beam relationship

or, in the case of a multi-storey industrial building, exposed within a service floor sandwiched between the production floors. A trussed girder is usually made up of angle and tee sections of various sizes for all members, connected by welding. The top and bottom flanges, or booms, are normally formed by a tee section or a tee split from a beam section. When the span is great or the loading is very heavy it may be necessary to fabricate the girder from channel and beam sections as shown at (C), figure 112, or even to fabricate the booms and struts as plate girders in order to attain adequate strength, especially in compression members liable to buckling. Trussed girders are triangulated in various ways, but for building work the arrangement shown, producing an 'N' girder, is most common. The struts or compression members are vertical and the sloping ties, lying in the plane of diagonal tension, are in tension.

The economic depth of trussed girders is from one-sixth to one-tenth of the span.

Lightweight lattice girders prefabricated in standard depths and lengths from hot- and cold-rolled sections are widely used for roof and floor framing in structures where the loading is relatively light (see chapters 6 and 9 and Part 1).

Space frames

These are described on page 353. Although more commonly applied to roof construction, the space frame can with advantage be applied to multi-storey frame construction. A light, rigid member with considerable stiffness results from forming a three-dimensional frame, and triangular section frames

can show substantial savings over conventional girders — some have quoted savings as much as 20 per cent. Used for floor framing, wide spans can economically be covered giving an open floor through which all services can be freely run.

Vierendeel girders

A Vierendeel girder may be used as an alternative to a floor height trussed girder when the diagonal ties of the latter would cross door or window openings, or for some other reason would be inconvenient (figure 111 D). The Vierendeel girder has no diagonal members, the shear normally carried by these members being transferred to the bearing by the stiffness of the chords and vertical members and by the rigid joints connecting them, as shown in figure 112 D. In such circumstances the chords would lie in the planes of the floors and the vertical members would appear as columns. These girders are expensive in steel and more costly than a trussed girder.

Stanchions

Universal and compound stanchions

The Universal beam and broad flange sections, used alone or plated if necessary, are commonly used for the stanchions.

Stanchions are usually positioned with the web in line with the main beams which they support so that the beam is connected to the stanchion flange. This is because smaller bending stresses due to the eccentric bearing will be set up in the stanchion than if the beam is connected to the web.

192

In the case of cleated joints, regulations require a minimum eccentricity of 100 mm to be assumed at the beam connection and when the beam is connected to the stanchion web rather than the flange, greater bending stresses always result, in spite of the smaller total lever arm, due to the small value of the minimum section modulus (see figure 113). With broad flange sections the difference is less marked and with direct welded joints a connection to the web will, in fact, produce less bending stresses in the stanchion because of the extremely small eccentricity at the joint.

As most stanchions fail by buckling in the direction of the least dimension, and since strength in this respect depends upon the value of $\frac{l}{r}$ (effective length divided by the least radius of gyration), those sections with the widest flanges are most efficient for use as stanchions since they give closer values for the radius of gyration in the direction of both principal axes. This is shown clearly on the graph in figure 114[1]. In shorter stanchions, where buckling is less critical, it can be seen that strength is to some extent proportional to weight, but with increase in height strength reduces more rapidly in sections having a high ratio of minimum to maximum radius of gyration. It will be seen that with broad flange beams of equal dimensions in both directions, it is possible to obtain a section differing little in its resistance to buckling about both axes.

Circular stanchions or columns

Solid and hollow circular columns have the same r value in all directions and occupy less floor space than any other equally strong stanchion under axial load, but they present difficulties in making connections to them and in a multi-storey frame vertical continuity suffers. In addition, the solid steel column is heavy and expensive.

In terms of efficiency, that is in terms of strength to weight ratio, as distinct from size, the solid circular column is less efficient than a Universal column section with a wide flange. For example, a 100 mm diameter solid steel column, weighing 61.7 kg per metre run, has an r value of 25 mm while a 152 x 152 x 37 kg broad flange section has a greater minimum r value of 38.7 mm. Where the effective length of a stanchion can be reduced by adequate restraint in the direction of least r,

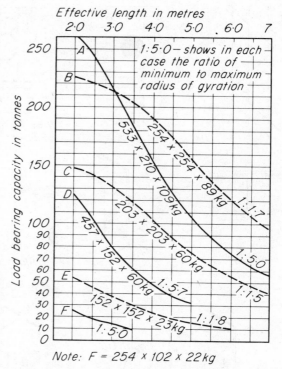

Note: F = 254 x 102 x 22 kg

114   Effect of wide stanchion flanges

the efficiency of the Universal section will be further enhanced.

The caps and bases of solid circular columns are of thick steel plate bored to fit over the ends of the column, which are turned to produce a small shoulder. The plates are 'shrunk' on to the column by being heated and then forced on to the turned ends where they cool and shrink thus tightly gripping the column (figure 115 A).

A hollow circular column has a greater r value than a solid circular column of the same diameter, that for the former being approximately 0.35 x mean diameter and for the latter 0.25 x O/A diameter. The tube section, with its smaller area of material and load-carrying capacity but greater r value, has advantages, therefore, in cases where buckling rather than direct stress will be the critical factor.

Cap and base plates and connecting cleats are normally welded to hollow circular columns (B). Tubular columns have been used in Great Britain for two-storey frames where the total height of column can be in one length to preserve continuity.

[1] See also Part I, page 64.

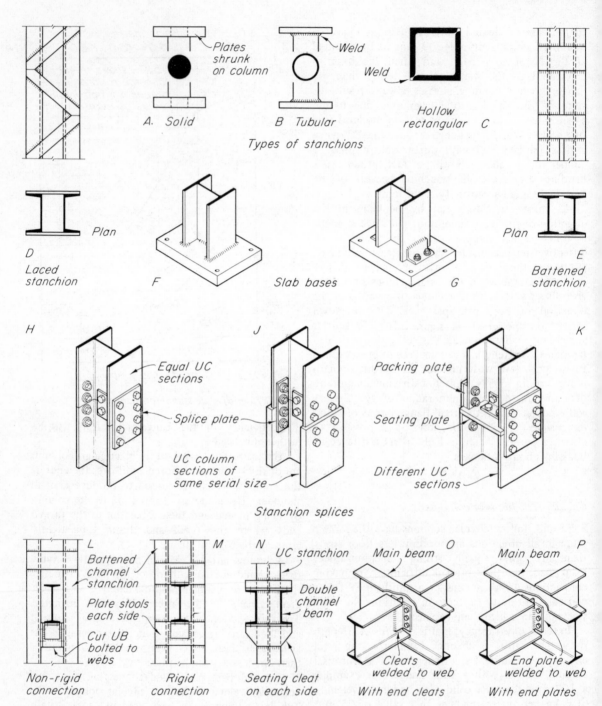

A. Solid

B Tubular

Plates shrunk on column

Weld

Weld

Hollow rectangular C

Types of stanchions

D Laced stanchion

Plan

F Slab bases G

Plan

E Battened stanchion

H

Equal UC sections

Splice plate

J

UC column sections of same serial size

K

Packing plate

Seating plate

Different UC sections

Stanchion splices

L

Battened channel stanchion

Plate stools each side

Cut UB bolted to webs

Non-rigid connection

M

Rigid connection

N

UC stanchion

Double channel beam

Seating cleat on each side

O

Main beam

Cleats welded to web

With end cleats

P

Main beam

End plate welded to web

With end plates

Cantilever beam to stanchion connections

Beam to beam connections

115   Steel frames

Hollow rectangular stanchions

A hollow square section permits connections to be made to it more easily than to circular columns, and has an equal r value about both normal axes slightly greater than that for a hollow circular column of the same dimension, being 0.41 x mean breadth. As with hollow circular columns this section is particularly useful for long, slender but lightly loaded stanchions and will show an economy in material over the normal Universal section. Connecting cleats are normally welded to these stanchions which may be made from two angles or two channels welded toe to toe (figure 115 C) but are now normally selected from the standard range of rectangular hollow sections.

Lattice or braced stanchions

Stanchions formed from pairs of beam or channel sections braced together by diagonal or horizontal plates are called respectively laced and battened stanchions, and are used in cases where very tall but relatively lightly loaded stanchions are necessary, requiring a maximum r value in both directions, or in cases where the load is too great for even the largest rolled section (see figure 115 D and E). These sections can easily be arranged to give equal r values about each axis by adjusting the distance between the pair of basic members. For heavy beam sections the distance between webs will be about three-quarters of the depth of the beam, and for channel sections placed back to back the distances between webs will be about half to two-thirds of the depth of the channel according to the particular section used. Limits on the spacing of lacing bars and battens are laid down in BS 449.

Connections

Connections between the members of a normal steel frame are made by means of welding and bolts with angle cleats and plates.

The subject of connections is introduced in Part 1, chapter 6, where types of bolts, the nature of welding and the basic types of connections used are described and reference is made to their detailing relative to fabrication. Welded connections are discussed in more detail under 'Welded construction' on page 200 of this volume.

Some of the connections made at different points in a steel frame have been described in Part 1. These will now be considered further and others will be described.

Stanchion bases

For reasons given in Part 1 the foot of a stanchion must be expanded by means of a base plate which will act as an inverted cantilever beam and is designed accordingly.

There are two types of base: (i) the slab or bloom base and (ii) the gusseted base.

Slab or bloom base This consists of a base plate thick enough to resist the moments caused by the bearing pressure. When the area of base required is such that its projection beyond the stanchion is small (figures 105 and 106, Part 1) these moments will be small and the thickness of the base plate will be relatively small. As its area increases so must its thickness. It is fixed to the foot of the stanchion by welding (figure 115 F) or by a pair of angle cleats when circumstances make it advisable to send stanchion and slab separately to the site (G). These fastenings are required mainly to secure the slab to the stanchion.

This base requires less labour in fabrication than the gusseted base (see below) and there is a saving in depth of the cover necessary to give a clear floor compared with that required when there are large gusset plates, but it sometimes requires more material, particularly in the case of large bases in which the slab may be very thick.

Gusseted base This consists of a base plate stiffened by gusset plates which act as ribs (figure 120 J). The size of the base plate as that of a bloom base depends on the safe bearing pressure which the concrete can resist. The breadth of the base generally varies from two to three times the breadth of the column, and the height of the gusset plates from one-and-a-half to three times the breadth of the column. Gusseted bases are commonly used when the stanchion transfers high bending moments to the foundation. The holding down bolts (figure 120 K) in these cases must be well tied down to the foundation block.

The underside of base plates and slabs need not be machined when the base will be grouted to a concrete foundation, provided the underside is true and parallel with the upper face. A bloom base is always used for solid circular columns.

Stanchion caps

A cap must be provided to the stanchion if the beams rest on top of it. This is illustrated in figure 105 and described on page 143 of Part 1.

Stanchion splices

Joints in the length of stanchions, generally termed splices, are necessary for erection and fabrication purposes because it is difficult and expensive to handle lengths much greater than about 10.50 m (see figure 115 *H*). Stanchions should be fabricated in lengths as long as possible and the limit of 10.50 m for an individual length results in stanchion lengths of two to three storeys in height, depending on the floor-to-floor height of the structure. The joints should be made as near as possible to the beam level, but in order that splice plates shall not obstruct beam connections they are usually placed about 300 to 450 mm above floor level.

Any change in the section is made at these points (*J*) and where the whole of the end of the upper length does not bear on the lower length a horizontal seating plate of the same overall size as the lower stanchion must be placed between the two lengths (*K*). Packing plates will be necessary between the upper smaller section and the splice plates. It will be seen by reference to figures 109 *A* and 115 *J* that the use of Universal related sections of the same serial size reduces the need for seating plates. When the ends of each adjacent length of stanchion are machined and bear directly one on the other, transmission of load will be by direct compression so that the splice plates serve only to secure the two lengths. If the ends are not machined and are not in full contact the splice plates and bolts must be sufficient to transmit the entire load.

Beam to stanchion connections

Direct compression connections are only possible where the beam rests on top of the stanchion (figure 105, Part 1). This is normally not possible at lower beams where loads have to be transmitted by shear connections as shown in figures 106 and 107, Part 1. In order to facilitate erection these joints are provided with seating cleats or stools welded to the stanchion in the shop, so that when the beam is hoisted into position it can bear on the stool while being fixed. Load transmission from beam to stool is by direct compression and through the stool to the stanchion by the welds in shear. Angle cleats are also provided to the web or top flange of the beam to prevent canting of beam. It is simpler to have these connected to the web and it also leaves the top flange clear at the junction with the stanchion, but sometimes it is not possible to arrange this and flange cleats must be used. A top flange cleat produces a stiffer joint than web cleats.

Bearings for cantilever beams passing through a double member stanchion can be provided by stools formed from pieces of beam section fixed between the stanchion members on which the beams rest and to which they are fixed by stool cleats as shown at (*L*), figure 115. Greater rigidity at the connection may be provided by plate stools connected to the flanges of the stanchion members and secured to the flanges of the beam with angle cleats (*M*). The top plates and cleats are fixed after the erection of the beam. Where the beam itself is a double member, projecting seating cleats are fixed to the flanges of the stanchion to take the beam members (*N*).

The maximum stress permitted in stanchions depends on the ratio of effective length to radius of gyration. The former varies with the number and disposition of the beams connected to the stanchion, as this affects the degree of fixity and restraint given to the stanchion. The restraint given by various arrangements is indicated in Appendix D of BS 449.

Beam-to-beam connections

The direct compression connection, in which one beam bears directly on the top flange of the other, is the most economical (figure 108, Part 1), but in multi-storey frames this is generally not suitable because of the considerable depth of the pair of beams one on top of the other. Shear connections are therefore usual with web connected to web. Angle cleats or end plates are welded to the web of the secondary joists in the shop and the connection is completed on the site by bolting to the main beam (figure 115 *O, P*).

When the main beam is much deeper than the secondary beam, for example in the case of a deep beam section or a plate girder, erection is facilitated by using deep seating stools made up of plates and angles or of a piece of cut beam section. These have the added advantage of forming web stiffeners at the point of application of the load which is desirable in deep beams. Where the flanges of both beams are to be level, that of the secondary beam must be notched to clear the flange of the main beam.

Tension joints are generally avoided but may have to be used when the secondary beam must pass under the main beam. The connection depends entirely upon the nuts of the bolts by which the secondary beam is hung from the other. Direct connection of flange to flange should be avoided whenever possible, the better method being to hang from the flange of the main beam by means of angles and gusset plates.

Wind bracing

Adequate lateral resistance against wind pressure may be provided in a number of ways. These may take the form of
(a) rigid frames
(b) diagonal bracing (X or K)
(c) shear walls
(a) In many cases where the beams are reasonably deep and satisfactory connections to the stanchions can be made the structure may be stiff enough as a whole to resist excessive lateral deflection. Connections can be made by means of top and bottom flange cleats of sufficient size and bolts in sufficient numbers. The cleats must be stiff enough to transmit moments without excessive distortion and often would become uneconomically thick, especially when the stanchion is narrow and necessitates the use of long cleats to accommodate the required number of bolts. The thickness of the cleats can be reduced by the use of plate or gusset stiffeners.

A more satisfactory method is to weld a steel fixing plate on to the end of the beam, making use of high tensile steel friction bolts for fixing (see figure 107 A, Part 1); when greater stiffness is required the plate and bolts may be extended beyond the beam. Since the columns in such a frame carry bending moments they will be larger than with other methods. Where wind loading is not excessive, that is in buildings which are not too tall, this method can produce an economic solution. Very tall structures with relatively small base areas, however, require other methods.
(b) In this method the principle of triangulation

is applied to certain frame panels up the height of the building to produce stiff vertical elements capable of resisting the wind pressure which is transmitted to them from the external walls by the floors of the structure acting as rigid diaphragms. The remainder of the frame is then required to carry only the gravity load (see figure 116 A).
(c) Shear walls are stiff walls capable of resisting the lateral wind loading. In principle they act in the same way as the braced frames in (b), wind pressure being transmitted to them by the floors (B). They can, in fact, be formed within the panels of the frame itself. Research has shown that even lightweight wall panels and 102.5 mm brick panels with door openings in them increase to a considerable degree the stiffness of the structural frame.

Lift shafts and stair wells in tall buildings are commonly formed of reinforced concrete, producing a stiff annular core or cores for the full height of the structure, and these may be used to provide the necessary resistance to lateral wind pressure. The floors, as in (b) and (c), transmit the wind pressure from the external walls to the core (figure 116 C). A development of this for very tall buildings, known as the 'hull-core' system, utilises a rigid or braced framework for the perimeter structure to form stiff walls interacting with each other to produce an external annular structure, called the hull. This acts together with the internal core through the floors to produce a very stiff total structure to resist wind forces (D).

Composite construction

In reinforced concrete structures it has been usual to keep in mind during the design stage the interaction of the different parts, but in the case of steel structures it has been usual to design a frame in complete isolation from the rest of the structure, leading to an analysis of its behaviour which is now known to be false, at least in the case of building frames. Considerable data is now available regarding the composite action of the various parts of a building structure, and BS 449 permits allowance to be made for concrete casing on beams and stanchions.

Due to its early failure under the action of fire, structural steel members in multi-storey frames must be adequately protected from the effect of fire, and this is fully discussed in chapter 10. Solid concrete casing is common and, while it is

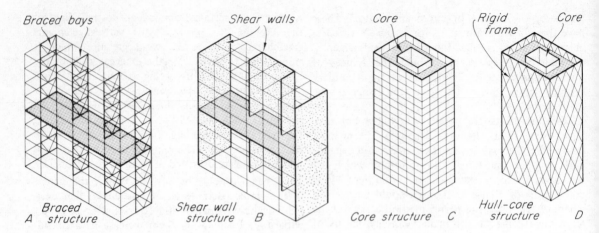

Braced bays   Shear walls   Core   Rigid frame   Core

A   Braced structure   Shear wall structure   B   Core structure   C   Hull-core structure   D

116   Wind bracing

used primarily for the fire protection of the steel[1], BS 449 permits allowance to be made for the stiffening effect of the casing on beams and stanchions resulting in higher permissible stresses for the steel. In addition, it also permits some of the stanchion load to be carried by the concrete, with a limit on the total axial load carried of twice that permitted on the uncased section. These allowances do not apply to stanchion sections greater than 1000 mm ẋ 500 mm nor to beams greater than 1000 mm in depth or 500 mm in breadth nor, in both cases, to box sections. In a multi-storey steel-frame building, having solid concrete casing to the members, a saving is steel can, therefore, be effected by taking the casing into account, even allowing for the somewhat heavier reinforcement required. This results in smaller overall sections and, therefore, a saving in the concrete casing content. In the case of stanchions, the more lightly loaded a stanchion is, the greater will be the benefits deriving from this procedure.

Composite stanchions can also be formed with the concrete placed internally, relative to the steel. Multi-storey stanchions of two opposing channels have been used, the channels being filled with concrete vibrated after erection, designed to carry 28 per cent of the load, and showing a saving of 50 per cent of steel compared with normal stanchions.

BS 449 lays down conditions in which the frictional resistance between the top flange of a steel beam and the floor slab it supports may be considered sufficient to provide adequate lateral restraint to the beam. As this restraint is more often than not available in practice solid concrete beam casing is not essential for this purpose and lightweight protection such as asbestos spray and hollow casings of various types can be used with economic advantage. In some circumstances hollow casings to the stanchions may show economies, but in others it may be more advantageous to use concrete casings to carry some of the stanchion load.

Concrete floor slabs may be used to act together with the steel supporting beams. A concrete slab

[1] The London Building (Constructional) By-laws lay down certain minimum thicknesses of solid casing to steelwork exposed to the weather, which may be greater than that required for fire protection. By-law 8.02 requires that all structural steelwork, other than internal steelwork, shall be protected from the weather by

(i) concrete not less than 75 mm thick; or

(ii) brickwork or stone or similar material, properly secured, if the steel is protected from the effects of corrosion by:

(a) 50 mm of concrete on the flange faces, 40 mm on the flange edges and 25 mm on all projecting rivet heads, bolts or splice plates so that the total thickness of encasement to the steelwork is not less than 100 mm; or

(b) such material as the District Surveyor may approve as being suitable, having regard to the particular circumstances of the case, and the thickness of the brickwork, stone or similar material, is not less than 90 mm; or

(c) a Class IIA, Class IIB or Class IIC enclosure.

Where the structural steelwork may be adversely affected by moisture from the adjoining earth, it shall be solidly encased with concrete at least 100 mm thick.

198

adequately bonded to a steel beam will act in the same way as in a reinforced concrete T-beam. This is particularly useful when the steel is encased, as the lever arm of the composite section will be greater by about half the thickness of the slab. The concrete casing by itself does not, however, provide sufficient bond between the steel and concrete to transmit the shear stresses, and this is essential if composite action is to be achieved. The greatest shear stress is at the neutral axis, which is always near the top of the combined section, so that a satisfactory bond can be obtained by welding shear connectors to the top flange of the beam. That most commonly used now is the shear stud fixed by a special electric welding gun (figure 117*A*). These may be used with both *in situ* cast and precast slabs. Alternatively, with precast slabs, high strength friction grip bolts may be used to produce sufficient friction between beam and slab to transmit the shear stresses (*B*). The problem of tension arising in the top of the member and compression in the bottom over the supports in continuous composite beams may be dealt with as follows:

(1) The concrete may be reinforced longitudinally with rods

(2) The beam may be strengthened by welding on top flange plates, ignoring the concrete entirely, as well as plates on the bottom flange to strengthen that.

Concrete is also used to stiffen high-tensile steel beams in order to permit them to be stressed to their working limit. In normal building structures these beams cannot always be stressed to their limit because the amount of deflection might be unacceptable, high-tensile steel having an elastic modulus approximately equal to that of mild steel. In this system the beam is deflected to the same extent that it would be under full working load and the concrete is cast as a casing round the tension flange. After this has gained strength, the deflecting force is removed, and as the beam reverts to its original unstressed state, it compresses the concrete. Under working load the tensile stresses in the bottom flange are resisted by the compressive stresses in the concrete casing and this has the effect of reducing the deflection of the beam, thus permitting the high-tensile steel beam to be stressed to its limit so that its high strength may be fully utilized.[1]

Prestressing may be applied to mild steel beams

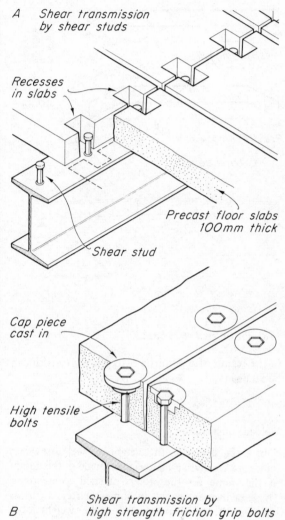

A   *Shear transmission by shear studs*

Recesses in slabs

Precast floor slabs 100mm thick

Shear stud

Cap piece cast in

High tensile bolts

B   *Shear transmission by high strength friction grip bolts*

117   Composite beams

in order to allow the use of reduced cross sections resulting in savings in weight and cost. The post-tensioning cables may be passed through a box tension flange with anchor plates at the ends (figure 118 *D*) or they may be external to the beam at the sides (*E*) or below the tension flange (*A, B, C*). In the latter examples it will be seen that the beam may be bent up so that the cable is straight or the cable may be strutted off the tension flange to form a trussed beam. Depth/span ratios

[1] This type of beam is produced under the proprietary name of *Preflex*.

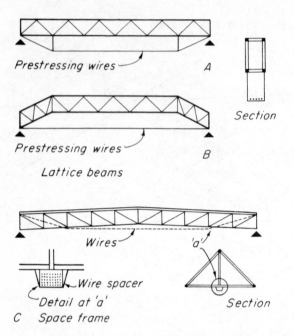

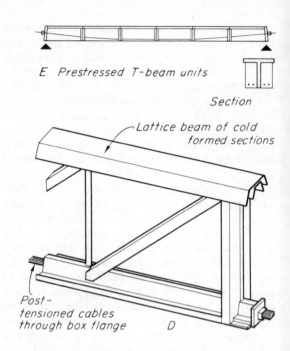

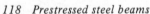

118   *Prestressed steel beams*

in the region of one-thirtieth to one thirty-fifth can be attained[1].

Welded construction

Substantial savings in the weight of steel arise when joints are welded and allowances made in the design of the frame for the effects of rigid connections. These savings are stated in some quarters to be as high as 20 to 25 per cent of the weight of a comparable frame in bolted construction. The economies are due to (a) the elimination of bolt holes; (b) simpler connections with no bolt heads or, in many cases, angle cleats; (c) the possibility of developing fully rigid end connections leading to the use of lighter members; (d) in the case of built-up members, such as plate girders, the use of single plates, welded directly one to the other without the addition of angles. Although not usual with multi-storey frames, where the frame is exposed the appearance is enhanced by the cleaner and neater joints, and less maintenance costs are involved in re-painting, because of the smaller sections and the absence of bolted connections.

The use of welding permits the better use of tubular and cold-rolled steel sections, both of which,

when used in appropriate circumstances, tend towards further reductions in weight. Flexibility in design arises from the fact that built-up sections may easily be fabricated, thus freeing the designer from the restrictions of the standard rolled sections. In addition, the material in such members may be proportioned to the moments it is called upon to resist in the different parts, thus producing maximum structural efficiency; greatest advantage may be taken of this in single-storey rather than in multi-storey frames.

Apart from the application of welded joints throughout a total structure, as in figure 108, Part I, the shop fabrication of the component parts, as indicated in Part 1, is now invariably carried out by means of welded connections, the site connections being bolted.

Methods of welding

Two methods are normally employed in structural work: (i) oxy-acetylene welding, (ii) metal arc

---

[1] See *The Structural Engineer,* November 1950, February 1954 and February 1955, for papers on the subject of prestressed steel beams and girders.

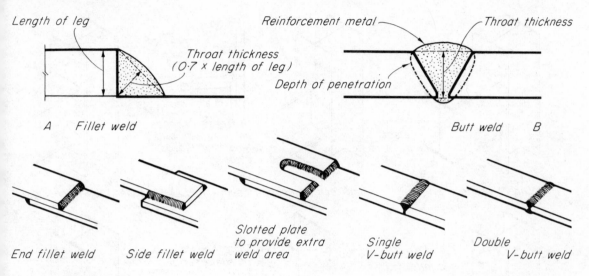

119   Types of welds

welding. In the former, the flame from an oxy-acetylene torch heats the two surfaces to be joined at their point of contact, and molten metal from a steel filler rod held in the flame at the same time fuses into the two surfaces. In the second process, an electric current passes through the metal filler rod, which acts as an electrode, and the material to be joined. An arc is produced, which heats and melts the surfaces and the end of the filler rod.

Types of welds   There are two types of weld: a butt weld and a fillet weld. The fillet weld is that in which the two surfaces joined are basically at right-angles one to the other (figure 119 *A*), the butt weld is that in which the two surfaces to be joined butt against each other (*B*). There are varieties of each of these and some are shown in figure 119. The strength of a weld is based upon its area, made up of the effective length and the throat thickness[1].

Types of welded connections

Varying degrees of rigidity may be produced in welded joints according to their design, but unless especially designed to produce full fixity, a welded joint will generally possess no more rigidity than a normal bolted connection. Welded joints may be in the form of direct or indirect connections. A direct connection is one in which the members are in direct contact one with the other and are joined to

each other directly by welds without the use of cleats or plates (figure 120 *A*). In the case of beams, particularly, this necessitates accurate cutting to length and complicates the erection process, so that more often than not an indirect connection is used, involving flange and web cleats (*B, D*), which overcomes these disadvantages. Stanchion bases and caps, and stanchion splices, may be formed with direct or indirect connections. Figure 120 *E, F, G,* show splices formed by direct connections. Temporary angle cleats are bolted to the webs to enable the lower and upper stanchion lengths to be held in position while the necessary site welds are executed. Splices are sometimes formed as shown at (*H*). The welded base (*J*) is a gusseted form referred to on page 195. When used to transmit bending moments to the foundation the holding-down bolts must be well secured to the base slab as shown at (*K*).

Indirect beam to stanchion and beam to beam connections may be made in a number of ways. A non-rigid connection may be made by the use of web cleats, fillet welded at the toes to each of the members, with a space between the members themselves (*D*). To facilitate erection, it is an advantage if a seating cleat is incorporated in the joint. Cleats

[1] General requirements and permissible stresses for welded construction are laid down in BS 449. For details of methods and forms of joints see also BS 693 and BS 1856.

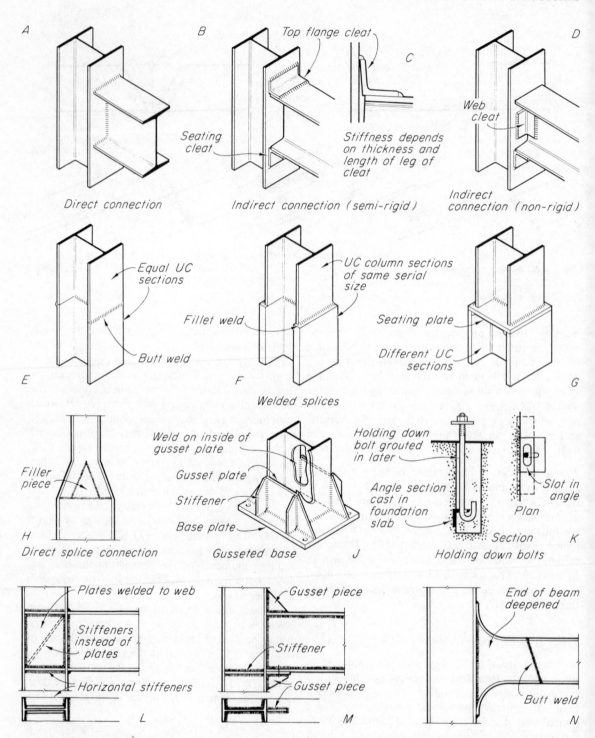

A

Direct connection

B

Top flange cleat

Seating cleat

C

Stiffness depends on thickness and length of leg of cleat

Indirect connection (semi-rigid)

D

Web cleat

Indirect connection (non-rigid)

E

Equal UC sections

Butt weld

F

UC column sections of same serial size

Fillet weld

Welded splices

G

Seating plate

Different UC sections

H

Filler piece

Direct splice connection

J

Weld on inside of gusset plate

Gusset plate

Stiffener

Base plate

Gusseted base

K

Holding down bolt grouted in later

Angle section cast in foundation slab

Slot in angle

Plan

Section

Holding down bolts

L

Plates welded to web

Stiffeners instead of plates

Horizontal stiffeners

M

Gusset piece

Stiffener

Gusset piece

N

End of beam deepened

Butt weld

120   Welded connections

are normally shop-welded to the various members and in this type of joint there is little advantage in making the site joint by welds rather than by the usual bolts.

Semi-rigid joints between beams and stanchions can be obtained by the use of flange cleats fillet welded at the toes, one acting as a seating cleat, the other as a top cleat, with a space between the end of the beam and the stanchion flange (B). The rigidity of such a joint varies mainly with the type of tension flange connection, that is, with the top cleat, the stiffness of which varies with the thickness and length of the legs of the angle. The stiffness increases with the increase of thickness and with the decrease in length of the vertical leg (C). By means of such variations it is possible to attain either a very rigid joint or a relatively flexible joint.

In fully rigid connections it may be necessary to increase the shear strength of the stanchion web. This may be accomplished (i) by welding in horizontal stiffeners at the level of the beam flanges (L and Part 1, figures 107 and 108), (ii) by the further addition of web plates (L), (iii) by means of a triangulated system of stiffeners which may be designed to act either in conjunction with the stanchion web, or entirely alone.

The use of gussets or brackets at the junction of the beam and stanchion, or deepening the end of the beam (M, N) has the effect of reducing the shear stress at the connection and can eliminate the need to strengthen the stanchion web. It also increases restraint at the joint.

Semi-rigid and fully rigid framing

The majority of steel frames have been designed on what is known as the simple design method, in which all the beams are assumed to be simply supported at the connections with the stanchions, in disregard of the obvious and proved partial fixity existing at the ends of beams with bolted connections. BS 449 permits the alternative methods of simple, semi-rigid and fully rigid design. In the semi-rigid method, in certain conditions of end fixity, beams with bolted connections may be designed as partially fixed at the ends and can achieve a saving in steel of from 2 to 5 per cent. An allowance in design may be made for the end restraint of beam to stanchion connections where the beam is efficiently connected to a stanchion, efficiency of the connec-

tion being based on the thickness of the top cleat and the number and size of bolts.

The fully rigid method in which the interaction of beams and stanchions is calculated for all conditions of loading gives greatest rigidity and greatest saving in steel, but is not so readily applied to multi-storey as to single-storey frames, because of the large number of joints. The development of fully rigid multi-storey frames depends upon the extensive use of welding as a joining technique or on the use of friction grip, or torque, bolts (see Part 1, page 143).

Cold-formed steel sections

These are made from steel strip cold formed to shape in a rolling mill, press-brake or swivel bender, those formed by rolling being known as cold-rolled sections, the others as pressed-steel sections.

Pressed-steel sections are largely used for flooring and roofing units and wall panels, and for metal trim such as skirtings and sub-frames, the lengths of which are limited by the maximum width of the press-brake or bending machine (figure 121).

The press-brake is a machine which has a long horizontal former rising and falling with a pressure of 150 tonnes. This former presses the steel strip into a suitable shaped horizontal bed to form folds in sequence as shown in figure 121 A, by means of which simple sections are shaped, usually up to 3 m long, although some machines produce lengths up to 7 m. The swivel bending machine folds the strip by means of clamps (B), in lengths up to 2 m, the folds or bends being made in sequence as before. These methods can produce sections economically in small quantities. The press-brake can handle steel strip in thicknesses up to 20 mm but the swivel bender only up to 3 mm.

For structural members of greater length, cold-rolled sections are used. These are formed into the required shape by passing metal strip between six to fifteen progressive sets of forming spindles or rollers, each pair of which adds successively to the shaping of the strip, the final pair producing the finished section (C). The basic sections rolled are plain angles and channels, lipped channels, and zeds (see figure 122). Outwardly lipped channels are commonly called top-hat sections. The length is limited only by considerations of transport. The maximum width of strip which can be formed is

203

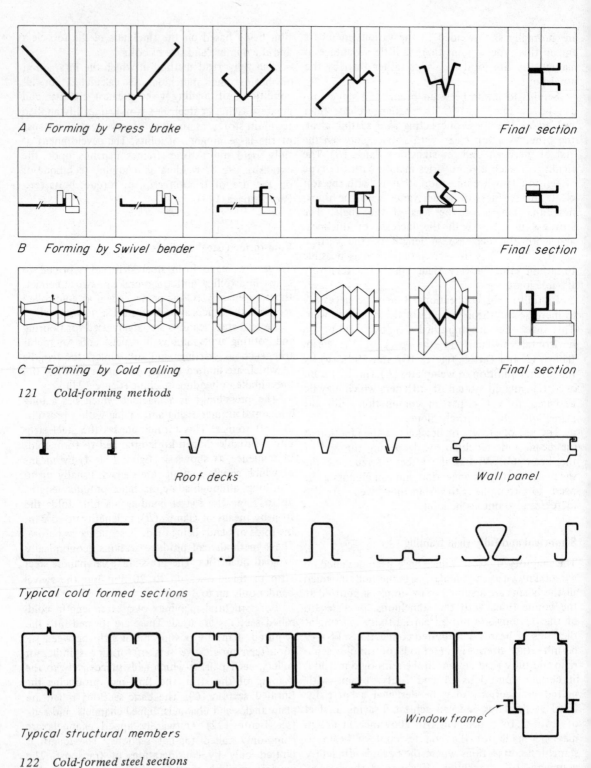

A   Forming by Press brake   Final section

B   Forming by Swivel bender   Final section

C   Forming by Cold rolling   Final section

121   Cold-forming methods

Roof decks   Wall panel

Typical cold formed sections

Typical structural members

Window frame

122   Cold-formed steel sections

204

rather more than 1 m, in thicknesses from 0.3 to 8.0 mm.

There is virtually no limit to the shape that can be rolled so that the designer can choose a shape best fitted for any particular purpose although, as the manufacture of the rollers for sections outside the basic range is expensive, the economic advantage of choosing such a section must be considered carefully if the quantity required is small[1]. Where possible the section should be made to fulfil more than one function by shaping it, for example, to avoid the casings often necessary with hot-rolled sections and to permit the direct attachment of claddings and windows. The strength of individual members may readily be varied by varying the gauge of the steel while maintaining the same overall sizes. These possibilities can lead to increased efficiency in the structure and to substantial economies in the weight of steel used.

As indicated in Part 1 cold-rolled sections are most efficiently used with structures of moderate loads and span, in which circumstances they can be cheaper than hot-rolled members. Erection of the structure is often cheaper and easier because of its light weight and rigidity. Cold-rolled sections are used for roof trusses, lattice beams and rigid frames, where they are probably used to greatest advantage, but they have been used for two-storey frames throughout, and for three-storey frames in which the bottom storey stanchion lengths are in hot-rolled sections in order to keep the overall size the same as the cold-rolled lengths above. As an alternative to the use of hot-rolled sections, a heavier section with the same overall size can be obtained by spot welding cold-rolled reinforcing sections on the inside.

Channels placed back to back and box sections are suited to axial load, the latter having considerable torsional strength. There appears to be little economic advantage in using cold-rolled sections for columns as far as cost of material and fabrication are concerned, but there are considerable savings in erection costs, which may be 10 to 15 per cent cheaper than with hot-rolled sections. When fire-resisting casing is not required external columns can, as mentioned above, be shaped to act as window mullions to accommodate windows by direct fixing as shown in figure 122. I-section beams formed from channels placed back to back are only likely to be economic for light loads over short spans, where the smaller hot-rolled sections

might not be used to their limits. The wider choice of section possible with cold-rolled sections gives them an advantage in these circumstances. Greater savings over hot-rolled work are found by using cold-rolled lattice beams over intermediate spans with light loadings although, as indicated above, the most economic field is in roof structures where the advantages of the low dead/live load ratio and ease of handling and erection show most clearly (see chapter 9, page 337).

Connections are made by various types of welds, self-tapping screws, bolts and cold rivets (see Part 1, pages 147-149) and sections can be formed to push-fit into each other, thus avoiding the use of gusset plates. For example, top-hat flange sections can be used for beams into which the bracing members fit, so that the node connection is direct. Details of construction in cold-formed steel are given in Part 1, chapter 6.

Structural considerations are similar to those in the design of light alloy structures. The design of thin wall structures requires special consideration, due to the possibility of local instability and, as in aluminium sections, lips to the edges may be provided to give increased stiffness to the section[2].

The need for protection from corrosion is important, because there is not so much margin for wastage of metal as in hot-rolled sections. Phosphating followed by paint dipping and stoving is usually adopted for internal work and where the structure is to be exposed to the weather, hot-dip galvanizing can be used. Mild-steel strip is widely used in the production of cold-rolled sections but high-tensile steel with rust-inhibiting qualities is likely to be used to an increasing extent because of its structural advantages.

THE REINFORCED CONCRETE STRUCTURE

Reinforced concrete, because of its particular characteristics, can be formed into walls as well as into

[1] The method is only economic for the production of large quantities. Up to 1500 m may need to be run off before the cost of setting up the machine is covered, depending on the complexity of the section to be rolled.

[2] The design of cold-rolled steel structures is covered by Addendum No. 1 (1961) to BS 449 – *The Use of Cold-Formed Steel Sections in Building.* BS 2994 (1958) 'Cold Rolled Steel Sections', gives the properties of a range of sections.

beams and columns to form a skeleton frame, and floor slabs can be designed without projecting beams to carry them. A reinforced concrete structure may, therefore, consist solely of walls carrying slab floors, slab floors and columns only or a combination of columns, beams and loadbearing walls, each being used to fulfil most satisfactorily the functions required at various points (see figure 125 *A, D*). Staircases and lift shafts often must be enclosed in solid walls, and it is logical to make these of reinforced concrete capable of both enclosing the areas and carrying the floor loads, rather than to surround the areas with beams and columns and then enclose with non-loadbearing panel infillings. Such enclosures, being monolithic in form, result in very broad annular columns running right through the building which can be used to provide resistance to wind pressure on the structure.

Greater flexibility in planning and design is possible with reinforced concrete than with steel.

At the beginning of its structural life reinforced concrete is fluid or plastic in character and this gives rise to two important factors concerning the nature of the structure for which it is used:

(a) the ease with which a monolithic structure may be obtained, producing a rigid form of construction with the economies inherent in this form (see page 206), and

(b) the ease with which almost any desired shape may be formed either for economic, structural or aesthetic reasons. For example, the material may be disposed in accordance with the distribution of stresses in the structural members, placing most where the stresses will be at a maximum and reducing it where they will be at a minimum.

These two factors together, monolithy, giving particular distributions of stresses, and variation in the disposition of material according to the stress distribution, produce characteristic concrete forms which are most obvious, as far as building structures are concerned, in single-storey structures. Nevertheless, these characteristic forms do find a place in multi-storey frames, and it should be made clear at this point that the designer's freedom to cast concrete in almost any shape is limited by the cost of the formwork or shuttering into which the concrete must be poured. This forms a large proportion of the total cost of a reinforced concrete structure as can be seen from the following approximate percentage break-down:

206

Concrete	40%	Materials	28%
		Labour	12%
Shuttering, including erection and stripping	32%	Materials	12%
		Labour	20%
Reinforcement	28%	Materials	20%
		Labour	8%

Shuttering costs for beams alone may be as high as 40 per cent of the total cost of the beam. It will be seen that the percentage labour content in shuttering is far greater than that in steel fixing or concreting, so that economies in shuttering will have a significant effect in reducing the cost of the concrete work. Such economies are the outcome of simple structural forms repeated a number of times, making the construction of the shutters a simple matter and enabling them to be used repeatedly to the maximum extent. Complicated shapes, particularly if curved, appearing only once in a structure lead to high shuttering costs.

In situ cast structures

Up to comparatively recent times multi-storey reinforced concrete frames have always been erected as *in situ* cast structures for which all the constituent concrete materials have been brought to the site, mixed and placed in formwork erected in the position the concrete will finally occupy in the completed structure. Such frames are invariably of monolithic construction by which full continuity throughout columns, beams and slabs is attained. The advantages of monolithic or fully continuous construction are:

(a) reduced deflections in the members

(b) reduced bending moments distributed more uniformly throughout the structure than in discontinuous structures. The reduced moments result in lighter members. The greater uniformity in distribution will, in members of uniform section sized to the maximum bending moment, have the effect of involving less waste of material at the points which are less highly stressed

(c) in the case of beams there is a less rapid increase in dead weight with increase in span because, due to the stress distribution, a great deal of the extra material is required over the supports which will take its weight directly.

Against these advantages must be placed the following disadvantages:

(a) the adverse effect of differential foundation settlement, which has been described on page 81

(b) adverse effect of temperature movement. Movement due to temperature changes has a similar effect upon a continuous structure to that of foundation movement, and close attention is necessary at the design stage to the maximum possible movement and to the use of expansion joints at appropriate points to prevent accumulative movement throughout the whole structure.

Cross wall construction in *in situ* cast concrete is known as 'box frame' construction. This, together with *in situ* cast external wall construction is dealt with in this chapter rather than in that on 'Walls and piers' because of its total monolithic character enabling the walls and floor slabs to act together. For ease of reference reinforced concrete floors have been considered in a separate chapter, although they do, in fact, form an integral part of most reinforced concrete structures.

Precast structures

A precast concrete component may be defined as a component cast in formwork in a position other than that which it will finally occupy in the completed structure and which, after removal from the forms and maturing, requires to be placed and fixed in position.

The technique of precasting concrete for structural purposes was originally applied to the manufacture of floor and roof slabs, but the process has now developed to such an extent that whole building structures can be erected from factory produced precast components involving columns, beams, floor and roof slabs, wall panels and cladding.

In terms of site work the great advantage of the precast structure is that the speed and simplicity of erection compares favourably with that for a steel frame and this, allied to the cheapness of concrete, makes it an extremely valuable method of construction. In addition to the saving of time and labour on the site factory production makes possible a closer control of the concrete than is often possible on the site, particularly in the case of small jobs,

and leads to a saving in materials and an improvement in quality. Formwork and its support is greatly reduced, the site is less obstructed and, in cases where the concrete is to be exposed, the production of satisfactory surface finishes is facilitated. The difficulties arising from the shrinkage of fresh concrete are eliminated because all maturing takes place before the components are built into the structure.

The principal disadvantage is that the continuity and rigidity of structure attained in an *in situ* cast structure are more difficult to achieve in the precast form and account must be taken of this at the design stage.

As with normal *in situ* work, precast work should be designed to produce the maximum repetitive use of a minimum amount of shuttering. Individual components should be simple in form and they should be as large as methods of transport and erection will permit in order to reduce the number of joints in the structure. As far as transport is concerned the limit on the size of a factory cast component is in the region of 18 to 21 m by 2.4 m overall. For multi-storey work a crane is invariably employed for erection purposes and, whereas a 250 kg component is about the heaviest which can be manhandled, when a crane is to be used on a job the size and weight of component should be related to the capacity of the crane likely to be used, since it will most economically be employed when hoisting at its maximum capacity. Individual joints can as easily be made in the case of heavy as in the case of light components when the units are supported by crane, but with smaller, lighter units more joints must be made and the crane must make a greater number of lifts at greater cost.

Precast concrete itself tends to be more expensive than *in situ* concrete because of factory overheads and transport costs, but against this must be placed, in terms of the structure as a whole, the savings in time and labour on the site, so that the costs of precast and *in situ* structures are generally about the same, unless the units are precast a considerable distance away from the site making transport costs high.

Precasting can, of course, be carried out on the site and some large contracting firms do, in fact, carry out much of their casting work in this way, although other comparable firms make a practice of carrying out all such work in a factory, even though this may involve transport over long distances.

Site casting does reduce the amount of handling and avoids transport costs but a large amount of site space is required for the casting beds. Provided that the quality of control and supervision usually available in the factory, and similar means of efficient vibrating and cleaning of shutters are available on the site, good results are possible. When structural components must be of such a size as to prohibit transport from a factory there is no alternative to site casting. Precasting of facing material or of components requiring a high degree of surface finish is probably best carried out under factory conditions.

Choice of structure

The choice of a particular reinforced concrete structural system as most suitable in any given case will depend largely upon the nature and purpose of the building. For example, the structure of a building to accommodate heavy, evenly distributed loads. might most economically be developed as flat slab construction, whereas one in which considerable concentrated loads caused by machinery would occur could most economically be formed with a beam and column system in which the various elements could be designed more easily with regard to the local loading at any point. In the case of flats requiring a high degree of fire resistance and a measure of sound insulation in the separating walls, a box-frame structure is suitable and can be economic whereas for an office block, requiring large areas which can freely be divided up in different ways by non-loadbearing partitions, this would be unacceptable.

Linked with these considerations will be that of resistance to wind. A box-frame structure provides ample resistance in a transverse direction but will require stiffening longitudinally by lift or stair enclosures or by solid walls. The frames to heavily loaded structures may have beams sufficiently deep to provide the necessary rigidity at the joints with the columns without any other form of bracing. In other cases, where solid walls must be provided on plan for functional reasons, perhaps fire division walls, it may be economical to transfer the wind loads to these entirely by the floor slabs so that the columns may be relieved of lateral pressure. When a precast concrete frame appears to be suitable the

decision must be made whether to use diagonal bracing within the frame or solid walls at suitable points to provide wind bracing.

As pointed out in chapter 3, the choice of structure is closely linked with the soil conditions on the site and the economic design of the foundations, and the consideration of all three must take place at the same time.

The most economic structure is not always that in which the amounts of steel, concrete and shuttering are all kept to a minimum, and the most economical beam is not necessarily one in which the 'economic percentage' of steel is provided, that is the amount of reinforcement which permits the safe working stresses in the steel and concrete to develop at the same time. In fact, it is normally only in slabs and sometimes in rectangular beams that this is possible. For economy of shuttering it may be desirable to maintain a constant depth for a continuous beam, although the bending moments vary considerably at different points, rather than attempt to reduce the amount of concrete. In order to standardize shuttering in this way to reduce its cost, considerable variations in the concrete mix and steel content, as a means of standardizing the size of beams and columns, can often be justified. In some circumstances it may be necessary to increase the steel content in order to restrict the depth of the beam for reasons of headroom, or in order to keep floor to floor heights to a minimum, and the extra cost may well be counterbalanced by savings in other directions.

The materials of reinforced concrete

Concrete itself has considerable compressive strength but is weak in tension, its strength in this respect being about one-tenth only of its compressive strength. CP 110: Part 1: 1972 gives the crushing strengths of various mixes and the allowable working stresses. In structural members in which both compressive and tensile stresses occur under load, the full compressive strength of the material cannot, therefore, fully be developed, and in order to overcome this deficiency a material, strong in tension, is introduced in the tensile zones to reinforce the concrete at those points. Steel is used for this purpose at the present time but investigations have been made for some time into the use of glass fibre. So far, however, a number of problems have not been solved.

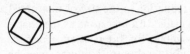

Square twisted bar

Ribbed and twisted bar   *Stretched and twisted ribbed bar*

Note: Twisted bars may be of cold worked mild steel or of H.T. steel   *Ribbed bars*

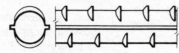

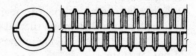

123   Deformed bars

Reinforcement

Steel is used, either as mild steel or high-tensile steel bars or wires, because it can easily be produced in forms suitable for the purpose, and it possesses to a large degree characteristics which are essential in any material to be used for reinforcement. Assuming adequate tensile strength these are:

(a) a surface which will satisfactorily bond with the concrete so that when the steel is stressed it will act together with and not pull away from the concrete

(b) a coefficient of lineal expansion much the same as that of concrete, so that under changes of temperature undesirable stresses will not be set up

(c) a relatively small elongation under stress to avoid excessive deflection.

Steel reinforcement is obtainable in a variety of forms of bars and fabric.

Bars   Circular section rolled mild steel bars or rods are most commonly used for all forms of reinforced work. High-tensile steel rods can also be used. Due to the higher working stress, less steel will be required to provide the same strength as mild steel, but in developing its higher strength it stretches more than mild steel and the cracks in the concrete around it will be larger. This might be undesirable in circumstances where corrosive fumes, for example, could attack the steel. The problem of the formation of large cracks is closely linked with that of bond between the steel and concrete. The better the bond the less risk there is of large, concentrated cracks developing, and as a means of increasing bond and limiting cracking to fine, well distributed cracks, *deformed bars* can be used. The greater bond stress obtainable makes it possible to stress the steel to higher limits and thus develop its strength to a

maximum. Greatest advantage is obtained when a large number of small diameter bars are used rather than a few larger bars, because the surface area in contact with the concrete is thereby increased. The use of these bars eliminates the necessity of end hooks, thus economizing in steel and simplifying work on the job.

Deformed bars are produced in a number of ways. Firstly, as high-tensile steel bars rolled with projecting ribs or corrugations along the length or, secondly, from mild steel bars which are cold-worked to increase the ultimate tensile strength and raise or eliminate the yield-point of the steel, the amount of increase depending upon the nature of the basic steel and the amount of cold-working. Both stretching and twisting are used as methods of cold-working and may be applied to circular ribbed bars or to square bars which become deformed by the twisting process and thus afford better bond. Some examples are shown in figure 123.

The importance of eliminating the yield-point lies in the fact that when mild or high-tensile steels are used for reinforcement, both of which have yield-points, the bond between reinforcement and concrete begins to break down when this stress is reached, so that in practice design is based on this value. The use of cold-worked mild steel or high-yield-point steel makes it possible to work to much higher stresses, particularly if deformed bars are used, and in this respect tests have shown that within certain limits such steel used as reinforcement can be stressed to its ultimate strength.

Fabric   The use of this form of reinforcement is an economic way of reinforcing large areas such as floor and roof slabs. It is produced in two main forms: as a mesh of wire or rods electrically welded at the points of crossing or as expanded steel sheets.

Mesh fabric is manufactured either from hard (cold) drawn steel wire or from small cold twisted steel bars, both of which, due to drawing or cold-working, have greater strength than mild steel, that of the wire being considerably greater. It is supplied as square or rectangular mesh in pieces or in rolls and its use avoids the necessity of tying together separate bars.

Expanded metal fabric of steel sheets slit and stretched to form a diamond-shaped mesh is supplied in pieces and for reinforcing purposes has a mesh of 75 mm or more.

Aggregates

Various materials are employed as aggregate, the selection depending upon the purpose for which the concrete is used. They may be divided into (i) heavy, (ii) lightweight aggregates.

Heavy aggregates These include the natural sands and gravels and crushed stones covered by BS 882, and crushed brick. These are normally used where strength and durability are required, although many lightweight aggregates are now being used for structural concrete.

Lightweight aggregates These were for many years used for reinforced concrete floor, roof and wall slabs, and are now used also for general reinforced concrete construction. Satisfactory materials are foamed slag, expanded or sintered clay or shale, and sintered fly ash. Expanded slate and imported pumice are also satisfactory. These aggregates can be used with the addition of sand to provide a satisfactory grading to give the necessary strength and impermeability, the resulting concretes having maximum densities from 1440 to 2000 kg/m^3 with crushing strengths from 13.8 to 62 N/mm^2. The modulus of elasticity of these concretes is less than that of gravel concrete so that deflection in beams and slabs tends to be higher.[1] Depths of members, therefore, need to be greater. Multi-storey structures have been constructed in lightweight reinforced concrete in the USA for many years and are being built in Great Britain, making use of both *in situ* and precast concrete[2].

Fire resistance When a high degree of fire resistance is required, the type of aggregate used is important as this largely affects the behaviour of concrete under the action of fire. The classification of aggregates in respect of fire resistance is given on page 371 and it will be seen that one natural stone, limestone, is included in the non-spalling types.

CP 110 permits a reduction in the size of reinforced concrete members made with limestone as the coarse aggregate. For example, for fire-resistance periods of four hours and two hours, columns may be reduced from 450 mm and 300 mm to 300 mm and 225 mm minimum overall size respectively.

Size of aggregate This should be as large as possible consistent with ease of placing round reinforcement. For heavily reinforced members the nominal maximum size is usually limited to 6.4 mm less than the minimum space between the bars or 6.4 mm less than the minimum cover to the bars, whichever is the smaller size. For general purposes 19 mm is usually satisfactory and normally used, and for members such as the ribs and the topping of hollow block floors 9.5 mm is normal. Where the reinforcement is widely spaced, as in solid slabs, the aggregate size may be as great, or even greater than, the minimum depth of cover to the steel, provided that the aggregate is not of a porous nature.

THE *IN SITU* CAST CONCRETE FRAME

For small span structures a rectangular grid layout, similar to that for a steel frame, with one-way spanning floor slabs, can be satisfactory, but with large spans or heavy loading a square grid with two-way spanning slabs is more economical because of the resulting reduction in thickness and dead weight of the slab. Codes of Practice restrict the thickness of floor slabs to a fraction of the span[3] as a precaution against excessive deflection so that there is a limit to the possible reduction in dead weight of slab for any given span. Up to the point at which deflection ceases to be the factor governing slab thickness, no advantage, therefore, is gained by using a two-way spanning slab in place of a one-way span, because the thickness of slab and, consequently, its dead weight, must be the same in both, and there will be little difference in the amount of steel required. After this point has been

---

[1] See table 66, page 169, *MBC: Materials*.

[2] See 'The Use of Lightweight Concrete for Reinforced Concrete Construction', by A. Short, MSc, AMIStructE, in *The Reinforced Concrete Review,* March 18, 1959.

[3] See table 15, page 239

reached advantage can be taken of the economies resulting from the use of a two-way spanning slab on a square grid.

In the case of one-way spanning slab construction the transverse, or tie-beams, necessary in a steel frame to provide lateral rigidity to the frame are not essential to an *in situ* cast reinforced concrete frame, since each floor is cast as the frame rises and can provide rigidity to the frame. Where such transverse beams are omitted, lateral stiffness against wind pressure must, of course, be provided by the floor slab which should be made strong enough to fulfil this function. Figure 124 shows ways of framing in this manner. (*A*) shows the floor beams running parallel with the main external walls resulting in a flat ceiling for the length of the structure, an advantage in certain types of buildings, such as offices, where movable partitioning is likely to be changed in position from time to time. With no transverse beam projections such partitioning can be standardized to the floor to ceiling height and be freely placed in any position. The supporting columns may vary in position along each beam relative to those carrying the other beams, although, unless essential for planning reasons, this would not be done becuase of the variations caused in beam shuttering and possible variations in foundation loading. When the width of the building necessitates two lines of internal columns these may be placed the width of a corridor apart and the floor slab between them thickened to form a stiff longitudinal beam (*B* and figure 149). This will act with the columns as a rigid inner structure to resist the wind pressure transferred to it through the outer spans of floor.

Figure 124 *C* shows the floor beams running at right-angles to the main external walls which are free of beams, thus permitting lightweight infilling panels to run from floor to slab soffit on elevation. If the beams are made sufficiently deep, internal columns may be omitted giving wide, unobstructed floor areas where these are necessary. (*D*) shows cantilever beam frames, the advantages and disadvantages of which are, in principle, the same as those constructed in steel (see page 189). The longitudinal beams necessary with steel can, however, be omitted as shown, provided the floor slabs give adequate lateral stiffness. The propped cantilever principle (see page 189) may be applied to a reinforced concrete structure and, in some cases, it may be economic to omit the cantilever beams

and design the floor slab itself to cantilever over longitudinal beams. The soffit can be sloped up to a shallow outer edge beam which will be supported by the outer 'props'.

As already mentioned at the beginning of this section, the rectangular grid layout can be economical for small spans and lightly loaded structures, but when larger spans and heavy loads are involved the square grid with two-way spanning slabs shown in (*E*) is likely to be cheaper. Although normal beams carrying a simple solid slab of this type will show economies over one-way spanning slab construction, in the case of wide-span grids certain variations will result in greater economies by further reducing the dead weight of the floor slab. The normal slab and deep beams may be replaced by a drop slab in which the beams are replaced by a thickening of the slab to form wide, shallow bands over the lines of the columns as shown in (*F*). The effect of widening the beam to a band, is to shorten the span of the slab with a consequent reduction in its thickness, dead weight and amount of reinforcement.

The floor slab, which is an integral part of the structure and has a significant effect upon the economics of the building as a whole, may be constructed in various ways. Types of floors are discussed in chapter 6.

As stated earlier the nature of concrete makes it an adaptable material. It permits considerable latitude in the form of structural members for structural or other reasons such as planning and lighting, and readily permits the interaction of beams and floor slabs to produce more economical members. Some examples of this adaptability are shown in figure 125. The use of beam and column framing with load-bearing concrete walls in the same structure is logical when the latter can fulfil an enclosing, as well as structural, function (*A*) and is often adopted to provide or assist in the resistance to wind pressure (*D*). (*B*) and (*C*) show ways in which members may be formed to fulfil dual functions, to accommodate services or to distribute material according to the stress distribution in a member.

Where the floor slab is cast monolithic with a beam, as is usual in most cases, the T-beam is commonly used (see figure 126). The necessary tension reinforcement in a beam can be accommodated in a relatively thin rib of concrete which needs to be only wide enough to accommodate the steel and to provide for shear stresses. By combining

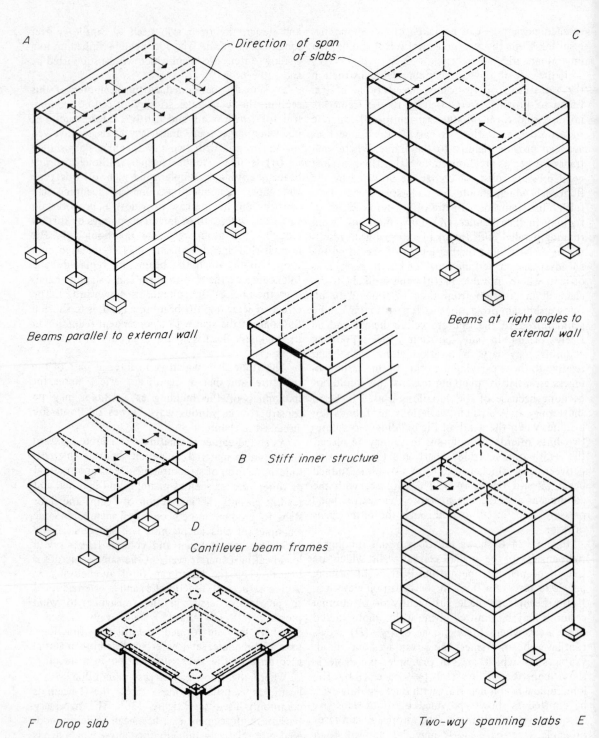

A

Direction of span
of slabs

C

Beams parallel to external wall

Beams at right angles to
external wall

B   Stiff inner structure

D   Cantilever beam frames

F   Drop slab

Two-way spanning slabs   E

124   Reinforced concrete frames

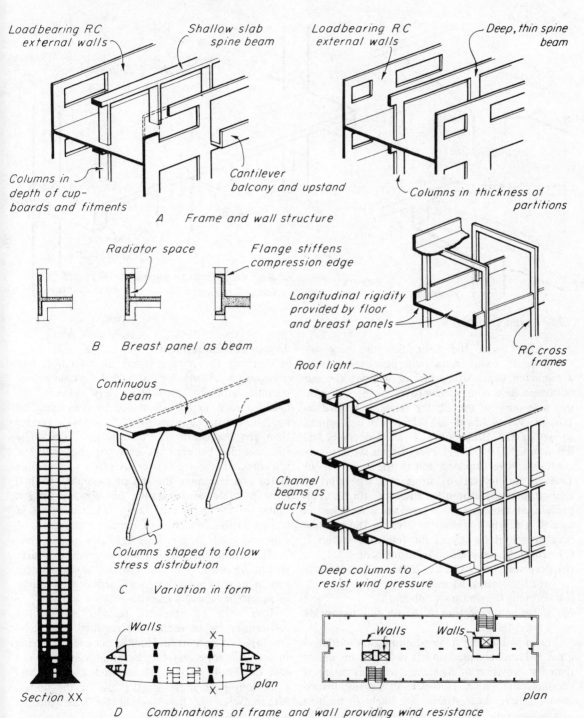

Loadbearing RC external walls

Shallow slab spine beam

Loadbearing RC external walls

Deep, thin spine beam

Columns in depth of cupboards and fitments

Cantilever balcony and upstand

Columns in thickness of partitions

A   Frame and wall structure

Radiator space

Flange stiffens compression edge

Longitudinal rigidity provided by floor and breast panels

B   Breast panel as beam

RC cross frames

Roof light

Continuous beam

Channel beams as ducts

Columns shaped to follow stress distribution

Deep columns to resist wind pressure

C   Variation in form

Section XX

Walls

plan

Walls   Walls

plan

D   Combinations of frame and wall providing wind resistance

125   Adaptability of reinforced concrete

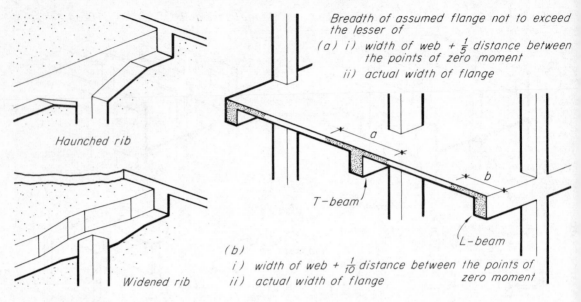

Haunched rib

Widened rib

Breadth of assumed flange not to exceed the lesser of

(a) i) width of web + $\frac{1}{5}$ distance between the points of zero moment

ii) actual width of flange

T-beam

L-beam

(b)

i) width of web + $\frac{1}{10}$ distance between the points of zero moment

ii) actual width of flange

126 T- and L- beams

such a rib with the floor slab the necessary resistance in compression may be obtained with a minimum depth of beam and without the use of compression reinforcement. Where the slab is on one side only of the rib the beam is termed an L-beam. The width of slab which may be assumed to act as the flange of a T- or L-beam is laid down in CP 110: Part 1: 1972. Beams are normally continuous over supports and in the case of T- or L-beams the reversal of stresses at the point of support presents a problem since the rib, in compression at those points, is generally insufficient in area to resist the compressive stresses. The problem may be solved by any of the following methods:

(a) providing compression reinforcement
(b) deepening the rib by means of a haunch, which increases the area
(c) widening the rib for its full depth
(d) widening the bottom of the rib only to provide a lower flange.

For most building frames compression reinforcement is invariably used and is generally no dearer than the provision of haunches, which is the most common alternative. However, in heavily loaded frames, where shear stresses are likely to be high at the supports, haunches may be preferable. The flaring or widening of the rib and the provision of a lower flange are rarely used as they complicate the shuttering (see figure 126).

214

Layout of reinforcement

In detailing the reinforcement in a member the arrangement of the bars should be as simple as possible with sufficient space left between the bars for each to be surrounded by concrete. The minimum distance between bars must be greater than the maximum size of aggregate used. The space needed between the bars, together with the thickness of external cover, is often a governing factor in determining the size of a member. CP 110 gives a table of maximum bar spacings and a table of the amount of cover required, which is related to the condition of exposure and grade of concrete, and ranges from 15 mm to 50 mm.

The minimum number of different bar sizes should be used, and the use of the largest size consistent with good design will reduce the number of bars to be bent and placed.

Bars must be extended beyond any section sufficiently far to enable the required grip to be developed or be hooked. Hooked bar ends occurring in a tensile zone may cause cracking, and to avoid this the hooks should be omitted and the bars made longer to allow for the loss of the hooks. Alternatively the bars should be bent into a compression zone.

Tension bars continuous round a re-entrant angle, as in a cranked slab, should have a radius large enough to reduce the outward pressure of

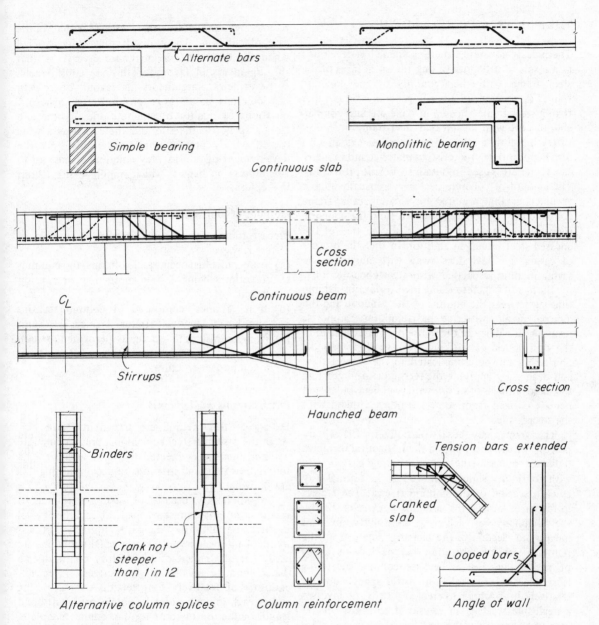

Alternate bars

Simple bearing   Monolithic bearing

Continuous slab

Cross section

Continuous beam

C_L

Stirrups   Cross section

Haunched beam

Binders   Tension bars extended

Crank not steeper than 1 in 12   Cranked slab

Looped bars

Alternative column splices   Column reinforcement   Angle of wall

127 Reinforcement

the steel to that which the concrete can resist in shear and tension. If this is not possible the tension bars should be linked by stirrups to compression bars, or better, the bars should be separate and should extend beyond the intersection sufficient for bond (figure 127).

Links must be provided to avoid the possibility of buckling of compression reinforcement and the bursting out of the concrete. To make the reinforcement stiff during concreting, and to hold stirrups in position, bars in the corners of beams are provided and the stirrups are continued round the tension side. At all points of intersection the bars must be wired together to prevent displacement during concreting.

Some typical details of reinforcement layout are shown in figure 127.

215

THE PRECAST CONCRETE FRAME

The precast concrete structural frame has developed as a result of attempts to link the advantages of the steel frame with the economy of the concrete frame. A precast frame will generally be cheaper than a steel frame encased in concrete and comparable in cost with an uncased steel frame. In multi-storey buildings a steel frame must be encased with some appropriate fire-resisting material and in many cases concrete is used to obtain sufficient protection. The amount of concrete necessary is usually almost as much as that required for a comparable frame constructed of reinforced concrete, and the concrete frame would show a saving in cost over that of the encased steel frame. In addition to this, the process of casing the steel does away with much of the saving in time associated with steel construction[1].

The *in situ* concrete frame involves a considerable time lag between the pouring of the concrete and the removal of all shuttering and temporary supports, resulting in a delay in the re-use of shuttering and in the obstruction of working areas for long periods. Shuttering can be complicated in the case of slabs and beams. By applying the technique of precasting the disadvantages of *in situ* work can be avoided and benefit derived from some advantages linked with the steel frame.

The frame may be (i) partially or (ii) wholly precast. In the first method the horizontal members only are precast, the columns being cast *in situ* with continuity simply achieved in the normal way as each section of the column is cast. The factors in favour of casting the columns *in situ* are, firstly, the simplicity of achieving continuity by this means and, secondly, the fact that solely from the point of view of shuttering there is little in favour of pre-casting. This is because column shuttering is simple in form, takes up little space, requires relatively little labour to erect and strip and involves a negligible wastage in re-use. In contrast, in the case of beams and slabs, bending stresses are set up in the shuttering while the concrete is wet, necessitating heavy forms and a considerable amount of propping. This is avoided when the shuttering is supported by the ground or a production bench and the formwork can be lighter and cheaper. Floor areas are obstructed by props for considerable periods while the concrete attains sufficient strength to support its own self-weight. Considerable wastage in horizontal shuttering occurs in stripping and

re-erecting[2], and this work takes longer than in the case of columns. These disadvantages are avoided when the horizontal members are precast, so that the arguments in favour of this are considerable.

Nevertheless, arguments in favour of *in situ* cast columns based only on shuttering disregard questions of quality of concrete and of time and labour spent on the site and there is now a wide use of wholly precast frames, especially since the development of multi-storey columns precast up to five storeys in height, which minimize site labour in erection.

Methods of fabrication

Precast frames can be fabricated in a number of different ways:
(a) from individual beams and columns, the columns sometimes being cast in more than one storey height
(b) from 'frames' composed of column sections and beam lengths forming a single cast unit
(c) from precast units acting as permanent structural shuttering to cast *in situ* concrete to form a composite structure.

Precast beams and columns

Details of this method are shown in figure 128. As in the case of a steel stanchion a precast concrete column must be connected by some means to the foundation slab, and methods of accomplishing this are shown at (*A*).

Column joints, usually of such a nature as to ensure continuity, must be made at each floor level in the case of storey height columns or at less frequent intervals if multi-storey height columns are used. The attainment of satisfactory continuity at the column connections necessitates either the exposure of relatively long lengths of reinforcing rod at each joint in order to obtain sufficient bond length or the use of some form of connecting plate. The section of column left open for the jointing of the rods must then be boxed in and concreted

[1] See, however, footnote to page 185

[2] It has been shown that average comparable figures for the re-use of horizontal timber shuttering are 4 to 6 times for *in situ* work and 20 times for precast work, although with the use of plywood for shutters, re-use is considerably greater in both cases. For large precast contracts the use of steel forms can be economic and results in more than 100 times re-use.

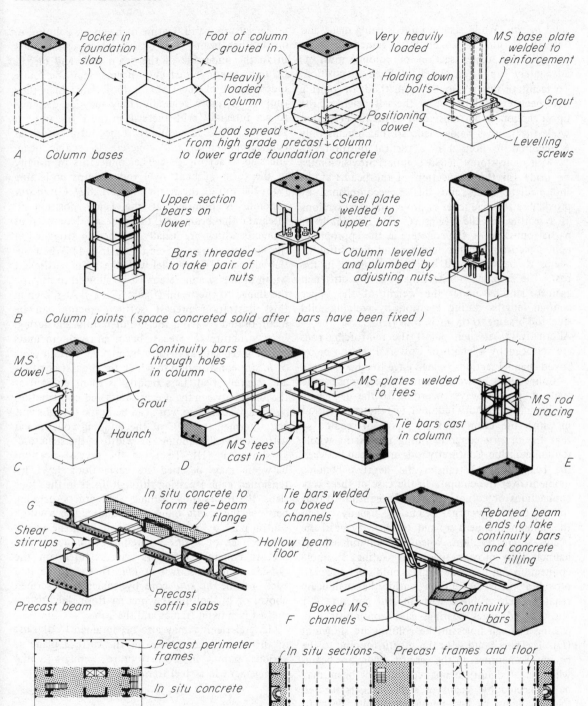

A Column bases

Pocket in foundation slab

Foot of column grouted in

Heavily loaded column

Very heavily loaded

Holding down bolts

Positioning dowel

Load spread from high grade precast column to lower grade foundation concrete

MS base plate welded to reinforcement

Grout

Levelling screws

B Column joints (space concreted solid after bars have been fixed)

Upper section bears on lower

Bars threaded to take pair of nuts

Steel plate welded to upper bars

Column levelled and plumbed by adjusting nuts

C

MS dowel

Grout

Haunch

D

Continuity bars through holes in column

MS plates welded to tees

MS tees cast in

Tie bars cast in column

E

MS rod bracing

G

Shear stirrups

In situ concrete to form tee-beam flange

Hollow beam floor

Precast beam

Precast soffit slabs

F

Tie bars welded to boxed channels

Rebated beam ends to take continuity bars and concrete filling

Boxed MS channels

Continuity bars

J

Precast perimeter frames

In situ concrete

Precast floor

H

In situ sections

Precast frames and floor

128 *Precast concrete structures*

solid *in situ*. While such a joint is being made and until the *in situ* cast concrete has gained sufficient strength, the upper section of column must be adequately supported and held fixed in position. To facilitate this a number of methods of jointing have been evolved to permit the self-weight of the upper section to be transferred to the lower section while the joint is being made so that bracing only is required to hold it in a vertical position (*B*).

· With multi-storey height columns provision must be made for the connection of the beams at the intermediate levels and this is accomplished by providing either haunches or projecting steel sections as a seating for the beams (*C, D*). The whole or partial omission of the concrete at the appropriate points according to the number and disposition of beams to be connected has been adopted in the past. The exposed reinforcing rods are stiffened against bending under the weight of the upper column lengths during erection by welding mild steel rod bracing to the main rods (*E*).

Alternatively, at such points the reinforcing rods are replaced by a length of Universal section or by boxed channel sections welded edge to edge (*F*).

Connections between beams and columns can be made in various ways according to the degree of rigidity and continuity required. (*F*) shows a method in which the beam ends are boxed or rebated so that the *in situ* concrete filling makes the whole joint monolithic, Continuity rods are placed between the beam ends to transfer the negative bending moment over the column. In the case of three-way connections or corner columns the continuity rods for the centre or corner beams are usually cast in the column and bent up out of the way until the beams are in position. When the columns can be haunched to provide the beam seating, a simple connection can be made using a mild-steel dowel to provide a positive beam fixing within the small beam rebate which is filled with *in situ* concrete (*C*). Site welded steel bearings and continuity bars passing through holes in the column are shown at (*D*). The bars bond with the *in situ* topping to the beams. Sometimes, in the case of secondary beams where great rigidity is not required, the beam may be connected by means of steel brackets secured to the column by high-tensile steel bolts. Beam to beam connections may similarly be made or, more commonly, by means of rebates formed in the main beam to accommodate the ends of the secondary beams.

To ensure a full bearing of beam on column, or on another beam, steel bearing plates can be cast in to the underside of the beam ends and on the column bearing surfaces. Beams may be quite independent of the floor or roof slab which they support or, more generally, they may be designed to act integrally with the slab in the form of T- or L-beams. This necessitates some form of shear connector and *in situ* concrete to enable beam and slab to act together. The connectors are usually in the form of bent steel rod stirrups projecting from the top of the beam (figure 128 *D, G*). *In situ* concrete is required at the beam position to integrate the beam and the precast floor or roof elements which are usually notched or troughed at the ends to receive it (see figure 150 *A*). When the slab elements run parallel with the beam, sufficient width of *in situ* concrete must be allowed to provide the flange to the beam (figure 128 *G*). As shown in this illustration precast soffit elements can be incorporated to avoid the use of normal shuttering.

A structure of precast beam and column units can be given lateral rigidity by making certain parts of *in situ* construction to which the precast portions are securely tied. Less rigidity is then required at the joints between the precast units and construction is simpler. The *in situ* work can be in the form of end and intermediate bays of the building constructed with cross walls running the width of the structure, as shown at (*H*). To these the precast portions between may be tied at each floor level by tensioned cables passing through ducts in the floor slabs, the floors thus forming wind girders spanning between the *in situ* blocks. In building types with a central core of services, stairs and lifts, the whole core can be of *in situ* cast reinforced concrete and precast concrete members can be used on the outside walls as shown at (*J*). These can be tied back to the *in situ* core by cables as described above, or by bolting the precast floor slabs to the precast perimeter frames and the core.

Lateral rigidity may also be obtained by the use of diagonal braces placed in the vertical plane at various points in the structure, similar to the arrangement in a steel frame.

Precast frame units

These can be formed in various ways but each type consists essentially of a pair or more of columns linked by a beam and so formed that beam and column connections do not occur at the same point, thus overcoming the difficulties of assembly which arise when they do coincide. A pair of columns linked by a beam is easier to brace temporarily while the joints are being made than separate columns. Such units are, therefore, easier and quicker to erect than separate beams and columns. These units are suited to a layout in which there are no lateral beams, the floor and roof slabs spanning directly between lines of support running parallel to each other.

Figure 129 shows types of such units. In (A) the column joint is located at the top of the beam and in (B) at the points of contraflexure in the columns where the bending moment is at a minimum. When the perimeter columns of a building are closely spaced to act as window mullions, frame units can be formed of two columns, a head beam and a cill beam. The method of linking the units varies according to the treatment of the elevation. In (C) the head and cill beams will be hidden by cladding and the columns will be exposed. The head beam, that is the floor beam, can project on each side and meet its neighbour at the centre of the adjacent bay without detriment to appearance. The cill beam acts as a brace to the frame and may or may not provide support to cladding. When the whole of the frame is to be exposed on elevation care is needed in the arrangement of the joints. The head and cill beams are kept within the line of the columns and when the units have been erected in alternate bays they are joined by separate beams at cill and head which are bolted to them (D).

These illustrations show storey height frames with a pair of columns, but they can be constructed three or four columns wide and with only a top beam, provided the beam is substantial enough to withstand the hoisting stresses and the weight is within the capacity of the crane to be used on the job. Two- or three-storey height frames can also be used. These multiple frames reduce erection time and labour by the reduction in the number of joints to be made.

Column connections are illustrated in figure 129 A, B, C, details 1, 2, 3. They may also be made by means of high tensile steel bars passed through holes formed through the height of the columns. These are connected at the joints by screwed couplers and tightened by a torque-controlled spanner. Connections between beam ends are made by coupling plates or rebated ends and dowels.

Rectangular frames may be formed of half-columns at each side and half-beams at head and foot. The half-columns are channel shaped so that when erected a void is formed between them which is filled with *in situ* concrete, which can be reinforced if necessary. The half-beams are bolted together.

Composite structure

Reference is made on page 218 to the use of precast soffit elements as permanent shuttering to portions of *in situ* cast work required to form T- and L-beams at the junctions of precast beams and slabs. The use of reinforced precast concrete units as permanent shuttering, designed to act with *in situ* concrete to form a composite structure as illustrated in figure 129 E, is a means of obtaining the continuity and rigidity inherent in *in situ* cast work without the use of normal formwork. It also reduces the amount of precast work which factory overheads and transport costs tend to make more expensive than *in situ* work[1].

For economy the units should be shallow, but for ease of handling their thickness should not be less than one-fortieth of the length. In order to obtain units of reasonable length, therefore, the section should be of such a shape as to give stiffness to a thin member (see figure 129E) or be stiffened in some other way. In the case of beams the precast element should extend no higher than the soffit of the floor slab, so that in some instances the precast unit may be only a shallow strip carrying the tensile reinforcement similar to the prestressed element shown. To stiffen this during transport, stirrups and any top reinforcement for the beam may be introduced. Increased stiffness is given if diagonal bars are used to form the reinforcement into a lattice girder similar to the elements of this type used for floor slab construction shown in figure 147 G.

Satisfactory bond between the precast units and the *in situ* cast concrete is essential in order to transfer shear stress, and although research has shown that a roughened surface on the precast unit is adequate, this does assume good site supervision and workmanship in forming the junction between the two. A definite mechanical bond is usually ensured in practice by means of projecting wire

[1] See also references to 'ferro-cement' shuttering on page 326.

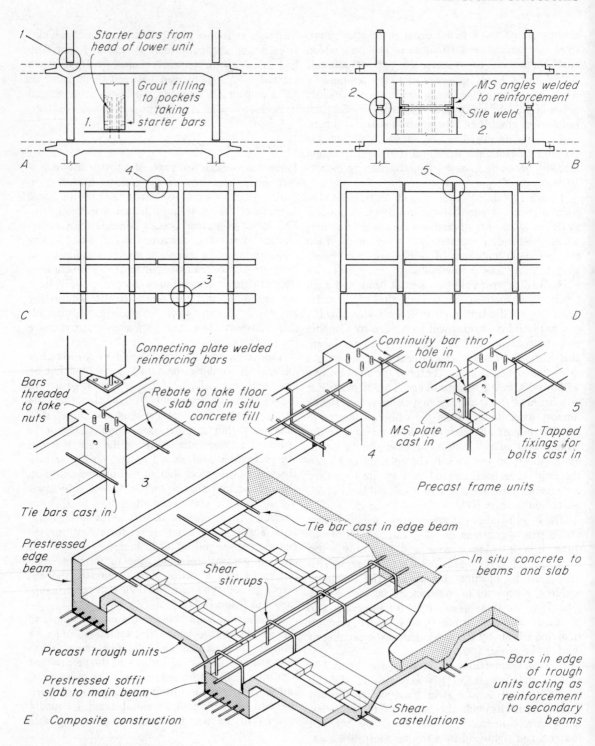

1 Starter bars from head of lower unit

Grout filling to pockets taking starter bars

1.

A

2 MS angles welded to reinforcement

Site weld

2.

B

4

3

C

5

D

Connecting plate welded reinforcing bars

Bars threaded to take nuts

Rebate to take floor slab and in situ concrete fill

3

Tie bars cast in

Continuity bar thro' hole in column

MS plate cast in

4

Tapped fixings for bolts cast in

5

Precast frame units

Tie bar cast in edge beam

Prestressed edge beam

Shear stirrups

In situ concrete to beams and slab

Precast trough units

Prestressed soffit slab to main beam

Shear castellations

Bars in edge of trough units acting as reinforcement to secondary beams

E   Composite construction

129   Precast concrete structures

stirrups or castellations. When composite beams or slabs are continuous the negative moments over the supports must be resisted by reinforcement placed in the *in situ* cast concrete. Figure 129 *E* shows a floor for heavy loads which incorporates a wide shallow beam in which the tensile zone is a precast prestressed slab, the compression zone being of *in situ* cast concrete. The side shuttering of the beam is formed by the edges of the precast trough units forming the lower part of the floor slab. These trough units are very thin but their form, which provides the shuttering for main and secondary beams, makes them stiff enough to carry the live loads during the casting of the *in situ* concrete.

The combination of precast and *in situ* cast concrete is particularly economical when allied to prestressing in the range of 6 to 9 m spans, for which normal prestressed concrete is not generally economical. In a normal prestressed concrete beam the concrete throughout is of high quality, but in a composite beam the lower precast and prestressed section only need be of high quality concrete, thus effecting an economy due to the smaller volume of high quality prestressed concrete to be manufactured and transported to the site.

Precast units and *in situ* concrete may be combined in columns as well as in beams and slabs. By using a precast concrete casing with an *in situ* cast core, time and labour can be saved by the elimination of normal shuttering and a good finish is obtained when the surface of the columns is to be exposed. It also permits the construction of the next floor to proceed more quickly while still maintaining full monolithic junctions with the floors and beams above and below.

THE REINFORCED CONCRETE WALL

In situ cast external wall

The reinforced concrete loadbearing wall used as the enclosing wall to a building is the alternative to its use as a dividing element in the concrete box frame described below. The wall areas over openings act as beams and those areas between openings as columns, thus no projections occur internally (see figure 125 *A*). These openings may be wide, since with normal cill heights there is ample depth of wall between window head and cill above to act as a deep, thin beam and the wide, narrow window is a characteristic of this form of construction. Alternatively, the whole height of the wall may be regarded as a beam pierced by any necessary openings for windows.

Sufficient width of wall must, of course, be left between openings to act as columns taking all the vertical loads. The problems of appearance and thermal insulation are the same as with the plain concrete wall, but the danger of cracking due to possible unequal settlement is reduced because reinforcement is present to resist any tensile stresses set up.

The concrete box frame

This is a form of cross-wall construction in which the walls are of normal dense concrete and, with the floors, form box-like cells as shown in figure 130. As in the case of brick or block cross-wall construction, it is suited to those building types in which separating walls occur at regular intervals and are required to have a high degree of fire resistance and sound insulation. The most common building type in this category for which it is suitable is the multi-storey flat or maisonette block. The advantages listed on page 131 in respect of brick and block cross-wall construction apply also to the box frame.

In concrete walls of normal domestic scale, about 2.4 m high and 100 mm thick, failure is almost wholly related to the strength of the concrete and very little to the slenderness of the wall. Reinforcement, therefore, may be nominal in amount or may be omitted altogether provided that the concrete is sufficiently strong to resist the stresses set up under load. For multi-storey blocks in the region of ten or eleven storeys high the mix would be designed to give a strength of around 15.5 N/mm^2 at 28 days, although for the two lowest storeys a stronger mix might be necessary as well as the inclusion of reinforcement.

Cracking due to the shrinkage of concrete is normally overcome by the inclusion of shrinkage reinforcement. Such cracking generally occurs only if the shrinkage is resisted by some restraint, such as that offered by changes in the plane of a wall or by a previously poured lift of concrete which has been permitted to take up its shrinkage before the next lift is poured on to it. Provided that concreting can proceed without undue delay and that the walls are in simple, straight lengths, shrinkage reinforcement in the walls may safely be omitted.

Although the junctions of walls and floors in a

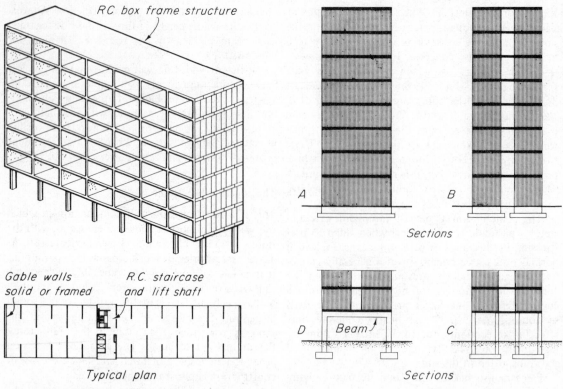

RC box frame structure

*Gable walls
solid or framed*

*R.C. staircase
and lift shaft*

Typical plan

Sections

A   *B*

D   *Beam*   *C*

Sections

130   *Concrete box frame*

box frame are monolithic, if the walls are not reinforced the structure can only provide rigidity in the length of the building to the extent of the precompression set up in the walls by the floor loads and self-weight of the walls, as explained in the case of normal cross-wall construction (see page 132). Additional stability must normally be given by staircase and lift shafts of reinforced concrete, or by the inclusion of longitudinal walls at certain points in the plan. The box-walls themselves provide rigidity in the transverse direction.

Many box frames have been constructed with the end, or gable walls similar in form to the internal cross walls. The solid external concrete wall suffers certain disadvantages (page 125 and Part 1, page 85) and to it must be applied thermal insulation and, generally, some external facing for the sake of appearance and to ensure weather resistance. As both can be applied as satisfactorily to a frame as to a solid wall and since the latter, used as a gable wall, is more expensive than the frame, it appears logical to use a frame in this position[1]. With regard to ther-

mal insulation it is desirable to apply insulation to the ends of the box walls as heat losses at these points can be high, leading to condensation on the internal faces of the walls adjacent to the exposed ends. Depending upon the relationship of infilling panels and wall ends the insulation may be applied to the inner faces or end of the wall as shown in figure 73.

Creasy[2] gives graphs showing that for low blocks the box frame is not so economical as either a reinforced concrete frame or brick cross-wall construction (see figure 74). In high blocks, however, it is cheaper than the frame when plan requirements permit the walls to be spaced 5.00 m or more apart but less economic when the walls must be placed

[1] It has been said that the cost of mounting and demounting the shuttering for the gable walls is about six times the cost of that for the internal walls.

[2] See 'Economics of Framed Structures', by Leonard R. Creasy, BSc (Eng), MICE, page 256, *Proceedings of the Institution of Civil Engineers*, Vol 12, March 1959.

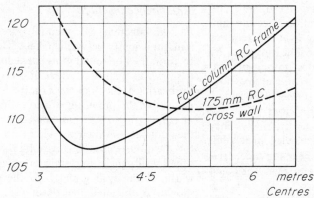

Based on ten-storey construction with 2·7m storey height and 9m depth of building

131   Concrete box frame — relative cost of structure

closer together (see figure 131). It should be noted that the optimum spacing of the walls is 5.00 to 5.50 m with the cost curve rising a little more rapidly with a decrease than with an increase in the spacing. It should also be noted that the graphs show the relative costs of the structural frames only, and disregard the effect of other elements which have a bearing upon the total cost of the building as a whole. For example, the cost of infilling panels of bricks or blocks at the separating wall positions, to provide the necessary degree of fire resistance and sound insulation, must be added to the cost of the reinforced concrete frame to make it comparable with the box frame.

In its simplest and most economic form all the box walls run in a straight, unbroken line from back to front of the building and are supported directly by a strip foundation (figure 130A). They may, however, be pierced by openings or be in completely separate sections on the same line, or staggered relative to each other provided that each section is in the same position throughout the height of the building (*B*). If the upper floors are to be supported on columns at ground level the necessity of beams and the disposition of the columns will depend upon the arrangement of the walls above. Straight, unbroken box walls can act as deep beams spanning between the supporting columns with any necessary reinforcement placed in the tension and shear zones. If the walls are broken extra columns must be introduced to enable each wall section to act as a beam (*C*) or, alternatively, a separate beam must be introduced to pick up the sections and transfer the loads to the columns (*D*).

Large precast panel structure

In this form of construction the loadbearing elements are large panels not less than storey-height, used with precast floor and roof units (figure 132 *D*). Window openings may be cast in the external panels which are usually finished with an exposed aggregate or tooled or profiled surface and incorporate thermal insulation, either sandwiched between two leafs or applied to the internal face. Internal panels can be made smooth enough to make plastering unnecessary.

The method is most suitable for residential buildings since the dense concrete panels can provide, as well as the strength for loadbearing, the degree of fire resistance and sound insulation required at the separating walls. Cellular, cross-wall and spine-wall plan forms may be used. The advantage of the cellular plan is its inherent stability and the fact that all walls may be loadbearing so that the floor panels may be two-way spanning.

Types of panels

External wall panels are commonly either of solid or of sandwich construction although waffle slabs are also used. The latter, however, have a number of disadvantages.

Sandwich panels have a layer of insulation incorporated either symmetrically or asymmetrically in the thickness of the slab (figure 132). In the former both internal and external leaves are loadbearing with transverse ties strong enough to ensure that both act together. The restraint thus offered to thermal and moisture movement in the external leaf

223

can cause it to warp. This is overcome by an asymmetrical positioning of the insulation since the thinner non-loadbearing leaf, usually external, requires only to be attached to the loadbearing leaf by lighter forms of ties. These ties, either of hot-dipped galvanized mild steel, or of suitable non-ferrous metal, must have ends formed to ensure a mechanical anchorage between the leaves.

Solid external panels are insulated internally. They are simpler to produce than sandwich panels but usually require a vapour barrier near the inner face. Certain types of these panels are in cavity or cored form.

Internal loadbearing wall panels are solid or cored and between 125 and 225 mm thick with nominal reinforcement. Adequate sound insulation can be achieved with a thickness of 175 mm if plastered on both sides and rather thicker if not.

Floor panels may be of solid or cored construction. The former may be reinforced as two-way spanning slabs and they also provide better air-borne sound insulation. The latter are lighter in weight, but can span in one direction only.

Casting panels

Horizontal casting is used for complicated panels which present some difficulty in casting, such as sandwich-panels for external walls, those with openings in them and those which are to have an integral or applied surface finish. When cast these are preferably removed from the moulds by means of pivoting mould beds or by vacuum pads, in order to avoid damage. The former method avoids the need for reinforcement to resist lifting stresses.

Vertical casting is preferable for wall and floor panels required to have a fair face both sides since this has the advantage of eliminating face trowelling. The moulds can be arranged in batteries with ten or more compartments, the division plates being of thick steel or concrete panels or of ply facing on both sides of a steel frame. The system of using two concrete panels, initially cast horizontally with a very smooth, true face, as the mould faces to reproduce a run of similar units was developed by the Building Research Station. The first panel is cast between the initial pair and has a true face on each side. After curing the three panels are spaced apart to provide the mould for two further panels, the five then being used to produce four more, and so on[1].

Structural connections

In situ concrete is commonly used to form the structural joints between panels. The method used to form the horizontal joint between the wall panels is shown in figure 132 *A, B*. It is preferable to limit the bearing of the floor slabs on the heads of the wall panels by the provision of projecting nibs or horns at about 150 to 225 mm centres along the edge of the floor panels, which provide the necessary support for the floor slabs. This permits the load from the upper wall panel to be transferred across the whole width of the wall directly to the panel below as shown. A threaded bar or dowel projecting from each end of the lower panel provides, by means of nut and washer, temporary support and a means of levelling the top panel. The joint is filled with *in situ* concrete and after this has set the gap above it is dry packed with cement mortar. When this in turn has set the nuts are run down to ensure contact between the upper slab and the packing thus offsetting the initial shrinkage of the mortar.

Vertical edge joints may be made as at (*C*), the ends of the panels being rebated to take a concrete filling, or recessed to hide the joint as in the right-hand detail. Shear keys are often formed on the edges of wall and floor panels by means of projections or indentations (see figure 150 *A*).

Methods of weatherproofing external joints are illustrated in (*B*) and (*C*), but reference should be made to Part 1, chapter 2, where this subject is discussed in detail and alternative methods are shown.

Addendum No. 1 to CP116 to which reference should be made[2] requires the whole structure to be designed against progressive collapse due to accidental loading and all members to be adequately tied together by the provision of steel connections of various forms. Steel reinforcement acting as a peripheral tie is required at each floor and roof level (figure 132 *B*) to which ties in both directions at each level should be anchored. Ties in the direction of the span may be encased in the floor or roof panels (*D*) and transverse ties may be placed in the *in situ* concrete in the joints between the transverse wall panels (*A*).

[1] See BRS Current Paper CP14/68, 'Developments in the production of concrete panels', for a more detailed description of these methods and other methods of producing concrete panels.

[2] Addendum no. 1 (1970) to CP116:1965 and CP116: Part 2:1969

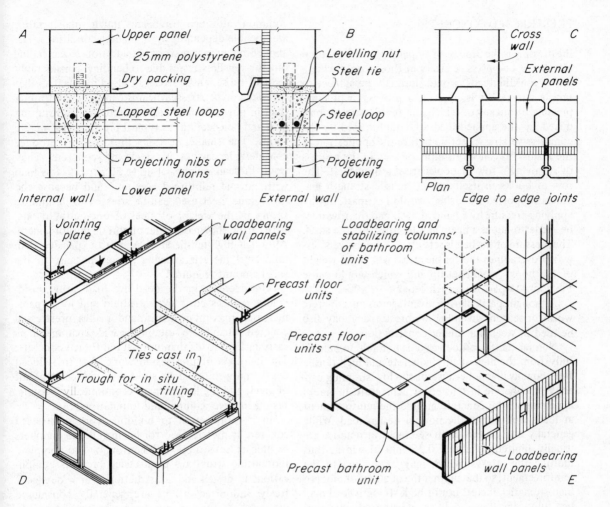

132   Large precast concrete panel construction

Vertical continuity may be obtained by forming *in situ* cast reinforced concrete columns within rebates formed between wall panels or by bolting wall panels toegther, one method of which is to use horizontal steel plates welded to the reinforcing bars projecting in rebates in the bottom corners of each panel (*D*). This method is similar to that used for precast columns shown in figures 128 and 129.

With cross and spine wall plans overall rigidity can be provided by *in situ* cast lift and stairwells, but this has the disadvantage of mixing precast and *in situ* work on the site. Fully precast construction uses bathrooms precast as reinforced concrete boxes complete with floor and ceiling and lift and stairwells precast in storey or half-storey heights which, when erected on each other, form structural 'columns' running the full height of the building. A number of these vertical units along the centre of the block form a structural spine to the remainder of the structure which is fabricated from large precast floor panels and storey height loadbearing wall panels (*E*).

PRESTRESSED CONCRETE

Prestressing is the process of imparting to a structural member a compressive stress in those zones which, under working loads, would normally be subject to tensile stresses. It is, in fact, a process of precompressing by means of which the tensile stresses produced by the applied load are counteracted by the compressive stresses set up before the application of the load. This can very simply be seen in the process of removing a row of books from a bookshelf. The row of books in itself has no tensile strength and unless supported by a shelf would fall apart, but by applying pressure by a hand at each end the row may be made to act as a beam and be lifted off the shelf. The pressure of the hands sets up a compressive stress which overcomes the tensile stress which the weight of the books would set up and which would cause the books to part from each other.

Although particularly advantageous in concrete work for reasons given below, prestressing may also be used in steel and timber construction.

Normal reinforced concrete is not able to benefit fully from the high-quality concrete and high-tensile steel now available because of the low straining capacity of concrete in tension, which results in cracks appearing in the concrete around the reinforcement at loads well below the normal design load. While generally not dangerous these cracks in practice are usually limited to about 0.25 mm in width, thus limiting the stresses which may be applied to the reinforcement, so that neither the qualities of modern high-strength concrete nor of high-strength steel may be fully developed[1]. In a prestressed member, however, the concrete is at all times under compression so that there is a complete absence of cracks. In the event of an overload, provided that this is within the elastic limit, the cracks formed will close again after removal of the load without harm to the structure. The high compressive strength obtainable in present day concrete can, therefore, be fully used, while at the same time the high-tensile qualities of modern steel may also be fully utilized because the steel is not used as normal reinforcement to take the tensile stresses which the concrete is unable to resist, but, as will be seen later, is used solely as a means of producing the compressive stress in the concrete[2].

Reduction in the depth of beams and slabs, thus producing higher stresses, is therefore possible without giving rise to crack formation, and depth/span ratios of 1:20 for beams and 1:40 for slabs are common, although for beams much smaller ratios are possible depending upon loading conditions. The applied compressive stress, in addition to cancelling out the tensile stresses due to bending, considerably reduces those tensile stresses caused by shear so that the webs of prestressed beams can be much thinner than in normal reinforced concrete beams, resulting in I- and box-sections as typical prestressed concrete forms. The smaller sections thus possible produce considerable savings in steel and concrete. They result in dead weights of up to 50 per cent less than with normal reinforced concrete, and because the high grade steel used can be stressed to its limit, a saving in the weight of steel of one-tenth to one-fifth can be shown over that of normal reinforcement. Although high-tensile steel is more expensive than mild steel there is a saving in cost because of the small amount required.

The decrease in the dead/live load ratio considerably reduces costs over medium and long spans, increases maximum spans, and makes prestressed concrete much more suitable for wide span members carrying light loads than normal reinforced concrete. The lightness of prestressed work, due to reduced depth and web thickness and to the reduced amount of steel, and the longer spans economically possible, results in lower column and foundation costs.

In its application to building work prestressed concrete is mostly used for beam and slab members in precast construction. When applied to complete monolithic structures prestressing presents complications in design and construction. For wide spans, freely supported beams are generally considered preferable to continuous beams in order to avoid similar complications (see page 232). For spans below 6 m, normal reinforced concrete construction is generally cheaper than prestressed concrete. Be-

---

[1] The use of special types of reinforcement in normal reinforced concrete to give increased bond and to limit cracking to fine, well distributed cracks is discussed on page 209.

[2] Concrete with a 28 day crushing strength of 40 N/mm^2 is generally used, although for pretensioned work 30 N/mm^2 concrete is sometimes used. Steel wire with ultimate tensile strengths up to 2310 N/mm^2 and high-tensile bars with ultimate strengths up to 1110 N/mm^2 are used.

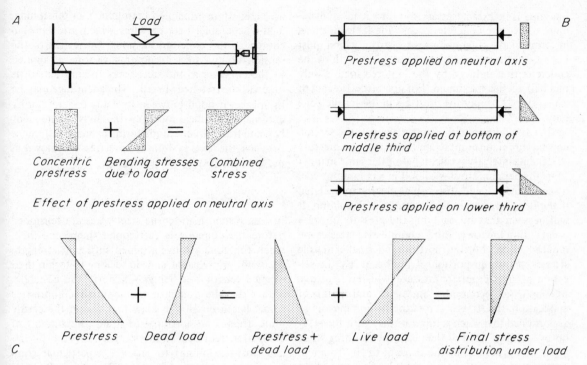

A

Load

Concentric prestress · Bending stresses due to load · Combined stress

Effect of prestress applied on neutral axis

B

Prestress applied on neutral axis

Prestress applied at bottom of middle third

Prestress applied on lower third

Prestress · Dead load · Prestress + dead load · Live load · Final stress distribution under load

C

133 Principles of prestressing

tween 6 and 9 m prestressed work may or may not prove more economical according to the particular job, having regard to such factors as the reduction in size and numbers of columns and foundations likely to result from the use of prestressed work. In this range the composite form of construction described on page 219 is likely to be the cheapest. For spans greater than 9 m prestressed work will usually show economic advantages over reinforced concrete, especially when the imposed loading is light, as in roof construction. Creasy[1] shows that for multi-storey frames as conventionally planned, prestressed concrete is not, in most cases likely to be cheaper than reinforced concrete.

Columns, being compression members, are normally not prestressed. However, in tall columns particularly, where bending stresses may be high due to wind pressure or an eccentric load such as applied by a travelling crane, prestressing can usefully be applied. Tensile stresses in walls, even loadbearing walls, are normally not such as to justify prestressing, but in tall retaining walls where bending stresses may be high, prestressing can be economical. As indicated in chapter 4, when applied to retaining walls prestressing, by preventing crack formation, has advan-

tages in terms of the water resistance of the wall (see page 148).

Principles of prestressing

The prestress, or precompression may be induced in a beam entirely without the use of steel by means of external jacks, in the same manner that hand pressure is applied to a row of books, provided that sufficiently solid abutments are available as in the case of a bridge (see figure 133 *A*). The principles of prestressing can usefully be considered on the basis of this method.

The pressure will be of a uniform intensity over the whole section if applied on the neutral axis of the beam, and if of equal intensity to the tensile stresses induced by the imposed load will cancel them out. As will be seen, this results in a final compressive stress in the upper fibres, assuming a beam of uniform cross-section, of twice that set up by the imposed load. This precompression of the compress-

[1] See 'Economics of Framed Structures', by Leonard R. Creasy, BSc (Eng), MICE – *Proceedings of the Institution of Civil Engineers,* Vol 12, March 1959.

ion zone is neither necessary nor does it make maximum use of the concrete in carrying its load, since, in terms of the final compressive stress, which must not exceed the maximum permissible strength of the concrete, that induced by the imposed load is only one half of this maximum. For greatest efficiency it is necessary to apply the prestress in the tensile zone only.

The distribution of the prestress across the section depends upon the point of application of the pressure (B), and its intensity is calculated in the same manner as are the bending stresses caused in a column or wall by an eccentric load[1]. This will be clearly appreciated if the beam is visualized as a 'horizontal' column. It will be seen that by applying the pressure at some point within the lower third, compressive stresses are induced in the bottom portion and smaller tensile stresses in the top portion of the beam. By the selection of an appropriate pressure, which is kept to a minimum to economize in steel, and point of application, the stresses across the section may be so apportioned that when acting together with those set up by the dead load of the beam the resulting stress at the top is zero while the stress at the bottom represents the maximum permissible compressive stress of the concrete (C), thus making maximum use of the strength of the concrete. Since the forces due to prestressing and the dead load act simultaneously the upper fibres of the concrete are not, in fact, subjected to the tensile stresses set up by the prestressing, nor the lower fibres to the excess compression indicated.

When the live load is applied additional compressive stresses are set up in the top and additional tensile stresses in the bottom fibres, and these forces, acting together with the residual forces from the combination of dead and prestressing loads, result in a compressive stress in the top fibres and a smaller compressive, or a zero, stress in the bottom. Greatest economy is obtained if the maximum compressive stress in the bottom fibres, due to dead and prestressing loads, is equal to the maximum stresses set up by the live load, thus producing zero stress at the bottom and a maximum permissible concrete stress at the top when the beam is under load.

Methods of prestressing

The method of applying the precompression by means of jacks, which presupposes sufficiently strong abutments, is of limited use and rarely practicable for normal building works. The alternative method used consists in principle of stretching, or tensioning, high-tensile steel bars or wires which are then anchored to the concrete member. On release of the tension on the steel a compressive force is applied to the concrete as the steel seeks to contract to its original, unstretched, length. The anchorage may be by means of bond between steel and concrete or by external mechanical means at the ends of the member, and these two methods form the main difference between the two systems of prestressing known as pre-tensioning and post-tensioning.

Pre-tensioning

In this system high-tensile steel wires are tensioned before the concrete is cast round them, and then, when the concrete has attained sufficient strength, the wires are released and, in seeking to regain their original length but being bonded to the concrete, induce in the concrete the required compressive force. Based on Hooke's law that within the elastic limit stress is proportional to strain, the amount of elongation required in the steel wires (both in pre- and post-tensioning) to produce a particular compressive force in the concrete can be easily calculated (see page 231 for the effect of certain stress losses).

As strong abutments are required between which to stretch the wires pre-tensioning is invariably applied to precast units and is usually carried out in a factory, although a prestressing bed set up on a site as a means of avoiding factory overheads might prove more economical for a very large contract. Factory production is generally preferable since the need for close control of the concrete preparation and its placing and of the stressing of the steel is more likely to be satisfied under factory conditions than on the site.

Although pre-tensioning can be applied to individual members formed and stressed in their own moulds, the most usual method is that known as the 'long line' system in which the wires are stretched within continuous moulds between anchorages 120 m or more apart. The wires pass through templates at each end which position them correctly and the ends are gripped in anchor plates (figure 134 A). Spacers are placed at various intervals along the mould according to the required lengths of units. The anchor plates are then jacked away the calculated distance to stretch the wires, the concrete is

---

[1] This is explained in Part 1, pages 55 to 56.

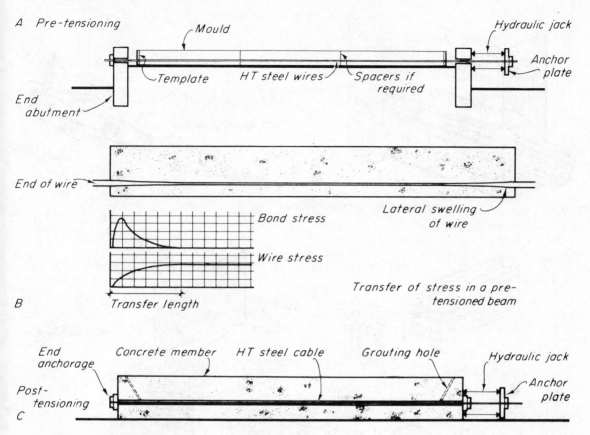

A Pre-tensioning

Mould

Hydraulic jack

Anchor plate

Template HT steel wires Spacers if required

End abutment

End of wire

Bond stress

Lateral swelling of wire

Wire stress

Transfer of stress in a pre-tensioned beam

B Transfer length

End anchorage Concrete member HT steel cable Grouting hole Hydraulic jack

Anchor plate

Post-tensioning
C

134 Methods of prestressing

poured and after it has hardened sufficiently the wires are released and are cut between each unit.

At the extreme ends of pre-tensioned members the bond between steel and concrete is not fully developed, and for a short length, varying from 80 to 120 times the diameter of the wire according to the quality of the concrete and the roughness of the surface, the wires contract considerably in their length with a consequent loss of stress in the wires, the stress at the cut end being zero. At the same time this contraction is accompanied by a lateral swelling which forms a cone like anchor (B). The length in which this occurs is termed the transfer length and requires reinforcement for shear in the form of stirrups. The lateral swelling of the released wires tends, of course, to occur throughout their length, thus further increasing the bond between wires and concrete.

Small diameter wires are used so that the greatest surface area is obtained to increase the bond, and the usual diameters lie between 2 and 5 mm. These wires have ultimate tensile strengths ranging from 1540 N/mm^2 for the larger diameters to 2310 N/mm^2 for the smaller. It is essential that the wires be thoroughly degreased and allowed to rust slightly in order to produce a satisfactory surface. Careful control of the concrete mix and vibration are used to produce high quality concrete, and some form of curing is normally applied to accelerate the hardening.

Post-tensioning

In this system the concrete is cast and permitted to harden before the steel is stressed. The steel, which is usually in the form of high-tensile steel cable or bar, if placed in position before concreting, is prevented from bonding with the concrete either by being sheathed with thin sheet steel or tarred paper or by being coated with bitumen. Alternatively, the prestressing steel can be introduced after the concrete

229

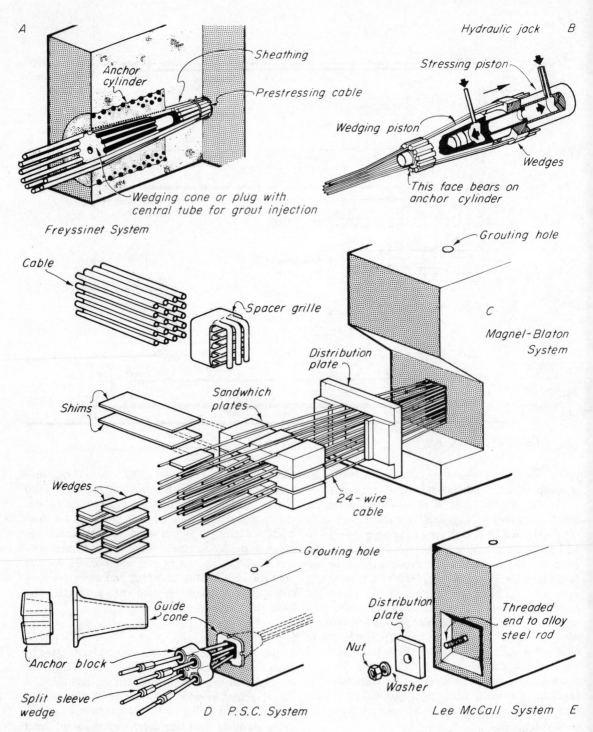

A

Anchor cylinder

Sheathing

Prestressing cable

Wedging cone or plug with central tube for grout injection

Freyssinet System

Hydraulic jack   B

Stressing piston

Wedging piston

Wedges

This face bears on anchor cylinder

Grouting hole

C

Magnel-Blaton System

Cable

Spacer grille

Distribution plate

Shims

Sandwhich plates

Wedges

24-wire cable

Grouting hole

Guide cone

Anchor block

Split sleeve wedge

D   P.S.C. System

Distribution plate

Nut

Washer

Threaded end to alloy steel rod

Lee McCall System   E

135   *Post-tensioning systems*

has set by casting in bars or duct-tubes at the appropriate positions which are extracted before the steel is inserted. It is also possible to place the wires outside the concrete and protect them. The cable or bar is anchored at one end of the concrete unit and stressed by jacking against the other end to which it is then also anchored (figure 134 *C*). The steel is subsequently grouted under pressure through holes at the ends of the unit to protect it from rust and to provide bond as an additional safeguard.

There are a number of methods of anchoring and jacking the prestressing steel, some of which are illustrated in figure 135.

The Freyssinet system uses a cable of eight to eighteen wires positioned round a central open spring forming a hollow core, and an anchorage device cast into the end concrete member consisting of a concrete cylinder with a central conical hole and a conical concrete plug grooved on the outside to take the cable wires which are laid between the cone and the cylinder (*A*). The special double-acting Freyssinet jack incorporates a stressing piston and a wedging piston (*B*). The wires, led through grooves spaced round the head of the jack, are wedged to the stressing piston which is operated until the required extension of the wires is obtained. Then the wedging piston is used to force the plug into the concrete cylinder to anchor the wires. The wires are then released, the cable grouted through the conical plug, the wires cut flush and the face of the anchorage protected with a pat of mortar.

The Magnel-Blaton system differs from the Freyssinet system in the form of anchorage used and in the manner of stressing the wires (*C*). The wires are stressed in pairs by a normal single-acting jack bearing on the anchorage and are secured by steel wedges to grooved steel plates, each of which anchors eight wires. These plates are arranged in layers, the number depending upon the size of the cable, and bear on a steel distribution plate. The wires in the cable are held about 5 mm apart throughout their length by spacer grilles.

The Gifford-Udall system and the *PSC system* both stress the wires one at a time. They are anchored individually, in the former system by means of a pair of conical half-wedges driven into a steel barrel accommodated in an anchor plate, in the latter, by a single-piece split sleeve driven into the tapered hole of an anchor block (*D*). The prestressing wires in

these post-tensioned systems are usually from 5 to 7 mm in diameter.

Secondary reinforcement is usually required in the concrete immediately behind the anchorages, and vertical stirrups at the ends of the beam to distribute the local loading from the anchorage of the cables.

The Lee-McCall system uses alloy steel rods instead of cables. The rods are from 13 to 29 mm in diameter and are anchored, after stressing by jack, by means of a special nut screwed on to the threaded end of the rod, the thread of nut and rod being so designed that the load is transferred by degrees to the nut in such a way that stress concentrations are largely eliminated (*E*). It is possible with this system to re-stress the rods at any time before grouting in, so that the loss of prestress due to shrinkage and creep in the concrete, which occurs in the early life of a prestressed member, may be wholly restored if desired.

After prestressing a concrete member a gradual reduction in the prestressing force commences and continues for a considerable period. This is due to the shrinkage of the concrete, the creep of the concrete and the creep of the steel. The creep of a material is the increase in strain, ie lengthening or shortening, which continues to take place after the stress on the material has become constant, so that it will be evident that the creep in the concrete of a prestressed member which is under compression from the stressing wires, will result after a time in the shortening of the beam, whilst the creep of the steel in tension will result in a lengthening of the wires, which, together with shrinkage of the concrete, leads to a loss in the initial prestress. In determining the initial prestress an allowance must be made for these losses, together with those due to elastic shortening of the concrete as it is stressed and, in the case of post-tensioning, to anchorage slip.

The distribution of prestress over a section is discussed on page 227 in terms of the point at which maximum stresses are set up by the external loads, that is, the point of maximum bending moment. At other sections the dead and live load moments will be less and the stresses due to prestressing will be excessive, in large beams particularly. A reduction in the moment of resistance of the section at these points can be made by varying the eccentricity of the prestressing wires.

This can be accomplished in two ways (i) by using straight cables and varying the section of the

231

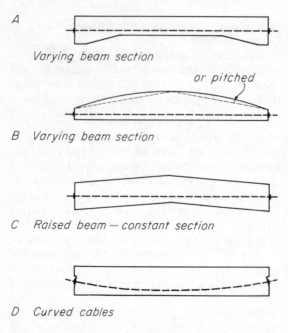

A Varying beam section

or pitched

B Varying beam section

C Raised beam – constant section

D Curved cables

136 Prestressed concrete beams

in any case, since variations in section along the length of a beam increase formwork costs, a constant cross section is preferable except for very large spans. The curving upwards of the prestressing cable gives a vertical component which helps to resist the shear forces in the beam and enables high shear loads to be taken.

Pre-tensioning is most suitable for the production of large numbers of similar units, particularly if they are of a cross-section too small satisfactorily to accommodate the relatively large post-tensioning cables. In pre-tensioning the wires must be straight so that shear resistance from curved-up wires is not obtainable. Generally speaking, the method is not suited to prestressing on the site. Beams range generally from 4.5 to 23 m in length, the maximum length depending upon transport and handling. Beams up to 30 m or more can be made as 'specials'.

Post-tensioning is invariably used for prestressing on the site and for large members. In most cases it is not economical for members less than 9 m long because the cost of the anchorages relative to the length is high, while the cost of jacking is the same as for a long beam. It may be cheaper to use reinforced concrete for a large number of small units if, for some reason, pre-tensioning is not suitable. The general range of spans is 15 m upwards. Post-tensioning has the advantage, particularly with members carrying heavy shear loads, that the cables can be curved upwards to provide added shear resistance.

beam as shown in figure 136 *A, B,* or by raising the centre of a beam of constant cross section, as in (*C*), or (ii) by curving the wires upwards from their lowest point as in (*D*). As the shear forces tend to increase as the bending moment decreases, reduction of the section as in (*B*) may not always be desirable and,

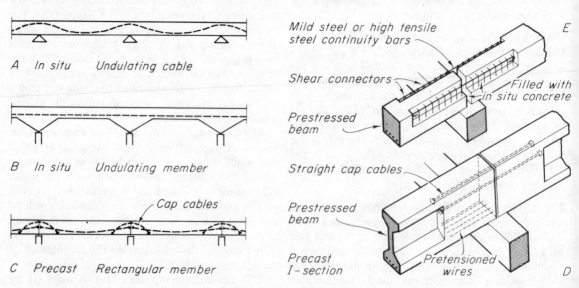

A In situ   Undulating cable

B In situ   Undulating member

C Precast   Rectangular member

Mild steel or high tensile steel continuity bars

Shear connectors

Prestressed beam

Filled with in situ concrete

Straight cap cables

Prestressed beam

Precast I-section

Pretensioned wires

137 Prestressed concrete beams – continuity over support

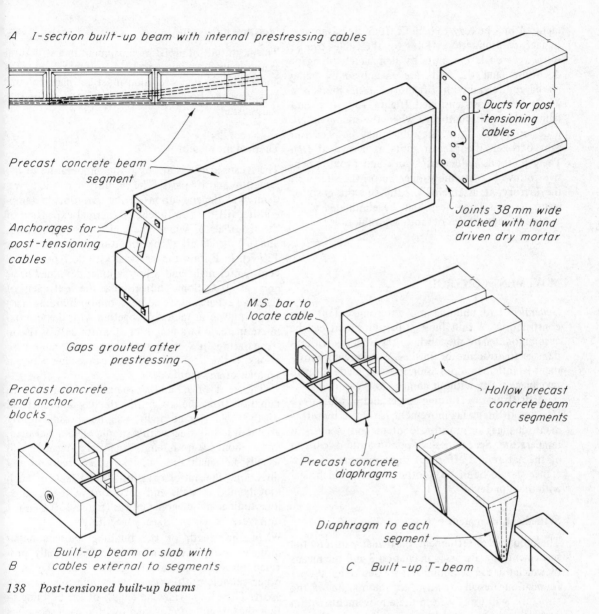

A   I-section built-up beam with internal prestressing cables

Precast concrete beam segment

Ducts for post-tensioning cables

Anchorages for post-tensioning cables

Joints 38 mm wide packed with hand driven dry mortar

MS bar to locate cable

Gaps grouted after prestressing

Hollow precast concrete beam segments

Precast concrete end anchor blocks

Precast concrete diaphragms

Diaphragm to each segment

Built-up beam or slab with
B   cables external to segments

C   Built-up T-beam

138   Post-tensioned built-up beams

When continuity over supports is essential the necessary stressing in the zones of negative moment can be effected in a number of ways, illustrated in figure 137. In *in situ* cast members either the cable must be undulating, which gives rise to friction losses (A) or the member itself must be 'undulating' in form (B) to vary the point of maximum prestress. Precast members can be prestressed in the normal way and continuity be provided over the supports by cap cables, curved in the case of rectangular sections or straight in the case of I-sections (C, D), or by continuity bars set in *in situ* filling in rebates in the beam ends (E).

By the use of post-tensioning it is possible to build up a beam from a number of precast concrete units or segments placed end to end like the row of books mentioned earlier. These units can be produced on the site, but being small are often manu-

233

factured in a factory to benefit from the advantages of factory production. Holes for the cables can be formed through the units by light steel tubing or duct-tube, and the units are assembled by being placed end to end with stiff mortar in the joints, the whole being post-tensioned (figure 138 *A, C* and figure 211). Alternatively, when the members are wide, or in slab form, the cables may be placed in gaps between the precast units as shown at (*B*). These methods reduce site work and avoid expensive formwork. Assembly may be on the site or in the factory; if assembled on the site the costs of transport of a large beam can be eliminated while gaining the advantages of factory production.

MOVEMENT CONTROL

Generally All buildings move to some extent after construction. Within limits this movement can be accommodated by the fabric of the building without damage to structure or finishes. When greater movement is anticipated provision must be made for it to take place freely without damage to the building.

Apart from overturning forces and possible overstressing of materials movement is caused by settlement, changes in moisture content and changes in temperature. Space does not permit a full discussion of the nature and effects of these movements. Reference should be made to other sources which deal with these in detail[1].

Settlement movement

The problems of settlement, particularly differential settlement, are discussed in chapter 3 and the means by which this can be minimized are described. Proper foundation design, relative to the nature of the structure, will usually keep these movements within acceptable limits, but in some cases it will be cheaper to provide for movement in the structure, This is done on subsidence sites as described on page 106.

Settlement joints are often provided at the junction of parts of a building which vary considerably in height or in loading (figure 47 *A*), or, where floors run through, provision for movement between the parts is made by a flexible or hinged bay at the junctions (figure 139 *A*). Claddings and windows in such bays must be designed to permit free movement.

234

Moisture movement

The magnitude of moisture movement in a structural frame is likely to be small. It is primarily in walls and claddings that provision must be made for this and reference should be made to page 176 where this is discussed.

Thermal movement

This is caused by variations in the temperature of the structure and its parts. The magnitude of the movement will depend upon (i) the variation in temperature, (ii) the coefficient of thermal expansion of the materials of which the building is constructed, and (iii) the length of the structure or its part (figure 139 *B*). In Britain the seasonal variation, between a cold winter night and a hot summer day, may be as great as 50°C. Some indication of the coefficient of thermal expansion of some common building materials is given in table 14, together with the approximate increase in a length of 30 m for a 28°C rise in temperature. It will be seen that considerable variations in the coefficient values can occur between samples of any material.

The roof of a building, being exposed fully to radiation from the sun during the day and radiation to the cold sky at night, will be most seriously affected. In buildings with a simple rectangular plan, up to about 30 m in length, thermal movement will usually be small and can take place freely in any direction. Precautions against the effect of such movements on walls and partitions in the form of longitudinal and diagonal cracks (figure 139 *C*) must, however, be taken[2] (see page 366).

When the length of the building exceeds much more than 30 m expansion joints are usually provided, placed at intervals not greater than 30 m or at other suitable positions not exceeding this distance apart. These subdivide the length of the building so that the amount of movement within each section is limited, and they provide space for expansion so that damage to the structure is avoided (*D*). The actual spacing and widths of the joints can be calculated

[1] See *Principles of Modern Building*, Vol 1 (HMSO) and *Building Research Station Digest* 12 (Old Series).

[2] See *Building Research Station Digest* 12 (Old Series), 'The Design of Flat Concrete Roofs in Relation to Thermal Effects', for a detailed consideration of this.

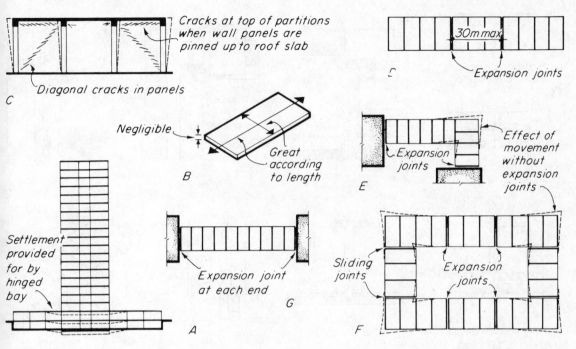

139 *Movement control*

Material	Coefficient of thermal expansion per $^\circ C \times 10^{-6}$	Approximate increase in a length of 30 m for a $28^\circ C$ rise in temperature (mm)
Concretes	10 - 14	10.00
Mild steel	11 - 13	9.90
Aluminium alloys	23	19.30
Brickwork	5 - 7	5.33
Limestones	2.4 - 9	4.82
Sandstones	7 - 16	9.90
Granite	8 - 10	8.12
Slates	6 - 10	6.86
Glass	9	6.86
Asbestos cement	12	9.90
Wood:		
along grain	6	4.82
across grain	46	35.50
Plastics (glass reinf'd)	12	9.90

Table 14   *Thermal movement of building materials*

from the coefficients of expansion and a selected rise of temperature.

When the form of the building is such that expansion at some points is restrained, movement will be greater at the unrestrained areas. This can be limited by the provision of expansion joints at the points of restraint as shown at (E) and (F). The design of the end joints in (F) must permit a sliding action. Expansion joints should similarly be placed where sudden changes occur in plan (G), and where floors and roofs are weakened by large openings in the structure.

Expansion joints should not be limited to the roof slab. They should also be formed in the external walls extending some distance down and inwards to enable the stresses set up by expansion to be distributed. It is common practice to carry expansion joints through the whole of the structure from top to bottom, particularly in monolithic reinforced concrete buildings. In a framed building the simplest method is probably to use double columns and beams as shown in figure 140 L, but when double members are undesirable the joint may be formed in the centre of a bay with cantilevered floor slabs or beams, or on the line of a column with sliding bearings to the floor structure as shown at (L). In suitable steel-framed structures joints are sometimes provided in the roof and top storey only, reliance being placed on the flexibility of the upper stanchions.

Typical details of expansion joints are shown in figure 140. As it is often difficult to remove the

235

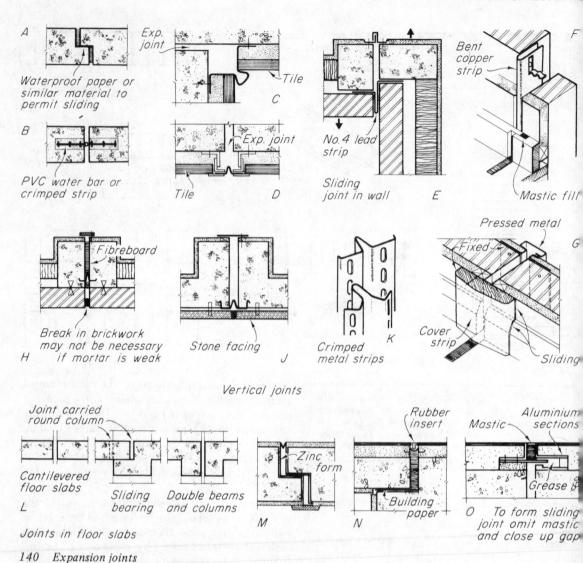

A — Waterproof paper or similar material to permit sliding

B — PVC water bar or crimped strip

C — Exp. joint / Tile

D — Exp. joint / Tile

E — No. 4 lead strip / Sliding joint in wall

F — Bent copper strip / Mastic fill

H — Fibreboard / Break in brickwork may not be necessary if mortar is weak

J — Stone facing

K — Crimped metal strips

G — Pressed metal / Fixed / Cover strip / Sliding

Vertical joints

L — Joint carried round column / Cantilevered floor slabs / Sliding bearing / Double beams and columns / Joints in floor slabs

M — Zinc form

N — Rubber insert / Building paper

O — Mastic / Aluminium sections / Grease / To form sliding joint omit mastic and close up gap

140   Expansion joints

shutter board forming a narrow gap between two *in situ* cast concrete members bitumen-impregnated soft fibreboard is frequently used for this purpose and left permanently in position (H). With wider gaps the board may be withdrawn and the space filled with non-extruding resilient filling (N), or left open (J) as circumstances require. Waterproof resilient filling or crimped 0.56 mm copper strip (K) is used to form a flexible weather-resisting barrier in joints in external walls and between columns (A and J).

Expansion joints in floors should be detailed in the same way to prevent water used for cleaning passing to the ceiling below (M, N). Alternatively a grease seal may be used (O). In positions where a sliding joint is required (figure 139 F), this can be detailed for floor slabs as shown at (O), for walls as (E) and for roof slabs in principle as figure 232 D.

For further consideration of expansion joints in walls and roofs reference should be made to pages 176 and 366 respectively[1].

[1] For a comprehensive treatment of the subject of expansion joints see *Joints de Dilatation dans la Construction en béton et en béton armé*, by Adolf Kleinlogel, Editions Eyrolles.

6 Floor structures

The broad categories of floor structures have been discussed in Part 1. An indication of the economic application of these relative to span and superimposed loading is given in table 18 of that volume and the types of floor construction suitable for short spans and for longer spans with light loading are described. In this chapter those types of floor appropriate to wide span and heavily loaded buildings will be considered, together with some special types of construction. As pointed out in Part 1, in large-scale and multi-storey buildings the floors are normally main structural elements closely related to the general structure of the building and as such they must be considered at the design stage in relation to it.

UPPER FLOORS

Choice of floor

The main factors influencing the choice of floor type, together with some indication of the manner in which each may be relevant, are discussed below. Reference should also be made to *Choice of appropriate structure* in Chapter 5 and to pages 208 and 210.

Nature of the building structure

With a steel frame, providing in itself all the necessary lateral rigidity, a precast concrete floor or a cellular steel floor laid quickly on the steel beams but contributing little to the rigidity of the frame could be suitable. With an *in situ* reinforced concrete frame an *in situ* concrete floor cast in with the frame and designed to provide lateral rigidity would be logical and would permit the omission of the beams parallel to the floor span.

In a reinforced concrete frame with beams on all four sides of the floor panels the latter may be designed as two-way reinforced slabs to minimize thickness and weight. With a steel frame, for which a rectangular grid layout is generally more economical than a square layout, one-way spanning floor construction will usually prove more suitable, particularly over short spans.

The height of the structural frame will also have a bearing upon the choice of floor type. Creasy[1] states that 'where the columns are carrying five or more storeys they have a relatively greater influence on the cost of the framework, and a hollow-slab form of construction compares favourably with the solid slab from the reduction in dead weight to be carried by the columns. For buildings of less height this advantage is offset by an increase in the construction costs and the solid slab becomes more economical'.

Loading

Flat slab construction is economical for heavy uniformly distributed loading, which with normal beam and slab construction might require very deep beams. Because of the smaller overall thickness the total floor to floor height is reduced, and this in turn will effect overall economies by a reduced height of structural frame and reduced area of external cladding. Where heavy concentrated loads must be carried a diagonal beam floor might be selected because such loads are dispersed throughout all the members of the grid, thus avoiding concentrations of high stress and resulting in a reduction in beam depths.

Span

This may sometimes be fixed by plan requirements, but as pointed out in Part 1 it may often be reduced to fall within the economic range of a cheaper type of floor system (see table 17, Part 1).

With increase in the span (and in the load) there will normally be an increase in the thickness and weight of a solid floor slab. Any increase in the weight of the floor will impose a greater load upon the structural frame and the foundations, with a consequent increase in the cost of these elements. The desirability of reducing the dead weight of the floor to a minimum, for reasons of structural economy, has resulted in the large variety of floor systems now available. Each system is most economical over a limited range of spans and loading, having regard solely to the relationship between weight of floor, the distance it will span and the load it will carry.

[1] 'Economics of Framed Structures', paper by Leonard R. Creasy, BSc, MICE, in *Proceedings of the Institution of Civil Engineers*, Vol 12 March 1959.

Other factors, however, are also involved. For example, the weight of floor beams and floor ribs may be reduced by an increase in their depth[1]. But such an increase may increase the total height of the structure, the area of external cladding and the cubic contents of the building, while the total floor area remains the same. Thus possible economies effected in the structural frame and foundations by the reduction in floor weight must be related to the extra cost of the greater height and the greater area of external facing and a balance struck between the two.

Degree of fire resistance required

This is frequently a determining factor in the choice of floor. Many buildings, because of their high fire load, must be divided into fire-tight compartments by walls and floors. These must have a degree of fire resistance sufficient to withstand the complete burn-out of the contents of any compartment and prevent the spread of fire to other parts of the building. The considerations involved in respect of forms of construction and the nature of the materials used with regard to their action under fire are discussed in chapter 10.

Provision of services

The large number of services and extensive equipment required in many types of building necessitates early consideration of the means of housing them both vertically and horizontally. Horizontal runs can often conveniently be accommodated in the topping or screed to the floor but if the services are extensive this can result in a screed 75 to 150 mm thick, adding to the depth and weight of the floor. A common alternative is to run the services below the floor slab and beams and to conceal them by a suspended ceiling. They may often economically be run freely in the actual floor depth by the choice of an appropriate type of floor construction.

In extreme cases, such as laboratories and industrial processing plants, the number of services and the amount of room required for their distribution may necessitate a considerable depth of space below the floor. This depth may be used to advantage in the adoption of a deep, light supporting structure to the floor. When plant is involved, the depth required may be so great as to result in what is, in fact, a service floor sandwiched between the production or processing floors.

Degree of sound insulation required

Weight of structure is important in connection with insulation against airborne sound. The greater the weight the greater the insulation provided. The degree of insulation provided by a boarded timber floor, as pointed out in Part 1, may be acceptable in the first floor of a house but is inadequate in most other buildings. Many types of light concrete floors can provide insulation against airborne sound sufficient for some buildings, but for others their inherent insulation value must be increased by the provision of a suspended ceiling underneath or a floating floor on top. A solid concrete floor of sufficient thickness and weight can give a reasonable degree of insulation against airborne sound but has little effect on impact sound. Apart from the increase they make in airborne insulation, floating floors are widely used to reduce the transmission of impact sound. For a detailed consideration of sound insulation see *MBC : Environment and Services.*

Cost, speed of erection, adaptability

Consideration of cost must enter into the choice of the floor, but not solely in terms of the floor itself. As with all other building elements the effect of each type of floor upon the remainder of the building must be examined and the economic consequences assessed and compared. One aspect of this has been considered briefly under *Span* on this page. Speed of erection may be an over-riding factor in some circumstances and in others adaptability to non-rectangular panels which, due to site or other limitations, are sometimes unavoidable.

Strength and stability

Reference is made in Part 1 (page 198) to the need for adequate strength and stiffness in floors in carrying their dead and superimposed loadings. In their design these loadings must be calculated or assumed on the basis of values laid down in regulations or codes of practice, and maximum permissable deflections must be established.

Floor loading

The dead load is usually based on the weights of materials specified in BS 648. Code of Practice 3:

[1] See pages 59 and 60, chapter 3, Part 1.

chapter V: Part 1 requires the dead load of any partitions not definitely located in the design of the building to be allowed for as a uniformly distributed load per square metre of floor of not less than one third of the weight per metre run of the partitions. If the floor is to be used for office purposes this load must not be less than $1 kN/m^2$.

The superimposed loads to be assumed in the design of a floor vary with the different uses to which the parts of the building may be put. Values for these are given in table 1 in CP3 : chapter V : Part 1 in the form of (i) average uniformly distributed loads per square metre of floor area and (ii) concentrated loads to be applied over any square with a 300 mm side.

A floor slab must be designed to carry whichever of these loads produces the greater stresses, the concentrated loads being applied in those positions which result in the maximum stresses or, where deflection is the primary design criterion, which result in maximum deflections. Where the slab is capable of effective lateral distribution of its load consideration of the concentrated load is not required.

Beams are designed to carry the distributed loads with the exception of beams, ribs and joists spaced at centres not greater than 1 m, which must be designed as floor slabs.

In the case of a single span of a beam supporting not less than 46 m^2 of floor at one general level, the imposed load may, in the design of the beam, be reduced by 5 per cent for each 46 m^2 supported, subject·to a maximum reduction of 25 per cent.

The London Building (Constructional) by-laws, 1972, adopt the provisions of the Code for dead loads and minimum imposed loads, but also provide discretionary powers to the District Surveyor in respect of these. The Building Regulations, 1972, similarly adopt the provisions of the Code, but with certain provisos.

Deflection

In order to minimize the cracking of plaster and other ceiling finishes the deflection of floor slabs and beams must be limited in accordance with the nature of the finishes. In Codes this is covered by laying down a fraction of the span as the maximum permitted deflection in the case of steel and timber, and in the case of reinforced concrete by laying

Reinforced concrete beams and slabs (limiting deflection to 1/250 span	CP 110 : Pt 1 : 1972

Rectangular beams up to 10 m span	Maximum values of span to effective depth
Simply supported	20
Continuous	26
Cantilever	7

These figures may need to be modified according to the area of reinforcement provided and its service stress.
For spans greater than 10 m when deflection must be restricted further to avoid damage to finishes and partitions other values given in the Code should be used.

Solid slabs: ratios and modifications as for rectangular beams. The ratio for two-way spanning slabs is to be based on the shorter span.

Steel beams: maximum deflection 1/360	BS 449 : Pt 1 : 1970

Timber beams and joists: maximum deflection 1/300	CP 112 : Pt.2 : 1971

Table 15   *Limitation of deflection*

down such values of the ratio of span to depth as will limit deflection to a given fraction of the span. These are shown in table 15.

Where the floor structure is used to transmit wind forces to strong points (see page 184) it must be stiff enough to fulfil this function and buckling must be taken into account in its design.

UPPER FLOOR CONSTRUCTION

As indicated in Part 1 upper floors may be constructed of timber, reinforced concrete or steel. The choice of a particular type will depend largely upon the factors already discussed.

The advantages and limitations of these different types of floor are discussed in Part 1 (pages 199-201) where the construction of timber floors is described in detail, together with that of a simple *in situ* cast solid concrete slab floor. Concrete floor construction will now be considered further and types of steel floor and special floor construction will be described.

Reinforced concrete floors

These fall into two broad categories, *in situ* cast and precast, in each of which there is a wide variety of types. Many types of precast floor involve the use of *in situ* topping of concrete acting structurally with the precast components to form a composite construction.

In situ cast floors

These floors, being a wet form of construction as indicated in Part 1, require temporary support until the concrete is strong enough to bear its loads. The length of time during which the support must be left in position depends upon the type of cement used in the concrete and the temperature of the weather when the work is in progress. Table 28 on page 425 gives a guide to the number of days after pouring at which the formwork may be removed or 'stripped'.

Solid concrete floor slab This type is commonly used when the slab is to act as a membrane supported on columns without beams, as in the flat slab and plate floors which are described later, or where high degree of lateral rigidity is required to be provided by the floor. In buildings up to four storeys in height in its simplest form it may often prove more economic than hollow block construction[1]. It gives maximum freedom in design on plan and section since it can easily be made to cover irregular plan shapes and can easily be varied in thickness at different points according to variations in load or span. It is a heavy floor but highly fire resistant.

The simplest form of solid floor is the one-way spanning slab described in Part 1 which, as explained there, is economic only over small spans up to 4.60 m.

For large spans or heavy loadings a two-way spanning slab should be used in which the reinforcement is designed to act in both directions, the proportion of load taken by each set of reinforcement depending upon the ratio of long to short side of the floor panel. The most economic application is to a square grid. The distribution bars in the upper diagram of figure 166, Part 1, would, for a two-way spanning slab, be replaced by main reinforcing bars. The economies of this type of floor arise from the reduced thickness and weight of the slab which result from two-way spanning. In practice,

however, the minimum thickness of a slab is limited to a given fraction of the span in order to avoid excessive deflection (see page 239). In many cases it is also limited by the requirements of fire resistance. There is, therefore, an advantage in using a two-way spanning slab only when the thickness of the slab ceases to be governed by these considerations.

Normal *in situ* solid floor slabs are not generally prestressed, this being applied most economically to rib or beam members.

A beamless floor consists of a reinforced slab resting directly on reinforced concrete columns with which it is monolithic. There are no projecting main or secondary beams, the slab acting as an elastic diaphragm bearing on point supports. The columns may or may not have flared heads. CP 110 covers the design of both types under the term 'flat slab', but when designed without column heads the construction is commonly called a 'plate' floor and for the sake of distinction that term will be used here.

Plate floor The slab, or plate, is reinforced at the bottom in each direction over the whole of its area with concentrations of reinforcement along the lines of the column grid. These form wide 'column bands' within the plate thickness. A mat of mesh or rods in each direction is formed in the top of the plate over the columns. It is most economic when the grid repeats uniformly, although this need not be the same in each direction. However, because of the large area of intersection of the wide column bands it is possible, within certain limits, to displace columns from the regular grid. This gives a flexibility which is useful in terms of planning (see figure 141 *A*). Any such column layout must, of course, be the same throughout the height of a multi-storey building. A regular spacing of columns of up to approximately 5.50 m in each direction produces the most economical grid.

The system is most efficient for light and medium loadings as in flats and offices. With domestic loading and a grid spacing up to 4.5 m, the plate thickness would be from 125 to 150 mm. For 3.50 kN/m² loading over a span of about 5 m a plate thickness of approximately 200 mm would be necessary. For maximum economy the thickness of the slab must be kept to a minimum consistent with deflection requirements and those of shear resistance at the columns. A minimum of three bays in each direction

[1] See reference to Creasy on page 237

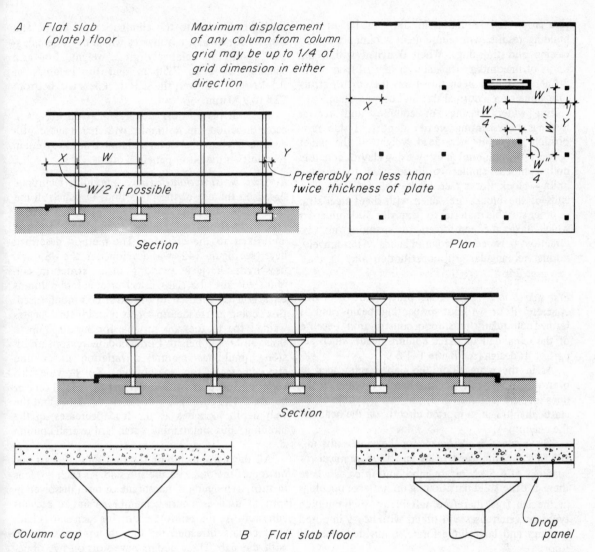

A   Flat slab
    (plate) floor

Maximum displacement
of any column from column
grid may be up to 1/4 of
grid dimension in either
direction

X   W

W/2 if possible

Section

X

W'
4

W'

W"

Y

W"
4

Preferably not less than
twice thickness of plate

Plan

Section

Column cap   B   Flat slab floor   Drop
                                                                     panel

141   Flat slab floors

is, therefore, desirable, together with a half-bay projection of the plate beyond the external columns in order to provide restraint on the outer panels of the plate[1]. Nevertheless, it is possible to use it over two bays only and also possible to reduce the plate projection beyond the external columns to twice the plate thickness, or even to eliminate this projection altogether. But the latter necessitates an uneconomic amount of reinforcement in the slab.

Wallings approaching the weight of normal cavity panel walls with a light-weight inner skin must be carried on the outer column bands between or close to the external columns. Lightweight panels, or

curtain walling requiring only lateral support from the floors, may be carried by the plate projecting up to half the grid beyond the external columns.

Practical advantages arising from the use of the plate floor are simplification of shuttering and reinforcement, reduction in dead weight compared with beams and slab and a flat soffit throughout facilitating the use of standard height partitions, particularly useful in office blocks. In many cases

[1] In respect of slabs designed by the 'empirical' method, Code of Practice 110, requires an arrangement of panels in at least three rows in two directions at right-angles and a ratio of length of panel to its width not exceeding 4:3.

241

also an overall reduction in the total height of a building results, with consequent economies in carcassing and finishings. When designed within the limits of maximum efficiency savings of from 15 to 20 per cent have been shown on the cost of structure alone over a normal slab and beam system.

For wider spacings of columns and heavier loadings necessitating greater depths of slab it is possible to reduce the dead weight of the panel between the column bands by using clay or concrete hollow blocks similar to those used for normal hollow block floors (see figure 143 *D*). The open ends of the blocks are sealed with sheet steel strip or other suitable material to form ribs 300 mm apart which take the reinforcement spanning in two directions between the 'column bands'. Alternatively, normal rectangular grid construction may be used (see page 246).

Flat slab floor   This is the other form of *in situ* concrete floor without projecting beams and is termed colloquially 'mushroom construction' because of the expanded or flared column heads which are part of its design (see figure 141 *B*).

As in the plate floor, the slab is reinforced in both directions with 'column bands' running on the lines of the column grid, the slab acting as an elastic diaphragm supported directly on the heads of the columns.

The system is designed for heavy evenly distributed superimposed loads and is economical for loadings of 4.5 kN/m^2 or more and in cases where there is little solid partitioning on, or large openings in, the slab. It is, therefore, suitable for such building types as warehouses and others with heavy imposed loadings and large imperforate, undivided areas of floor.

For maximum efficiency the columns should be on a regular grid of about 6 to 7.50 m in approximately square bays. In order to provide adequate resistance to the compression stresses in the bottom of the slab over the points of support, and to increase the resistance to shear and punching stresses at these points, the heads of the columns are expanded to give the typical 'mushroom' cap. This will be square or circular depending upon the shape of the column. In some circumstances it is necessary to thicken the slab over this cap as shown in (*B*) to form what is termed a drop panel. The main advantages over normal beam and slab floors, as with plate floors, are the reduction in floor to

floor height due to the elimination of beams and simpler formwork, advantages which must be balanced against its relatively great weight. For grid spacings of 6 to 7.50 m and for loadings of 4.5 kN/m^2 and over, the slab thickness will be from 225 to 300 mm.

As with plate floors, not less than three bays in each direction are desirable, with half-bays cantilevering beyond the outer column bands to obtain restraint on the outer panels.

Lift-slab construction   This is a building technique based on the use of the plate floor in which all the floor slabs, together with a roof slab, are cast at ground level and subsequently raised into position and fixed to the columns. The method described here (see figure 142) was developed in the USA and uses hydraulic jacks working on the structural columns to raise the slabs. Steel or concrete columns on a grid of about 6 m are most commonly used. Box columns, made up of two mild steel angles welded toe to toe, are suitable for buildings up to three storeys in height. For economic reasons multi-storey buildings require a variation in column strength from top to bottom. To provide this variation I-sections with cover plates welded on can be used. The cover plate thickness decreases and the web depth increases as the load decreases up the building, thus maintaining a standard overall column size.

All the slabs, including the roof slab, are cast one on top of the other on the site slab. As each hardens in turn, a sprayed-on membrane of resin dissolved in spirit is used as a curing membrane and to prevent adhesion. As the reinforcement for each slab is laid, steel collars threaded on the columns are cast in with the slab. These collars have slots on two opposite sides as shown in figure 142 *A*, to accommodate the ends of two screwed steel hoisting rods connected with the lifting jacks.

When all the slabs have been cast and have sufficiently matured, a hydraulic jack is set on top of each column and, in the case of a building of up to three storeys, each slab in turn is hoisted to its appropriate level. All the jacks are synchronized through a single control panel set up on the top slab. The slabs, in their final position, bear on and are welded to shear blocks site welded to the columns on two or four sides after the slab has been levelled (figure 142 *B*). An alternative to this, which speeds erection, is the use of a rebated steel wedge (which

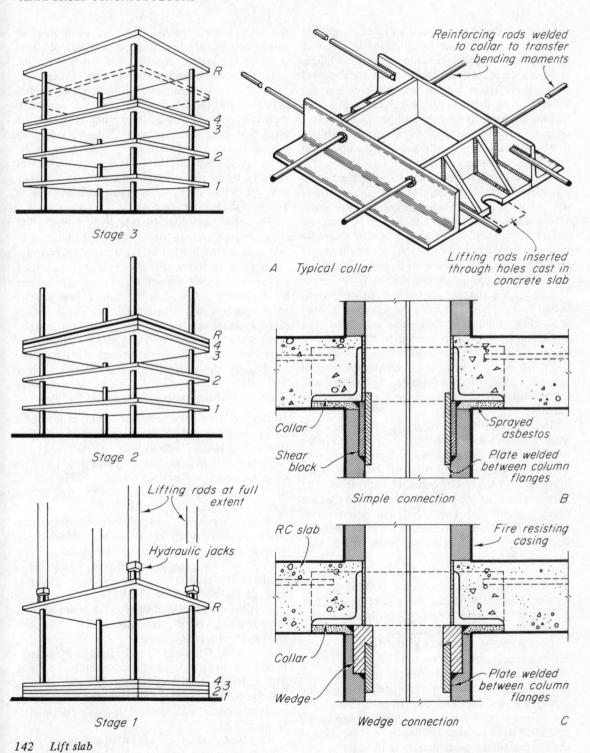

Stage 3

Stage 2

Stage 1

Lifting rods at full extent

Hydraulic jacks

R

4 3
2 1

A   Typical collar

Reinforcing rods welded to collar to transfer bending moments

Lifting rods inserted through holes cast in concrete slab

Collar

Shear block

Sprayed asbestos

Plate welded between column flanges

Simple connection   B

RC slab

Fire resisting casing

Collar

Wedge

Plate welded between column flanges

Wedge connection   C

142   Lift slab

replaces the shear block shown in (*B*) bearing on top of a shop welded shear block set flush on an I-section (*C*). On concrete columns the shear block, in either method, is welded to a cast-in prefabricated steel insert. When welded to the columns the collars act as shear heads and bearing plates to transmit the loads to the columns. Bending moments are transferred by means of reinforcing rods welded to the collars and by the weld between the collar and shear block or wedge.

In multi-storey buildings the columns are erected in sections. All the slabs are cast after the erection of the lowest section. When mature, all the slabs, except those to be fixed to the lowest section, are hoisted to the top of the section and temporarily secured or 'parked' by means of steel wedges placed between the collars and shear plates shop welded in the appropriate positions to the columns. The roof slab is hoisted first and 'parked' in order to stiffen the columns and enable the rest to be hoisted in batches of three or four slabs at a time. The remaining slabs are then hoisted to their positions up the lowest column section and permanently fixed to the columns, after which the second section of columns is butt-welded to the first and the whole operation repeated up the height of the building.

This process is illustrated in figure 142 in which Stage 1 shows all slabs cast and the roof slab lifted to the top of the first column length and about to be 'parked' ready for the lifting of the floor slabs. Stage 2 shows the second column lengths fixed after the roof slab has been 'parked' at the top of the first column length and slabs 1, 2 and 3 have been fixed. Slab 4 is carried temporarily by slab 3. Stage 3 shows the roof slab lifted and finally fixed ready for the raising and fixing of slab 4.

Extensive slabs are cast in sections. A gap is left between adjacent sections which is filled in with *in situ* concrete after the slabs have been fixed, picking up the reinforcement left projecting for this purpose from the edges of each section.

In the course of hoisting, the load is applied through the jacks at the top of the columns. During the early stages of lifting in each section the columns are fixed only at the foot, and braced at the top by the roof slab, thus the critical condition for the columns, in terms of buckling, occurs as the remaining slabs are raised in batches off the stack at the bottom of the column section. In order to keep the column sizes within practical limits the height of

the column sections must, therefore, be limited and this results usually in a maximum length of about 9 m for the lowest section and about 6 m for the upper sections where the column area reduces. The difficulty of stabilizing the lifting rods when projecting above the columns at the end of a lift, especially in high winds, is another factor in the limitation of column heights. Temporary bracing during erection is provided by guy wires or by angles at the tops of the columns, and permanent stability by *in situ* concrete elements such as stair and lift wells, or by steel framed elements with adequate diagonal bracing.

The advantages claimed for this system are that it eliminates elaborate formwork and the hoisting of materials to great heights, and that it enables concrete placing and the fixing of reinforcement to be carried out at ground level. Electrical and other services can be positioned and fixed in the slabs at ground level before pouring and lifting the slabs. These factors taken together produce a simple economical system, particularly if the building is designed specifically for this form of construction. Where span and loading make it advisable the alternative forms described under 'Plate floor' or ribbed slabs with broad slab-depth beams may be used instead of the solid flat plate.

In situ tee-beam or ribbed floor   This is illustrated in figure 143 *A* and consists of a series of tee-beams cast monolithically side by side to produce a relatively thin slab with ribs on the underside. The basis of a tee-beam is described on page ·211, and by applying the principles to slab construction a floor lighter than a solid slab results.

It is an expensive floor to construct with normal shuttering and proprietary steel forms, hired by the general contractor or used by the makers as subcontractors, are generally employed. These forms produce ribs at 600 mm centres 90 or 100 mm wide at the bottom and slightly wider at the top. The depth is adjusted according to the load and span. The thickness of the top slab is 25 to 75 mm reinforced with rods or mesh.

Hardwood fillets can be embedded in the bottom of the ribs to provide a fixing for lathing or battens. The underside may then be enclosed with a ceiling and the resulting voids between the ribs can accommodate services and recessed lighting fittings (see figure 152 *B*).

This type of floor, being cast monolithic with the main supporting beams, may be used to stiffen the structural frame in buildings up to three or four

REINFORCED CONCRETE FLOORS

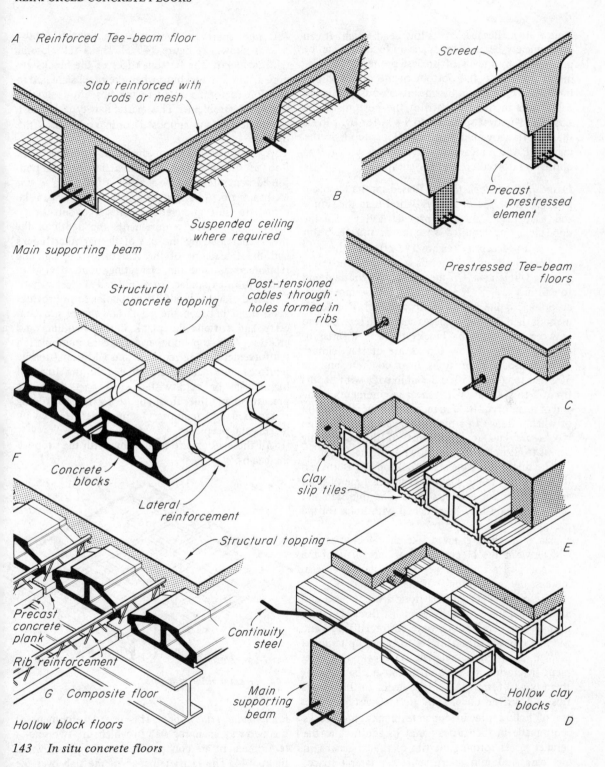

A   Reinforced Tee-beam floor

Slab reinforced with
rods or mesh

Suspended ceiling
where required

Main supporting beam

Screed

Precast
prestressed
element

B

Structural
concrete topping

Post-tensioned
cables through
holes formed in
ribs

Prestressed Tee-beam
floors

F

Concrete
blocks

Lateral
reinforcement

Clay
slip tiles

C

Structural topping

E

Precast
concrete
plank

Rib reinforcement

Continuity
steel

G   Composite floor

Main
supporting
beam

Hollow clay
blocks

Hollow block floors

D

143   In situ concrete floors

storeys high. Because of its low dead weight it can economically be used for spans up to 9 m. It can be prestressed by means of precast pre-tensioned elements bonded to the bottom of the ribs as shown in figure 143 *B* or by means of post-tensioned cables run in holes formed at the bottom of the ribs (*C*). To cover spans up to 14 m for light loads these methods would require a floor depth in the region of 450 mm overall. For spans less than 9 m normal reinforced concrete is cheaper.

In situ hollow block floor Based on the tee-beam principle this type of floor is lighter than the simple solid slab floor, and as it provides a flat soffit the applied ceiling required with an *in situ* tee-beam floor is not necessary (figure 143 *D*). Like the solid concrete floor, in most forms it requires shuttering over its whole area. On this the blocks are laid end to end in parallel rows, about 75 to 100 mm apart, according to the width of rib required. Reinforcement is laid in these spaces and concrete poured between and over the blocks to form a series of tee-beams. The hollow blocks are of clay, similar to those used for partitions, or of concrete, and are 300 mm long and 250 or 300 mm wide with depths from 75 to 200 mm. When shear requirements at the end of spans necessitate an increased concrete section or when a flange to a main tee-beam is required the blocks are laid to stop short of the supporting beams to allow for this, as shown in figure 171 *B*.

The thickness of the structural topping is not less than 25 mm for the shallowest of this type of floor and increases for greater depths. The thin slab and the blocks only may be punched with holes for the passage of services.

Clay slip tiles, grooved like the blocks, can be laid between the blocks to cover the soffit of the ribs and give a uniform key for plaster. They also increase the fire resistance of the floor (*E*).

The main types of this floor are patented and are designed and erected by specialist firms. They are normally designed as one-way reinforced floors and the majority are most suitable for spans up to about 6.70 m. They can be adapted to two-way reinforcement by closing the ends of the blocks and spacing them out to form ribs at right-angles to the normal ribs as described under *Plate floor*. One type makes use of hollow precast concrete blocks, the blocks being made in such a way that in addition to the parallel spaces forming the ribs of the tee-beams, in this case 450 mm apart, narrower lateral spaces

450 mm apart are formed at right-angles to the ribs as shown at figure 143 *F*. These take lateral reinforcement. The bottom edges of the blocks are lipped, the lips touching when the blocks are laid in position to form a soffit to the ribs, thus reducing shuttering problems. This particular type is most economical for superimposed loadings of 3.50 kN/m^2 or more.

A composite form of the *in situ* hollow block floor is shown at (*G*). In this a thin precast reinforced concrete plank forms the soffit of an *in situ* tee-beam rib. The widths of the planks can be varied to suit grid requirements. To suit heavy load or wide span requirements the depth of the ribs is increased by the use of deeper filler blocks and the thickening of the structural topping. The reinforcement and the projecting bond steel is in lattice form expanded from cold rolled steel strip or is formed of welded rods. Its particular form provides good bond between the plank and the *in situ* concrete and stiffens the plank during handling and erection. The top boom can serve as compression reinforcement when required and additional tensile reinforcement can be introduced in the form of high-tensile or cold-twisted steel bars. Like the precast rib and filler floor (figure 147) no shuttering is required and temporary support is provided by ledgers and props.

All these floors are monolithic with the supporting beams.

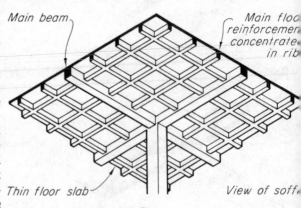

144   *Rectangular grid floor*

Rectangular grid floor This is a development of the two-way spanning slab in which the two sets of reinforcement are concentrated in ribs, as shown in figure 144. The restricted span of the slab over the ribs results in a considerable reduction in its thickness, and consequently in the total dead weight of the floor. It is most economic over relatively wide spans where the weight of a solid slab would be excessive. The optimum spacing of the ribs is controlled by the minimum practical thickness of the slab they carry. The latter is governed by fire-resistance requirements, depth required for effective accommodation of reinforcement and the maximum permitted ratio of span to depth for the slab.

The methods used for constructing this type of floor are referred to under *Formwork*, page 417.

In situ diagonal beam floor   This type of floor, like the rectangular grid floor, is a single layer grid construction (see page 326). It consists of two intersecting sets of parallel beams equally spaced and set at 45 degrees to the boundary supports as shown in figure 145. The beams are basically all of the same depth and cross-section and are rigidly connected at their intersections and to continuous edge beams. The support and end restraint provided by the shorter and stiffer corner beams to those of longer span, together with bottom slabs provided in the corners to resist uplift due to the negative bending moments in these areas (*A*), results in a considerable reduction in the bending moments in the longer beams compared with those in a normal slab and beam floor of the same span. Thus a depth/span ratio of 1:30 is possible. In addition to bottom slabs the short beams across the corners are sometimes, for greater efficiency, made slightly deeper and wider than the other beams.

For greatest efficiency the grid layout should give a sub-division of three panels along the shorter side of the structural bay. Where internal columns are necessary these are preferably located on the intersection of the beams (*B*). Crossed cantilevers, springing from the heads of the columns within the thickness of the floor, may be necessary if the layout results in beams of excessive length as shown in (*B*). Cantilever beams are also required if the columns are located within a panel instead of at an intersection of beams (*C*). Ideally, this type of floor should be square on plan, or subdivided into squares, to give maximum efficiency.

The floor slab need not be integral with the beams. Although it is generally of *in situ* concrete it may be of precast concrete units, of corrugated asbestos-cement sheets with a topping, or of timber

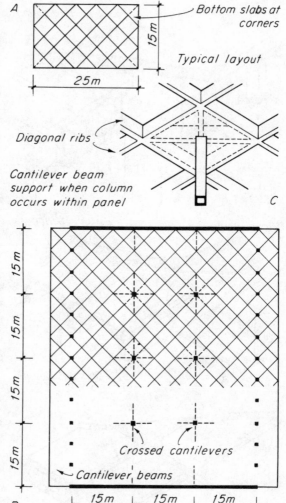

145   *Diagonal beam floor*

joists with boarding, since the beam structure is designed independent of the slab.

This floor is most economic when used to carry heavy superimposed loads, particularly concentrated loads, over spans of 15 m or more. In Great Britain it is very generally called the 'Diagrid' floor, although this particular name is, in fact, the trade name of Truscon Ltd.

Prestressing can be applied to this type of floor because of its beam structure. When prestressed the beams are generally precast in panel lengths, supported on temporary runners and props, and post-tentioned together as shown in the roof example in figure 223 *C*.

247

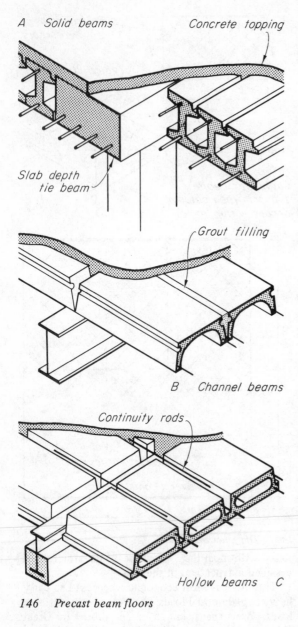

A   Solid beams   Concrete topping

Slab depth
tie beam

Grout filling

B   Channel beams

Continuity rods

Hollow beams   C

146   *Precast beam floors*

Precast floors

Most types of *in situ* concrete floor require shutter-ing over the whole of their area which must be kept in position for a considerable period until the con-crete has gained sufficient strength. The use of pre-cast floors reduces or eliminates shuttering and results in shorter construction times as indicated in Part 1, where the advantages and disadvantages of

this form of construction are outlined. The two basic categories of precast beam and precast rib and filler block construction will now be considered in greater detail.

Precast beams   These are illustrated in figure 146. They provide the simplest form of precast concrete floor and are widely used because of the speed with which they can be erected and used as a working platform. The floor consists of a number of beams placed side by side and, to a very large extent, spanning between supports independent of each other. The cross-section varies with the individual maker, some having a closed soffit, others an open soffit requiring an applied ceiling. All are designed to produce a more or less hollow floor to reduce the dead weight.

Solid beams   In this type some form of I-section is normal (*A*). The beams are laid with edges touching or, in some cases, overlapping slightly, so that the bottom flanges form a flat soffit. If continuity over supports is required portions of the top flanges must be removed to allow the insertion of reinforcement and *in situ* concrete.

Hollow and channel beams   The solid units are relatively narrow and heavy, and developments in the form of channel and hollow beam sections of greater unit width, and much the same weight, enable floor areas to be laid more quickly (*B, C* and figure 166 Part 1). Widths of the units vary from about 290 to 420 mm and depths according to span and load. Each beam is reinforced in the bottom corners and in some types in the top corners also. The sides are splayed or shaped to form a narrow space between the beams. This is filled with grout to assist the units to act together in some measure, the adjacent faces of the beams being grooved or castellated to provide mechanical bond. The grout is normally taken up to finish flush with the top of the beams, a structural concrete topping being used only when extra strength is required for heavy loads over long spans. Continuity over supports is obtained by the insertion of rods in the joints prior to grouting (*C*).

In all types of precast beam floors the units are made to the correct lengths between bearings with the ends designed to suit the type of bearing. The use of foamed plastic cores to form the voids in hollow beams, instead of pneumatic tubes, gives greater design freedom at the ends, which can be solid, because the cores are not withdrawn (see figure 150 *A*). Notches, holes and fixings for applied ceiling and floor finishes can be incorporated. The underside of the units can be left smooth or keyed for plaster in types having a flat soffit forming a continuous ceiling.

The beams are delivered to the site, hoisted and placed in position on their supports and the joints grouted up. A minimum of *in situ* filling is required and no temporary shuttering or supports. Spans up to approximately 6 m are possible although the economic limits for most types are 3.50 to 4.80 m.

Precast ribs and fillers   This type of floor consists basically of precast reinforced concrete ribs spanning between the main supports and carrying hollow blocks or slabs to fill the spaces between the ribs (figures 147 and 151 *E*). In most systems a flush soffit is produced. The *in situ* topping may be simply a screed (figure 147 *A*) or may be structural, to act with the ribs (*B*), depending on the load range for which the system is designed.

The ribs, which are placed at centre spacings ranging from about 250 to 600 mm, are manufactured in the required lengths for each job. In the case of heavy floors they may be pierced at intervals to permit the passage of lateral distribution rods.

The fillers may be hollow clay tiles or pots, hollow precast concrete blocks or wood wool or lightweight concrete slabs. The pots or blocks are usually rebated at the bottom edges and sit on projecting flanges or lips at the bottom of the ribs so that a flush soffit results (*A, B*). In those systems designed as a composite construction with a structural topping, the filler blocks are shallower in depth than the ribs, or are deeply splayed, so that the latter project above the fillers and bond with the topping when this is cast as shown in *B*. The sides of the ribs are usually castellated to provide mechanical bond. In systems with lightweight slabs resting on the tops of the ribs, bond is obtained by means of steel stirrups or dowels left projecting from the top of each rib (*C*).

No shuttering is required for this type of floor, although in those systems with structural toppings a few props may be necessary under the ribs until the *in situ* concrete has matured. Continuity over supports is obtained by means of rods placed between the ribs.

Prestressed floors   Prestressing is now applied to many types of precast floors. This is advantageous where wide spans are involved because it reduces the thickness and dead weight and increases the economic

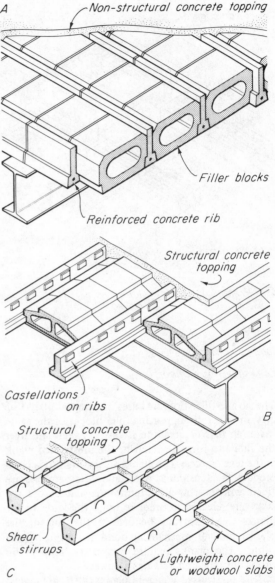

147   *Precast rib and filler floors*

span of the floor.

Pre-tensioning is most commonly adopted so that no stressing is carried out on site and it is applied to all the precast floors so far described — solid, hollow and channel beams and rib and filler types.

The economic advantage of combining prestressed precast elements with *in situ* concrete has been referred to on page 221. Many types of prestressed floor are based on composite construction of this nature, two of which are illustrated in figure 148.

249

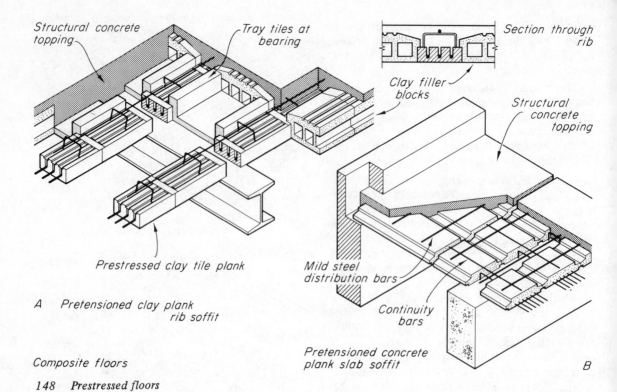

Structural concrete topping

Tray tiles at bearing

Section through rib

Clay filler blocks

Structural concrete topping

Prestressed clay tile plank

Mild steel distribution bars

Continuity bars

A   Pretensioned clay plank rib soffit

Pretensioned concrete plank slab soffit

B

Composite floors

148   Prestressed floors

The rib and filler floor (A) uses prestressed planks of clay or concrete supporting filler blocks of the same material. *In situ* concrete is cast between and over the fillers to form a rib having the prestressed plank as the tension zone. The planks are grooved to accommodate the tensioning wires and these grooves are filled with mortar after tensioning and before the jacks are released in order to bond the wires to the planks. Steel stirrups bedded in the top of the planks provide mechanical bond with the *in situ* filling. The planks require temporary support by props until the concrete has matured.

The prestressed concrete planks in (B) are placed close together side by side without filler blocks to form permanent slab shuttering. The edges are grooved to provide a dovetail key for the *in situ* concrete structural topping in addition to the natural bond between the rough top·of the plank and the topping. The planks are 355 mm wide by 50 to 100 mm thick in lengths up to 7.50 m, and the thickness of the topping is varied according to the span and load. Steel distribution rods, and continuity rods over supports, are laid on top of the planks before casting the topping. As with the other

system no shuttering is required and only a temporary central prop for spans over 2.40 m.

Prestressed precast tee-sections may be used in composite construction either in rib and filler block combination or placed close together to form a flat soffit and filled over with solid *in situ* concrete (figure 149). The latter produces a heavy slab and is useful in circumstances where stiffness for lateral rigidity is required, as, for example, in the spine beam in certain reinforced concrete frames as shown here (see also figure 124 *B* and the relevant text).

Many of these prestressed floors can be used for spans of 9 to 12 m and over. Generally speaking they will not often prove cheaper than the normal reinforced types for spans less than 6 m, although in certain circumstances some types may prove cheaper with spans as small as 4.50 m.

Large precast floor panels   It is normal to employ some form of crane on contracts of any size, and their economic use depends on operating them at maximum capacity. In all forms of precast concrete work regard is paid to this fact. This has led logically to the use, in suitable circumstances, of relatively

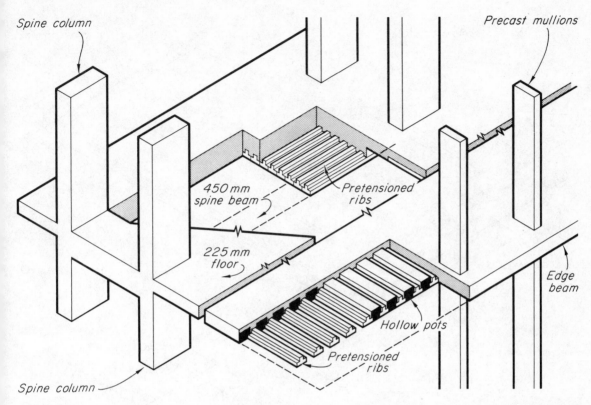

Spine column

Precast mullions

450 mm
spine beam

Pretensioned
ribs

225 mm
floor

Hollow pots

Edge
beam

Pretensioned
ribs

Spine column

149   Prestressed floors

large precast floor panels. The proprietary hollow beam floor has been developed in the form of precast multiple units, or wide slabs, as shown in figure 150 *A* which can be as wide as 2.70 m, the width for any particular span being dependent on the lifting capacity of the crane available. Advantages, in addition to the economic use of the crane, are greater speed of erection, reduction in weight compared with an equivalent area of normal hollow beam floor, reduction in grouting and the simplification of trimming large holes by casting during manufacture.

Prestressed planks are produced in widths up to 1.20 m, and 50 mm thick reinforced planks up to 2.4 m wide (figure 150 *B*). The latter incorporate lattice reinforcement similar to that to the precast rib soffit in figure 147 *G*.

Wide multiple tee-beam or ribbed slabs (figure 166, Part 1) are also produced, especially for wide spans.

Flat plate floors on precast concrete columns on a grid of about 5.50 m have been precast in large panels. The column bands are cast as panels span-

ning between column head plates and the whole of the centre panel is cast in one piece. Rebates in the top edges of the panels form channels or strips about 450 mm wide in which *in situ* concrete is placed to bond with the steel reinforcement left projecting from the edges of each panel. The column head plates need most accurate setting and levelling and it is usually more economical to carry these out as *in situ* work.

Openings and services in concrete floors

Apart from small holes for pipes in solid floors and holes within the width of an individual section of a hollow beam (figure 151 *A*) or of a filler block, openings in concrete floors require to be trimmed. Illustrations of methods commonly adopted are given in figure 151.

In *in situ* solid concrete and hollow block floors small openings can be trimmed by the use of appropriate reinforcement placed within the thickness of the slab round the opening (*B*), but for larger openings such trimmers and trimming beams are too

251

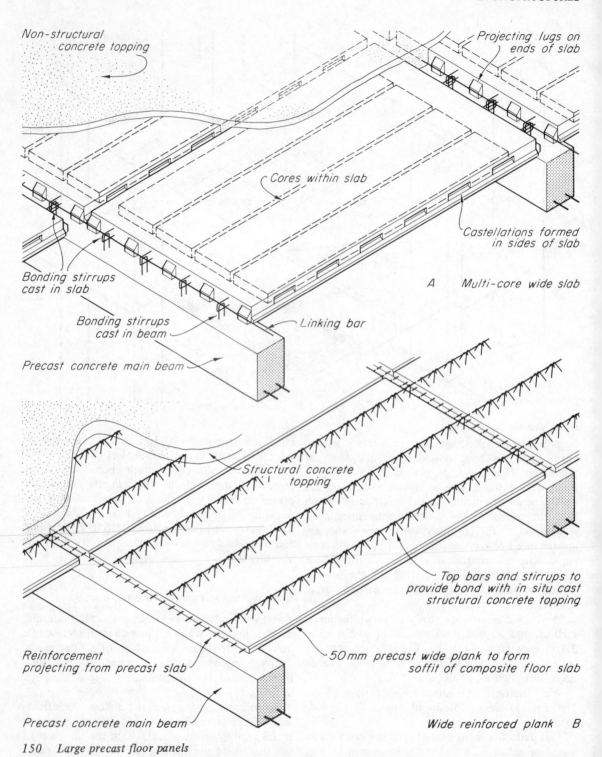

Non-structural
concrete topping

Projecting lugs on
ends of slab

Cores within slab

Castellations formed
in sides of slab

A   Multi-core wide slab

Bonding stirrups
cast in slab

Bonding stirrups
cast in beam

Linking bar

Precast concrete main beam

Structural concrete
topping

Top bars and stirrups to
provide bond with in situ cast
structural concrete topping

Reinforcement
projecting from precast slab

50 mm precast wide plank to form
soffit of composite floor slab

Precast concrete main beam

Wide reinforced plank   B

150   Large precast floor panels

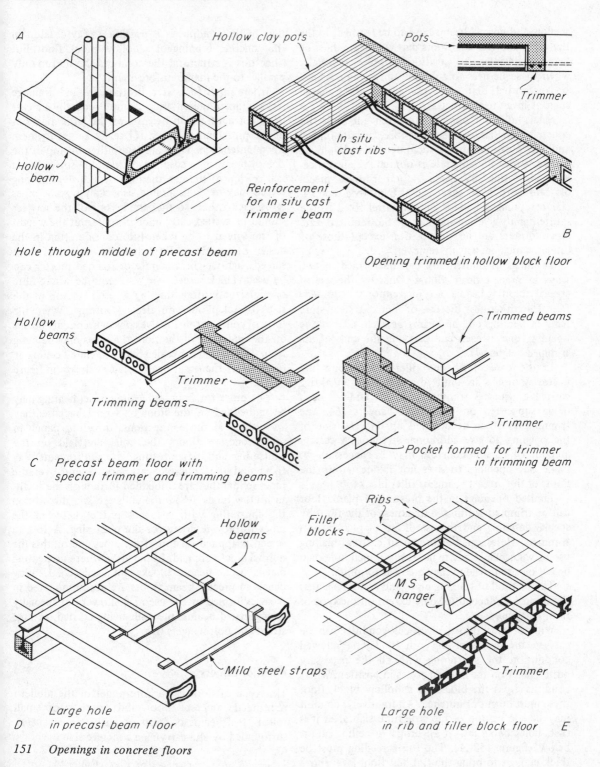

A

Hollow clay pots

Pots

Trimmer

Hollow beam

In situ cast ribs

Reinforcement for in situ cast trimmer beam

B

Hole through middle of precast beam

Opening trimmed in hollow block floor

Hollow beams

Trimmer

Trimming beams

C   Precast beam floor with special trimmer and trimming beams

Trimmed beams

Trimmer

Pocket formed for trimmer in trimming beam

Hollow beams

Mild steel straps

Ribs

Filler blocks

M S hanger

Trimmer

Large hole in precast beam floor

D

Large hole in rib and filler block floor   E

151   Openings in concrete floors

shallow and normal beams have to be formed. In the flat plate and flat slab floors particular care must be taken in deciding the positions of holes, especially when these lie near to columns and column bands. It is essential that early consideration at design stage be given to this matter.

Small holes for services are often cut after the concrete floor has been cast but this is uneconomical and the positions of all such openings should be settled before the concrete is poured. Small vertical holes are formed by means of timber boxes, pieces off cardboard rolls, sheet metal bent to a rectangular or cylindrical shape or gas barrel fixed to the shuttering to give an aperture of the required size. Most formers are left in position except those of timber, which are usually removed.

Openings of limited size can be formed in two ways in precast beam floors. One, by the use of special trimming beams carrying reinforced trimmers (C). The other, by the use of cranked steel strap hangers bearing on the beams at each side of the opening, one at each end carrying the ends of the trimmed beams (D).

In the case of rib and filler block floors the necessary blocks are omitted and the shortened ribs carried on special trimmer ribs, the ends of each being supported on mild steel hangers (E). The trimming ribs can be doubled up on each side of the opening to give additional strength. A certain amount of *in situ* concrete may be necessary to make out openings to sizes not falling within the limits of the precast beams or filler blocks.

Limited openings in the prestressed plank floor can be trimmed within the thickness of the floor by special detailing and the use of cut planks to form trimmers. Holes can be formed in casting, making use of wide planks where necessary. Small holes can be cut on site provided they are near to joints.

The method of running pipes and conduits across the floors will depend to some extent on the type of floor being used.

With *in situ* floors, electrical conduits can be cast in the soffit of the slab by laying screwed conduit on the shuttering with all the necessary outlet, junction and draw-in boxes in position. Where conduits cross the blocks of a hollow block floor these must be cut or notched. Alternatively, conduit may be laid on top of the structural slab after it is cast, holes being left at all drops to ceiling points below (figure 152 A). The top screeding must be thick enough to bring the finished floor level above

that of the conduit. It is possible to lay conduit in the concrete topping of a hollow block floor, but since this is structural the conduits should run only parallel to the main reinforcement.

Where pipes for water, heating, gas and drainage services are required they cannot generally be cast in the structural slab because of the large diameter of the pipes and the need for ready access to them for maintenance. They are laid either on top of the slab or fixed to, or suspended from, the soffit. When laid on top of the slab the thickness of the finishing screed may be considerable, especially if some pipes cross each other, so that the layout of the services should be settled early enough to permit the weight of the screed to be taken into consideration in the design of the floor. To reduce the weight of thick screeds lightweight concrete or foamed mortar can be used. Thick screeds are also required where fibre or metal underfloor ducting is used on top of the slab for telephone and electric wiring[1]. When the form of construction involves ribs or secondary beams on the underside of the floor the pipes may be hung from the slab and concealed by a suspended ceiling at the level of the rib or beam soffit as shown in figure 152 B.

The pipes for floor and ceiling panel heating may be embedded in the floor or screed because they are welded at the connections. In ceiling panels in solid concrete floors the coils are laid on the shuttering and, after testing, the reinforcement is fixed and the concrete cast around and over the pipes so that they are embedded flush with the soffit. In hollow block floors the coils are laid directly on the shuttering with special slip tiles between the pipes to provide a key for the plastering. A 19 mm screed is applied on top of the pipes and on this the normal blocks are laid. In the areas where no heating panels occur deeper blocks are used (C). Heating panels in precast beam floors are accommodated in recesses formed by the use of shallower but specially strengthened beams, the coils being carried on steel straps and bolt hangers (D).

Filler joist floors

This type of floor is the forerunner of the modern reinforced concrete floor and consists of small rolled steel beams or joists at fairly close centres, surrounded by and carrying a concrete slab as shown

[1] See *MBC: Environment and Services*, chapter 14.

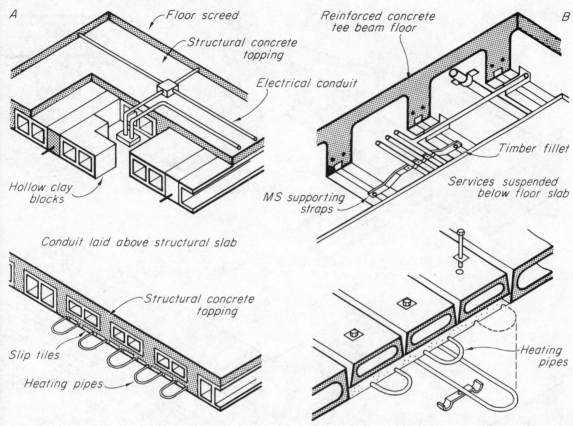

A

Floor screed

Structural concrete topping

Electrical conduit

Hollow clay blocks

MS supporting straps

Reinforced concrete tee beam floor

B

Timber fillet

Services suspended below floor slab

Conduit laid above structural slab

Structural concrete topping

Slip tiles

Heating pipes

Heating pipes

C   Heating panel in hollow block floor

Heating panel below precast beam floor   D

152   *Services in concrete floors*

in figure 153 *A*. The steel, or filler, joists are cleated direct to the main steel beams, or, more commonly, are supported on a continuous steel seating angle bolted to the web of the main beam, every second or third filler only then being cleated. Where headroom is not restricted it is preferable to run the fillers over the supporting beams, extending over at least three spans. The maximum spacing of the fillers is related to the superimposed loading on the floor and to the thickness of the concrete unless the concrete is reinforced to span as a slab, or function as an arch, between the filler joists. Slabs arched between the filler joists are simply constructed by using ribbed metal lathing curved between the joists and acting as centering and a key for plastering (*B*). The thickness of the concrete at the crown must be not less than 50 mm.

Other methods, designed to eliminate shuttering, use hollow clay blocks or tubes spanning between the bottom flanges of the filler joists. They have lips which protect the underside of the fillers and provide a key for plastering. The minimum thickness of concrete over the blocks is 32 mm for filler spacings up to 450 mm and 50 mm for greater spacings. One system is illustrated at (*C*).

The filler joist floor tends to be heavy and is generally used in cases of severe loading conditions as in factories, loading platforms and colliery buildings where it can be economical. Compared with a solid reinforced concrete floor it has the advantage that holes may be cut subsequent to its completion anywhere between the filler joists. This can prove useful in certain classes of building involving periodic alterations in equipment and services.

Steel lattice joist floor

Lattice, or open web, joists with flanges of light steel channels, angles and flats, or of cold-formed

255

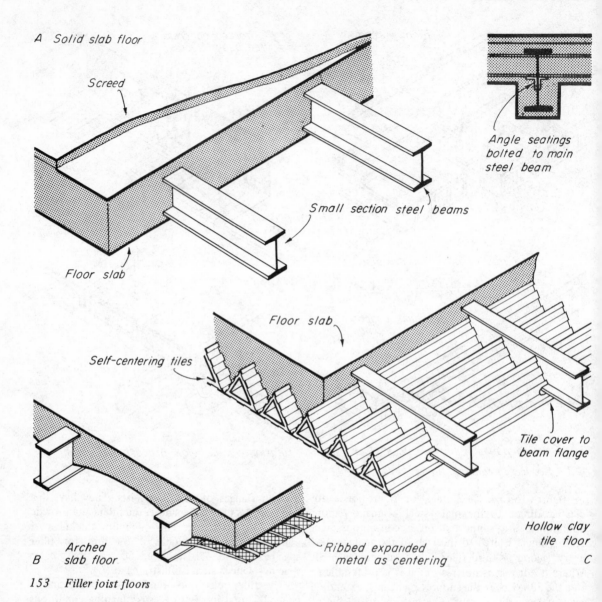

A Solid slab floor

Screed

Floor slab

Small section steel beams

Angle seatings
bolted to main
steel beam

Floor slab

Self-centering tiles

Tile cover to
beam flange

Arched
B   slab floor

Ribbed expanded
metal as centering

Hollow clay
tile floor
C

153   Filler joist floors

sections, and with bent rod, angle or tube lacing to form the web, can be used instead of normal timber joists or Universal beams. Continuous timber nailing fillets can be fixed on the flanges to provide fixing for floor and ceiling finishes (figure 154 A). The open lattice web gives complete freedom in running services through the thickness of the floor.

For a given floor loading and normal joist spacing, greater spans can be covered than with timber joists of the same depth. For example, with normal domestic loading 225 mm by 50 mm timber joists

at 400 mm spacing will span approximately 4.90 m whereas a lattice joist of approximately the same depth would span 8.50 to 8.80 m. Over the 4.90 m span a lattice joist of about 175 mm deep could be used.

When a building programme is sufficiently large the use of shallow lattice joists at 400 to 450 mm centres can be competitive for domestic work, having regard to the ease with which services can be run through the floor thickness. But, generally speaking, their most economic use is at 600 mm centres

256

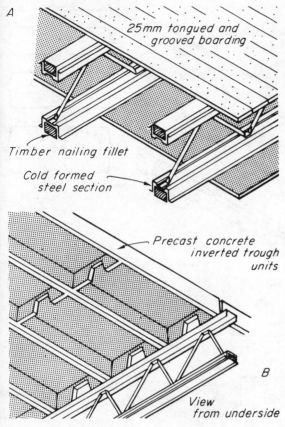

A

25mm tongued and grooved boarding

Timber nailing fillet

Cold formed steel section

Precast concrete inverted trough units

B

View from underside

154   Steel lattice joists

using an increased thickness of floor board. For this spacing, 25 mm tongued and grooved boarding is required. Greater .spacing of the joists is possible when light concrete slabs are used to span between them as shown in figure 154 *B*, and a lighter floor is generally obtained than if rolled steel beams are used to carry the slabs. The concrete slabs may be formed from light, precast trough panels or by stretching stout building paper over the joists, stiffened with light mesh reinforcement, to form shuttering for concrete about 25 to 75 mm thick.

Cellular steel floor

Steel in the form of cellular units working in conjunction with a concrete topping can be used to form a structural floor slab as illustrated in figure 155 *A*. The steel cellular units consist of top and bottom profiled sections spot-welded together along their length, the sections varying in depth to suit different spans and loadings. At 600 mm intervals

along the top section slots are punched, through which steel strip bonding anchors are fed to provide the bond necessary to permit steel and concrete to function together as a composite loadbearing floor slab. The minimum thickness of concrete topping is 45 mm. The steel units are fixed to supporting steel beams by welding or, if the beams are of reinforced concrete, to steel plates embedded in the concrete. Alternatively, cartridge hammer fixing to concrete may be used.

The continuous cells of the units form a series of ducts at 200 mm centres running parallel to the span. By the use of header ducts laid at right-angles over the units services of all types can be brought from main distribution points to any point of the floor or the ceiling below. The header ducts with the various outlet accessories are, in principle, the same as in the underfloor duct systems used on normal reinforced concrete floors and referred to on page 254.

The floor can also be used for air ducting. The welded joints between the top and bottom sections of selected cells are closed with a special sealer and the cells are connected to the air supply by header ducts on the underside of the steel units.

The units are 600 mm wide and up to 8.20 m in length. They are relatively light in weight, can be quickly hoisted in bundles and fixed in position, and when fixed provide a platform for materials and other trades before the concrete topping is poured. The floor is lighter in weight than most forms of *in situ* and many forms of precast concrete floors. The unprotected underside under standard test has a fire rating of half-an-hour and a false ceiling of suitable material, or asbestos sprayed direct on to the soffit, must be used to achieve a higher degree of fire resistance.

Profiled steel floors

A form of construction used on the Continent employs corrugated steel sheeting as the main supporting element with a topping of concrete to distribute the load. The corrugated sheeting, specially protected against corrosion, spans between secondary steel beams and is riveted at each end to angles welded to the beam webs. The spacing of the beams depends upon the gauge of metal and depth of corrugation used for the sheeting.

A layer of cork aggregate concrete is placed on the sheeting and covered with roofing felt and on

257

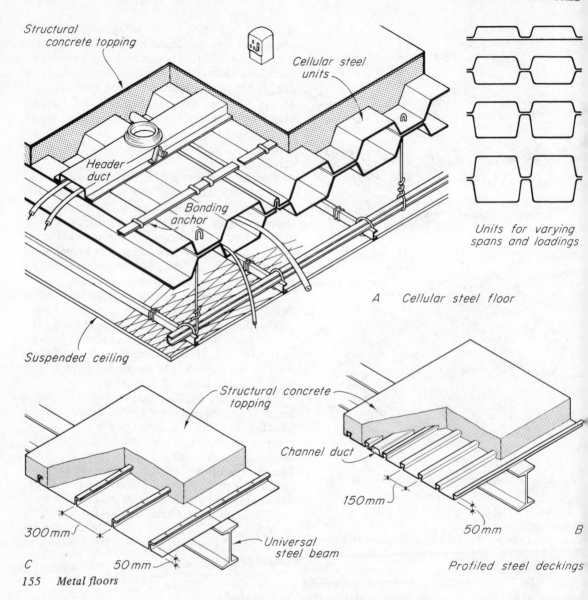

Structural concrete topping

Cellular steel units

Header duct

Bonding anchor

Suspended ceiling

Units for varying spans and loadings

A   Cellular steel floor

Structural concrete topping

Channel duct

150mm

50mm   B

300mm

50mm

Universal steel beam

C

Profiled steel deckings

155   Metal floors

this is poured a 38 mm thick layer of normal concrete reinforced in both directions. The concrete is not bonded to the steel to form a composite construction but serves to distribute the applied loading over the corrugated sheeting below.

Another form uses cold-rolled steel decking with wedge-shaped folds projecting upwards, which key into the concrete to form the reinforcement to a composite construction (figure 155 B). The sheeting is spot welded or gun-fixed to supporting steel beams and acts as a working platform during erection

which requires no props and as permanent shuttering to the floor. The open folds form a key for soffit plastering. Channel ducts can be formed between the folds with header ducts at right angles laid in the thickness of the concrete topping to take electrical services. (C) shows another type using shallow, individual cold-formed channels riveted to each other to form the decking and reinforcement to a composite floor. This requires some propping during concreting. Similar provision can be made for electrical services.

258

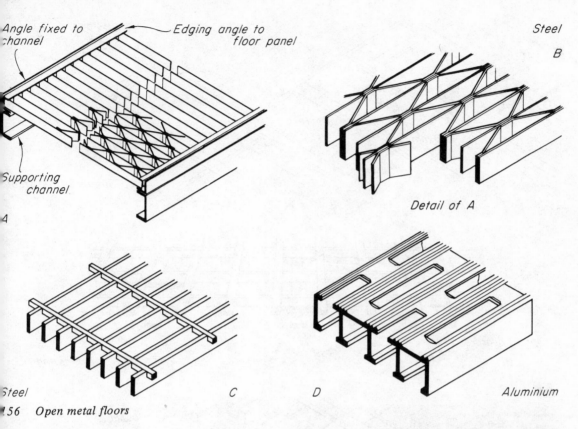

Angle fixed to channel

Edging angle to floor panel

Steel

B

Supporting channel

A

Detail of A

Steel   C   D   Aluminium

156   Open metal floors

Open metal floors

Open metal flooring, some examples of which are shown in figure 156, is used mainly in industrial buildings, particularly for service and operating platforms to machines, where the passage of light and air is required to be maintained.

This type of flooring is made up in steel or aluminium alloy in panels of varying widths and lengths as required. It can be formed of parallel flats spaced apart and braced either by similar flats (A, B) or by bars intersecting at right-angles at intervals along the length of the panel (C). The junctions of all members are welded or riveted and depths range from 19 to 63 mm. Clear spans are up to 2.4 or 2.7 m.

Open flooring pressed from 16 gauge mild steel sheet is also available, produced in sections or planks 225 mm wide and in 1.35 m and 1.80 m lengths. The depth is standard at 38 mm so that variations in loading must be allowed for by variations in the support spacing.

In aluminium alloy this type of floor is also produced as a 150 mm wide ribbed extrusion in depths from 19 to 50 mm, with rectangular or square holes punched in the top plate (D). Clear spans are much the same as for the other types of floors. For all types the supporting structure is formed from various rolled-steel sections to which the floor panels are fixed by means of clips and bolts (A). An insulated clip and stainless steel or cadmium plated bolts are used with the aluminium extrusions to isolate the two metals and avoid possible electrolytic action.

Perforated cast-iron plates of a similar nature are used externally for walkways and for fire escape stairs. It is a heavier form of construction than the steel panels described above, but is used externally because of the greater resistance of cast iron to corrosion (see figure 179).

Spring floors

Gymnasium and dance floors, particularly the latter, should be resilient. This resilience is imparted by the use of a sub-floor of timber bearers, with or

259

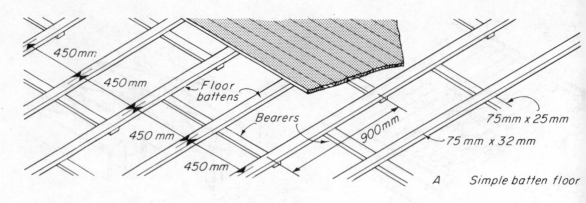

450 mm

450 mm

*Floor
battens*

Bearers

450 mm

900mm

75 mm x 25 mm

75 mm x 32 mm

450 mm

A   *Simple batten floor*

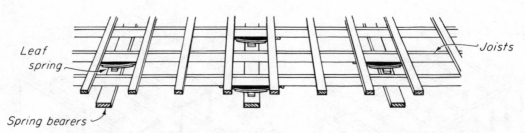

*Leaf
spring*

Joists

Spring bearers

C   *Leaf-spring floor*

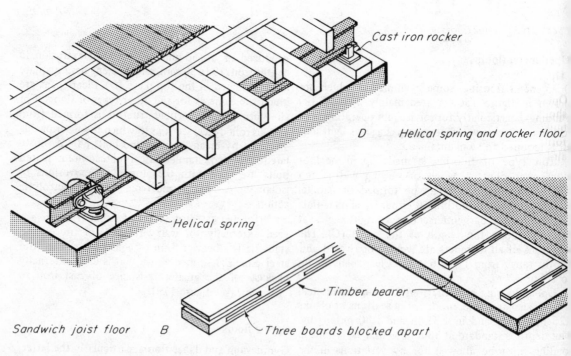

Cast iron rocker

D   *Helical spring and rocker floor*

Helical spring

Sandwich joist floor   B

Timber bearer

Three boards blocked apart

157   Spring floors

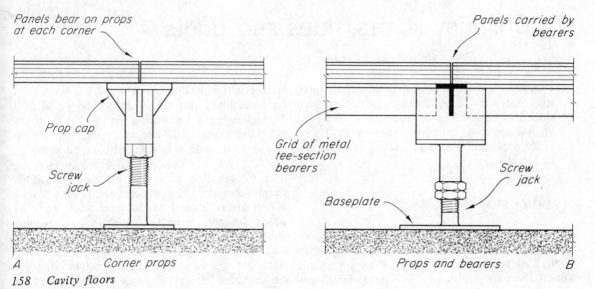

Panels bear on props at each corner

Panels carried by bearers

Prop cap

Grid of metal tee-section bearers

Screw jack

Screw jack

Baseplate

A   Corner props   Props and bearers   B

158   Cavity floors

without springs or rockers as shown in figure 157, on which narrow strip flooring is fixed.

In its most economical form the sub-floor may consist of thin floor battens at 400 to 450 mm centres carried on bearers fixed to the main floor in a staggered arrangement (A). This gives a bearing at every 900 mm to alternate floor battens and at 450 mm to the intermediate battens. Although not producing as much resilience as a floor incorporating springs it is cheaper than a true spring floor.

Another form without springs, developed in Sweden for gymnasium floors, uses bearers made up of timber boards in 'sandwich' form in which the boards are blocked apart at staggered intervals (B). Resilience is given because the blocks carrying the upper board bear on the middle board at a point midway between the lower blocks: not directly on them.

Fully resilient floors incorporate special springs or rockers to permit the sub-floor to deflect when in use. In some forms the springs are of a leaf type placed at intervals between the joists and continuous parallel wood bearers to form 'spring joists' on which the strip flooring is laid (C). Other forms incorporate spring and rocker fitments carrying steel beams on which timber bearers are laid (D). Others make use of rocking bars only. Devices can be incorporated to lock the springs when a rigid floor is required.

Cavity floors

These are structural sub-floors raised above the normal floor slab to form a cavity to accommodate services, and have developed with the increase in services in areas such as laboratories and computer rooms.

The floor may be of normal timber strip incorporating access panels and carried on small timber joists or metal sections which are raised above the main floor on metal struts or props but more usually it is formed entirely of removable panels about 600 mm square supported either directly on props at their corners (figure 158 A) or at their edges on a grid of timber or metal bearers which is supported on props (B). The props in all cases are adjustable in height to permit levelling of the floor and give a minimum depth of cavity from 75 to 100 mm.

MOVEMENT CONTROL

Settlement movement

When settlement joints are provided in the structure, as described in the previous chapter, the junctions of the floor slabs must be designed to permit relative movement by rotation. These would be detailed as pivoted joints.

Thermal movement

Typical details of expansion joints in floors are shown in figure 140 and reference is made to these on page 236.

261

7 Chimney shafts, flues and ducts

The construction of flues and fireplaces to serve heating appliances of a domestic scale has been described in Part 1. The construction of the large flues and chimney shafts necessitated by higher capacity appliances and large heating plants is discussed in this chapter.

CHIMNEY SHAFTS

A chimney shaft is a free standing structure containing a flue which by virtue of the heating apparatus it serves is generally larger than a domestic flue[1]. Flues are bounded by flue linings which protect the surrounding structure of the chimney and assist the formation of convection currents by which gases are taken to the outlet of the flue for dispersal to the upper air.

Size of chimney shafts

The height and cross sectional area of a given flue clearly relate to the basic need to provide a suitable volume of warmed air and gases whose temperature difference to that of the outside air is sufficient to allow convection currents to form within the flue. These currents must be strong enough to overcome the frictional resistance of the flue (and any other airways connected to it through which the gases must pass) and the inertia of cool air present in the flue in different atmospheric conditions.

The design of large chimneys and flues involves calculations in which the factors of cross-sectional area, height, plant type and size and fuel are related. Useful approximations of the probable size of chimneys in relation to a given plant size and height can be made by reference to simplified tables or graphs.[2]

The temperature differential between the inside and outside of the flue is a critical factor and may be affected by:

(i) *Atmospheric conditions* High atmospheric temperatures and heavily polluted atmospheres will cause sluggish action and possibly down draughts

(ii) *Materials of construction* Flue linings of low insulation value will not maintain high enough temp-

eratures within the flue to promote strong convection currents. Flues surrounded by chimneys of high thermal capacity will take a lot of heat from the gases before the surrounding material warms up enough to allow gases to reach high temperatures

(iii) *Size of flue* Too large a flue will dissipate the heat of the flue gases too rapidly resulting in sluggish operation. This problem occurs when several boilers are connected to the same stack, some of which are not in use in summer. The remaining boiler(s), often providing hot water only, does not require such a large flue and cannot produce sufficient hot gas for the flue to function. In these circumstances separate flues are preferable, grouped together if possible. in a single chimney stack to engender a concentration of flue gases at maximum load sufficient to make the smoke plume rise well into the upper air and disperse the effluent.

Height of flues

Apart from theoretical factors affecting the design flues, there are statutory requirements for the minimum heights of flue outlets. The minimum height for domestic chimneys and for small plant is determined by section L of the Building Regulations 1972 and for industrial and large central heating plant by the second edition of the Clean Air Act Memorandum 1956 and section 6 of the Clean Air Act 1968. The following factors are taken into account in relation to the determination of chimney heights:

Nature and quantity of effluent Conventional fuels (gas, oil and coal) are assessed in terms of their sulphur content, the quantity of sulphur dioxide in their flue gases and the sizes of the furnace installation.

Speed of gases through flue The memorandum recommends flue gas velocities of up to 15 metres

---

[1] It should be noted that flues of large dimension can be accommodated inside buildings or within elements of buildings such as party walls, the chimney being formed as a duct within the structure of the building (see page 263).

[2] See Chimney height nomograms included in the Clean Air Act 1956 and *Memorandum on chimney height 1963.*

per second irrespective of flue efficiency in enabling the plume of smoke to rise well above the flue terminal before dispersion[1].

Neighbouring sources of air pollution This is affected by the general character of the district. In view of the reduced efficiency of flues in polluted atmospheres, the memorandum gives five classifications of districts ranging from country districts to large cities with varying degrees of pollution.

Presence of other buildings and tall trees These can cause downwash of air around chimneys and consequent premature deposit of effluent on the ground in the vicinity. If the height of the chimney as determined by the previous factors is less than two and a half times the height of any local object some modification may be necessary to achieve proper dispersion of effluent in the upper air.

Materials and construction of chimney shafts

Generally, the basic parts of a chimney shaft comprise: foundations, structural shell, flue lining (where considered necessary), branch flue connection to furnace(s) and terminal.

Foundations

Chimney shafts standing alone are clearly subject to wind loads and foundations must distribute asymmetrical loading to the subsoil within the allowable bearing pressure. Settlement associated with clay subsoils must be expected since the heat from the base of the chimney will in time be transferred to the subsoil. Due to the concentration of load it is common for chimney foundations to be piled.

Structural shell

This part of the shaft must perform as a stable structure cantilevered from the base and having overall stability when subjected to wind pressure from any given direction. The wind load is assessed in accordance with the method laid down in CP 3 : chapter V : Part 2, which is described briefly on page 116. The actual pressure on the shaft varies with its shape and the code gives coefficients to take account of this.[2]

In many cases the structural shell will be designed to give support to lining materials with or without the presence of an air space and certain openings

may be needed for access to air spaces in addition to those required for branch flues.

Chimney shafts are normally constructed of brickwork, concrete or steel with or without some form of lining. Table 16 summarises forms of contemporary construction.[3]

Masonry shafts

These are now less used due to their high cost and the difficulty in getting experienced bricklayers. When used the design of brickwork chimneys, as with masonry walls, is based on the use of the material in compression only and calculated on the assumption that the chimney will fail section by section up its height. Thus the designer must check the stability of each section in order to arrive at the overall stability of the structure. This results in a structure of increasing thickness towards the base as shown in figure 159 *A* which illustrates a traditional example with typical proportions[4] (*B*) and (*C*) show methods of lining such shafts.

Large chimneys may be constructed of masonry within buildings provided precautions are taken to protect the building from damage by heat or through corrosion of any structural steel. Such chimneys will generally require less thickness of shaft since they are given lateral support at each floor level and are not subject to wind loading. In such cases the main factor governing the thickness will be the

---

[1] Anthony Collins in 'Chimneys and flues: design and construction' *RIBA Journal* October 1973 suggests a minimum speed of 6 metres per second and a maximum of 18 metres per second on a sliding scale relating to plant size. He also suggests that to enable flue gases to continue in the same direction after leaving the flue they should travel at about one and a half times the wind speed likely to be encountered at a given site. The speed of flue gases can be calculated by reference to inlet temperatures, area and shape of flue and the frictional resistance of the flue. Draught can be produced by natural convection or by inducing it, using motorised fans.

[2] The Building Regulations rely on CP 3 chapter V: Part 1: 1967 for calculation of dead and superimposed loads and on CP 3 chapter V: Part 2 for calculation of wind loads.

[3] This information is reproduced from 'Chimneys and flues: design and construction' by Anthony Collins, *RIBA Journal* October 1973.

[4] These are based on the London Building (Constructional) By-laws 1972 (By-law 12.13).

Simple: single skin construction in one material (principally for gas appliances)

1 riveted steel plate 2 welded steel plate 3 cast iron	4 asbestos cement pipe 5 precast concrete flue blocks for building into walls	note: numbers 1 to 4 can be guyed or unguyed

Composite: double or multi-skin construction using one or more materials (suitable for oil and coal fired furnaces)

structure: materials and system of construction	lining materials		
masonry construction of brick, stone or concrete block	1 sand and cement render 2 clay pipe flue liners 3 firebrick set in masonry 4 firebrick insulated with rock wool 5 firebrick with ventilated air space 6 Moler brick set in masonry 7 Moler brick with ventilated air space 8 acid resistant bricks set in masonry	9 acid resistant bricks insulated with rock wool 10 acid resistant bricks with ventilated air space 11 Moler concrete (precast, gunned, or *in situ*) set direct to masonry 12 Moler concrete (precast or *in situ*) with ventilated air space 13 refractory concrete (*in situ* or gunned) set direct to masonry	14 refractory concrete with ventilated air space 15 flexible metal with unventilated air space 16 flexible metal with insulating fill between liner and chimney (usually existing) 17 insulating concrete (Rentokil)
precast reinforced concrete, cast in complete plan sections, with main reinforcement threaded and grouted in as building proceeds	1 precast Moler concrete 2 precast refractory concrete		
precast reinforced concrete, panels with the main reinforcement cast into the vertical and horizontal joints between the panels as the work proceeds (Trustack)	1 Moler concrete precast in tubular form with unventilated air space		
reinforced concrete cast *in situ*	1 firebrick with ventilated air space 2 firebrick with rock wool insulation	3 Moler brick with ventilated air space 4 acid resisting brick with ventilated air space 5 acid resisting brick with rock wool insulation	6 refracrory concrete (precast) or *in situ* with ventilated air space 7 refractory concrete with rock wool insulation 8 Moler concrete with ventilated air space
steel plate*, exposed externally with or without supporting guy ropes *see BS 4076:1966	1 firebrick 2 Moler brick 3 acid resisting brick	4 refractory concrete (*in situ*, precast, or gunned) 5 Moler concrete (*in situ*, precast, or gunned)	note: all these linings should be positioned no further than 25 mm from the steel to prevent the buildup of dust, which will cause the lining to bulge
steel plate covered with aluminium: air space between the two materials: with or without supporting guy ropes	1 structural steel acts as lining		
as above but with the air space filled with rock wool insulation	1 structural steel acts as lining		

Table 16 *Chimney shaft construction*

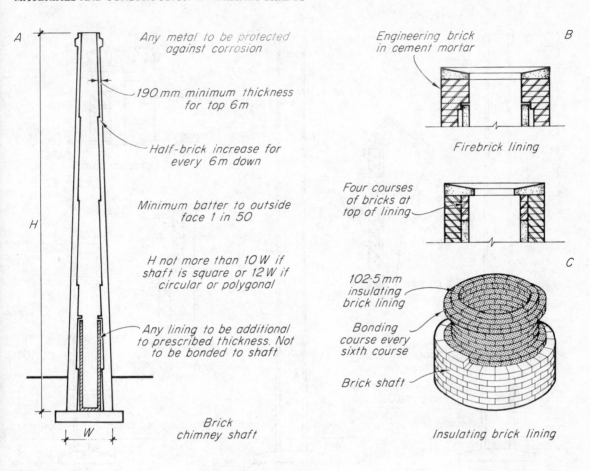

Any metal to be protected against corrosion

190 mm minimum thickness for top 6 m

Half-brick increase for every 6 m down

Minimum batter to outside face 1 in 50

H not more than 10 W if shaft is square or 12 W if circular or polygonal

Any lining to be additional to prescribed thickness. Not to be bonded to shaft

Brick chimney shaft

Engineering brick in cement mortar

Firebrick lining

Four courses of bricks at top of lining

102·5 mm insulating brick lining

Bonding course every sixth course

Brick shaft

Insulating brick lining

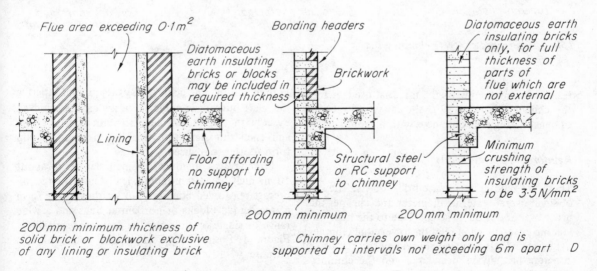

Flue area exceeding 0·1 m²

Diatomaceous earth insulating bricks or blocks may be included in required thickness

Lining

Floor affording no support to chimney

200 mm minimum thickness of solid brick or blockwork exclusive of any lining or insulating brick

Bonding headers

Brickwork

Structural steel or RC support to chimney

200 mm minimum

Diatomaceous earth insulating bricks only, for full thickness of parts of flue which are not external

Minimum crushing strength of insulating bricks to be 3·5 N/mm²

200 mm minimum

Chimney carries own weight only and is supported at intervals not exceeding 6 m apart

159   *Brick chimney shafts and large chimneys*

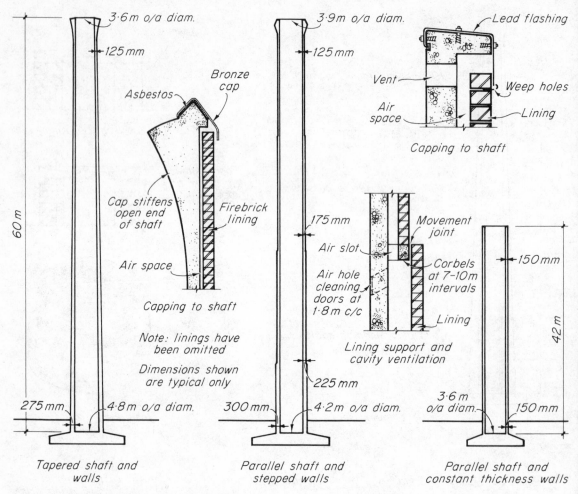

160   Reinforced concrete chimney shafts

resistance to crushing of the material used to form the chimney. Figure 159 *D* shows typical methods of constructing such chimneys within buildings[1].

Reinforced concrete shafts

These are tending to replace brick shafts. They are of small overall external diameter and cheaper than brick over about 35 m height. Due to the high wind loading experienced in relation to very tall chimneys of such buildings as power stations *in situ* reinforced concrete has always been used and the chimneys designed as special structures. Typical reinforced concrete shafts are shown in figure 160. Reduction

in thickness may be achieved by stepping or tapering the shaft but considerable economy can be achieved by constructing the whole shaft with constant bore and shell thickness thus enabling a rising or slip form shutter to be used.

Chimneys of medium or small size up to about 50 m high are often economically constructed of precast reinforced concrete units either in the form of segmented blocks or horizontal rings and a wide range of single or multiple flue designs are available. Figure 161 shows typical blocks used for this purpose.

[1] These are based on the London Building (Constructional) By-laws 1972 (By-law 12.05).

266

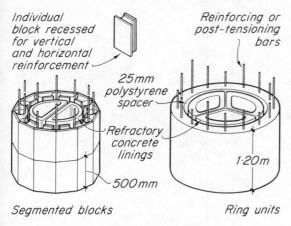

Individual block recessed for vertical and horizontal reinforcement

25mm polystyrene spacer

Refractory concrete linings

500mm

Segmented blocks

Reinforcing or post-tensioning bars

1·20m

Ring units

161   Precast concrete chimney shafts

Steel shafts

These may be either self-supporting, guyed or stabilized by enclosing framed structures. Steel chimneys can be erected quickly and may be lined or unlined depending on the fuel being burnt (see *Linings*).

Self-supporting steel chimneys have been erected up to 135 m in height but a more normal height is around 75 m. The cantilevered chimney shell is secured to a concrete base by bolts and is usually stiffened by gusset plates at the base (figure 162 *A*). A steel chimney weights far less than a concrete chimney and resistance to overturning will largely be provided by the mass of the concrete base. Steel chimney shells are either riveted or welded structures particular care having to be paid to the latter form with regard to the resonant vibrations caused by wind eddies which, if they match the natural frequency of the sway of the chimney, may cause the sway to develop to the point of collapse. Unlike reinforced concrete or riveted steel, welded shafts have little structural damping within the structure to resist oscillation and in order to break the wind vortices affecting such chimneys helical strakes are often fitted to the upper part of the stack (*B*)[1].

Guyed steel chimneys (*C*) are built of 5 mm to 19 mm steel plate depending on their size. Chimneys of up to 30 m are guyed at two thirds of their height and over 30 m at 0.4 and 0.8 of their height. The guys have tightening bolts in the linkage and terminate at anchor plates buried in concrete anchor blocks in the ground. As the guys stabilise the chimney against sway from wind pressure and overturning the concrete foundation only needs to be large enough to resist dead weight.

Where there is not space for guying chimneys and a fully cantilevered construction may not be desirable due to site conditions, steel chimneys can be stabilised by erecting around them singly or in groups a framed structure to which the chimney shaft(s) is attached by suitable bracketing (figure 163 *A*).

Steel chimneys are cheaper than either masonry or reinforced concrete but are comparatively short lived. Unless lined or clad it is necessary to protect the metal against corrosion due to condensation caused by the heat loss from such chimneys. For this purpose proprietary metal paints are available and PVC coatings.

Steel construction is also suitable for ejector chimneys which provide induced draught by means of a small high pressure blower discharging into a venturi to produce suction. A typical example of an ejector chimney providing a draught equal to a chimney 60 m high by 3 m diameter using natural draught is shown at (*B*), figure 163.

The performance of steel chimney shells may be improved both in functional performance and longevity by building them in double skin form with a steel inner shell and an outer loadbearing shell of heavier steel or by clothing them in aluminium and interposing insulation, such as rockwool, between these skins (figure 163 *C*). The performance of such chimneys depends on the workmanship in making the joints gas proof since the steel lining is impervious to vapour and can accommodate high gas velocities and pressures.

Asbestos cement chimneys are only suitable for gas appliances and if free standing may be guyed (see later under Gas flues).

Resin bonded reinforced glass fibre chimneys have the advantage of lightness of construction and resistance to acid corrosion, less maintenance and longer life than steel. Guy ropes are normally required or a steel supporting tower. The main disadvantage of such chimneys is the limited working temperature which is in the region of 235°C. The chemical composition of the resin bond is critical and it is advisable to fit a temperature indicator alarm and some means of admitting cold air in case of fire.

---

[1] BS 4076:1966 Appendix B gives useful design information in connection with this problem. The BS covers general requirements for steel chimneys, linings and claddings and supporting structures.

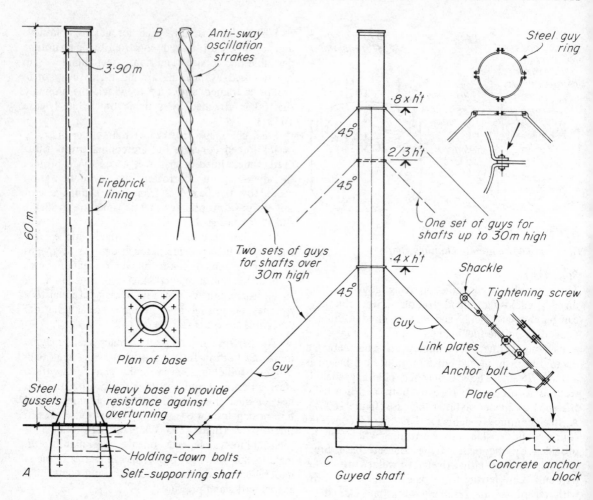

162   Steel chimney shafts

Flue linings

Flue linings protect the structural shell from the effects of heat and corrosive agents. If suitably chosen linings provide thermal insulation around the flue and thus help to maintain the temperature of the flue gases thereby reducing the production of corrosive agents and promoting strong convection currents by maintaining a high temperature differential between the inside of the flue and the outer air.

Types of linings in common use are: firebricks, refractory concrete, Moler bricks and concrete, acid resisting bricks. For small installations linings used include: asbestos cement pipes, clayware pipes,

flexible metal tubing and proprietary insulating concrete. Special linings of resin bonded glass or plastic may be used under controlled temperature conditions.

The properties and characteristics of the main group of lining materials used for larger chimneys are as follows:

Firebricks   These have a 28-32 per cent alumina content and are suitable for temperatures up to 1200°C. The bricks are moulded in radial form to suit the chimney and they are laid in fire cement or mortar of ground fireclay. Such bricks are not good insulators but are effective anti-corrosion liners, gas tight and of long life. Firebricks must be

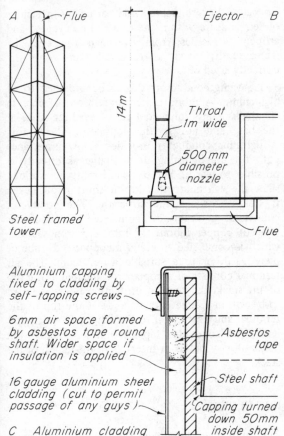

A Flue

B Ejector

Throat 1m wide

500 mm diameter nozzle

14 m

Steel framed tower

Flue

Aluminium capping fixed to cladding by self-tapping screws

6mm air space formed by asbestos tape round shaft. Wider space if insulation is applied

16 gauge aluminium sheet cladding (cut to permit passage of any guys)

Asbestos tape

Steel shaft

Capping turned down 50mm inside shaft

C   Aluminium cladding

163   Steel chimney shafts

laid in a manner which allows them to move freely relative to the surrounding chimney shell (figure 159 *B*). Touch headers should be built against the outer shell at 3 m intervals.

Refractory concrete   This is similar to firebrick but can be cast *in situ* or applied by gun.

Moler bricks   Made from diatomaceous earth in solid form, these bricks are good insulators and effective within the temperature range 150-800°C. The material may be made to given shapes and due to its low coefficient of expansion may be built tightly against the enclosing chimney shell or if permitted, bonded into masonry structures (figure 159 *C*, *D*), or used as permanent shuttering for concrete shells. The normal thicknesses of moler brick linings are between 76 mm and 114 mm and they are set in mortar made from ground moler bricks and Portland or aluminous cement − depending on the anticipated flue temperature. Moler

bricks have low crushing strength but do not need excessive support due to their light weight.

Moler concrete   This is made from diatomaceous earth with aluminous cement and may be cast *in situ*, gunned or precast. The performance of moler concrete is similar to the bricks of this material in the temperature range 150-980°C.

Acid resisting bricks   These are highly vitrified fire or clay bricks set in acid resisting cement. They produce an impervious lining used when flue gases are likely to be very acidic or at or near their dew point (150°C or below). Such bricks are not good insulators and do not stand up to sharp changes of temperature and they should be used in combination with other materials or with an enclosing air space about the lining. Clay acid resisting bricks withstand temperatures of up to 540°C and those of firebrick up to 1100°C.

Linings should be taken to the top of brick and concrete shafts to prevent damage to the chimney shell through sudden increase in temperature. They must be protected from the weather by an adequately oversailing capping which also allows for thermal movement (see figures 159 *B*, *C* and 160). Corbel supports are usually spaced at about 7 m for heavy linings such as firebricks to 13 m for lighter linings such as moler bricks. Where air spaces are relied upon to give a shallow temperature gradient through the chimney construction and afford removal of diffused gases, air holes must be provided in the corbels with adjacent cleaning doors (see figure 160). In steel chimneys brick linings are usually supported on internal steel angles with a 25 mm gap between firebrick and metal shell which is filled with loam. Alternatively moler bricks or precast moler concrete liners may be used self supporting and built against the steel shell depending on the compressive stress at the base and the amount of movement between the shell and lining anticipated in any particular location.

Branch flues

Where possible branch connections should sweep in to ensure a good gas flow (figure 164). Two branches should not enter the stack directly opposite each other unless a splitter plate is provided to divert the gases and avoid turbulence which could cause unstable draught conditions. Generally openings should not occupy more than 25 per cent of the

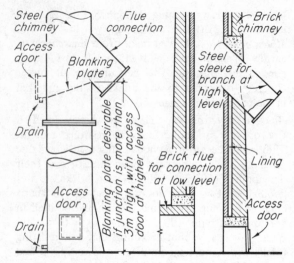

164   Branch flue connections

circumference of the chimney unless the structure is supported and stiffened as necessary. In large installations involving a number of boilers branch flues are collected into one horizontal header flue having a cross sectional area of about one and a third times that of the largest or main flue. Such flues are usually made accessible for cleaning and should be kept as short and well insulated as possible. Brick or concrete construction is mostly used with insulating linings as described for flues previously. Steel or cast iron horizontal headers should be insulated to prevent overheating of the boiler house. Some types of plastic insulation used for this purpose should be separated from the hot metal by an air space for which purpose the flue is wrapped in expanded metal before applying the insulating composition.

Terminals

Due to the cooling effect of the outside air, there is greater danger of corrosion at the terminals of flues. Terminals to large chimneys must be constructed of acid resisting materials such as acid resisting aluminium bronze, chemically pure lead, high alumina cement or thick mild steel insulated internally with aluminium. Furthermore the upper external surfaces of chimneys should be protected with surface coatings of acid resistant paint or be constructed of high alumina concrete. Guy ropes used to stabilize steel chimneys should be secured at least

3 m below the terminal to avoid corrosion. Inversions of temperature within the flue can be avoided and effluent velocities from oversized flues can be increased by using a truncated cone with the correctly sized outlet.

Lightning conductors should be provided to all high chimneys, reinforcement in reinforced concrete chimneys being connected to an earthing plate at the base and not to the conductor at the top. A lightning conductor provides a low impedance path for the lightning to discharge as directly as possible to earth. Sharp bends must be avoided since the lightning will take the shorter route even if this is through the structure. The conductor consists of an air terminal connected to a conductor tape of copper (about 25 mm x 5 mm section) which is connected to either a copper rod, tube or plate earth or to a suitable water main. The air terminal consists of a copper rod located at the top of the stack, the zone of protection being assumed to be within a cone with its apex at the top of the terminal and base equal to the height above ground level. The conductor tape should have expansion loops of easy radius at intervals in the length of the stack.

GAS FLUES

The subject of chimneys for gas-fired appliances has been introduced in Part 1 in respect of domestic scale appliances and it is here extended further.

Compared with the burning of solid fuels which require 100 per cent excess air to promote adequate natural draught in a flue, gas (and oil) fuels require only 20 to 30 per cent excess air resulting in smaller flues. Gas has a low sulphur content and produces a reasonably low condensate of sulphuric acid on thin walled cold flues such as steel or asbestos cement, which in single skin form are only suitable for gas flues.

The burning of gas fuels produces a large quantity of water during combustion and this together with the weak draught resulting from the relatively low temperature of the gases leaving the boiler leads to a considerable accumulation of condensate at the base of the flue. Flues to gas boilers require to be properly drained by incorporating condensation trays or vessels in the base of the chimney or header flue as shown in figure 165.

Baffles or draught diverters are often incorporated

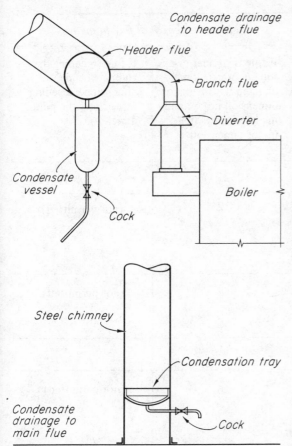

appliances when they have a rated input of under 45 kW. Incinerators may be class I or II according to size whether fired by gas or not. Gas appliances with an input rating of over 45 kW are classified as 'high rating' and their flue requirements are not dealt with in the Regulations.

The Regulations limit the maximum length of certain flues according to situation and the type of appliance they serve and this is indicated in table 17.

The Regulations also control the minimum size of flue required for different appliances, as shown in table 18, and the cross-sectional proportions of rectangular flues. In regard to the latter the smaller dimension, B, must not be less than 63 mm (this also applies to the diameter of circular flues) and the larger dimension must not be more than:

6B if serving one gas fire only
5B if serving one other type of appliance only
1½B if the flue is a main flue
2B if the flue is an appliance ventilation duct.

The Regulations are explicit as to the positioning of flue pipes and the relationship of outlets to walls, roofs and openings in them.

165 Gas boiler flues — removal of condensate

in the design of gas boilers or should be provided at each boiler flue outlet to allow free removal of the products of combustion and to divert any down draught away from the combustion chamber. In addition, this 'break' in the flue system enables a quantity of air to mix with and dilute the flue gases thus lessening the risk of condensation within the main flue. Further dilution may be provided by a ventilating grid at the base of the main flue situated in the boiler room and this should be as large as practicable up to the cross sectional area of the main flue.

Flue size and outlet position

Flues to gas fired appliances are covered by Parts M and L of the Building Regulations 1972 under which such appliances are given the designation Class II

Materials and construction of gas flues

In discussing suitable materials and methods of constructing flues it is important to note that the Regulations differentiate between a 'chimney' as including any part of the structure of a building forming any part of a flue other than a flue pipe and a 'flue pipe' which means a pipe forming a flue but does not include a pipe built as a lining into a chimney or an appliance ventilation duct. The word 'flue' means a passage for conveying the discharge of an appliance to the external air and includes any part of the passage in an appliance ventilation duct which serves the purpose of a flue.

Class II appliances may discharge into a flue in a masonry chimney (see Part 1, page 231) or into a flue pipe. In the former the flue must be lined or be constructed of dense concrete blocks made of and jointed with high alumina cement, except flues which serve one appliance only within the limits of appliance type and flue length given in table 17 which may be of normal brick or concrete construction without a lining.

Flue pipes may be formed of the same materials

271

Situation of flue	Type of appliance	Maximum length of flue in metres	
		If flue is circular or square, or is rectangular and has the major dimension not exceeding three times the minor dimension	If flue is rectangular and has the major dimension exceeding three times the minor dimension
(a) Flue formed by a chimney or flue pipe which is internally situated (that is to say, otherwise than as (b) below)	Gas fire	21	12
	Heater installed in drying cabinet or airing cupboard; or instantaneous water heater	12	(not permitted)
	Air-heater or continuously burning water heater	6	(not permitted)
(b) Flue formed by a chimney having one or more external walls; or by a flue pipe which is situated externally or within a duct having one or more external walls	Gas fire	11	6
	Heater installed in drying cabinet or airing cupboard; or instantaneous water heater	6	(not permitted)

Table 17   *Length of flues to gas fired appliances*

as used for those serving solid fuel appliances (Part 1, page 229) with an addition of sheet steel[1].

Asbestos cement pipes may be used either singly or in groups surrounded by suitable non-combustible material. There is considerable saving in the space required for grouped flues of asbestos cement over normal brickwork and they may be used up to a limit of 30 m, the weight of the flue pipes being separately supported at each floor level. Individual pipes must be securely fixed, with the sockets upwards, and each joint must be filled with asbestos rope and pointed with plastic asbestos. Suitable terminals are required and need careful design.

Asbestos cement is not resistant to attack by corrosive flue gas condensates. If condensation is anticipated internal protection should be applied in the form of bitumastic paint or brush-on acid-resistant cement finish. Alternatively, asbestos cement pipes with vinyl acetate protection already applied may be used. The height above the appliance at whcih condensation may occur depends on a number of variable factors. When a very tall flue is being considered the local Gas Board should be consulted in case internal coating of the flue or other precautions may be advisable. Asbestos cement pipes cool very easily and long flues on the outside of a building invariably result in condensation

[1] BS 715, *Sheet metal cylindrical pipes and fittings for gas-fired appliances.*

272

Situation	Type of appliance	No	Input rating (kW) not to exceed	Minimum area (mm²)
Not specified	Gas fire	1	Not specified	12 000 mm² *
Not specified	Class II appliance (not a gas fire)	1	Not specified	Not less than the outlet to the appliance
In the same room or space	Main flue serving class II appliances (not gas fires)	2 or more	13 18 30 35 45	3750† 5750† 7000† 9000† 11 500†
In different storeys	Any class II appliance	2 or more	Not specified	40 000‡

* Opening in any local restrictor unit not to be less than 6000 mm²
† Or the area of the larger of the outlets of the appliances, if greater
‡ Subject to the requirements of Regulation M 10

Table 18   *Minimum size of flues to gas fired appliances*

trouble and may even fail to function as a flue. In such circumstances an insulated flue such as described below would be better.

Metal pipes may be used for flues, but preferably only in positions where the flue can be seen and easily replaced if necessary, because of the liability of corrosion. Protected steel or cast-iron pipes and sheet metal with welded or folded seams are suitable.

Increased use is being made of factory made chimney systems, suitable for gas, oil or solid fuel installations. These systems comprise flue units and components and fittings for both internal and external situations and are generally restricted to heights of between 12 and 15 metres although special support systems can be designed for heights in excess of this.

The flue units are normally formed of a lining (such as stainless steel) and an outer casing of corrosion resistant material with about 25 mm of non-combustible high temperature insulating fill between the two concentric tubes. Most systems may be installed in traditionally constructed dwellings with timber floors and roofs provided account is taken of their self weight. The principal components are the chimney sections, support plates (or brackets externally), fire stops, special flue pipe connectors and terminals. These factory made chimney systems claim to save up to 50 per cent of the cost of a traditional brick chimney and are, of course, much quicker to erect.

The requirements relating to the proximity of combustible material to gas flue pipes are given on page 231 of Part 1.

Table 19 summarises the application of materials which are suitable for flues to gas appliances[1].

Balanced flues

Certain gas appliances are designed to operate with 'balanced flues'. A 'balanced flue' or 'room-sealed' appliance is so designed that the combustion chamber is enclosed and sealed from contact with the air in the room. Air for combustion passes into the casing from outside the building and the products of combustion are discharged directly through the wall at the same position. The air pressure is the same on both inlet and outlet and the warm gases are discharged without being affected by wind

[1] This list has been extracted from that given in the 'Gas Handbook for Architects and Builders' published by the Gas Council.

Material	Condition	Suitability	Protection against condensation
Brick	Corrosion resistant lining	All appliances	Inside face may be lined with acid resistant tiles embedded in an acid resistant jointing material Acid resistant lining may be introduced consisting of suitable clay flue lining, earthenware pipe or asbestos cement flue with protective coating of acid resisting material. See below under 'asbestos cement or lead lined aluminium tubing' Inside face may be rendered with acid resistant cement
Precast concrete	Protected, length/breadth ratio not exceeding 3:1	All appliances	May be wholly of acid resistant cement May be composite, with inside wall made of acid resistant material. Joints should be made with an acid resistant joint material
Asbestos cement	Protected	All appliances	Inside face may be coated by manufacturer or on site with an acid resistant compound. Suitable coatings have been prepared from: (a) vinyl acetate polymer (b) a rubber derivative base compound
Earthenware pipes	Glazed or unglazed but of low porosity	All appliances	
Metal	Protected or corrosion resistant	All appliances — generally used for connection to chimney or flue pipe	Mild steel acid resistant vitreous enamelled Galvanised iron† Aluminium† Stainless steel Protected cast iron Lead lined aluminium
Double walled flue with a 6 mm to 16 mm air space, metal or asbestos cement‡		All appliances	

continued . . .

Material	Condition	Suitability	Protection against condensation
Brick	Unprotected	All appliances – subject to height limitation given in table 17	
Precast concrete	Unprotected, length/breadth ratio not exceeding 3:1	All appliances – subject to height limitation given in table 17	
Precast concrete	Unprotected, length/breadth ratio greater than 3:1 but not exceeding 5:1	Radiant and convector gas fires – subject to height limitation given in table 17	
Asbestos cement	Unprotected	All appliances – subject to height limitation given in table 17	
Bitumen asbestos	Protected	All appliances. External use only and provided that temperature of products entering flue does not exceed 93°C	

Table 19 *Materials for flues to gas appliances*

† Used only when accessible for renewal
‡ Stainless steel with insulation is available

pressure or gusts (see figure 166 *A*). Such appliances may also be connected to common flue systems or appliance ventilation ducts using natural or mechanical ventilation[1].

Shared flue systems

The use of shared or common flues reduces considerably the amount of space required for gas flues in multi-storey buildings and facilitates the positional planning of gas fired appliances away from outside walls resulting in functional planning and economic advantages.

One system links up conventional appliances by short branch flues to a main flue. Others use a main flue or duct serving room-sealed appliances.

[1] Regulations L20 and M10 of the Building Regulations 1972 cover these cases of one or more appliances connected to the same flue.

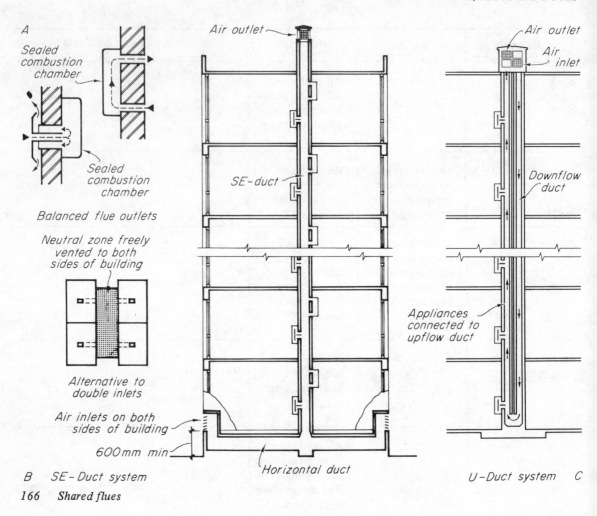

A
Sealed
combustion
chamber

Sealed
combustion
chamber

Balanced flue outlets

Neutral zone freely
vented to both
sides of building

Alternative to
double inlets

Air inlets on both
sides of building

600 mm min

B   SE–Duct system

166   Shared flues

Air outlet

SE-duct

Horizontal duct

Air outlet
Air
inlet

Downflow
duct

Appliances
connected to
upflow duct

U–Duct system   C

SE-duct and U-duct systems

The Gas Council has developed two systems of providing common ventilation ducts into which room-sealed appliances can be discharged: the SE-duct system and the U-duct system. The former consists of a main duct with air inlets at the base drawing air from the perimeter of the building at ground level and terminating in a specially designed terminal above roof level (see figure 166 B) whilst the latter consists of a twin duct in the form of an elongated 'U' with both air inlet and outlet at roof level, thus avoiding the necessity to build horizontal ducts beneath the building (figure 166 C). In both systems the air pressure at the inlet and the outlet of the appliances must be equal to allow the residual

heat from the flue gases to act as the motive power to move gases up the flue. A number of precast concrete units for which the Gas Council holds patent rights is available comprising duct blocks, storey height duct units, terminals and intakes. Tables 20 and 21 show the types and sizes of SE-ducts for different numbers of appliances in flats.

Branched flue system

It has been normal practice on the Continent for many years to obviate the provision of a separate flue throughout the whole height of a building for each room heating appliance by connecting

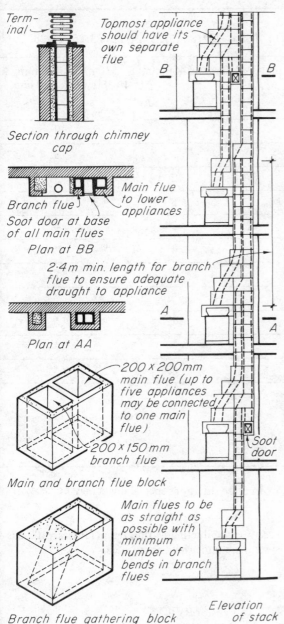

Term-
inal

Topmost appliance
should have its
own separate
flue

Section through chimney
cap

Branch flue

Main flue
to lower
appliances

Soot door at base
of all main flues

Plan at BB

2·4 m min. length for branch
flue to ensure adequate
draught to appliance

Plan at AA

200 x 200mm
main flue (up to
five appliances
may be connected
to one main
flue)

200 x 150 mm
branch flue

Main and branch flue block

Main flues to be
as straight as
possible with
minimum
number of
bends in branch
flues

Branch flue gathering block

Elevation
of stack

Soot
door

167 *Branched flues*

individual appliances to a common main flue by
short branch flues, the principle of operation being
to provide the necessary draught in the branch flue
and to evacuate the gases through the main flue.
A branched flue system saves considerable flue
space on the upper floors of tall buildings (figure

167). The system, also called the shunt system,
may be constructed from precast units of refractory
concrete formed with two internal apertures for
branch and main flues.

Fan diluted flues

Conventional flue systems operate on the natural
draught principle and because the pressure difference
is so small the flue has to have relatively large
dimensions to keep the flow resistance low. However, if an external source of power is introduced,
such as a fan, the area of the flue can be considerably
reduced (see *Ejector chimney,* page 267). In fact,
the only limits on area reduction are the noise
level, air velocity, and the power consumption
which must be kept to an economical level. This is
the principle of the mechanical extraction system
which, apart from the reduced flue area, has the
added advantage of being able to draw combustion
products downwards from the appliance and horizontally under floors where necessary. It is, therefore, useful where natural draught flues are not
practical. A development in this field is the fan
diluted flue (figure 168). This system has been
specifically· developed to solve the problems of
ground floor shop premises in mixed tower blocks
of offices, shops and flats. The system operates
with fresh air being drawn in through a duct by fan,
mixed with the products of combustion, and finally
being discharged to the atmosphere with a carbon
dioxide content of not more than one per cent.

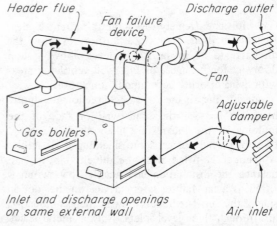

Header flue

Fan failure
device

Discharge outlet

Fan

Gas boilers

Adjustable
damper

Inlet and discharge openings
on same external wall

Air inlet

168 *Fan diluted flues*

Rating of appliance (kW)	Number of storeys 3/4	5/6	7/8	9/10	11/12	13/14	15/16	17/18	19/20	21/22	23/25	26/28	29/30
Continuous appliances with kW ratings of:	Duct type: 1 - 200 mm x 300 mm; 2 - 230 mm x 400 mm; 3 - 330 mm x 480 mm; 4 - 380 mm x 560 mm												
2.93	1	1	1	1	1	1	1	1	1	1	2	2	2
4.93	1	1	1	1	1	1	1	1	2	2	2	3	3
5.86	1	1	1	1	1	1	2	2	2	3	3	3	3
8.79	1	1	1	1	2	2	3	3	3	3	3	3	4
11.71	1	1	1	2	2	3	3	3	3	3	4	4	4
14.66	1	1	2	2	3	3	3	3	4	4	4	*	*
17.59	1	1	2	3	3	3	3	4	4	4	*	*	*
20.52	1	2	3	3	3	3	4	4	*	*	*	*	*
23.45	1	2	3	3	3	4	4	*	*	*	*	*	*
26.38	1	2	3	3	4	4	*	*	*	*	*	*	*
29.31	1	3	3	3	4	4	*	*	*	*	*	*	*
Instantaneous water heaters with rating of: 28.58	1	1	1	1	2	2	2	3	3	3	3	3	3

Table 20   *SE-ducts: required sizes*

* In these situations special ducts may be necessary

Gas with its extremely low sulphur content is ideal for this method. Many local authorities allow the discharge to be made at low level, above a shop doorway for instance, or into well ventilated areas with living or office accommodation above.

Ideally, the air inlet and discharge louvres should be positioned on the same wall or face of the building. If the louvres are likely to be subjected to strong wind forces, some shielding is desirable. A damper is fitted near the diluent air inlet to balance the installation and an air flow switch is fitted as a fan failure safety device on the suction side of the fan. The boilers draw in their combustion air through floor level inlets. Normal metal sheet, asbestos pipe or rectangular asbestos ducting can be used for the ducting as flue temperatures with this system are rather low, about 65°C.

DUCTS FOR SERVICES

Although some services may be run within the floor structure as described on page 254 there are, as pointed out in *MBC: Environment and Services*, advantages in accommodating them in ducts formed for this purpose in the building fabric. The services are concealed and protected but they are accessible without the necessity of breaking open a floor screed, for example. Inspection and maintenance

278

Rating of appliance (kW)	Number of storeys									
	3/4	5/6	7/8	9/11	12/14	15/16	17/19	20/22	23/29	30
Instantaneous water heaters (28.58 kW) and continuous appliances with ratings of:	Duct type: 1 - 200 mm x 300 mm; 2 - 230 mm x 400 mm; 3 - 330 mm x 480 mm; 4 - 380 mm x 560 mm									
2.93	1	1	2	2	3	3	3	3	4	4
4.39	1	1	2	2	3	3	3	3	4	4
5.86	1	1	2	2	3	3	3	4	4	*
8.79	1	1	3	3	3	3	4	4	*	*
11.72	1	2	3	3	3	4	4	*	*	*
14.66	1	2	3	3	4	4	*	*	*	*
17.59	1	2	3	3	4	4	*	*	*	*
20.52	2	2	3	3	4	*	*	*	*	*
23.45	2	3	3	4	*	*	*	*	*	*
26.38	2	3	3	4	*	*	*	*	*	*
29.31	2	3	4	4	*	*	*	*	*	*

Table 21   *SE-ducts: required sizes*   * In these situations special ducts may be necessary

are thus made simpler and alterations and additions are facilitated. Ducts also have the beneficial effect of separating operations (see Part 1, page 26) since the installation of the services may be independent of the construction of the building fabric.

Services normally require vertical and lateral distribution within a building and in multi-storey buildings this usually results in three types of duct: main, vertical and lateral.

Main ducts

These link the various service lead-in points with the main controls and provide the primary horizontal distribution of services. They are usually located below the ground floor in the form of either a *subway*, a *crawlway* or a *trench*.

A subway, in which a man may walk, should be at least 2 m high and wide enough to provide a clear working space of 686 mm between pipes or pipe racks (figure 169 *A*). In order to minimise the overall width it is desirable to group large and small pipes on opposite sides of the subway.

A crawlway must be not less than 1.067 m high with the same minimum working space as a subway (figure 169 *B*). Access panels should be provided in the top at frequent intervals. Crawlways are used preferably only where the services require little maintenance.

Subways and crawlways are used where a large number of services must be distributed as in heavily serviced buildings such as laboratories and hospitals.

A floor trench is used when relatively few pipes are to be accommodated. Access to service pipes is from the top and the depth of the trench is generally not greater than 750 mm (figure 170 *A*).

Subways and crawlways are constructed with a concrete base and walls of either concrete or brick as shown in figure 169. If the site has a high water table some form of waterproofing as described for basements is essential. The floor should be laid to falls with a shallow channel on one side to convey

279

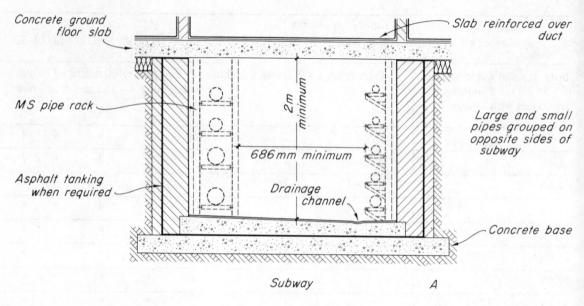

Concrete ground floor slab

Slab reinforced over duct

MS pipe rack

Large and small pipes grouped on opposite sides of subway

2m minimum

686mm minimum

Asphalt tanking when required

Drainage channel

Concrete base

Subway   A

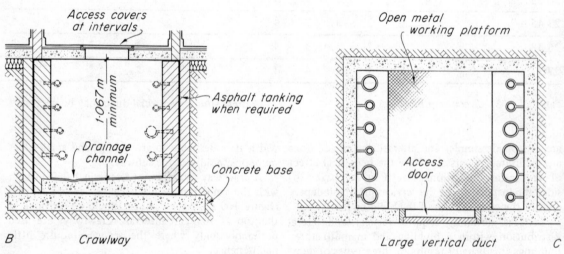

Access covers at intervals

Open metal working platform

1·067m minimum

Asphalt tanking when required

Drainage channel

Concrete base

Access door

B   Crawlway   Large vertical duct   C

169   Ducts for services

water from possible leaks and condensation to sumps or, in suitable circumstances, to a drain through a sealed gulley.

Trenches are commonly constructed of concrete with continuous access provided by trench covers which span from side to side of the trench. The covers may be in the form of metal trays, filled to match the floor finish, set in metal frames or in the form of precast concrete slabs over which the floor finish runs as shown in figure 170 *A*. Separating strips along the line of the slabs permits

the ducts to be opened up without damaging the adjacent floor areas. If manhole covers, filled as above, are used these must be placed at junctions and bends in the pipes and be long enough to permit lengths of pipe to be introduced and removed easily.

Vertical ducts

These lead from the main ducts to distribute the services to the various floors. Depending upon the number of pipes to be accommodated these may be

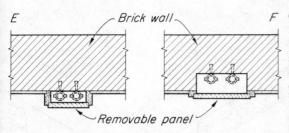

E   Brick wall   F

Removable panel

Surface casing   Chase or recess

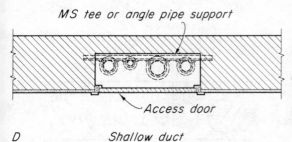

MS tee or angle pipe support

Access door

D   Shallow duct

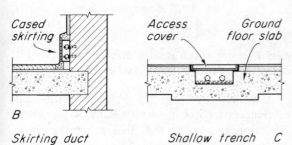

Cased skirting

Access cover   Ground floor slab

B

Skirting duct   Shallow trench   C

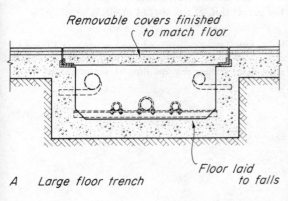

Removable covers finished to match floor

Floor laid to falls

A   Large floor trench

170   Ducts

casings on the face of a wall or chases cut in a wall for small pipes or cables (figure 170 *E*, *F*) or shallow ducts up to about 600 mm deep with access panels or doors the full width of the duct (*D*). A wide, shallow duct makes access to the pipes easier than a relatively deep but narrow one. Ducts large enough for a man to enter are often constructed for vertical runs of large numbers of pipes without branches crossing the risers and restricting access to them (figure 169 *C*). Access is provided either at each floor or at every other floor with internal steel ladders to working platforms of open metal flooring (see page 259).

Lateral ducts

These lead from the main or vertical ducts to provide horizontal distribution of services at various levels. They may be in the form of wall casings, such as a hollow skirting duct for small pipes (figure 170 *B*), or a shallow trench in a ground floor (*C*).

In concrete upper floors it is a simple matter to form ducts in the thickness of most types of one-way spanning floors so long as they run in the direction of the span (see figure 171). In a solid *in situ* floor the necessary width of duct is boxed out on the shuttering (*A*). In a hollow block floor (*B*) or precast rib and filler floor (*C*), one or more lines of blocks are omitted and in the case of a precast beam floor one or more beam sections are omitted. The service pipes pass through slots in the main supporting beams made, preferably, on the line of the neutral axis. If the latter is too low and the slot is formed at a higher level, compression reinforcement may be required in the 'bridge' over the slot. Holes through the webs of steel beams may need strengthening by means of a reinforcing plate welded around the slot. From such ducts lateral branches from the service pipes can be run at right-angles to the span in the thickness of the floor screed or rise directly to fittings when, for example, the ducts are placed relative to a run of laboratory benches.

In solid *in situ* slabs continuous over supporting beams it is possible to form shallow ducts at right-angles to the span on the points of contraflexure (*D*). These may be of a depth not exceeding about one-quarter of the thickness of the slab. If a deeper duct, or a complete opening through the floor, is required at any point at right-angles to the span it is necessary

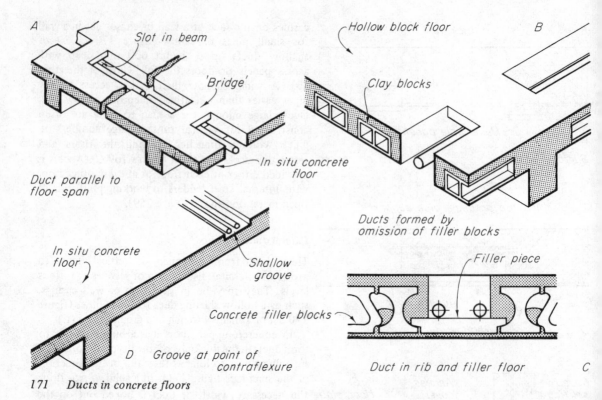

A   Slot in beam

'Bridge'

In situ concrete floor

Duct parallel to floor span

Hollow block floor   B

Clay blocks

Ducts formed by omission of filler blocks

In situ concrete floor

Shallow groove

Concrete filler blocks

D   Groove at point of contraflexure

Filler piece

Duct in rib and filler floor   C

171   Ducts in concrete floors

to form structural trimming beams on each side of the duct.

Ducts of various diameters can be formed in floor slabs or screeds by means of expendable fibre or cardboard tubes left in position. Alternatively, pneumatic cores in the form of long, inflatable tubes may be laid in position in an inflated state. The concrete is cast round the tube and, when set, the tubes are deflated and withdrawn. The construction of the wall of the tube is such that when it is deflated the tube twists and pulls away from the concrete. Pneumatic cores have the advantage that they can be laid in curved lines.

Fibre and metal ducting for telephone and electric wiring which is set within the floor screed is described in *MBC: Environment and Services*, chapter 14.

As mentioned on page 238 the lateral distribution of complex services in such buildings as hospitals, laboratories or some industrial processing plants often makes the provision of a service floor necessary.

Reference should be made to page 391 regarding the prevention of fire spread in ducts[1].

---

[1] See *MBC: Environment and Services*, chapter 17, and CP 413: 1973 *Design and Construction of Ducts* for information on the arrangement and support of pipes within ducts.

8 Stairs, ramps and ladders

STAIRS

The subject of stairs has been introduced in Part 1 where the functional requirements, basic design factors and the different types of stair are described. An introduction to concrete stairs is given and the construction of timber stairs is described in detail. These are considered further in this chapter.

Loading

Timber stairs for small domestic buildings, as pointed out in Part 1, are normally constructed on the basis of accepted sizes for the various parts. Other types must be designed having regard to the imposed loadings laid down in CP3: chapter V: Part 1: 1967, which requires the assumption of 1.5 kN/m² distributed load for dwellings not over three storeys high and a concentrated load of 1.8 kN over any square with a 300 mm side. The loadings for stairs in other building types must be the same as those for the floors to which they give access, but with a minimum value of 3.0 kN/m² and a maximum of 5.0 kN/m². The superficial area on which these loads are assumed to act is measured horizontally.

The loading on balustrades, assumed to be acting horizontally at handrail level, is to be taken as follows:

(a) Light access stairs, gangways and the like, maximum width 600 mm: 220 N/m run
(b) Light access stairs, gangways and the like more than 600 mm wide, stairways, landings and balconies, private and domestic: 360 N/m run
(c) All other stairways, landings and balconies:
                                                   740 N/m run

It should be noted that the last figure of 740 N/m run is also applied to parapets (acting at coping level) and handrails to roofs. Where crowds can panic much greater forces can be exerted and an allowance of 3000 N/m run must be made in respect of panic barriers.

Timber stairs

Newel and ladder type stairs are described in Part 1. As indicated in that volume by the use of laminated timber construction stairs of considerable span and width may be constructed. By the adoption of glulam techniques (see page 303) using horizontal laminae cranked strings may be formed in a similar manner to reinforced concrete, and cantilever treads may be built up of tapered form to permit treads of considerable width and overhang beyond the strings.

Geometrical stairs

Geometrical stairs in timber must be constructed with strings which are helical in form for circular stairs or have circular or wreathed portions linking straight sections in rectangular stairs.

The curved strings may be constructed on the traditional method using narrow vertical staves, or strips of timber, with radiating joints glued together and forming a core between two plywood laminates (see technique for circular column forms, figure 254) or by the use of continuous vertical board laminates in the form of normal glulam construction, the thickness of the boards depending upon the radius of the curve. The traditional method is used for small radii curves.

In both cases a vertical circular jig or former is required of the appropriate radius and strong enough to remain stable whilst the laminae are wrapped around it. As in concrete stairs, a helical timber stair may be constructed with a single wide laminated spine string over which the treads cantilever on each side. In this case the jig is provided with set-off steps to form a bed on which the string may be clamped to hold its helical shape as it is built up.

The steps may be constructed in normal closed or open riser form between pairs of strings or open riser construction with the treads bearing on recesses cut in the tops of the strings may be used. With a single spine string the cantilever treads may similarly bear on recesses in the string or may be supported on some form of metal plate or strap brackets screwed to the top of the string, similar to the examples shown in figure 181.

Handrails for geometrical stairs may be laminated in the same manner as for the strings.

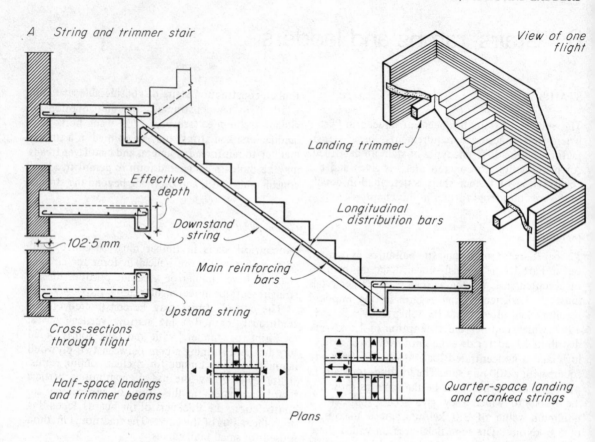

A   String and trimmer stair

View of one flight

Landing trimmer

Effective depth

Downstand string

Longitudinal distribution bars

102·5 mm

Main reinforcing bars

Upstand string

Cross-sections through flight

Half-space landings and trimmer beams

Quarter-space landing and cranked strings

Plans

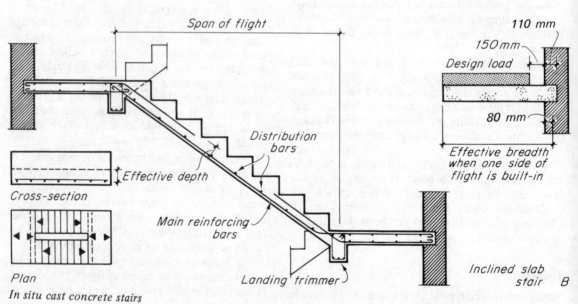

Span of flight

110 mm

150 mm

Design load

80 mm

Effective depth

Distribution bars

Cross-section

Main reinforcing bars

Effective breadth when one side of flight is built-in

Plan

Landing trimmer

Inclined slab stair   B

172   In situ cast concrete stairs

284

Reinforced concrete stairs

Concrete stairs are widely used because of their high degree of fire resistance and the relative ease with which a variety of forms may be produced. *In situ* cast stairs will be described first followed by a continuation of the material on precast stairs given in Part 1.

String stair

The strings may span between landing trimmers or be cranked to span beyond the landings to take a bearing at the perimeter of the stair (figure 172 *A*). This type of stair will be thinner than a slab type and therefore somewhat lighter in weight. With half-space landings as illustrated the inclined strings, which are reinforced as normal beams, can bear on reinforced concrete trimmers at the landings, and the flight slab will span between the strings. The landings will span between the trimmers and the enclosing staircase wall or frame. The strings may be upstand or downstand and, in the case of the latter, the effective depth will be from the soffit of the string to the internal junction of the treads and risers. The waist[1] thickness of the flight need only be about 75 mm. An upstand string is useful both from a functional and an aesthetic point of view. It prevents dropped articles and cleaning water from falling over the sides of the stair and it gives weight and smoothness of flow to a stair designed to appear as a slab flowing between floors. When an intermediate flight is incorporated, producing quarter-space landings, it is necessary to use cranked strings, and as these must run across the flight it is necessary to use downstand beams.

For freestanding stairs, a single substantial central string may be designed to carry the flight which cantilevers on each side. The flight may be cast *in situ* or be made up of precast elements bolted or tied into the *in situ* cast string, to give a smooth inclined soffit or a stepped soffit. Precast concrete treads or laminated timber are often used to produce an open riser stair in this form.

Inclined slab stair

Unless the span of the flight is very long, or strings are required for visual reasons, the stair can be designed without strings, the flight being designed to act as a slab spanning between the

trimmers (figure 172 *B*). In this case the span of the flight is the horizontal distance between the centres of the trimmers. The effective depth is the waist thickness of the slab, which is designed on the same basis as a floor slab. In the case of slabs designed to span in the direction of the flight, one side of which is built into a wall not less than 110 mm, CP 110: Part 1: 1972 permits a 150 mm wide strip next to the wall to be deducted from the loaded area, and the effective breadth of the slab to be increased by two-thirds of the embedded breadth up to a maximum of 80 mm.

Cranked slab stair

In this stair there are no trimmers and the top and bottom landings, together with the flight, are designed as a single structural slab spanning between enclosing walls or frame (figure 173 *A*). The appearance is clean and the thickness of the slab is not unduly great if the flight is not too long. This form of stair is useful when there are no side supports available for trimmer beams, as in the case of completely glazed sides to a projecting staircase. Should supports be available at the ends of the landings so that the latter may be made to span at right-angles to the direction of the flight, the landing slabs may be considered as beams supporting the flights, in which case CP 110: Part 1 requires that the effective span of the flight should be taken as the going of the flight plus half the width of the landing, subject to a maximum of 900 mm, at each end.

Monolithic cantilever stair

In this stair the flights and landings are cast *in situ* and cantilever out from a wall, either the enclosing wall to the staircase or a central spine wall as illustrated in figure 173 *B*. The soffit may be smooth or stepped, the latter resulting in a fairly thin slab, the compression zone of which is stiffened by the folded form. When the stair cantilevers from a central spine wall, it becomes completely self-supporting and, if projecting from the face of a building, may be fully glazed all round. This form of stair also provides a useful solution to problems of sound insulation when the stair, for this reason,

[1] See Part 1, page 237.

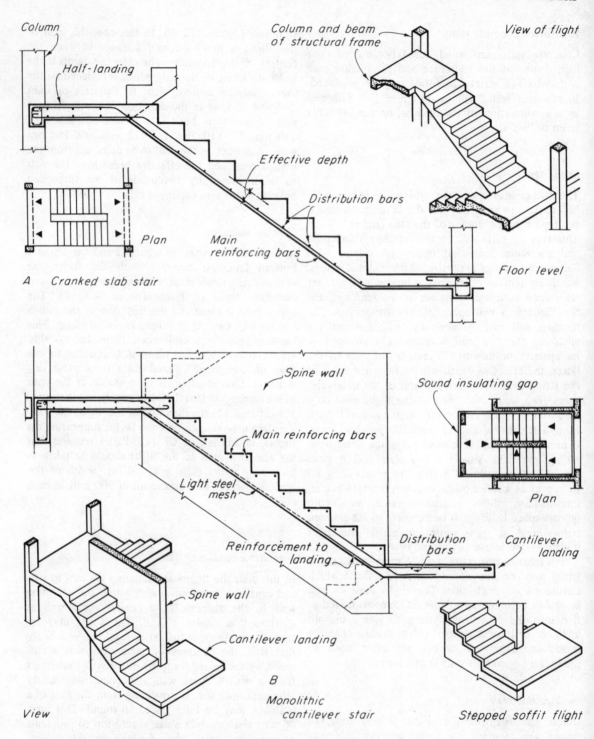

Column

Half-landing

Plan

A   Cranked slab stair

Column and beam
of structural frame

View of flight

Effective depth

Distribution bars

Main
reinforcing bars

Floor level

Spine wall

Sound insulating gap

Main reinforcing bars

Light steel
mesh

Reinforcement to
landing

Plan

Distribution
bars

Cantilever
landing

Spine wall

Cantilever landing

View

B
Monolithic
cantilever stair

Stepped soffit flight

173 In situ cast concrete stairs

is required to be separated from the surrounding structure, since it is possible to leave an insulating gap all the way round at all points (see plan). In these cases the half-space landing would be partially supported by the end of the wall and partly by the ends of the adjacent flights as shown, and reinforcement would be designed accordingly.

Continuous slab stair

This is a double-flight stair which receives support only at the floors above and below (figure 174 *A*). It consists structurally of a continuous slab, monolithic with the floors, which runs from one floor to the landing level, turns on itself and continues without any support to the next floor. This is not a cheap stair to construct since not only the normal stresses of bending and shear, but also those of torsion have to be resisted. The slab may be reduced in width to form a wide shallow beam carrying open riser cantilever treads. It may be placed centrally under the treads or eccentrically, so that they cantilever entirely over one side. With materials strong in tension, such as reinforced or prestressed concrete or laminated timber, the latter arrangement is practicable even with broad flights, although torsion stresses in the slab are increased.

In the types of stairs without trimmer beams the relationship of the end risers of the flights at a half-space landing affects the positions of the intersections of the sloping soffits and the landing soffit, as well as the form of the handrail turn. If the top riser of the lower flight is set back to line approximately with the second riser of the upper flight, it is possible to make the intersections of the sloping soffits and the landing coincide on the same line, which can be the face edge of the landing between the flights, without the landing being made excessively thick (see figure 174 *A*). This gives a clean appearance on the underside, simplifies detailing of applied finishes and permits a satisfactory handrail turn. This point is also illustrated and discussed in *MBC: Components and Finishes*, relative to balustrading.

Spiral stair

The newel type of spiral stair in its smaller form is usually constructed of precast concrete and is described in Part 1; the larger stairs are constructed on a large core as a monolithic cantilever stair. The core may be solid or hollow in the form of a duct.

The open well, or helical stair, figure 174 *B*, although visually very fine if well designed, is complicated in structural design and construction. A large proportion of steel is required to resist the bending, shear and torsion stresses, and the shuttering is expensive. The slab usually varies in thickness from top to bottom, increasing towards the bottom, and may vary in thickness across the width. There are two or three sets of reinforcement with top and bottom layers in each: continuous bars running the length of the spiral, cross or radial bars and sometimes diagonal bars laid tangential in two directions to the inner curve. The large amount of steel reinforcement and the complicated shuttering makes this an expensive stair to construct.

The helical stair may also be designed with closed strings with the flight slab spanning between. The effect of a helical stair depends upon the free flow of the curve from one floor to another. In many cases the limitations on the number of steps in a flight makes it impossible to design such a stair without intermediate landings, which interrupt the flow of the stair.

Precast concrete stairs

The concrete stairs discussed above are primarily *in situ* cast stairs. Although precast concrete has long been used for simple solid or open riser steps, or for small utilitarian spiral stairs, some examples of which are shown in Part 1, the precasting of large stairs has not been common. However, with the general use of cranes on building sites, many large stairs which once would have been constructed only *in situ* are now precast. These may be precast in the separate parts of strings and steps or they may be cast in complete flights and landings, depending upon the nature of the particular job and the size of the crane to be used.

Figure 175 shows a 1.37 m wide cantilever stair in which the steps are in the form of an 'L' with only the building-in end as a solid rectangle. This considerably reduces the weight of the stair. Masonry walls should be built in cement mortar for at least 300 mm above and below the line of a cantilever stair such as this.

Figure 176 shows a cut string and open riser stair in which the strings are stepped to take the treads, in this case of timber, which are screwed on. Precast concrete treads could be used and bolted to

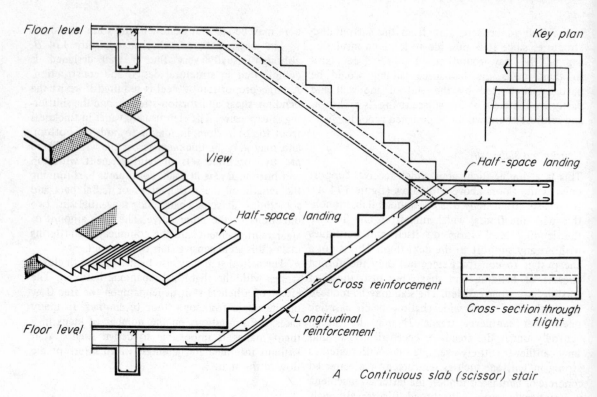

Key plan

Floor level

View

Half-space landing

Half-space landing

Cross reinforcement

Longitudinal reinforcement

Cross-section through flight

Floor level

A   Continuous slab (scissor) stair

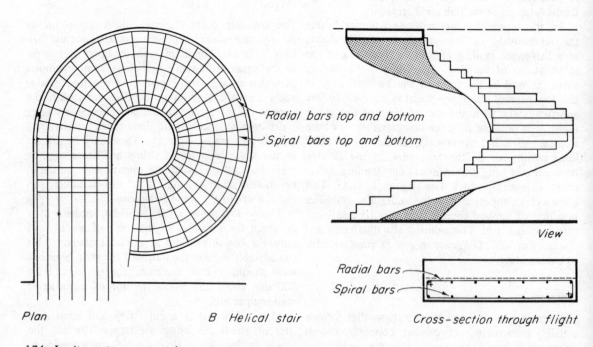

Radial bars top and bottom

Spiral bars top and bottom

View

Radial bars

Spiral bars

Plan

B   Helical stair

Cross-section through flight

174   In situ cast concrete stairs

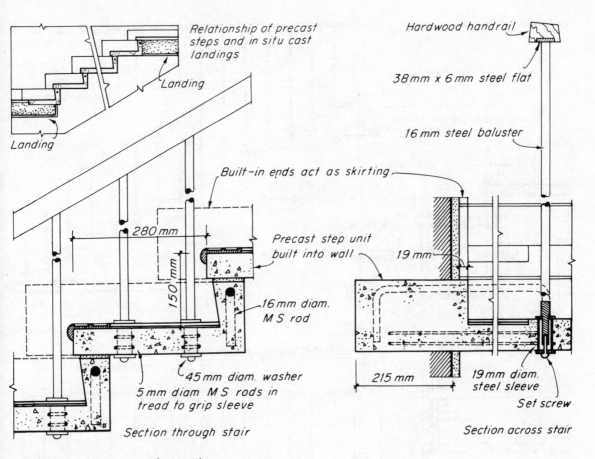

Relationship of precast steps and in situ cast landings

Landing

Landing

Hardwood handrail

38 mm x 6 mm steel flat

16 mm steel baluster

Built-in ends act as skirting

280 mm

150 mm

Precast step unit built into wall

19 mm

16 mm diam. MS rod

45 mm diam. washer
5 mm diam MS rods in tread to grip sleeve

Section through stair

215 mm

19 mm diam. steel sleeve

Set screw

Section across stair

175   Precast concrete cantilever stair

the strings in a similar way, or they could be secured by projecting rods and a small amount of *in situ* concrete cast in mortices or grooves left for the purpose.

A closed string type is shown in figure 177, in which the strings are precast and post-tensioned and on the inside of which are cast stepped bearings to take the ends of the precast treads. To position the treads and to avoid subsequent movement a stub is cast on each end which drops into an accommodating mortice in the stepped bearings on the string.

An alternative to the casting of the strings as separate elements is to cast them as a pair braced apart the required distance to form an open frame, so that the whole can be hoisted by crane and set in rebates formed on the edges of the landing. This is shown in figure 178 *A*. The steps, precast individually, are positioned and grouted in recesses formed to take them in the top faces of the strings.

When a slab stair without strings or trimmers is required this can be cast in elements consisting of a flight and parts of the top and bottom landings, the ends of the landings bearing on rebates in the staircase wall or frame. The half-space landings are completed by *in situ* concrete filling placed around rods left projecting for this purpose at the sides of the landing sections, as shown in figure 178 *B*. Such elements are heavy and require a crane of sufficient capacity for hoisting.

Metal stairs

Cast-iron escape stair

The oldest type of metal stair is probably the external cast iron and steel fire-escape stair. These are made up of standard strings, about 175 mm by 10 mm mild steel, to which are bolted perforated

289

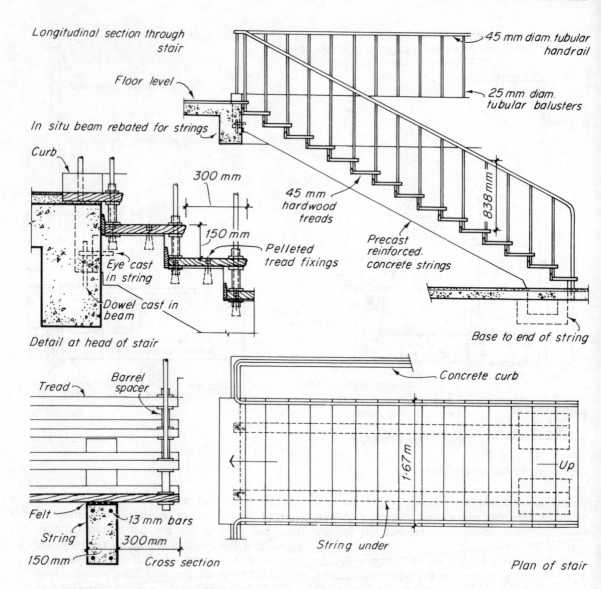

Longitudinal section through stair

Floor level

In situ beam rebated for strings

Curb

45 mm diam. tubular handrail

25 mm diam. tubular balusters

300 mm

150 mm

45 mm hardwood treads

Pelleted tread fixings

Eye cast in string

Dowel cast in beam

838 mm

Precast reinforced. concrete strings

Detail at head of stair

Base to end of string

Tread

Barrel spacer

Concrete curb

Felt

13 mm bars

String

300 mm

1·67 m

Up

150 mm

Cross section

String under

Plan of stair

176 Precast concrete cut string stair

cast iron or mild steel chequer-plate treads. Perforated cast-iron risers can also be fitted if required. The landings are formed of 13 mm cast iron or mild steel chequer plates and the stair and landings are usually carried on a structure of rolled-steel beams and channels. Details are shown in figure 179.

Spiral stair

Standard newel-type spiral stairs of a similar nature

can be obtained, but since the appearance of these leaves much to be desired, they are often specially designed and constructed when the finances of a job permit, especially for small internal stairs. These may be formed in cast iron, the steps being similar to those for the concrete stair shown in Part 1. They are threaded on to a metal newel-post. Alternatively, mild steel T- or channel-brackets are welded at the necessary points to a mild steel tube newel, as shown in figure 180, the timber treads

290

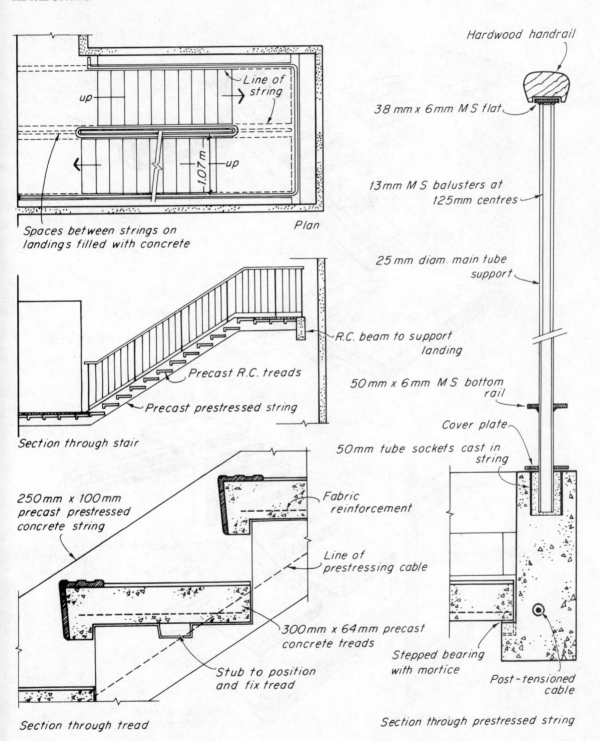

Plan

Spaces between strings on
landings filled with concrete

Line of
string

up

up

1.07 m

Section through stair

R.C. beam to support
landing

Precast R.C. treads

Precast prestressed string

Hardwood handrail

38 mm x 6 mm M S flat

13mm M S balusters at
125mm centres

25 mm diam. main tube
support

50 mm x 6 mm M S bottom
rail

Cover plate

50mm tube sockets cast in
string

250mm x 100mm
precast prestressed
concrete string

Fabric
reinforcement

Line of
prestressing cable

300mm x 64mm precast
concrete treads

Stub to position
and fix tread

Stepped bearing
with mortice

Post-tensioned
cable

Section through tread

Section through prestressed string

177   Precast concrete closed string stairs

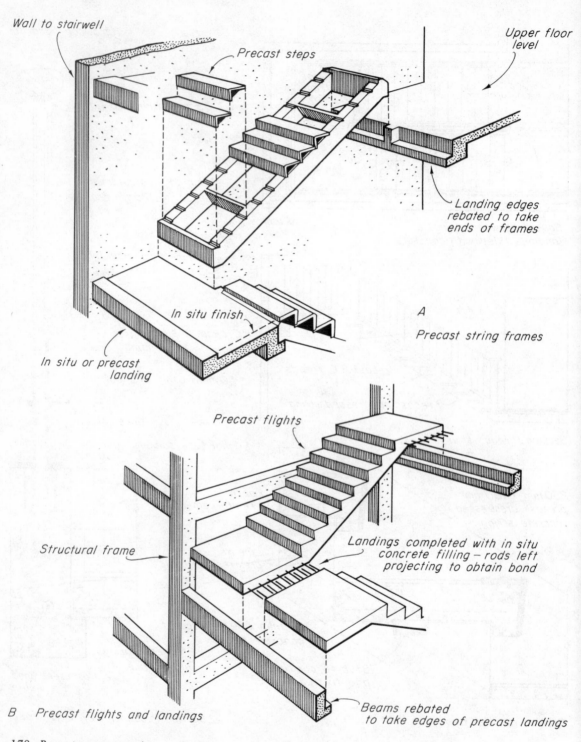

Wall to stairwell

Precast steps

Upper floor level

Landing edges rebated to take ends of frames

In situ finish

A
Precast string frames

In situ or precast landing

Precast flights

Structural frame

Landings completed with in situ concrete filling — rods left projecting to obtain bond

B   Precast flights and landings

Beams rebated to take edges of precast landings

178   *Precast concrete stairs*

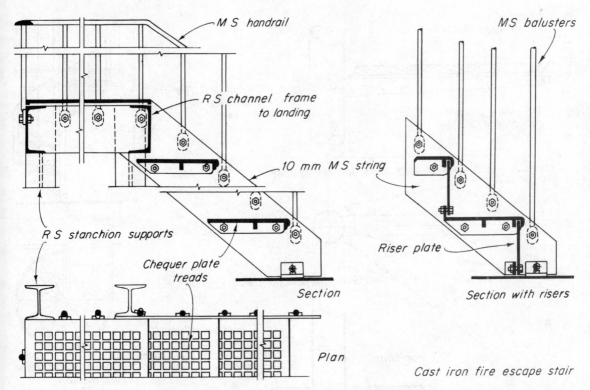

MS handrail

*RS channel frame
to landing*

10 mm MS string

RS stanchion supports

*Chequer plate
treads*

Section

Plan

MS balusters

Riser plate

Section with risers

Cast iron fire escape stair

179 Metal stairs

being bolted or screwed to the cantilever brackets. In the cast-iron type, the winder may be cast with a sinking in the top face to take a filling such as linoleum.

Helical stairs may be formed with cut or closed strings of mild steel plate (figure 181 *A*). If the diameter is large and the rise is considerable, this type of stair is likely to require intermediate support.

String stair

This can be constructed in various ways with strings of mild-steel tube, rolled-steel beams or channels and treads of steel, timber or precast concrete. These form open riser or closed stairs, examples of which are shown in figure 181. The treads are bolted to mild-steel plate seating brackets which are bolted or welded to the strings (*B* and *D*), or the brackets may be formed of small diameter bar in the form of an inverted flat 'U', the legs of which are welded to the string (*C*). Very careful detailing of the brackets is necessary in order to obtain a satisfactory appearance in the finished stair.

Figure 182 *A* shows a string stair in which the boxed channel string is raised above the treads to the level of a guardboard. The treads are suspended from the strings by square steel balusters, screwed to the outside face of the strings and connected to each other at the bottom by crossbars which support the treads. A stair of very light appearance resulting from the use of small section steel strip as a string is shown in figure 182 *B*. It should be appreciated that this construction is, of course, suitable for short flights only, such as that shown.

Where some degree of fire resistance is required in the case of stairs likely to be used internally for escape purposes, the above types of stair would not be suitable, since in many cases the risers are open and steel is exposed. Where, in such cases, it is considered desirable to use steel strings, these must be provided with some protective coating. Illustration (*E*) at the bottom of figure 181 shows a staircase with mild steel channel strings enclosed with 25 mm vermiculite plaster on expanded metal and carrying precast concrete steps. The risers return at the ends in line with the outside face of the

293

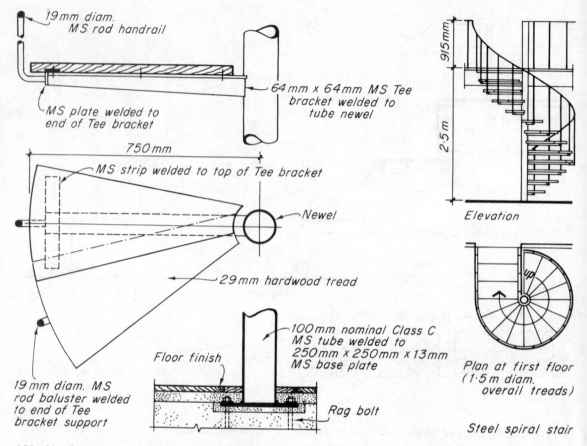

19 mm diam. MS rod handrail

64 mm x 64 mm MS Tee bracket welded to tube newel

MS plate welded to end of Tee bracket

750 mm

MS strip welded to top of Tee bracket

Newel

29 mm hardwood tread

19 mm diam. MS rod baluster welded to end of Tee bracket support

Floor finish

100 mm nominal Class C MS tube welded to 250 mm x 250 mm x 13 mm MS base plate

Rag bolt

915 mm

2·5 m

Elevation

up

Plan at first floor (1·5 m diam. overall treads)

Steel spiral stair

180 Metal stairs

strings and the treads cantilever a considerable distance at each side.

Pressed steel stairs

These are constructed in light pressed metal (figure 183). Each step is a pressing, consisting of the tread and riser, secured at each end to a pressed-steel closed string or a deep boxed channel member, according to the span of the flights. The treads are designed to take a filling of granolithic, terrazzo, or similar material, to form the finished surfaces. Alternatively, the tread (and riser as well if desired) can be covered with timber or marble so that the stair becomes simply the structural element. The landings are constructed from dovetail steel sheeting which gives a rigid structure and provides a good key for any filling. The soffits of flight and landing

are usually fitted with steel clips to which expanded metal can be fixed for plastering or the lining methods shown may be adopted.

Universal stairs

Staircases of any material are normally designed and constructed for each individual job because of the variation in floor to floor height in different buildings. Over past years standard or universal steps have been developed by means of which the rise and going of a flight may be varied within wide limits, and by the use of which staircases of any height and pitch can be constructed. This is accomplished by providing a sloping joint between tread and riser as shown in figure 184, so that, as the steps slide backwards or forwards for adjustment, every change in the going is accompanied by a proportional change in the rise. The proportion of

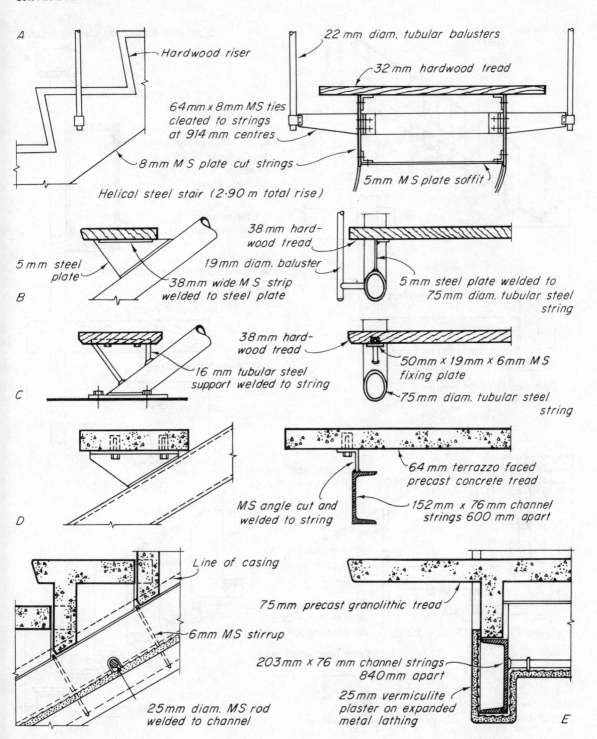

A

Hardwood riser

22 mm diam. tubular balusters

32 mm hardwood tread

64 mm × 8 mm MS ties cleated to strings at 914 mm centres

8 mm MS plate cut strings

5 mm MS plate soffit

Helical steel stair (2·90 m total rise)

5 mm steel plate

38 mm hard-wood tread

19 mm diam. baluster

38 mm wide MS strip welded to steel plate

5 mm steel plate welded to 75 mm diam. tubular steel string

B

38 mm hard-wood tread

16 mm tubular steel support welded to string

50 mm × 19 mm × 6 mm MS fixing plate

75 mm diam. tubular steel string

C

64 mm terrazzo faced precast concrete tread

MS angle cut and welded to string

152 mm × 76 mm channel strings 600 mm apart

D

Line of casing

75 mm precast granolithic tread

6 mm MS stirrup

203 mm × 76 mm channel strings 840 mm apart

25 mm vermiculite plaster on expanded metal lathing

25 mm diam. MS rod welded to channel

E

181   Steel string stairs

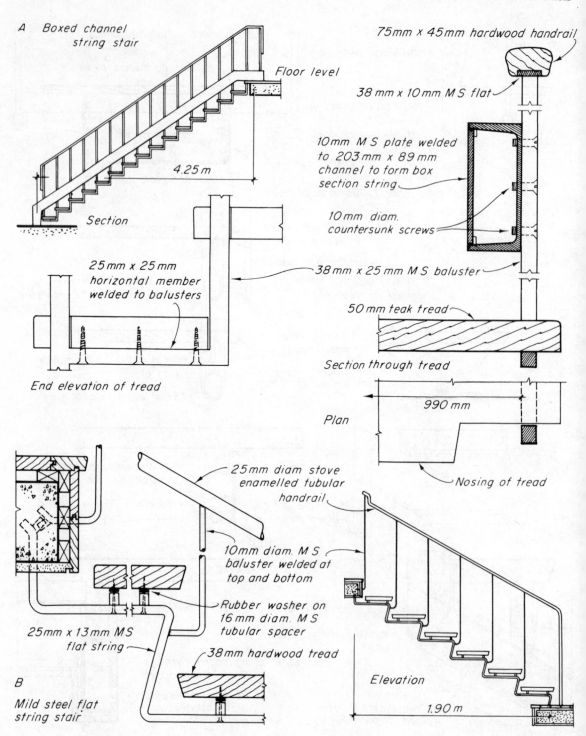

A Boxed channel string stair

Section

Floor level

4.25 m

25 mm x 25 mm horizontal member welded to balusters

End elevation of tread

75mm x 45mm hardwood handrail

38 mm x 10 mm M S flat

10mm M S plate welded to 203 mm x 89 mm channel to form box section string

10 mm diam. countersunk screws

38 mm x 25 mm M S baluster

50 mm teak tread

Section through tread

Plan

990 mm

Nosing of tread

25mm diam stove enamelled tubular handrail

10mm diam. M S baluster welded at top and bottom

Rubber washer on 16 mm diam. M S tubular spacer

25mm x 13mm MS flat string

38mm hardwood tread

B

Mild steel flat string stair

Elevation

1.90 m

182 Steel string stairs

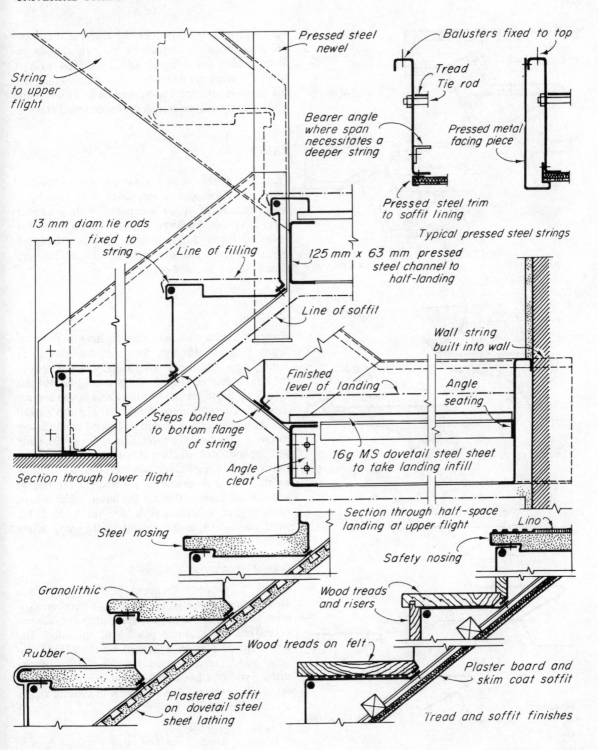

String to upper flight

Pressed steel newel

Balusters fixed to top

Tread Tie rod

Pressed metal facing piece

Bearer angle where span necessitates a deeper string

Pressed steel trim to soffit lining

Typical pressed steel strings

13 mm diam. tie rods fixed to string

Line of filling

125 mm x 63 mm pressed steel channel to half-landing

Line of soffit

Wall string built into wall

Finished level of landing

Angle seating

Steps bolted to bottom flange of string

16g MS dovetail steel sheet to take landing infill

Section through lower flight

Angle cleat

Section through half-space landing at upper flight

Lino

Steel nosing

Safety nosing

Granolithic

Wood treads and risers

Rubber

Wood treads on felt

Plastered soffit on dovetail steel sheet lathing

Plaster board and skim coat soffit

Tread and soffit finishes

183 Pressed metal stairs

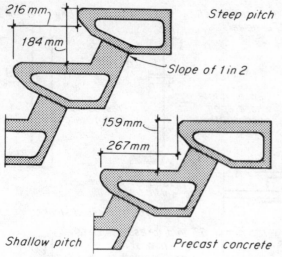

Steep pitch

216 mm

184 mm

Slope of 1 in 2

159mm

267mm

Shallow pitch   Precast concrete

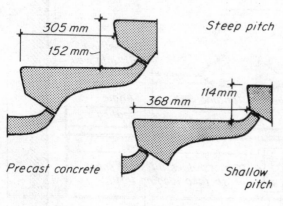

Steep pitch

305 mm

152 mm

368 mm

114mm

Precast concrete

Shallow pitch

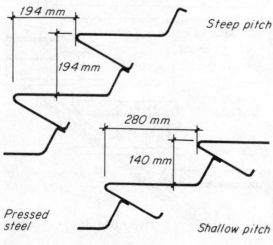

194 mm

Steep pitch

194 mm

280 mm

140 mm

Pressed steel

Shallow pitch

184 Universal stairs

rise to going is governed by the angle of the slope and is commonly governed by the rule that twice the rise plus the going should equal 584 to 610 mm.[1] The steps can be made of any suitable material and the same principle can be applied to a universally adjustable mould for casting *in situ* concrete stairs[2].

RAMPS

These are mainly used for the passage of vehicles rather than for pedestrians, since they take up a large amount of space compared with a normal stair. They can, however, be used with good effect both functionally and architecturally where the space is available. They are essential for access and circulation by disabled persons.

Car ramps

Ramps for cars generally should have a slope of about 1 in 10 although they may be as steep as 1 in 7, especially if the ramp is short. The radius to the centre line of curved ramps should be not less than 7 m, based on the turning circle of the average size car. Curved ramps should be slightly banked and the whole surface should be treated in some way to give a good hold for tyres. The width of the ramp will depend upon whether it is for one- or two-way traffic. With a minimum radius of 7 m, a minimum width of 3.65 m should be allowed for the former and not less than 7.30 m for the latter, which allows for a central separating curb 300 mm wide and a width of 3.35 m on the outside of the ramp where the radius is greater.

Pedestrian ramps

Ramps, if properly designed, probably provide a safer means of pedestrian movement between different levels than the normal stair, since they do not necessitate the accurate placing of the foot. The safe slope of a ramp is limited by the risk of slipping, which is influenced by the nature of the surface and whether or not the ramp is internally or externally situated. The following maximum slopes are considered safe:

[1] See Part 1, page 234.
[2] A step of this type was used many years ago by Alva Aalto (centre of illustration). In Great Britain the design of universal stairs is protected by patent.

For reasonable slip-resistant surfaces subject to wetting – 1 in 10.

(This is the maximum gradient permitted in the Greater London area if the ramp affords a means of escape.)

For reasonably slip-resistant surfaces usually dry – 1 in 8.

For highly slip-resistant surfaces – 1 in 6.

Ramps steeper than 1 in 6 are sometimes required in certain circumstances, as in factories or other industrial situations where, for example, a footway may be required to follow an inclined conveyor. The slope should never exceed 1 in 3. These steeply sloping ramps should always be provided with evenly spaced cleats across the ramp, spaced apart to suit natural walking. The cleat spacing will depend upon whether or not the pedestrians using them are likely to carry loads. Table 22[1] indicates suitable spacings for slopes up to 1 in 3. As cleats require accurate foot placement, ramps on which they are used lose their main safety features and should be avoided if possible.

The minimum width of a ramp to be used by one person should be the same as those for stairs. Greater widths will be required when two or more persons will pass. This will depend upon the nature of the building in which the ramp is situated and the amount of pedestrian traffic likely to use the ramp. A landing at least equal in width and length to the width of the ramp should always be placed at a change of direction in the ramp. The requirements for balustrades and handrails are the same as those for stairs.

Slope of ramp	Recommended spacing	
	If load is carried (mm)	No loads carried (mm)
1:6	355	455
1:5	330	430
1:4	305	405
1:3	280	380

Table 22   *Spacing of cleats on ramps*

A ramp should always be constructed with a good slip-resistant surface. Cement or granolithic surfaces may be finished with a wood float or swept with a stiff broom while still green. This exposes the particles of sand and provides a rough surface which does, however, wear smooth after a time, although the granolithic surface will give a more lasting result because a greater degree of roughness may be obtained initially. Abrasive grit materials may be added to the surface mix to increase the friction and reduce the wear. Slip-resistance is further increased by the provision of transverse grooves formed in the surfaces. If a suitable aggregate is included in the top, asphalt can provide a good slip-resistant surface. Wood, in a dry unpolished state, is reasonably slip-resistant, but can become slippery when wet. Metal surfaces are not altogether satisfactory as even when formed with figured surfaces these soon lose their pattern and become slippery.

FIXED LADDERS

These are usually of metal and are used as a means of access to roofs and other high places and sometimes as a means of escape.

Ladders should be steep enough to make the user face them when descending, but vertical ladders should be avoided if possible because they are less safe and harder to climb. Whenever possible the minimum advisable pitch of 60 degrees should be used. Ladders with a pitch of 75 degrees or steeper should have rungs. Those with a lesser pitch should have flat treads. Landings should always be provided at the top of ladders, and it should be remembered that when descending it is safer to step sideways rather than backwards on to a steep ladder. Single flights should generally not exceed 6 m in length. When longer, intermediate landings should be introduced. Ladders exceeding 6 to 9 m in height should be enclosed by a safety cage. Handrails should be provided on both sides of step-type ladders but the strings serve as hand rails for rung-type ladders. Details of recommended clearances, heights and sizes are given in figure 185. Reference should be made to chapter 10 for the requirements of the Greater London Council in respect of ladders used for escape purposes.

[1] This table is extracted from Industrial Data Sheet S3, 'Safe Access above Ground Level', issued by the Industrial Services Division, Department of Labour and National Service, Australia, from which much of the information on ramps and ladders has been obtained.

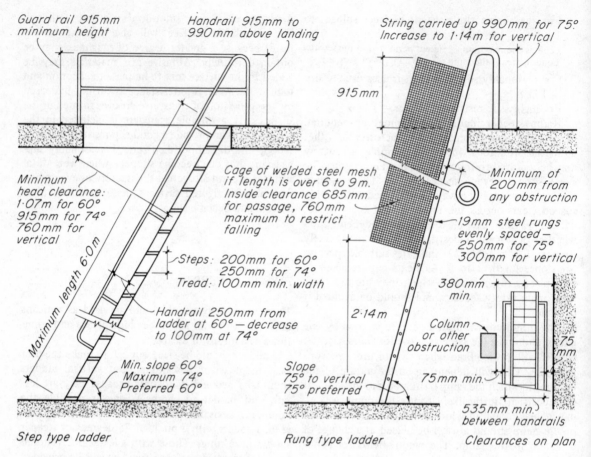

Guard rail 915mm minimum height

Handrail 915mm to 990mm above landing

String carried up 990mm for 75°. Increase to 1·14m for vertical

915mm

Minimum head clearance: 1·07m for 60° 915mm for 74° 760mm for vertical

Maximum length 6·0m

Cage of welded steel mesh if length is over 6 to 9m. Inside clearance 685mm for passage, 760mm maximum to restrict falling

Minimum of 200mm from any obstruction

19mm steel rungs evenly spaced— 250mm for 75° 300mm for vertical

Steps: 200mm for 60° 250mm for 74° Tread: 100mm min. width

Handrail 250mm from ladder at 60°—decrease to 100mm at 74°

2·14m

380mm min.

Column or other obstruction

75mm

Min. slope 60° Maximum 74° Preferred 60°

Slope 75° to vertical 75° preferred

75mm min.

535mm min. between handrails

Step type ladder

Rung type ladder

Clearances on plan

185 Fixed ladders

9 Roof structures

The functional requirements of the roof are discussed in Part 1 and reference is made to the economic and structural significance of the materials to
be used in its construction. These will now be
discussed in more detail.

Materials for roof structures

Steel, aluminium, reinforced concrete, timber and
plastics are all commonly used for the construction
of roofs. With all these materials constructional
forms have been developed in roof structures which
take advantage of the particular characteristics of
the materials. A brief comparison of those properties which are particularly relevant to their use in
roof structures is made below.

Steel

Strength   Steel has high strength in both compression and tension and a small amount of material
is able, therefore, to carry large loads. Working
stresses for mild steel of 165, 230 and 280 N/mm^2
for grades 43, 50 and 55 respectively are permitted
for all normal structural members by BS 449.

Elasticity   A structural material under stress should
not stretch or contract to an excessive degree. This
is particularly important in horizontal members
where large deflections due to loading must generally
be avoided. The ratio of stress to resultant strain,
known as Young's Modulus or the modulus of
elasticity[1], indicates the extent to which the material
will resist elastic deformation. If its resistance is
high the material is stiff, the deformation under
stress will be low and the deflection of a beam under
load will therefore be small. Since the minimum
depth of a beam is often dictated by deflection
rather than by the strength of the materials used, a
high modulus of elasticity permits either a shallower
beam section for a given deflection or a greater span
for a given depth of beam. This can be seen clearly
in the expression for the deflection of a beam,

$$d = \frac{\text{constant } (c) \times w \times l^4}{EI}$$

Steel has a modulus of elasticity of 200 kN/mm^2
indicating that it is a stiff material.

Ductility   Structural materials should be able to
withstand large deformations without suddenly
failing and cracking. In structural frames high stresses
are often induced over restricted areas at some
points and deformation will occur. Provided the
material is sufficiently ductile it will not crack, but
what is known as plastic flow will take place and
the load will be transferred to the surrounding
material, so that at no point is the failing stress
reached. Steel is a ductile material and, with a
yield point of up to 350 N/mm^2, undergoes considerable strain after the elastic limit and before
ultimate failure. This can be seen in the diagram of
stress/strain curves in figure 186.

Generally   The properties of steel are such that the
dead/live load ratio of steel members is small. That
is to say they are able to carry heavy live loads at
the expense of a comparatively small dead or self-
weight.

Aluminium alloys

Pure aluminium is quite soft and is alloyed with
other elements to make it a suitable structural
material. Aluminium alloys have the advantage of
being corrosion resistant and only about one-third
of the weight of steel, but they have the disadvantage of being more expensive in first cost.

Strength   The stress/strain curves of aluminium
alloys exhibit no sharply defined yield point (figure
186), so that there is no clear indication of the
elastic limit and no obvious point to which the
working stress can be related. A 'proof' stress is,
therefore, specified to aid this purpose. This is the
tensile stress which produces a non-proportional
extension of a defined amount of the original
length[2]. This 0.2 per cent. It will be seen from
figure 186 that the proof stress on which the working
stress is based can lie close to the ultimate stress
of the alloy. One of the commonly used structural
alloys has an ultimate stress of 310 N/mm^2, a proof
stress of 270 N/mm^2 and a working stress in
bending of 162 N/mm^2. While stronger alloys are

---

[1]   See chapter 3, Part 1.
[2]   See also *MBC* : *Materials,* chapter 9.

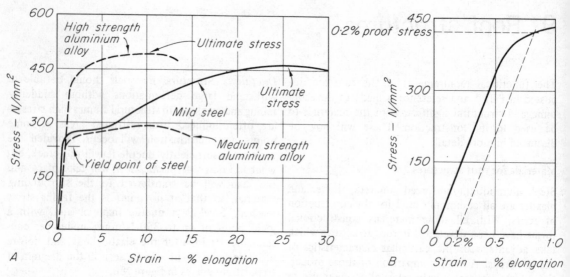

186 *Stress-strain curves*

available, unfortunately the aluminium-copper alloys which have greater strength have an inferior corrosion resistance.

Elasticity The modulus of elasticity of aluminium alloys, 69 kN/mm^2, is about one-third that of steel. An aluminium alloy beam will, therefore, under the same load and support conditions and neglecting self-weight, deflect three times as much as a similar steel beam. The stiffness of a beam is measured by its flexural rigidity (*EI*), and in order to maintain a given deflection when the modulus of elasticity decreases, the moment of inertia must be increased (see expression for deflection on page 301). This may be done by increasing the depth of the section to between $1^5/_8$ and $1^3/_4$ of the depth of the steel section. In addition, or as an alternative, the cross sectional area may be increased. The lack of elastic stability of thin members and the danger of local buckling must be guarded against when using aluminium sections. The web of a beam may buckle sideways if it is too thin or insufficiently stiffened, lateral buckling due to torsion may occur in an unrestrained slender beam, and local buckling may take place in a thin compression flange at the points of maximum bending moment. These dangers are avoided by using sections with greater flange and web areas or with a stiff cross-sectional shape such as box sections or sections with lipped edges as shown in figure 187.

302

Ductility The proof and ultimate stresses of aluminium alloys lie close together and there is very little elongation before failure occurs. In areas of high stress there will, therefore, be little plastic flow taking place to permit the load to be transferred to surrounding areas. This has to be borne in mind in the design of aluminium structures.

Generally In normal conditions aluminium is highly resistant to corrosion. It has a coefficient of expansion about twice that of steel, so that provision for greater temperature movements must be made in aluminium structures.

Aluminium alloys may be applied to structures in a similar way to steel but result in a considerably lighter structure. They are most economically used in cases where the weight of the structure itself

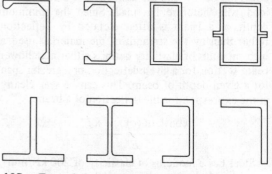

187 *Typical aluminium structural sections*

forms a substantial part of the load and the imposed loads are comparatively light. They are, therefore, obviously suitable materials for roof structures, particularly those of long span where the dead load rises rapidly and the dead/live load ratio is likely to be high. Because of their characteristics, aluminium alloys are best used in structures which are inherently stable and resistant to torsion and where the members can be kept short and directly loaded as in grid structures.

Reinforced and prestressed concrete

Strength   The ultimate compressive strength of normal concrete is from 20 to 60 N/mm^2, but its tensile strength is only approximately one-tenth of this. Methods of making good this deficiency by reinforcement are described in chapter 5, page 208, and the advantages of prestressing are described in the same chapter on page 226.

Elasticity   Concrete has a low modulus of elasticity ranging from 25 to 36 kN/mm^2, but due to creep effects the effective modulus for long term loading may fall to only one half of these values.

Generally   Over wide spans reinforced concrete used in beam and slab form has a high dead/live load ratio. It is most effectively used in structural forms which can take advantage of its monolithic character, particularly three-dimensional forms such as shells, doubly curved slabs and folded slabs. The prestressing of concrete brings about a considerable reduction in depth of spanning members and depth/span ratios from 1:30 to 1:120 are possible. The reduction in depth together with a reduction in the overall cross-sectional area, made possible by the increased resistance of prestressed concrete to shear, results in a considerable reduction in dead weight and results in satisfactory dead/live load ratios over wide spans.

Timber

Timber was one of the earliest materials to be used for structural purposes, but our knowledge of its behaviour and capabilities is comparatively new. Developments in methods of joining timber, particularly by means of metal connectors and glues, developments in the stress-grading of timber and research into improved methods of design have resulted in the more efficient use of the material.

Timber is a comparatively light material and the species used for normal structural purposes have weights approximately one-sixteenth that of steel.

Strength   Timber is an organic material and the knots and faults brought about during growth or seasoning constitute zones of weakness. The technique of stress-grading is a means of establishing the loadbearing capacity of a piece of timber either in terms of knots and other visible faults by which, on the basis of the strength of clear wood of the same species, a reduced allowable working stress is determined or by mechanically measuring the modulus of elasticity of the piece and allocating an allowable stress. The strength of normal structural softwoods in bending is approximately one twenty-eighth to one twenty-third that of steel. The basic stresses in flexure and compression parallel to the grain are 3-18 N/mm^2, depending upon the species and grade of timber, with values about half as much again in tension.

The problem of joints is an important factor in timber design, and earlier forms of joints had an efficiency of no more than 15 to 20 per cent relative to the timber entering the joint. Two methods of increasing their efficiency are by means of connectors and synthetic glues..

Timber connectors consist of various forms of metal plates and rings through which a bolt passes (see figure 206 and Part 1, figure 138). Their effect is to increase the strength of the joint, particularly in tension and shear.

Glues made from synthetic resins produce joints as strong and even stronger than the timber joined. A considerable number of these glues is available, all of which have different characteristics and some of which are immune to attack by dampness or decay. Glues can be used for lattice construction but are more advantageously used in building up laminated timber members, which can be much stronger than the same size section in solid wood. A further advantage of laminated construction is that different qualities of timber may be used in the same section, the better quality being limited to the more highly stressed zones. By means of gluing, continuity of structure is obtained and it is a simple matter to form curved members and portal frames with the greatest depth where the stresses are highest. The cost of manufacture, however, makes the use of 'glulam' members as these are called economic only

303

when it is clear that the requirements are beyond the range of solid timber.

Elasticity The modulus of elasticity of normal structural softwoods is from 4 to 12 kN/mm^2. Although some timbers such as Douglas Fir have a somewhat higher value the modulus of elasticity of timber is low compared with that of other materials. In spite of this, however, having regard to the light weight of the material, comparative analysis with other materials shows that in terms of flexural rigidity timber is the most efficient from the point of view of weight and cost of material.[1]

Generally The stiffness of timber relative to its weight and cost makes it particularly suitable for structures in which the load-carrying capacity is determined by its flexural rigidity (*EI*), such as structures which are large in relation to the load they carry. This includes roofs of all types, floors bearing moderate or light loadings and single-storey buildings, particularly those of large height and span. As a rough guide structures which are liable to fail through elastic instability, and for which timber is the most suitable material, are likely to be those with a load-intensity ratio, $\sqrt{P/l}$, less than about 1.50 where P is the compressive load in Newtons and l is the effective length of the member in millimetres.

In addition to its use in framed and laminated structures timber may be used in the form of stressed skin plywood panels built up as folded or prismatic slabs or in the form of planks built up as doubly-curved shells. Very high strength to weight ratios are attained in this way, with weights per square metre of floor area covered as low as 25 kg. As a structural material timber has the advantage of ease of working and fabrication. Because of its comparatively light weight, built up members can be easily handled. When properly used it is a permanent material and has satisfactory thermal insulating properties, which is a further advantage when it is used in stressed skin forms of roof structure. Although it burns freely in thin sections, when used in the sizes normal in lattice construction it remains structurally stable during a fire for a greater length of time than steel (see chapter 10).

Plastics

The large range of plastics now available has diverse characteristics and properties[2]. At present the widest application of plastics in building is still for non-

structural purposes, but considerable development of these materials for structural purposes has taken place in certain types of structure where advantage can be taken of their particular properties.

Plastics are light in weight and on an average weigh only about one-sixth of the weight of steel.

Strength The tensile strength of unreinforced plastics is only about 60 N/mm^2, but when reinforced with suitable material, such as glass fibre, a tensile strength of 160 N/mm^2 is developed, using a randomly oriented fibre of fibre volume fraction 30 per cent. The compressive strength is of the same order as that of the tensile strength.

Elasticity Thermo-setting plastics have a low modulus of elasticity, much the same as those for timber (3.50 kN/mm^2), but the modulus rises to 8.00 kN/mm^2 for a similar material to that described above.

Ductility Plastics are not ductile materials. Little plastic flow, therefore, takes place in areas of high stress and, as in the case of other materials of low ductility, this must be borne in mind in the design of structures incorporating plastics.

Generally Plastics have a high coefficient of expansion, about 8 times that of steel, but when glass fibre reinforced they have a coefficient about the same as that of steel. The light weight of plastics gives a favourable strength/weight ratio so that they are particularly suitable for roof structures, provided these are of the type in which the low stiffness of the material is overcome by the inherently stiff form of the structure. The types of structure, therefore, to which plastics can most advantageously be applied are space structures of the stressed skin type, in which the strength is derived more from the geometry of the form than from the properties of the material. Plastics have been used in folded plate form and in the construction of many geodesic domes and in grid structures combining plastic tetrahedra, or doubly curved sheets, with grids of metal rods.

[1] See paper by Philip O. Reece, 'A review of the Structural Use of Timber in the United Kingdom', Building Research Congress, 1951, papers in Division 1, Pt 2.

[2] See *MBC* : *Materials*, chapter 13.

Roof loading

(i) The dead load consists of the self-weight of the structure itself[1] and of the roof claddings, coverings and internal linings

(ii) The superimposed load consists of the weight of snow, any incidental loads applied during the course of maintenance work and, in the case of flat roofs used as roof gardens, play areas or for other purposes, additional loads according to the purpose for which the roof is used

(iii) Wind forces.

The dead load is usually based on the weights of materials specified in BS 648.

Superimposed loads on flat roofs CP 3: chapter V: Part 1: 1967 recommends for flat roofs and pitched roofs up to and including 10 degrees, with no access provided other than that necessary for normal maintenance purposes, an allowance for superimposed loads of 0.75 kN/m² measured on plan or a load of 0.9 kN concentrated on a 300 mm sided square, whichever produces the greater stress. When access in addition to that necessary for maintenance purposes is provided the figures are 1.5 kN/m² and 1.8 kN respectively. The figures include a load in respect of loose snow up to a depth of 600 mm. Loose freshly fallen snow weighs approximately 80 kg/m³, but compact snow may weigh 300 kg/m³ and in districts subject to heavy snowfall an allowance should be made for this.

Superimposed loads on pitched roofs The superimposed load allowance on roofs with a pitch greater than 10 degrees, to which no access is provided other than for maintenance purposes, should be:

(i) For a pitch of 30 degrees or less — as for flat roofs above

(ii) For a pitch of 75 degrees or more — no allowance.

For pitches between 30 and 75 degrees the load shall be obtained by interpolation. To allow for loads incidental to maintenance works it is recommended that all roof coverings, other than glass, at a pitch of less than 45 degrees should be capable of carrying a load of 0.9 kN concentrated on any 125 mm sided square.

Wind forces Some indication of the variations in wind pressure and suction over roof surfaces is given in figure 68 in this volume and figure 122 Part 1. The allowances to be made for wind pressures normal to the surface of flat and pitched roofs are assessed by the method given in CP 3: chapter V: Part 2: 1972 and briefly outlined on page 116. As in the case of walls greater suctions and pressures occur at gable ends, near the eaves and near the ridge of a pitched roof (see figure 122, Part 1). Fastenings for roof sheeting near these points should therefore be designed to take account of these greater forces which will be exerted[2].

TYPES OF ROOF STRUCTURE

The classification of roof structures into two- and three-dimensional forms has been described in Part 1 (pages 163-164) and as indicated there, these may be constructed of a number of different materials using different constructional techniques. At this point the various structural forms used for roofs of medium and large span will be reviewed, with particular reference to the principles on which they are based and which govern their structural behaviour, and to the practical considerations which have led to their development. Later in this chapter these are discussed in terms of constructional methods and details in different materials (see page 339).

Trusses and girders

Trussed roofs

The classification of roofs constructed of two-dimensional members as single, double and triple roofs according to the number of stages necessary economically to transfer the loads to the supports has been described in Part 1. The primary structural member in triple pitched roof construction, as explained in Part 1, may be in the form of a roof truss or a rigid frame, the former implying a double-pitched triangulated structure and the latter a structure with continuity between vertical and spanning members. Small span trussed and rigid frame roofs are described in Part 1, chapter 7.

Trussed roofs are widely used for single-storey and shed-type buildings (see figure 188). Considerations affecting the triangulation of the truss are

---

[1] The significance of the self or dead-weight of the structure, especially with increasing span, is referred to in Part 1.

[2] See Part 1, page 162, on pressure and suction distribution on roofs and on the effect of suction on lightly clad structures.

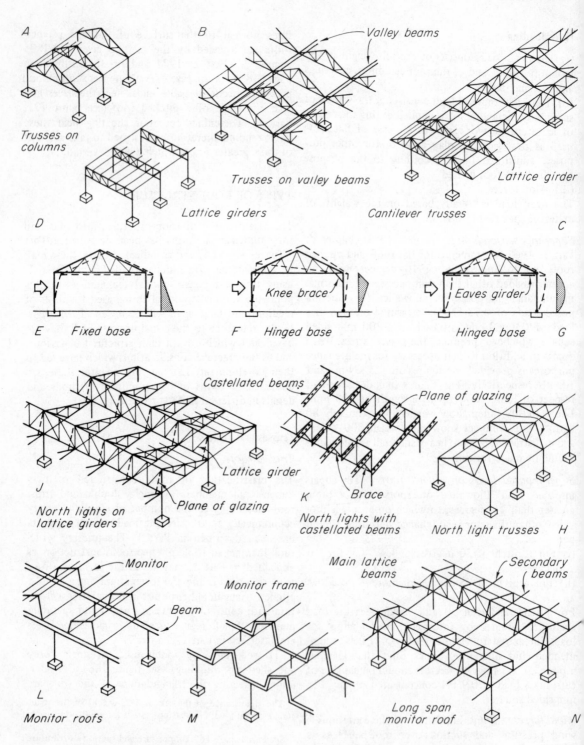

A Trusses on columns

B Valley beams Trusses on valley beams

Lattice girder

Lattice girders

Cantilever trusses

D

E Fixed base

F Knee brace Hinged base

Eaves girder

Hinged base G

C

J North lights on lattice girders

Castellated beams Plane of glazing

Lattice girder

Plane of glazing

Brace

North lights with castellated beams K

North light trusses H

L Monitor Beam

Monitor roofs

Monitor frame

M

Main lattice beams Secondary beams

Long span monitor roof N

188 Truss and girder roofs

306

discussed in Part 1, pages 67 and 180.

Light types of factory building requiring a clear internal height of about 3.65 m and a span from 9 to 12 m can be economically constructed with steel trusses spaced from 3 to 3.65 m apart bearing on the tops of stanchions as shown in figure 188 *A, E*. The normal fixing at the feet of the truss produces a reasonably unrestrained joint with the stanchion. The normal methods of fixing the column feet to the foundation pads[1] produces a comparatively rigid joint at this point. Side wind pressure will set up bending stresses in the columns as they react as vertical cantilevers, so that the stress distribution will be zero at the top of the columns increasing to a maximum at the foundation, which it will tend to rotate. Within the limits of sizes given above, the bending stress at the base will be comparatively small and the rotational tendency on the foundation will be slight. Thus columns of comparatively small cross-section can be used and, with foundation pads sized to take the vertical loads, the stresses on the soil due to wind-pressure can usually be kept within safe limits.

When the span is greater than 12 to 15 m and for functional reasons the columns must be high, the column section increases rapidly and large turning moments are applied to the foundation slabs. In these circumstances, to avoid uneconomic foundations a non-rigid or hinged joint between the column foot and the foundation may be introduced to relieve the latter of any rotational tendency, since no bending stresses can be transferred through such a joint (see also page 310). In order to provide the necessary rigidity against lateral wind pressure a knee brace may be introduced to provide a stiff joint between the columns and the roof truss, *(F)*. The stress in the columns will then be somewhat reduced and will be zero at the foundation and a maximum at the knee brace. Because of the rigid joints, some bending will be transferred to the feet of the truss which must be designed to withstand it, but the foundation slabs may be limited to the size required solely by the vertical loading. The introduction of rigid and hinged joints in this way results in the structure acting as a whole under side pressure of wind, with a tendency to 'uplift' on the windward side, which may be marked in the case of light structures.

An alternative method, which reduces bending in the columns and avoids bending stresses in the truss, is to introduce horizontal eaves girders instead of knee braces, *(G)*. These are lattice girders on a horizontal plane, running the length of the building and supported at the heads of the columns and the bottom ties of the trusses. Lateral rigidity is provided by these girders which pick up the wind pressure on roof and walls through the columns and transmit it to the ends of the building or to cross walls or braced frames. Intermediate foundations are thus relieved of any vertical component due to wind pressure. This is an economic method for high buildings provided the building is not too long or is divided at intervals by walls or braced cross frames which can transfer the wind forces from the girders to the soil. The columns are designed with unrestrained top and bottom joints so that the bending stresses are a maximum at the centre and zero at top and bottom. Induced bending stresses in foundations and truss are therefore avoided.

Roof trusses may be constructed in steel, aluminium alloy or timber, in spans up to more than 60 m when required. In the case of very large spans the pitch is kept low in order to avoid excessive internal volume and to reduce the area of roof to be covered and the weight of the structure.

Roof girders

Trussed or lattice girders are widely used for medium and large spans when a flat or low pitch roof is required (figure 188 *D*). Universal steel beams are not economic for spans much above 10.50 m, although this can be extended by the use of castellated beams (see figure 109). Reinforced concrete beams have an economic limit of about 9 m. Prestressed concrete, however, is very suitable for wide span roof beams since small depth/span ratios are possible.

Trussed girders may be designed with parallel chords or with the top chord double-pitched or curved where a low pitched roof is required[2]. This is normal for large spans, but in the case of smaller spans beams as a whole are sometimes pitched in the middle to give a low double-pitched roof suitable for low-pitch roof coverings.

Girders may be used as valley beams in multi-span trussed roof structures to permit the wider spacing of internal columns, the girders supporting a number of trusses (figure 188 *B*). Where very wide column spacing is required the depth of the girders

[1] Described in Part 1, chapter 6.

[2] See Part 1, figures 42 and 45.

307

will increase excessively the height of the building. In this case what is known as cantilever truss or 'umbrella' truss construction is used. In this the girders are made the full depth of the truss and are placed in the line of the ridge so that the truss cantilevers out on each side of the beam, the feet of adjacent trusses meeting at the valleys as at (C). The economic depth of trussed girders is from one-sixth to one-tenth of the span and, owing to the large depth of beam at the junction with the supporting columns, the beam to column joint can be comparatively rigid. Suitable materials for the construction of trussed girders are steel, aluminium alloys and timber.

Vierendeel girders without diagonal members, but with rigid joints between the chords and the vertical members, are only occasionally used in special circumstances in single-storey structures. These are described in chapter 5.

Beyond a span of about 12 m, depending upon the standard of lighting required, a reasonable degree of natural lighting through the walls is not likely to be achieved unless the structure is very high. In order to provide satisfactory lighting to the interiors of extensive single-storey buildings a number of roof forms have developed.

North light roofs

This type of roof may be in shell form, which is described later, or in trussed form. In lattice construction it is an asymmetrical truss, the steeper and shorter side of which is glazed and is sited to face north as shown at (H). As in the case of symmetrical trusses, the supporting columns may be placed at greater distances apart, say three bays, with the intermediate trusses picked up on valley beams. Where greater column spacing is required and the valley beams become excessively deep, lattice girders in the line of the ridge are used to form cantilever north light construction. To obtain wide spacings of the main lattice beams the construction may be in the form of trussed rafters spanning between the beams, with the plane of glazing running from the ridge, that is the top of the lattice beam, on to the 'back' of the adjacent trussed rafter (J). An alternative to this is the use of castellated beams, which are suitable for long spans carrying light loads, spanning between the top chord of one lattice girder and the bottom chord of the adjacent girder. The plane of glazing is the depth of the main beam (K).

Monitor roof

A monitor is a mono-pitch lantern light with glazing at the sides only. The side facing north is usually large in area and that facing south is small. This roof provides very even lighting at the working plane for a comparatively small volume of roof; good lighting may be achieved with quite low ceilings, the spacing of the monitors being arranged for any given height to provide an even distribution of light. When spans are not great and the columns may be spaced about 4.50 m apart, the monitor frames may be built off Universal beams or shallow lattice beams (L) or, alternatively, the monitor frames may be formed as integral parts of a cranked beam of welded steel or in situ or precast concrete (M). Where wide column spacing is essential, deep lattice beams are used spaced 6 to 7.50 m apart, according to the spacing required for the monitors, which support shallow lateral beams on the bottom chords spaced about 4.50 m apart. These secondary beams carry the monitor frames which straddle the top of the main beams (N). Instead of separate secondary beams and monitor frames, cranked welded steel beams may be used spanning from the top chord of one main beam to the bottom chord of the adjacent beam.

Rigid or portal frames

Over large spans, deep lattice girders and pitched roof trusses, particularly the latter, may result in excessive volume within the roof space of the building which, because of the obstruction by beams and ties, may not always be useful space. Further, with an increase in span the extra material necessary to provide adequate strength must be added to a simple beam or truss at the points where its own dead weight will increase the bending moments in the structure. The use of rigid frame construction overcomes these disadvantages to a very large extent.

The characteristic of the rigid frame is continuity of structure due to the stiff, or restrained, joints between the parts, and because of the nature of the stress distribution within such frames, less material is required at the centre of the spanning elements than in a comparable simply supported beam. With increase in span, the whole of the necessary extra material is not required to be placed in the beam element, so that the maximum economic span is much greater. The smaller depth of the beam elements results in comparatively unobstructed, usable space for the full height of the building.

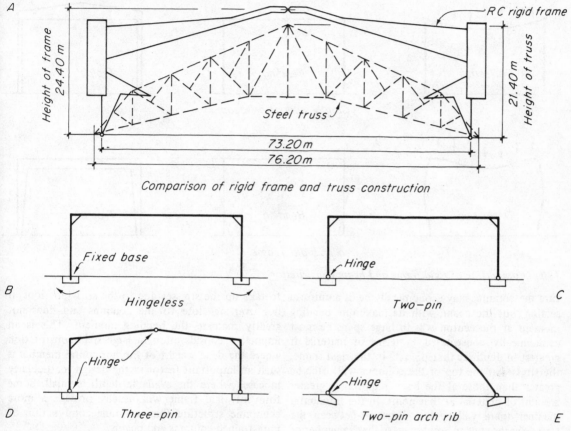

A

Height of frame
24.40 m

RC rigid frame

Steel truss

21.40 m
Height of truss

73.20 m

76.20 m

Comparison of rigid frame and truss construction

Fixed base

B

Hingeless

Hinge

Two-pin

C

Hinge

Hinge

D

Three-pin

Hinge

Two-pin arch rib

E

189   *Rigid frames*

The difference between these two forms of construction in this respect can be seen in figure 189*A*, which shows a rigid frame and a trussed roof construction spanning very much the same distance; the overall height of the rigid frame structure, the space within which is wholly utilized, is approximately the same as the overall height of the truss which encloses a large volume of space lying above the volume of the building below. In some types of building, this space is, of course, valuable for housing services.

The result of the continuity arising from the introduction of stiff or rigid joints between the parts of the frame is illustrated in figure 190, where a portal frame rigidly fixed to its foundations is compared with a beam structure simply supported on two columns. It can be seen that the bending in the beam of the portal frame is transferred through the rigid joints to the columns. The resistance to this bending offered by the column results, however,

in a reversed bending at the ends and a reduction of bending at the centre of the beam[1]. The nature of the deflections in each structure can be seen. There is little or no bending in the columns of the beam structure and only a single curve deflection in the beam, but there is considerable variation in curvature in the rigid frame. The points at which the direction of curvature changes, that is the points of contra-flexure, are points at which there is no bending moment and at which the bending stresses in the members change 'signs'. This can be seen in the bending moment diagrams, which also show that the stiff junctions of beam and columns in the rigid frame are zones at which the bending moments are large. In contrast, there are no bending moments at the unrestrained junctions between columns and beam in the beam structure.

These differences in stress distribution produce differences in form. In the beam and column struc-

[1] See *Beam action* Part 1, page 61.

309

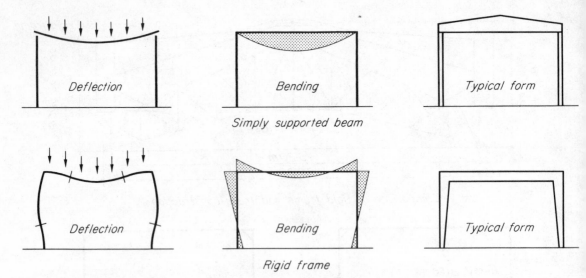

Deflection Bending Typical form

Simply supported beam

Deflection Bending Typical form

Rigid frame

190   *Comparison of rigid frame and beam construction*

ture the columns may economically be of a uniform section, but the beam, with its maximum bending moment at the centre, will in large spans be most economically constructed in terms of material if greatest in depth at that point[1]. In the rigid frame, the stresses at the top of the columns will often be greater than those at the base, requiring a greater amount of material at that point. In the horizontal member there will be less disparity between the stresses at the ends and mid-span, so that a member of uniform depth will be economic. The relative proportion of end and mid-span moments depends on the relative stiffness of beams and columns. If the columns are slender compared with the beam they will provide little fixity at the ends and the mid-span moment will approach that of a simply supported beam; the end moments will be small. In some cases where the bending moments at the ends or haunches are considerably greater than those at mid-span, the depth of the beam at the centre may logically be less than that at the ends.

It can thus be seen that continuity, because of the transfer of stresses from one part to another, results in all parts of the structure providing resistance to the stresses set up by the load with a consequent reduction in bending moment at particular points. As mentioned earlier, less depth is required at the centre of the spanning member of the rigid frame, thus reducing the bending moment further by the amount of dead weight saved at this point. Although more material has been added at the ends

to take up the stresses due to the stiff junctions, it lies over or close to the columns and does not greatly increase the bending moment. This is an important consideration in large span construction where the dead weight of the horizontal member is such an important factor in the design, particularly in cases where the available depth is small. Some form of rigid frame will usually provide a more economic structure than a simple combination of unrestrained columns and beams.

As with the knee-brace truss construction described above, the stiff joints between columns and beam in the rigid frame provide lateral rigidity and make it possible to introduce hinges when necessary. *The hinged joint* is also referred to as a non-rigid, unrestrained or pivoted joint. In structural frames a hinge implies a junction between two parts that can transmit a thrust and shearing force but not a bending moment, since it permits free rotation as explained in chapter 3 of Part 1. This, for the designer, simplifies the analysis of the structure by making it statically determinate. Direct stresses only exist at such joints and, since these can be resisted efficiently with greater concentration of material than bending stresses, the shape of the structure can be varied accordingly and produces what are now typical forms in rigid frames with hinged joints (figure 191). For practical purposes hinged joints may facilitate the site erection of prefabricated

[1] Although in overall terms costs of fabrication may make it cheaper not to vary the depth.

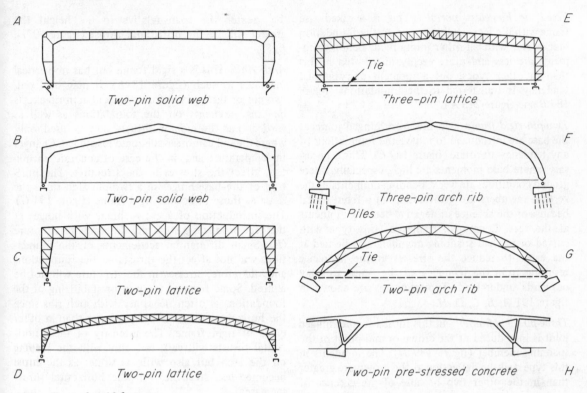

A — Two-pin solid web

B — Two-pin solid web

C — Two-pin lattice

D — Two-pin lattice

E — Three-pin lattice — Tie

F — Three-pin arch rib — Piles

G — Two-pin arch rib — Tie

H — Two-pin pre-stressed concrete

191   Typical rigid frames

frame components since they may be simply executed in comparison with the forming of continuous joints, particularly in concrete frames. In addition, pin jointed base hinges may be used as fulcrum points when lifting half-frames into position. They also serve, as explained already, as a means of relieving foundations of all tendency to rotate under the action of wind or other imposed loads on the frame. The actual form of hinge will depend upon its main purpose and upon the extent of freedom from restraint it is required to give. It need not necessarily be a true hinge or pivot provided the rigidity of the structure at the hinge point is low since the movements and degree of rotation are comparatively small. The 'split' hinges without metallic parts used in *in situ* concrete work are, therefore, feasible. Methods of forming hinged joints of various types are illustrated in figures 212, 213 and 214.

It has already been pointed out that rigid structures of all types are sensitive to differential settlement of the foundations and to movements due to changes in temperature. The effects of these movements must be borne in mind at the design stage,

the former by considering the superstructure and foundation design together, and the latter by the provision of expansion joints or hinges where necessary.

Rigid frames can be constructed in steel in lattice or solid web form; in aluminium alloy in lattice form; in timber in lattice or solid web form, and in concrete.

Types of rigid frames

The fundamental and constant characteristic of the single-storey rigid frame in all its forms is the stiff or restrained joint between the supporting and spanning members. Apart from this the form can vary in a number of ways. The spanning member may be horizontal, pitched or arched; the junctions of the vertical members with the foundations may be restrained or hinged; a hinge may be introduced in the middle of the spanning member, and the structure itself may be solid or latticed. The members may be regular in cross section or may vary in shape according to the distribution of stresses within them.

311

Fixed or hingeless portal This is a fixed-base frame with the feet rigidly secured to the foundation blocks and with all other joints rigid. Bending moments are less and more evenly distributed in this than in other types, but a moment or rotational tendency is transferred to the foundations (figure 189 *B* and figure 213 *A*).

Two-pin rigid frame In this form hinged joints at the base are introduced to relieve the foundations of any tendency to rotate (figure 189 *C*). This is necessary where base moments are high, especially where ground conditions are weak. Bending moments in the vertical members are greater than in a fixed portal because of the absence of negative bending moments at the feet. Long span frames of this type with curved or pitched spanning members may be tied at the eaves to reduce the stresses at the haunches resulting from the tendency of the frame to splay outwards under load. Typical forms are shown in figure 191 *A, B, C, D, H*.

Three-pin rigid frame In this form a further hinged joint is introduced at the crown or mid-point of the spanning member (figure 189 *D*). The moments in this type and the deflection at the crown are greater than in the other two because of the absence of rigidity at the centre and the resultant reduction in positive moments in the spanning members (compare (*F*) with (*E*), figure 192). It will, therefore, be less economical in material, unless the centre point is considerably higher than the eaves. The presence of the three hinges, however, makes this form of frame statically determinate and simpler to design. In precast concrete or lattice forms this type is usually easier to erect than the portal or two-pin frame.

Rigid frames impose horizontal thrusts of some magnitude on the foundations due to the tendency of the column feet to splay outwards as the spanning member deflects. This must be met by adequate resistance in the soil or, where necessary, by the use of inclined foundation slabs, piled foundations or ties between the foundations (*E, F, G*, figure 191). This is of particular importance in fixed and two-pin frames in which even a small horizontal movement will cause considerable redistribution of moments in the frame, with possible adverse effects on the structure as a result.

The horizontal thrust will vary with the stiffness of the frame and with the relative proportion of spanning member to vertical members. It will, therefore, be large with three-pin frames, and in all types

the greater the span relative to the height the greater will be the horizontal thrust (figure 192*D, E, F*).

Arch rib This is a rigid frame but has no vertical members as such (figure 189*E*). It may be fixed, two-pin or three-pin in form. A fixed arch rib exerts bending moments on the foundations as well as vertical and horizontal thrusts and is used only where the soil can offer adequate resistance. Changes in temperature and, in the case of concrete, shrinkage, affect the stresses in the structure. The function of the base hinges in a two-pin arch rib is the same as those in normal rigid frame (figure 191 *G*). The introduction of a crown hinge with hinges at the feet produces a statically determinate structure (*F*). Small differential settlements of the foundations will not affect the thrusts on the foundations and, therefore, stresses in the structure will not be altered. Some form of tying in or stabilizing of the foundations is often necessary with arch ribs since the horizontal thrust is often greater than in other types of rigid frame. The intensity of these horizontal thrusts will vary not only with the loading on the arch but also with its slope: as the curve becomes less steep so will the horizontal thrust increase[1].

Forces acting on a structure will set up a 'natural line of forces' through which the loads would, in theory, be transferred most economically to the bearings. For example, a rope or chain, uniformly loaded, will take up a catenary or near-parabolic curve through which all the forces act directly in tension, with no rotation or bending in the rope. The depth of the curve will depend upon the resistance at the ends of the rope. A similar 'thrust line' or 'curve of pressure' is set up in an arch under uniform load through which the load is transmitted to the abutments. It is, in fact, the bending moment diagram for the loading. If the arch form follows closely this parabolic curve it will be in direct compression at all points with no bending stresses in it (figure 192 *A*). Loading, however, is not always uniform and the arch may be subject to appreciable wind loads or other variable loads, so that in practice it may not be possible to eliminate bending entirely (*B*). But, by designing the arch to follow as closely as possible the bending moment diagram due to the particular loading, bending can be kept to a

[1] See pages 53 and 54, Part 1.

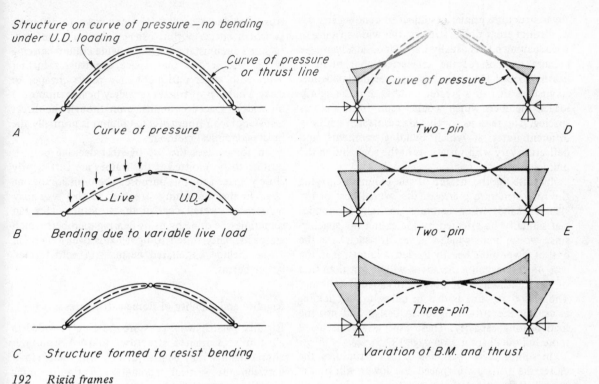

Structure on curve of pressure – no bending under U.D. loading

Curve of pressure or thrust line

A   *Curve of pressure*

Curve of pressure

Two-pin   D

↓ ↓ ↓ ↓ ↓

Live   *U.D.*

B   *Bending due to variable live load*

Two-pin   E

Three-pin

C   *Structure formed to resist bending*

Variation of B.M. and thrust

192   Rigid frames

minimum (*C*). In many circumstances, however, it may not be economical to shape an arch in this way, nor indeed may the arch form be so suitable as the normal rigid frame.

The magnitude of the bending moments induced in an arch or rigid frame which does not follow the 'thrust line' will vary as the modified shape diverges from the curve of pressure (figure 192 *D, E, F*). Any point of an arch rib or frame lying on the curve of pressure will be free of bending moment, as at the points of contraflexure or at all hinged joints which, as they are incapable of transferring bending, must always lie on the curve of pressure. Those parts lying furthest from the curve will be under greatest stress (*F*). Whilst the use of a crown hinged joint will automatically produce a zero bending moment at the joint and simplify analysis of the structure, the frame will be far stiffer and a more even stress distribution will occur if continuity is maintained and hinges are restricted to the base connections (*E*). This will be clear from the illustrations if it is borne in mind that the moments at various points in the frame are in direct proportion to the distance from the 'curve of pressure'.

It will be seen that the arch or near-arch form is usually subject to smaller bending moments than the normal rigid frame because its shape lies closer to that of the curve of pressure (*D*). It is, therefore, particularly useful for very wide spans in which economy in dead weight is a critical factor in the design. When a curved shape running right down to the ground level is not suited to the function of the building, the arch form is often used as the spanning member of a normal rigid frame so that the curve springs from a point at a reasonable distance above the floor level as in figure 191 *D*.

Spacing of main bearing members

The spacing of main bearing members, unless determined by other requirements, is fixed by the most economical combination of frames and purlins. For the common types of lattice truss and beam construction in all materials, the spacing lies between 3.70 and 7.50 m, although for very wide spans it may be economical to increase the spacing to as much as 15 m. For rigid frames of various types the spacing lies between 4.50 and 12 m. With *beam* and *frame* structures primarily subject to bending stresses as distinct from direct stresses, the wide spacing of wide span members usually has economic advantages because the cost of the smaller number of more heavily loaded members with deeper purlins is not so great as that of a greater number of more lightly loaded members which would result from a closer spacing. The reasons for this are connected with the different rates at which bending moments and deflection vary with variations in the span and in the unit loading.

Increase in the height of the columns carrying the roof structure increases the overall cost of the structure and, when columns are tall, economies can similarly be effected by increasing the spacing, since, within limits which will vary for each case, the cost of fewer more heavily loaded columns, as in the case of the spanning members, will be less than that of a greater number of more lightly loaded columns. The reason for this is that in tall columns buckling is usually the critical factor in design and not the load bearing capacity. These aspects are discussed more fully in chapter 5 on page 182.

In simple *steel truss and purlin* structures, the closer the trusses are spaced the lower will be the overall steel content of the structure. This is because variations in the spacing of the trusses has a much greater effect on the weight of the purlins than on the weight of the trusses. An increase in spacing necessitates the use of the next largest angle section for the purlin results in the increased cost of the purlins more than offsetting the saving arising from the smaller number of trusses and columns required. There is, however, a practical limit on the spacing of the trusses. This is reached when the trusses are so lightly loaded that the smallest standard steel sections are too large to permit the members to be reduced to the areas required for design purposes. This limit occurs at spacings just under 3 m. Within limits the weight of the purlins per m² of floor area is largely independent of the truss span and is nearly constant for any particular spacing of trusses. For spans ranging from 12 to 21 m a spacing of 3.80 m is often satisfactory in practice, the optimum span being about 15 m.

For maximum overall efficiency of the structure, however, particularly in the wider spans, the spacing of the main members will not be the same for all spans. When spans are large, fabrication costs of the main members increase, and for spans greater than about 21 m it is economically desirable to space the

314

trusses at wider centres, in some cases to distances at which lattice purlins can economically be used. In some circumstances where a wide column spacing is required, rather than space the trusses further apart with larger purlins, it may prove cheaper to carry a number of trusses on valley beams supported on widely spaced columns, even though the direct transfer of load from truss to column is normally the most economical method.

In some cases the use of steel decking permits purlins to be omitted if the frames are sufficiently closely spaced for this purpose. The omission of one stage in the supporting structure in this way, may reduce the overall cost of the structure. This can sometimes be done by using bearing members which are particularly suited to light loads over long span. These include castellated beams and cold formed lattice beams.

Rigidity and stability of framed structures

Rigidity against wind pressure must be provided in all forms of framed structure. In roof structures lateral rigidity may be provided by stiff joints either between the vertical members and the spanning member or between the frame and the foundations. The significance of these alternatives and the use of eaves beams have already been discussed (page 307). Wind pressure in a longitudinal direction would cause 'racking' or tilting over of the frames. This, in structures with only light purlins spanning between the trusses or frames, is resisted by bracing which may be placed either in the plane of the roof or, in the case of roof trusses, in the plane of the ties (see figure 193 *A*). This wind bracing should be placed at each end of the building and, in long structures, at intervals of 30 to 60 m (*B*). With large span trusses it is advisable to brace continuously at tie level along the length of the structure. Vertical bracing in the plane of the 'king rod' between the gable end and one or two end trusses gives increased rigidity; alternatively, vertical bracing can be provided in the wall panels. In large structures with greater spacing of trusses or frames, the purlins will be deep and may provide adequate longitudinal rigidity without separate wind bracing (*C*).

The stiff joints at the haunches of rigid frames or between roof trusses and columns makes the structure act as a whole in a lateral direction, and hinged joints at the feet result in a tendency for the whole

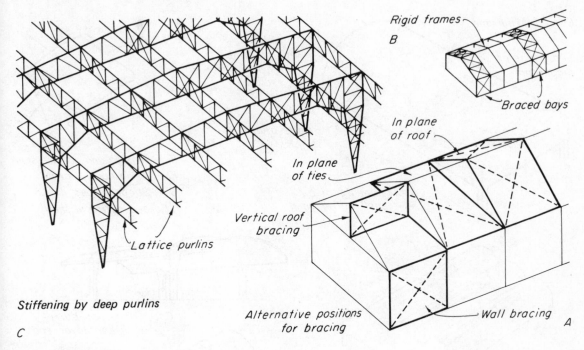

Stiffening by deep purlins

C

193   Wind bracing

structure to rotate about the leeward hinge and to lift up at the windward hinge (see figure 188 *F*). In large, comparatively light structures, this uplift may be so great that the dead weight of the structure is insufficient to anchor it down, and very large concrete foundation blocks may be required for this purpose, or tension piles may be used (see also page 184). The base must then be designed with long holding down bolts at the joint and heavy reinforcement in the foundation blocks.

The structures discussed so far are all two-dimensional structures. There are numerous forms of three-dimensional, or space structures, and these may be constructed of slabs or plates of solid material or of lattice framework, either form of which can be plane or curved in shape. They are discussed in the following pages.

Shell roofs

The term 'shell' is usually applied to three-dimensional structures constructed with a curved solid slab or membrane acting as a stressed skin, the stiffness of which is used to transfer loading to the points of support. They may be considered as single curvature shells, based on the cylindrical or parabolic form, and double curvature shells, based on the spherical and other more complicated forms. The term 'doubly curved shell' is commonly applied only to forms other than the spherical shell.

The main characteristic of a shell construction is the very thin curved membrane. This thin membrane is made structurally possible by providing restraint at the edges such that bending stresses in it are so small as to be negligible or are completely eliminated. The membrane is then subject only to direct stresses within its thickness (figure 194 *A*). Many examples of the shell form are to be seen in nature. For example, the bamboo rod, the crab shell, the bird's egg, all of which are exceedingly light but exceedingly strong. A blown egg has been known to support more than 45 kg distributed load. The efficiency of many modern shell constructions can be judged from the fact that they are relatively thinner than the shell of an egg. Long span concrete barrel vaults may be as thin as 57 to 63 mm thick for spans up to over 30 m. Short span barrels will be somewhat thicker than this and domes of 45 to 50 m span may be as thin as 90 mm. Thicknesses as

315

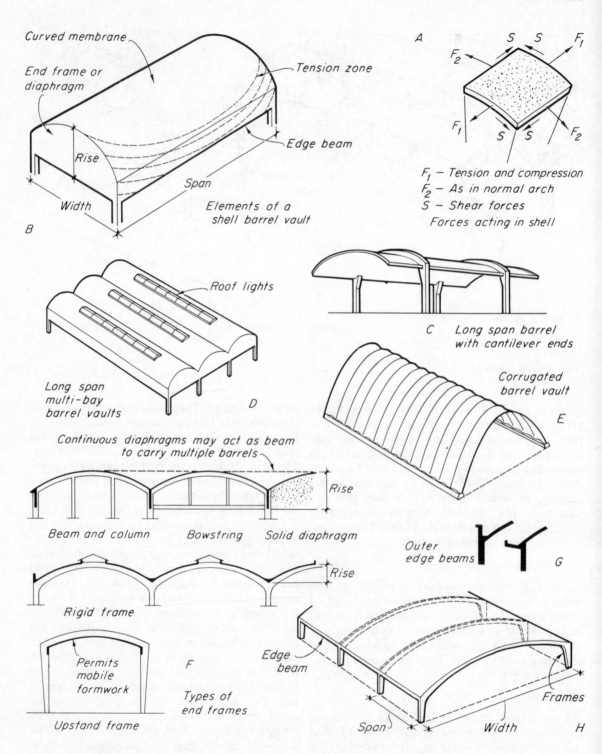

Curved membrane

End frame or diaphragm

Tension zone

Edge beam

Rise

Span

Width

Elements of a shell barrel vault

B

A

F_1 – Tension and compression
F_2 – As in normal arch
S – Shear forces

Forces acting in shell

Roof lights

Long span multi-bay barrel vaults

D

C *Long span barrel with cantilever ends*

Corrugated barrel vault

E

Continuous diaphragms may act as beam to carry multiple barrels

Rise

Beam and column   Bowstring   Solid diaphragm

Rise

Rigid frame

Outer edge beams   G

Permits mobile formwork

F

Types of end frames

Upstand frame

Edge beam

Span

Width

Frames

H

194   *Shell barrel vaults*

small as 38 mm are possible over spans of 30 m with some double curvature forms such as hyperbolic paraboloids.

In practice, for application to building work at least, the basic form of sphere or cylinder, for example, must be cut. The cut or free edges represent zones of structural weakness because direct stress can then only· be transmitted in a direction parallel to the cut edge. In order therefore to make full use of the structural properties of the remaining parts of the curved membrane, it is necessary to strengthen the edges by means of ribs or edge members which are called edge beams, or in the case of a dome, a ring beam. The latter will be subject to direct tensile stresses and, if the dome is supported on columns, to bending stresses as well. The edge beams of a cylindrical shell vault will be subject to bending stresses. In addition to edge beams, stiffening members are required at the open ends of cylindrical shells (figure 194 *B*). These end frames, as they are called, stiffen the cut end of the shell against buckling, the maintenance of the shape of the shell being essential in order to develop the membrane stresses within it.

Single curvature shells

These are barrel vaults of which there are two forms: long- and short-span barrels, typical examples of which are shown in figure 194 *D, H* respectively.

Long-span barrel vaults act primarily as a 'beam', the span of which is the length of the vault. The shell constitutes the compression member and the edge beams the tension members or flanges. Although the direct stresses in a shell are mostly compressive, shear forces are set up near the supports which give rise to diagonal tension (figure 194 *B*). In the case of reinforced concrete shells, this must be resisted by reinforcement placed at 45 degrees across the corners of the shell as shown in figure 217 *A*. The width of the barrel should be one-half to one-fifth of the span and the rise from the underside of the edge beams to the crown of the vault about one-tenth of the span for single span, and one-fifteenth for continuous span vaults. The depth of the edge beams is usually about half of the total rise, but in the smaller spans this may sometimes be reduced. In multi-bay buildings, the edge beam may be eliminated altogether, provided the necessary rise is obtained by increased curvature of the shell membrane. In these

circumstances the fold in the shell constitutes a beam as shown in the rigid end frame example in figure 194 *F*. In the case of a concrete shell it is a disadvantage to have an excessive rise to the shell as this prevents easy placing of the concrete. Various ways of forming edge beams and end frames are shown at *F* and *G*.

The width of a long-span barrel is usually not more than 12 m with a maximum practicable width of 15 m. The maximum economic span is about 30 to 45 m. When the end frames are as far apart as this, equilibrium without bending moment in the shell is generally not possible, but where such a span is necessary, a satisfactory solution can be obtained by a suitable choice of rise to span and depth of edge beams. In order to avoid excessive rise of the structure, the radius of curvature of the shell increases as the span and, therefore, the width increases. For small spans a radius of about 6 to 7.5 m is used, for spans from 15 to 30 m a radius of 9 m, and for spans over 30 m a radius of 12 m. Lighting openings may be formed in the crown of a shell vault provided they are kept clear of the ends and are not more in width than about one-fifth the width of the barrel in reinforced concrete shells and about one-third in timber shells. Circular lighting openings in reinforced concrete shells can be formed in various parts of the shell, but should not exceed about 1.20 m in diameter and must be kept clear of the bottom edges and the ends, particularly the corners, of the shell. The edges of all openings must be strengthened with edge ribs and in the case of a long opening in the crown of the vault, cross ribs at intervals along the opening are required.

Since a shell-barrel vault acts as a beam along its length, it is possible to cantilever the structure beyond the end frames. The external edge of the curved shell will usually require stiffening with a rib unless the projection is quite small (figure 194 *C*).

By prestressing the tension zones of a long span barrel vault (figure 217 *D*)·the rise may be reduced to one twentieth of the span.

Short-span barrel vaults are used when the clear span is beyond the practicable and economic limits of a long-span barrel indicated above, or where the interruption of roof space by the valleys of a succession of narrow long-span shells would be a disadvantage (compare (*D*), (*H*) in figure 194).

End frames in the form of arch ribs, rigid frames or bow-string frames, are usually spaced 9 to 12 m

317

apart, sometimes up to 18 m. The depth of the edge beams should be about one-fifteenth of the span. The total rise should not be less than one-tenth of the span or the chord width, whichever is the greater. Because the spans of the shells are generally less the edge beams are shallower than in long-span barrels. The width of short-span barrels is often great and the radius of curvature is therefore large. With shells of over 12 m radius it is usually necessary to prevent buckling of the curved membrane by introducing stiffening ribs running across the width of the vault at 3 to 6 m centres. These can be placed either above or below the curved shell and need not necessarily continue down to the edge beams, since buckling will occur in the upper compression zone of

the shell. Due to the interaction of the shell and the stiffeners, bending stresses often occur, and because of the thinness of the shell they may be very high. Provision for these must be made either by a local increase in the thickness of the shell or by special reinforcement.

North light and cantilever barrel vaults   Many variations of the simple barrel vault are possible, especially in reinforced concrete, and examples of these are shown in figure 195. These include asymmetrical forms such as north lights and cantilever shells (*A, B*) and double-cantilever shells (*C, D*). The total vertical rise of these should be not less than one-eighth of the span and the rise of the arc should be at least one-twentyfifth of the span. Edge beams

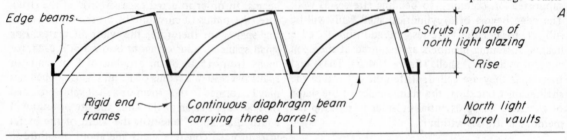

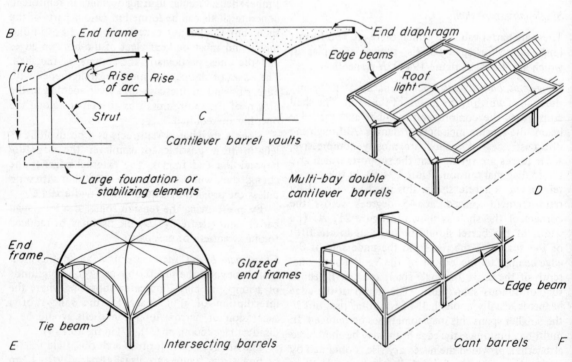

195   *Shell barrel vaults*

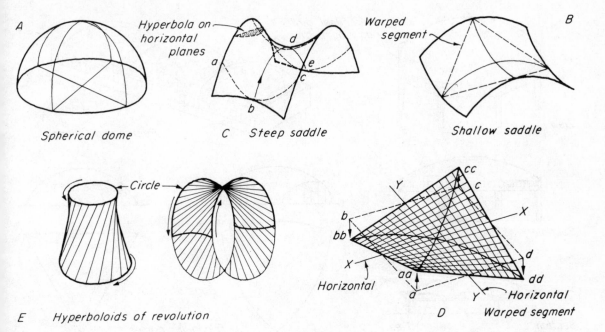

A   Spherical dome

C   Steep saddle

Hyperbola on horizontal planes

Warped segment

B   Shallow saddle

Circle

E   Hyperboloids of revolution

D   Warped segment

Horizontal

196   Double curvature shells

will normally need to be not less than one-eighteenth of the span in depth, but this may vary according to conditions. When two shells meet at a considerable angle, as at the middle of a double-cantilever shell, the edge beam may be omitted as indicated above, since the fold in the shell constitutes a beam. The upper edge beam of a north light shell is usually very small as it is normally supported by struts at 1.80 to 2.75 m centres in the plane of the north light glazing. These struts bear on the side of the valley gutter which is therefore designed as a deep 'L' edge beam to the bottom of the adjacent shell (A and figure 217 C). In cantilevered shells the end frames or diaphragms carry the whole of the roof load and their depth must therefore be substantial. In the case of a single-cantilever shell the foundations must be designed to prevent overturning, or struts or ties must be provided to ensure stability (B). End frames in all forms of shell may be placed either above or below the curved slab. Cylindrical vaults may intersect to form cross vaults or be 'canted' as shown at (E), (F). This permits glazed lights to be formed between the stiffening frames of adjacent barrels.

Barrel vaults may also be constructed as parabolic or elliptical shaped shells springing direct from the

foundations and stiffened with ribs at intervals. Alternatively, in concrete a corrugated shell can be used in which the stiffening effect of the ribs is provided by corrugations in the surface of the shell which, when repeated, forms a continuous corrugated surface as shown in figure 194 E.

Double curvature shells

Double curvature adds to the stiffness of a stressed membrane and is not limited to the normal rotational dome. Geometrical surfaces with double curvature are divided into two main groups:

(i)   Those in which the curvature is in the same direction in sections cut at right-angles, that is to say, either both concave or both convex, as in a sphere

(ii)   Those in which the curvature is opposite in sections cut at right-angles, as in hyperbolic paraboloids and hyperboloids of revolution. In these the surface appearance resembles a saddle.

Shell domes are included in the first group and the second group are commonly called doubly-curved slabs or shells (see figure 196 A, B).

Shell domes   The simplest form is the spherical dome which has been constructed in concrete over

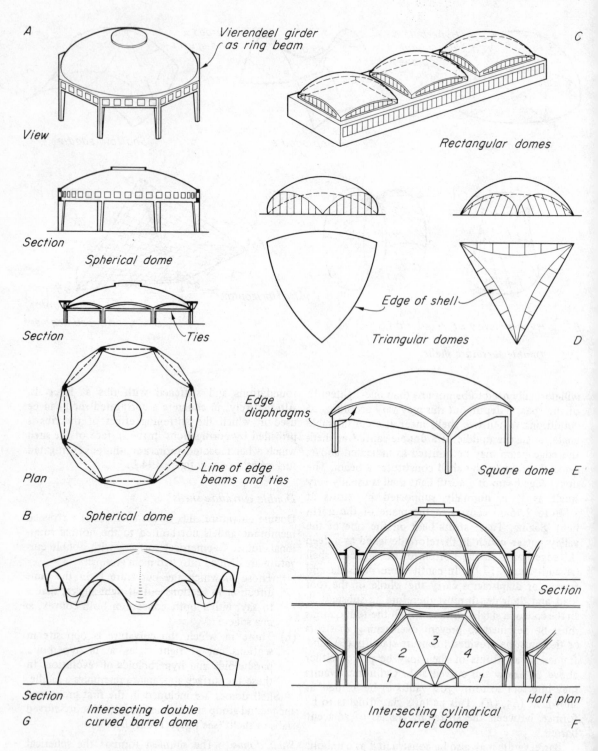

A

Vierendeel girder
as ring beam

View

Section

Spherical dome

Section   Ties

Plan

Edge
diaphragms

Line of edge
beams and ties

B   Spherical dome

Section

Section
G   Intersecting double
curved barrel dome

C

Rectangular domes

Edge of shell

Triangular domes   *D*

Square dome   *E*

Section

Half plan

Intersecting cylindrical
barrel dome   *F*

197   *Shell domes*

320

spans of 45 m or more. The shape may vary according to the plan shape to which the edges are cut. All cut edges must be stiffened with edge beams. To avoid horizontal thrust in circular shell domes it is necessary to construct the shell of approximately elliptical cross section with a rise of about one-sixth of the span, otherwise a ring beam or ties must be provided at the base of the dome to take up the thrust (figure 197 *A, B*). In the case of domes square or triangular shaped on plan (*C, D, E*), the edge beams must be designed to take up this thrust and to transfer it to the bearings.

The ring and edge beams serve also to resist stresses set up by temperature changes in the shell, which may sometimes need thickening near the beams and also round any openings formed in the shell.

Early concrete shell domes were formed by the intersection of a number of cylindrical shells resulting in a polygonal form. The finest example of this is probably the Market Hall at Leipzig, covered by three octagonal domes each 75.50 m span, with a shell thickness of 90 mm (*F*). Each dome is formed by the intersection of four cylindrical shells, the ridges at the intersections replacing the rigid frames in a normal barrel vault. (*G*) shows intersecting double-curvature barrels which result in a 'dome' form. Double-curvature barrels may also be used either as a means of providing increased stiffness to the shells in short-span vaults or in order to obtain curved edge beams in long-span vaults when prestressing is to be applied (see figure 217 *D*).

Doubly curved shells Of this group of three-dimensional geometrical surfaces that most used is the hyperbolic paraboloid, or saddle shaped form, so named because, when cut, some sections reveal hyperbolas and others parabolas as shown in figure 196 C. It is formed by moving the vertical parabola *abc* through the parabolic trajectory *bde*. This type of surface has greater resistance to buckling than dome forms because of its shape. In practice part of a hyperbolic paraboloid is most frequently used, in the form of a warped segment or parallelogram although the saddle-shape and conoidal forms are appropriate in some circumstances. The relationship of a warped segment to the saddle-shape is indicated at (*B*).

The basic construction of a warped segment or parallelogram is illustrated at (*D*). Points *a* and *c* of the horizontal plane *a b c d* are raised to a new higher position *aa, cc* and points *b* and *d* are depressed to *bb, dd*. The sides are divided into an equal number of sections and the corresponding opposite points are joined. The resulting net of straight lines defines part of a hyperbolic paraboloid surface. The characteristic feature of this surface is that although all cross sections cut parallel to the edges are straight lines, cross sections parallel to the diagonals are parabolas. That on the diagonal running through the raised corners, and all sections parallel to it, being concave upwards, and that on the diagonal running through the lower corners being convex upwards. That is to say, the upper surface of the segment curves in concave form between the higher points and in convex form between the lower points.

In spite of its complicated shape the stresses in a paraboloid shell can be more easily analysed than in most other surfaces. The shell may be considered as a series of arches and suspension cables intersecting each other at right-angles; the shell is in direct compression in directions parallel to the convex or 'arched' section and in direct tension in directions parallel to the concave or 'cabled' section (figure 198 *C*). Because of the great stiffness given by the double curvature, and as all the stresses are direct, the shell may be exceedingly thin. For spans in the region of 30 to 40 m a reinforced concrete shell will be from 38 to 50 mm thick. The forces exerted on the edges by the 'arches' and 'cables' are the same, and since they act also at equal angles to the edge but in opposite directions, they resolve into shear forces along the edge with no component perpendicular to the edge to cause bending stresses. Since the principal stresses are equal, the shell itself is in a state of uniform shear and, as the stress is the same over the whole surface, the practical design in terms of reinforcement in a concrete shell or the boarding in a timber shell is simplified.

Edge beams are required to carry the edge shear forces. The depth of these should be disposed equally above and below the shell membrane. They may be placed wholly above or below the shell but the cross section of the beam will need to be larger. The overall rise of the shell, that is the difference in height between the low and high corners, is important. If this is small the shell will be shallow and if too shallow it will tend to buckle. The ratio of the rise to the diagonal span is, therefore, a significant factor and this should be as large as possible and never less than one-fifteenth.

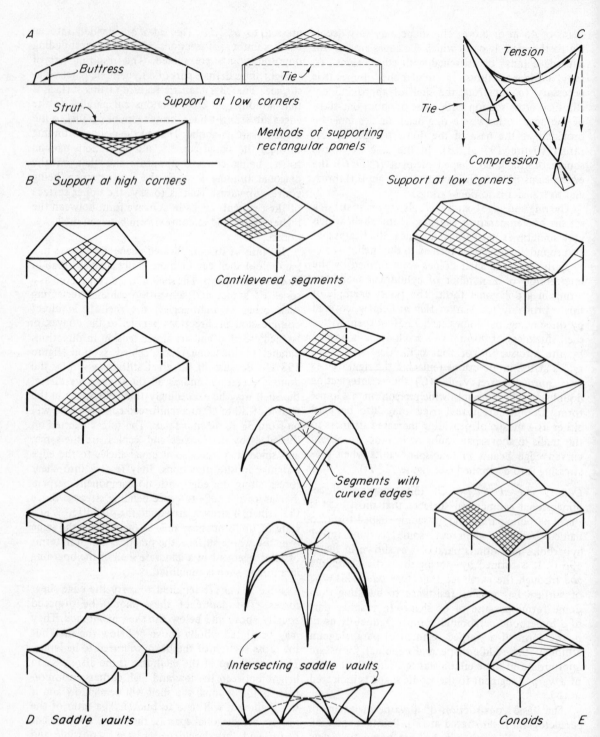

A

Buttress

Tie

Support at low corners

Strut

Methods of supporting rectangular panels

B   Support at high corners

C

Tension

Tie

Compression

Support at low corners

Cantilevered segments

Segments with curved edges

Intersecting saddle vaults

D   Saddle vaults

Conoids   *E*

198   *Doubly curved shells*

Doubly-curved panels may be used singly or in combination and the number of supports required will depend upon the way in which the panels are related to each other. In the case of a single rectangular panel only two supports are required. If the roof is supported at the two lower corners the edge beams will be in compression and a horizontal outward thrust will be exerted on the supports. This thrust can be resisted either by a heavy buttressed support or by a tie rod between the lower corners of the shell (figure 198 *A, C*). If the panel is supported at the two high points the edge beams will be in tension and an inward pull will be exerted on the supports. This must be resisted by a strut running between the high points (*B*). The tie is the cheapest of these three methods but the reduction in clear headroom it causes and the possibility that it may spoil the interior appearance are disadvantages. The buttressed supports may be costly if the columns are high because of the high bending stresses which will be set up by the outward thrust of the shell. Some provision must be made for tying down the unsupported corners against wind forces, particularly if the sides of the building are not enclosed (*C*). When the building is enclosed at the sides the edge beams are fastened to the heads of the enclosing walls.

Some combinations of warped segments are illustrated in figure 198. It will be seen in some cases that the edges of some panels lie alongside those of adjacent panels, so that the forces in each balance each other. In certain combinations of panels the inclined edge beams of adjacent panels can be integrated with the corner supports to form a rigid frame in order to obviate the use of a tie. The bounding edges of warped segments need not necessarily be straight, but may be curved in parabolic form. An example of this is shown. The development of the conoidal form, using both straight and curved edges is shown at (*E*) and the use of the full saddle shape at (*D*).

The hyperboloid of revolution referred to earlier is based on the circle and its development is illustrated in figure 196 *E*.

Shells can be constructed in reinforced concrete, timber and occasionally in steel plate.

Folded slab roofs

This form of construction is also called folded plate and, when there are a large number of facets, prismatic structure. It is another form of stressed skin or membrane structure in which the stiffness of the skin is used to distribute the loading to the points of support.

If a flat slab is folded or bent it can behave as a beam spanning in the direction of the fold with a depth equal to the rise of the folded slab (figure 199 *A*). When loaded, compression and tension stresses will be set up at the top and bottom of the section respectively and shear stresses in the slabs on each side of the fold. Each slab spans between the folds and must be thick enough to span this distance and to have sufficient stiffness to distribute the loads longitudinally. End frames or diaphragms must be provided to collect the forces in the slabs and transfer them to the supports (*A, C*). The shape of the roof may vary from a simple pitched roof of two slabs to a multi-fold or prismatic form involving several plates, some examples of which are given in figure 199.

The span and width of each bay governs the overall depth. As an approximate guide this should not be less than between one-tenth and one-fifteenth of the span, or one-tenth of the width, whichever is greater. The width of each slab is limited only by the requirements of adequate lateral stiffness which dictate the thickness. In practice, it is often cheaper to use a large number of narrow slabs rather than a few wider slabs. This is because, although a greater number of folds must be formed, thinner slabs can be used and the amount of material and the dead weight of the structure is less. When 'barrels' are formed of a large number of 'facets' or slabs (*D*), the influence of the rigidity of the folds will be proportionately greater than when fewer and, therefore, wider slabs are employed. One-twentieth to one-twentyfifth has been given as a reasonable value for the thickness/slope length ratio for the slab element. The slab thickness will generally be thicker than that in comparable cylindrical shells for spans in excess of about 4.5 m. Folded slab construction is competitive with these shells, however, because the flat shuttering is comparatively inexpensive. It is possible to use composite construction using repetitive precast concrete slab elements as permanent shuttering to an *in situ* structural topping (see figure 220).

There are various methods of retaining the folds in position at the points of support where the loads are taken down to the foundations. A solid diaphragm beam, a lattice truss or a rigid frame may be used, to which the ends of the slabs are rigidly secured.

323

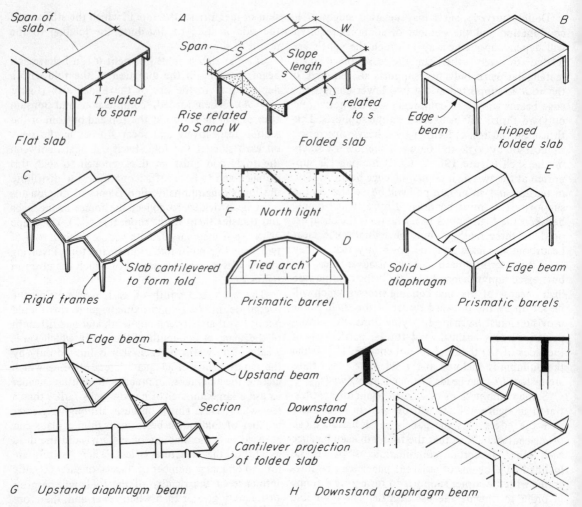

199   Folded slabs

When vertical supports down to the ground can be permitted, columns may be placed under each fold carrying a diaphragm beam which follows the shape of the folded slabs (figure 199 *A*, *G*). Diaphragms may be placed above or below the slabs as required (*G*, *H*). The diaphragm need not necessarily be vertical in order to fulfil its function: it may be sloped to form a hipped end, the angle folds then serving to transfer the loads to the supports (*B*). Wide span diaphragms may carry a number of folded elements when wide column spacing is necessary (*E*, *F*, *H*).

The free edges of folded slabs should be stiffened. They must either be supported along their length, be provided with edge beams or be cantilevered a short distance to form a fold (*B*, *G*, *C*).

The slabs may be perforated where light is required through the roof and, provided sufficient slab is left on each side of the perforations and adjacent to the folds, the slab can be reduced to a series of struts. It becomes then, in effect, a Vierendeel girder, the top and bottom flanges of which are the folds in the slabs. Folded slabs can be used over continuous spans, with depth/span ratios sometimes as low as one-fortieth, and as cantilevers (*G*, *H*).

This form of construction can be used to extend down to the ground in arch or rigid frame form as well as for simple span roofs supported on columns or walls (figure 200 *A*). A great variety of complicated and interesting forms can be obtained by the

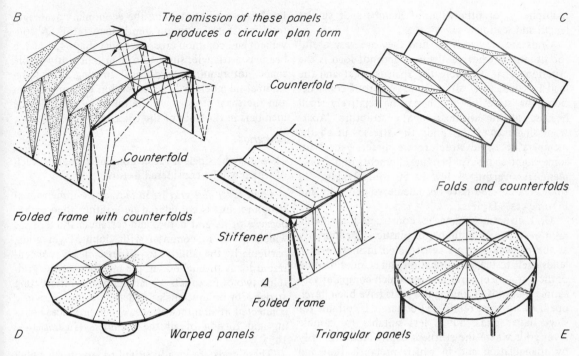

B

The omission of these panels produces a circular plan form

C

Counterfold

Counterfold

Folds and counterfolds

Folded frame with counterfolds

Stiffener

A

Folded frame

D

Warped panels

Triangular panels

E

200   *Folded slabs*

introduction of reverse or counterfolds at various points (*B, C*). Dome and vault forms can also be constructed. Circular shapes can be covered with folded slabs developed to form a horizontal roof structure as distinct from the dome form. This necessitates variation in the slab widths, the introduction of counter-folds, the use of triangular shaped slabs or the warping or twisting of the slabs, which then become hyperbolic paraboloids. Either the centre or perimeter supports may be omitted and the structure then becomes cantilever in form with tension and compression rings replacing the stiffening diaphragms. Two examples are shown at (*D*) and (*E*).

Folded slab structures derive most of their strength from their shape. The folded form gives great rigidity and makes possible the efficient use of material of high elasticity such as plastics and aluminium alloys, the use of which in other forms of construction often results in excessive deflections. Concrete, timber and plastics may be used in this form of construction.

Grid structures

Grid structures, apart from single layer flat grids, are three-dimensional or 'space structures'. Unlike shells or folded slabs, however, they are constructed not with solid membranes but with lattice or grid frameworks. In some systems, however, the grid is partially formed by the folds and edge junctions of bent or folded sheet panels of suitable material in which the skin strength of the sheet element forms a very large proportion of the total strength of the structure.

These structures usually provide a simple and economic method of covering very large areas without internal intermediate support. As they permit the prefabrication and standardization of the component parts, and as the dead weight is small compared with many other forms of structure, there are considerable savings in construction costs, the savings increasing with the increase in span. The economy of these structures as far as prefabrication and standardization is concerned is related to the simplicity, or otherwise, of the jointing technique and the

325

multiplicity, or otherwise, of members of similar length and section.

Grids are very stiff so that they are very useful for situations where deflection and not load is the criterion. That is, where the span is great and the load is small, as is often the case in roof construction. The structural depth can be relatively small because of the stiffness of the structure. Apart from single layer flat grids the stresses in all the members of a grid structure are direct, except for some slight transverse bending moments in diagonal members. Weights as low as 50 to 150 kg/m² of floor area covered have been achieved generally and in some cases far less.

Grid structures may be constructed in metal, reinforced concrete, timber or plastics. The majority at the present time are constructed in metal which lends itself to the prefabrication and standardization of the component parts, and for which comparatively simple methods of joining the parts have been developed. *In situ* cast reinforced concrete is suitable for single layer grids, but is less suitable for double layer grids where the stiffness is provided primarily by triangulation and in which prefabrication and standardization of the parts is logical. Although this can be done with concrete in the form of precast elements post-tensioned together, it is not an ideal medium for this purpose. Plastics may be incorporated in this type of structure in the form of three-dimensional folded or curved elements.

Grids can be applied in many arrangements to flat, curved or folded roofs. Their application to floors in the form of rectangular grid and diagonal beam construction has already been mentioned in chapter 6. The application to roofs can be broadly classified as follows:

(1) Space frames
(2) Flat grids
(3) Folded grids
(4) Folded lattice plates
(5) Braced barrel vaults
(6) Braced domes

Space frames

Although all the structures under consideration, apart from single-layer flat grids, are in fact space frameworks, the term 'space frame' is usually applied to a hollow section or three-dimensional lattice beam (figure 201 *A* and figures 221 and 222). The hollow shape confers great lateral rigidity on the beam whilst retaining the economic advantages resulting from optimum depth/span ratios. A convenient and common cross-sectional shape for such a beam is a triangle, since this is an inherently stable shape not requiring additional bracing. The longitudinal members will be in tension or compression and the shear stresses will be taken by the diagonal members in the sides of the triangle.

Flat grids

These may be single-layer or double-layer grids, each of which will be considered in turn.

A single-layer flat grid is, in fact, a two-dimensional structure, but is considered here rather than earlier because of its grid nature and certain characteristics which it has in common with double-layer grids, particularly the ability to disperse heavy concentrated loads throughout all the members of the grid. It has two or more sets of parallel beams intersecting at right- or oblique-angles and the beams are rigidly connected at all intersections, which produces bending and torsion of all the members (figures 144, 145).

These grids are ideally suited to structures which are to carry heavy concentrated loads. Because of the interconnection of the parts, a concentrated load is is distributed between all the members of the grid, decreasing the high stresses in the directly loaded area. The stress distribution in grid frameworks under heavy concentrated loads is therefore comparatively even.

The interconnected beams may be arranged in various layouts and the boundaries of the grid may be rectangular or circular. The rectangular grid, although widely used, is not the most efficient in terms of stress distribution since there are no members in the corners and these are in fact the most highly stressed zones. In theory, the beams could be arranged in such a way as to follow the trajectories of the principal stresses in the slab. Thus they would be where they are most needed. As the beams would have to be curved, such a layout is not likely to be economic, particularly as far as reinforced concrete is concerned, although this has been done using 'ferro-cement' pans[1] as shuttering. A close approx-

[1] These are thin slabs 19 to 38 mm thick, of cement mortar reinforced with superimposed layers of wire mesh and small bars. They are strong and light in weight and can also be used as permanent shuttering. See *The Structural Engineer*, May 1956, paper by Pier Luigi Nervi, 'Concrete and Structural Form'. See also page 354.

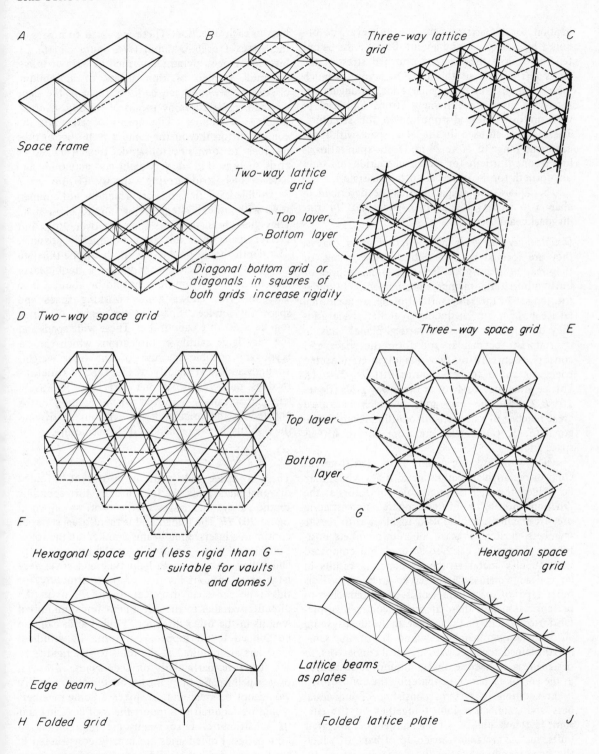

A

Space frame

B

Two-way lattice
grid

Three-way lattice
grid

C

D Two-way space grid

Top layer
Bottom layer

Diagonal bottom grid or
diagonals in squares of
both grids increase rigidity

Three-way space grid   E

F

Hexagonal space grid (less rigid than G —
suitable for vaults
and domes)

Top layer

Bottom
layer

G

Hexagonal space
grid

H Folded grid

Edge beam

Lattice beams
as plates

Folded lattice plate   J

201   Grid structures

imation to this arrangement can, however, be obtained by a diagonal grid layout. Because the beams tend to follow the lines of principal stresses, the stress distribution in this type of layout is much more even than in rectangular grids. Triangular or three-way grids are extremely strong and lead to very uniform stress distribution in the structure. These layouts are used for the larger spans within the economic range of 15 to 24 m. Depth/span ratios as low as one-thirtieth for rectangular grids and only one-fortieth for the others are often possible.

The square and diagonal grids are described in chapter 6 under 'Rectangular grid' and '*In situ* diagonal beam' floors.

Double-layer flat grids may be lattice grids, that is, they are formed by intersecting lattice beams, or they may be space grids. In lattice grids, each set of bottom horizontal members lies immediately under the top set in the same vertical plane, as in normal lattice beams, but in double-layer flat space grids they do not lie in the same vertical plane. Thus, in its simplest rectangular grid form the space grid consists essentially of two sets of interconnected triangular space frames (figures 201 *D*, 224 *A*). Lattice grids may be two- or three-way grids (figure 201 *B*, *C*), and the space grids, of which there are a great number of variations, in addition can be hexagonal (*E, F, G*). Lattice grids are not so stiff as space grids.

These structures are usually fabricated from circular or rectangular section metal tubes which may be welded together or joined by connectors at the junctions, although reinforced concrete precast compressive members can be used together with tensile members in steel. Precast concrete members posttensioned together can also be used, but compared with tubular metal structures concrete results in large dead weights. Double-layer grids, as well as other types of space frameworks, lend themselves to prefabrication and there are many commercial prefabricated systems on the market. In the ideal system, all the framing members would be of the same length, joined together with identical connectors, so designed as to require the minimum of fastenings at the end of each member entering the connector.

In addition to systems consisting of individual bars and connectors joined together on the site, some systems consist of individual prefabricated units, such as pyramids, made up of bars, of which the edge members of the flat bases are joined together on the site and the apices tied together by tie

bars in each direction. These are fixed to bosses on the apex of each pyramid (see figure 224 *B*, *C*). Instead of open pyramids formed of bars or tubes, pyramids formed of thin sheets of aluminium, plastics or plywood can be used, fixed in the same way along the edges by means of bolting, riveting, welding or gluing. The apices of all units are connected together in the same way by bar or tube members to form a bottom grid. Stressed skin space grids such as this are very light and economic and have a high load-carrying capacity (figure 225).

Double-layer flat grids are generally not competitive with other systems below about 21 m span. They have been used for spans of over 90 m but greater spans than this are economically possible. The depth should be one-twentieth to one-thirtieth of the span so that the depth of a double-layer grid for spans around 30 m would be about 1.0 to 1.50 m. This allows ample working space and space for services. Spans in the region of 90 m require a depth of about 3 m. These wide spans and the very large cantilever projections which can be used, are possible because of the light weight, strength and great rigidity of this type of structure. It does not collapse if one part fails. Relatively large areas can be removed when required or be omitted in the original design without destroying the stability of the remainder of the structure.

Folded grids

These are space grids in the form of bent or folded diagonal plane grids, the folds usually corresponding to the valleys and ridges of the roof, as shown in figure 201 *H*. The folded grid is therefore a series of continuous intersecting beams cranked at the folds. Folded grid roofs are usually of the multiple ridge and valley type or north light type and cover very large areas without internal support. Structures of this type can span distances up to 90 m in the direction parallel to the ridges and almost unlimited lengths in the other direction. Folded grids can also be applied to the hipped roof form. Longitudinal members are required at the folds to give rigidity to the structure, particularly on wide spans. Although over small spans they might be omitted, rigidity would not be great and, in any case, some members would be required to support the roof covering and its substructure. Edge beams are required at all boundaries. Folded grids are usually constructed in steel using ordinary rolled steel sections for the diagonal members and channels for the edge beams.

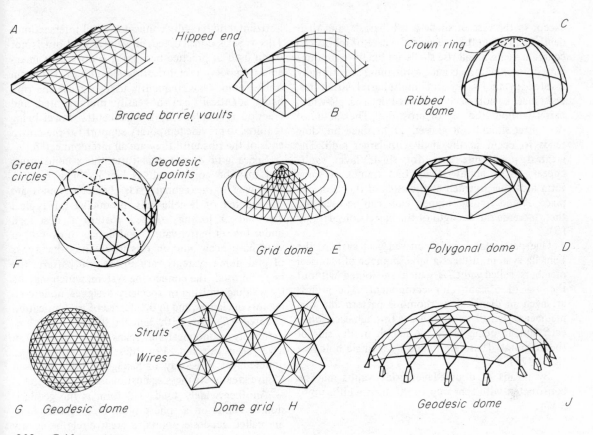

A

Hipped end

Braced barrel vaults

B

C

Crown ring

Ribbed dome

Great circles

Geodesic points

Grid dome   E

Polygonal dome   D

F

Struts

Wires

G   *Geodesic dome*

Dome grid   H

Geodesic dome   J

202   Grid structures

Folded lattice plates

In principle, this is the same as folded slab construction in reinforced concrete or timber, but the planes or plates are constructed as lattice beams, usually of steel, although occasionally of timber. The adjacent edges are interconnected at the folds, usually at intervals at the node points (figure 201 *J*). The roof profiles to which it can be applied and all other considerations are the same as for folded slabs. Folded lattice plate structures in steel can span greater distances than folded slabs in reinforced concrete on account of their lighter weight.

Braced barrel vaults

Although similar in form to reinforced concrete shells, braced barrel vaults, being an assembly of bars, are non-homogeneous (figure 202 *A, B*). The members can be arranged to follow directly the lines of maximum stresses. In practice, several types

of bracing are used each of which results in a different type of structural behaviour. The shape of the barrel is of great importance in relation to the stress distribution and although the cylindrical form is not the best shape, it is very useful from a practical point of view since prefabrication is facilitated by the fact that the slope of all the members is the same, the length of the bars can be the same and identical connectors can be used at all joints. In some cases the barrel is formed of curved members but more frequently the 'curves' are made up of short straight members.

The rise of the barrel should be between one-eighth and one-twelfth of the span. These figures relate to multi-barrel structures and the lower figure is only suitable for continuous spans or barrels which cantilever considerably at the ends. Greater rises should be used for single-width barrels, whether of single span or continuous span form. Where these rises cannot be obtained, edge beams may be used,

329

except in the case of single-width barrels, and their depth included in the rise as in the case of reinforced concrete shells. Small rise shells are likely to produce large horizontal thrusts and, while these are mutually resisted in the valleys of multi-barrel structures, resistance is not easily provided in a single-width barrel without the use of cross-ties. The radius of the barrel should not exceed 11 m since buckling tends to occur in the shell with larger radii. The greatest practicable span for single layer vaults appears to be about 36 m. End frames of some form must be provided at the ends of the vaults. In place of end frames, hipped ends can be used and these increase the strength of the barrel considerably (*B*).

One particular form of curved grid system, the Lamella system, utilizes a large number of identical members called *lamellas* which are joined without the use of a connector component. The grid is arranged in diamond or rhombus pattern and one member or lamella is continuous through each joint, each lamella being twice the length of the side of the diamond. These are joined by a single bolt (see figure 227).

By means of double-layer grids, vaults may be constructed with spans up to 90 m or widths up to 30 m.

Braced domes

These are constructed either with curved members lying on a surface of revolution, or of straight members with their connecting points lying on such a surface. Some forms consist of ribs running from the base to the crown or of inclined bars, also running from base to crown, but connected by a number of horizontal polygonal rings. Other types are constructed as space grids similar to those used for barrel vaults.

The first group includes ribbed domes, consisting of meridional ribs connected only at the crown or at a crown compression ring when there is a top opening (figure 202 *C* and figure 228). The other types in this group are polygonal in form and have horizontal polygonal rings and diagonal stiffening members and differ in the number of diagonals, the relationship of the polygonal rings to each other in terms of rotational position and in the number of sides making up the dome (*D*). In addition to these there is the stiffly jointed 'framed' dome which consists of continuous meridional ribs and polygonal

330

horizontal rings rigidly connected at all intersections. This type is constructed in welded steel but is not often used in practice because it is not easily amenable to modern prefabrication techniques. Three-pin ribbed domes are frequently used because the ribs, being identical, can be easily prefabricated and erection is simple, only a small central tower being required to provide temporary support for the crown ends of the ribs until they are all interconnected.

Space grid domes constitute the second group and may be constructed as lamella or as two- or three-way or hexagonal grids. Lamella domes are constructed of lamella ribs, producing the typical diamond or lozenge shaped pattern as described under braced barrel vaults, and domes of over 90 m span have been built in this way. The other types of grid dome systems vary in the grid pattern, the material used, the connecting system, which may be by welding, bolting or specially designed node connectors or clamps, and in the degree of prefabrication adopted (*E*). As with lamella domes, they are extremely economical and spans of well over 90 m have been covered. For these very large spans, double-layer three-way or hexagonal grids are used to give greater stiffness against buckling.

Another widely used grid form is the geodesic dome. Points on a sphere lying on a 'great circle' are called geodesic points, a great circle being any circle running round the surface of the sphere having the same radius as that of the sphere itself. Lines joining such points are called geodesic lines (see figure 202*F*). Since a sphere encloses the maximum volume with minimum surface area, a true geodesic dome would enclose the maximum volume with the minimum amount of structural material. This necessitates the use of curved members lying on great circles (*G*), but surface coverage is usually achieved by repetitive straight-edged patterns, using triangles and hexagons, in which the nodes only are geodesic, these being connected by straight members which are not geodesic (*H*, *J*). As with any other form of braced structure geodesic domes may be single-layer or double-layer, the latter being extremely rigid and suitable for very large spans. They may also be of stressed skin construction in which the covering acts as an integral part of the structural system, or of what may be termed formed surface construction, in which flat sheets of suitable material are bent and interconnected along their edges to form the main structural grid of the dome (figure 229).

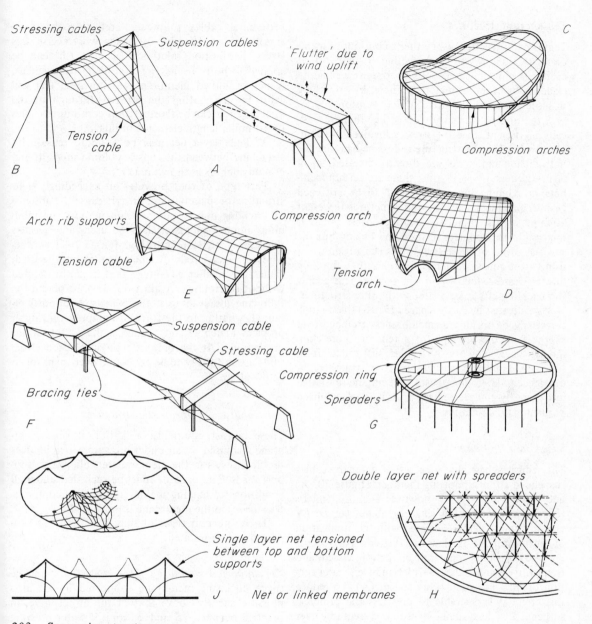

Stressing cables

Suspension cables

'Flutter' due to wind uplift

C

Tension cable

Compression arches

B

A

Arch rib supports

Compression arch

Tension cable

Tension arch

E

D

Suspension cable

Stressing cable

Compression ring

Spreaders

Bracing ties

F

G

Double layer net with spreaders

Single layer net tensioned between top and bottom supports

J   Net or linked membranes   H

203   Suspension structures

A pin-jointed triangular frame is rigid and this shape is therefore used, or hexagonal grids are broken down into triangles. It is impossible to cover a complete sphere entirely with hexagons, a certain number of pentagons being essential. Apart from exceptionally simple layouts, the members will be of different lengths and, further, a level alignment of the members at the base of such domes is not possible. This needs careful architectural consideration.

Double layer geodesic domes have been built over spans of more than 116 m.

Tension roof structures

Tension construction uses either a continuous membrane as structure and roof covering or a network of cables or pin-connected links to support a separate cladding or covering material. These, by stretching or suspending, are put into a state of tension.

Structures which have to resist bending are basically inefficient since the stresses involved are complex. Even when the structure is designed so that the individual components are directly stressed, as in lattice girders or space grids, there are certain members in compression which need to be stiffened against buckling. Shell or arch structures represent attempts to use materials in direct stress but it is impossible to eliminate all bending in any but the smallest forms of these structures. The advantage of using a tension system for a roof is that the only direct stresses which occur are tensile. This avoids the buckling effect associated with other structures.

As indicated in Part 1 (pages 17, 18) these structures may be formed as membranes stretched over supports on the principle of a tent or, more commonly, as a network of cables suspended from supports, known as suspension roof structures. A more recent development uses a membrane tensioned into shape by compressed air, known as air stabilized or pneumatic structures.

Suspension structure

This consists basically of a set of catenary cables suspended from supporting members (figure 203 *A*) and secured by guys as shown or anchor elements (*F*) or by compression rings or arches (*C*, *D*, *E*). One of the problems associated with the application of this technique is that of 'flutter'. This occurs because the roof is comparatively light in weight and frequently assumes shapes which induce greater suction or 'lift' at certain wind speeds (*A*). To reduce this effect it is desirable to prestress the principal suspended cable system by another system of cables curved in the opposite direction. In principle there are three methods: (i) a doubly-curved system of cables in which the main cables are prestressed by another set at right-angles (*B,E*), (ii) bracing ties which tension the two cables in concave form (*F*), (iii) spreaders which tension the two cables in convex form (*G*). The effect of suction will be to reduce the tension in the principal cables following the catenary paths and to increase the tension in the

restraining cables following 'arched' paths. Conversely, the application of loads will reverse this stress distribution. Methods (ii) and (iii) can be applied to both 'beam' (*F*) or grid (*H*) arrangement. A net or linked membrane as shown in (*H*) will assist in distributing the stresses throughout the system. These can be fabricated of continuous cables or of similar length elements or links.

A single layer net may be put into tension by stretching between the tops of column supports and ground anchors as shown in (*J*).

This type of roof provides an exceedingly light structure for spanning large areas because the absence of buckling due to compression permits the minimum amount of material to be used, and because the use of high-strength, high-tensile steel permits wires or linked rods of small diameter to be used. It has the further advantage that it can be erected without scaffolding. Against this must be placed the following disadvantages: the expensive supports required, whether in the form of compression ring or props and anchors and, secondly, the complications in cladding. At the present time suspension structures are not likely to be economic over spans much below 60 m.

Air stabilised or pneumatic structures

These consist essentially of a thin flexible membrane stabilised by air under pressure which induces tensile stresses in the membrane enabling it to support loads. The pressure must be such that under all conditions of loading the development of compressive stresses in the membrane is prevented.

There are two types of such structures: air supported and air inflated.

Air supported structure This consists of a single space enclosing membrane supported by air at a pressure slightly above that of the atmosphere, in practice between 15 and 25 mm of water pressure (figure 204 *A*). The membrane causes up-lift forces at ground level which must be counteracted by adequate anchorage. A constant air supply must be provided to make up for leakage through access points and to some extent through joints in the membrane. Air supply is maintained by low pressure fans and air loss through large access openings is usually minimised by the use of air locks or air curtains.

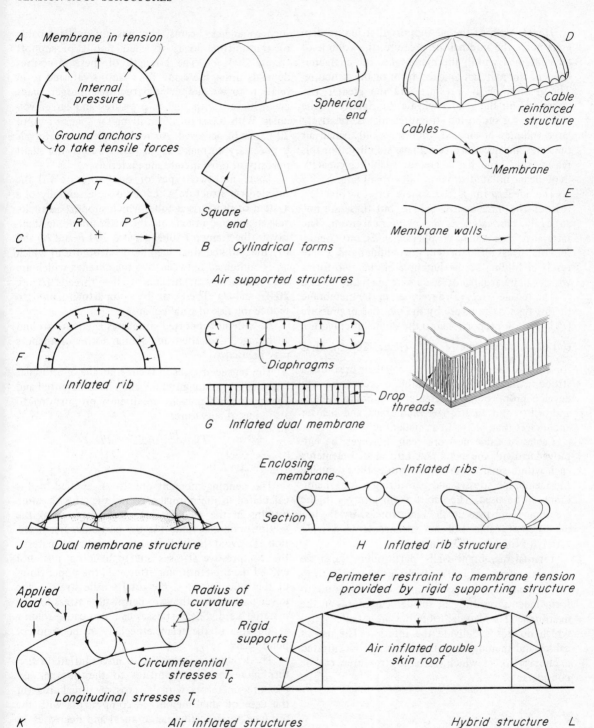

A Membrane in tension

Internal pressure

Ground anchors to take tensile forces

C

T

R P

Spherical end

Square end

B Cylindrical forms

D Cable reinforced structure

Cables

Membrane

E

Membrane walls

Air supported structures

F Inflated rib

Diaphragms

Drop threads

G Inflated dual membrane

J Dual membrane structure

Enclosing membrane

Inflated ribs

Section

H Inflated rib structure

Applied load

Radius of curvature

Circumferential stresses T_c

Longitudinal stresses T_l

K Air inflated structures

Perimeter restraint to membrane tension provided by rigid supporting structure

Rigid supports

Air inflated double skin roof

Hybrid structure L

204 Air stabilised structures

The strength of the air supported structure depends upon the internal air pressure, the volume of air contained within the membrane, the characteristics of the material and the form of the structure. The greater the air pressure and the greater the volume of air the greater will be the rigidity of the structure. Air supported structures enclose relatively large volumes of air and, therefore, only small air pressures are necessary to achieve stability. For this reason this type has a greater spanning capacity than the air inflated structures described later.

The ideal form is the sphere or, for practical reasons, the three-quarter sphere, but these are not so easily fabricated as the cylindrical form. The latter may have square or spherical ends, but circumferential stress differences in the cylinder and sphere result in folding at the junction of the two forms which can be avoided by a square end (figure 204B).

The tensile stress (T) developed in the membrane of this type of structure by the internal air pressure (P) is directly proportional to the radius of curvature (R) of the membrane: $T = \dfrac{PR}{2}$ (figure 204 C). Thus for equal areas covered and for equal air pressures a structure having a greater radius of curvature will develop proportionately greater membrane stresses and will require a stronger membrane and heavier anchorages than one with a smaller radius.

Cables or cable networks can, however, be employed to divide the membrane into smaller elements each with a small radius of curvature, thus reducing the membrane stresses and permitting a thin membrane to be used. The cables form grooves in the membrane and take up the main forces. By the use of different cable and net arrangements a wide variety of forms can be achieved.

Internal membrane walls, partitioning the space within the structure, can similarly influence the form of the membrane by anchoring it to the ground, to form indents resulting in smaller curvatures in the membrane (E). Vertical cable ties are an alternative which do not sub-divide the interior. The use of cables and membrane walls results in concentrated anchorage forces which, in large structures, can be considerable.

Air inflated structure   In this form of air stabilised structure air is used to inflate tubes and double membrane elements to form stiff structural members

such as arches, beams, columns and walls capable of transmitting loads to their points of support (figure 204 F). The strength of these structures depends upon the same four factors referred to in relation to air supported structures. Larger spans are achieved with higher pressures and larger volumes. With tubes of 0.5 m diameter spans of up to 20 m can be achieved. Air replenishment at intervals is necessary to make up air losses due to the slight porosity of most membrane materials.

There are two types of this structure (a) the inflated rib structure which consists essentially of a system of pressurised tubes which support an enclosing membrane in tension, the latter providing stability to the frame of tubes (figure 204 F and H) and (b) the inflated dual membrane structure in which air is contained between two membranes which are held together by diaphragms or 'drop threads' (figure 204G and J). These can be compartmentalised to reduce the risk of total collapse.

As with air supported structures adequate ground anchorage is required and similar forms of anchors may be used.

The tensile stresses in the membrane of a tubular beam or rib are longitudinal and circumferential and for any internal pressure are directly proportional to its radius of curvature:

$$T_l = \frac{PR}{2} \text{ and } T_c = PR$$

(figure 204 K).

The bending moment due to an applied load is calculated in the normal way and the stresses set-up by this in the member are established, using the section modulus value for the particular beam section. To avoid the folding of the membrane material the compressive stresses due to bending must not exceed the pre-tensioning stresses in the upper fibres of the tube, nor must the total tensile stress in the lower part exceed the safe permissible stress for the material used. Shear stresses can also cause folding at the sides of the beam, especially in the region of point loads.

The loadbearing capacity of an air inflated structure increases in proportion to the internal air pressure, increases with its cross-sectional area (in the case of dual membrane components with the distance between the membranes) and decreases, as with conventional components, with an increase in span. The arch and dome forms are stiffer and carry greater loads than straight beam and flat slab forms.

Hybrid construction Threr are many forms of combined or hybrid construction using air stabilised structures some of which integrate the two forms described, taking advantage of the superior insulating properties of the inflated dual wall construction and of the greater spanning capability of air supported construction. Others combine air stabilised construction with conventional forms of structure as in the example shown at (*L*) in figure 204[1].

CHOICE OF ROOF STRUCTURE

The most appropriate form of structure will depend upon the type of building, foundation conditions, spans to be covered, nature and magnitude of loads, lighting requirements and accommodation for services, the possibility of future alteration and speed of erection, as well as aesthetic considerations. For some types of buildings a few solutions only are possible, as in the case of an auditorium with rigid requirements of space, acoustics and volume and ventilation. In others, industrial buildings for example, particularly those of small and medium span, a wide range of structure is possible.

As indicated in Part 1, page 164, short span construction will usually be cheapest but the span of the structure is usually fixed by the type of building as this will dictate the miniumm areas of unobstructed floor space required. The minimum spans must therefore be compatible with requirements of clear floor area.

The clear internal height of the building is another factor governed largely by the use of the building. In industrial buildings sufficient clearance must be provided for the installation and maintenance of plant, and the widespread use of fork trucks requires minimum headrooms for efficient working. Much ancillary equipment is hung from the roof structure and the provision of sufficient headroom below this will govern the minimum internal height.

Lighting requirements may have a profound effect upon the form of structure used, particularly in the case of very wide buildings, in which the central areas cannot adequately be lit from the side walls. Roof lighting must then be provided and this will affect the shape of the roof. Generally speaking, with a height of 5.50 m reasonable natural light cannot be obtained from wall windows if the distance between them exceeds about 12 m. Various types of roof structure designed to provide adequate lighting for the interior have already been described, and the suitability of a particular form will depend not only upon lighting requirements but on other functional requirements.

Apart from considerations of lighting and the other considerations mentioned above, the roof structure must be thought of in relation to the heating requirements of the building. From this point of view the structure which, while fulfilling other functional requirements, restricts the internal volume of the building to a minimum and at the same time offers the minimum exposed roof surface area, will be the most efficient[2]. The reduction in volume reduces the amount of space to be heated and the reduction in roof surface area keeps heat losses through the roof to a minimum. From this point of view alone, therefore, the flat roof is the most acceptable form.

Although it may not always be acceptable for other reasons, it may be generally accepted that as far as lighting and heating requirements are concerned the larger the span the lower should be the pitch of the roof. To achieve this, where functional requirements demand large span roofs and wide column spacing, the use of forms of construction which are more expensive than others may be economically justified. When rigid frames are used to decrease the volume of enclosed space it is often economically advantageous to pitch the spanning members sufficient to permit the use of forms of sheet roof covering which require a fall of a few degrees.

The need to suspend equipment from the roof must be considered, particularly if the loads to be carried are heavy and are point loads. Shell roofs, which are essentially designed to carry comparatively light uniformly distributed loads, may not for this reason be suitable. Point loads of any magnitude should, wherever possible, as in other structures, be restricted to the main stiffening beams and end frames of shell structures. If, however, it is essential to suspend from the shell, all loads greater than about 50 kg should be spread by

---

[1] For a more detailed consideration of air stabilised structures see *Air Structures – a survey* (HMSO 1971) and *Principles of Pneumatic Architecture* by Roger N. Dent (Architectural Press 1971).

[2] See comparison of truss and rigid frame structures on page 310.

plates or cross reinforcement. In addition to suspended equipment, provision may have to be made for ventilating or extract ducts and, particularly if these are large, they will have to be related to the roof structure at an early stage.

The choice of structure will sometimes be weighted by considerations of adaptability and maintenance. Steel construction is generally more adaptable than reinforced concrete if alterations are necessary, and it is comparatively easy to strengthen the structure, although fixing to matured concrete can be made by metal fixing studs applied by cartridge hammer.

Concrete generally requires less maintenance than steel, which must be painted periodically. Steel lattice construction of all forms offers a large surface area for corrosion and is difficult to paint. Rigid frames built up of welded steel plates, on the other hand, expose less surface to the air and are easier and cheaper to paint. In highly corrosive atmospheres it may be more economical to use aluminium alloy instead of steel in order to reduce maintenance costs, even though the cost of the structure itself may be higher than in steel. Timber structures do not require protective coatings, particularly internally, and maintenance costs are therefore considerably less than for metal structures.

Very often speed and cost of erection will have a bearing upon the nature of the structure to be used. This is discussed in more detail later.

Span and type of structure

Although short-span construction will usually result in the cheapest structure, functional requirements often dictate medium or large spans. For these the most economic form must then be selected.

Of the wide variety of roof structures available, some only are economic over large spans, some only when there is considerable repetition of the structure and its parts on one particular job. As far as the materials of the structure are concerned, some, such as aluminium alloys, are only really economic when used in large-span construction. Care must be taken in design, therefore, not to use materials which can be exploited most favourably only over wider spans when, in fact, large spans are not justified by other requirements, nor to select forms of structure the economic use of which depends on repetition, when the job is too small to provide adequate repetition.

As with all types of structure the characteristics of the soil on which the building rests must be considered. In circumstances of weak soil requiring expensive foundations, it may be cheaper to use widely spaced, wide-span construction in order to limit the number of foundation points.

Research on the comparative costs of different forms of roof structure in different materials has shown that the variation in cost of structural frames alone of the same span and loading, and fulfilling the same requirements, is generally small. Bearing in mind that the cost of the structural frame may be as low as one-sixth of the total cost of the building[1], it will be appreciated that quite large variations in this element will have only a relatively small effect on the total cost. Nevertheless, in large buildings in particular, one-sixth of the total cost may be a substantial sum in which case consideration must be given to the problem of keeping this figure to a minimum.

For simple shed-type buildings where lighting is not important, the truss and column frame is probably the cheapest structure for spans up to about 30 m. The space above eaves level, however, is obstructed by the trusses and over the wider spans this type of structure encloses a considerable volume of space. For spans above 30 m 'umbrella' construction or north light roofs with lattice girders in the plane of the ridge can be used up to about 45 m span, at spacings not exceeding 18 to 21 m. The unit weight and cost of a steel folded lattice plate and an 'umbrella' roof of the same span and spacing of supports are about the same, but the folded plate construction has the advantage of unrestricted space under the roof.

In trusses or girders and in structural systems relying upon 'beam effect' such as space frames, economy in the use of any material will result when increased moments of inertia are produced by increased depth of structure. Structures which are more closely related to the space which they enclose, such as rigid frames or arch forms, will frequently be more expensive span for span than those incorporating deep beams, girders or trusses, but may compensate for this by reducing the volume of the building and the area of cladding and finishes.

Rigid frames in welded steel and reinforced con-

---

[1] *Factory Building Studies*, No. 7, 'Structural Frameworks for Single-Storey Factory Buildings', HMSO, 1960.

crete either 'square' or 'arched', although not always economic over smaller spans, are frequently used because of the advantages which have been mentioned above. Reinforced concrete rigid frames can be economic over 18 to 30 m spans. For spans of over 30 m, and certainly over 45 m, rigid steel frames with lattice structure are likely to be cheaper than those with solid web designed on the elastic theory, but not necessarily those designed on the plastic theory. For spans of 60 m or more, arch rib construction is likely to be the most economic.

Prestressed concrete is usually competitive in cost with uncased steel for solid web frames, so that where lattice frames are not suitable there is likely to be an economic advantage in using prestressed concrete.

Over spans of more than 24 m the saving in weight over steel in an aluminium structure may be as much as four-fifths. In spite of this, the cost of an aluminium structure may not be lower than that of steel, and aluminium only begins to be competitive with steel over very large spans of 60 m or more.

In suitable types of building advantage can be taken of the high strength to weight ratio of timber[1]. Bowstring trusses with laminated timber top chords can span up to 70 m or more, and lattice rigid frames have been constructed over spans of 30 m or more. Laminated timber 'bents' forming rigid frames and laminated timber arches are used over spans of more than 60 m.

Medium span structures in reinforced concrete can be competitive with steel, if constructed of standardized precast members, and sometimes when constructed of non-standard members cast on the site. Shell concrete construction can be economic in medium and wide span roofs, using long-span barrels up to spans of 30 to 45 m, above which short-span barrels would be used. There will not be much difference in the relative cost of steel and shell structures provided that the shell is repeated more than four to five times. Although shell construction may not always be competitive in cost with steel, nor usually in speed of erection, it is in other respects equal to steel and superior to it as far as maintenance is concerned.

Shell vaults can be prestressed, with a consequent reduction in the rise to span ratio, but the structural advantage of prestressing must be weighed against the increased cost and complication of placing of ducts, anchorages and cables and it is unlikely that this will be justified for spans much below 33 m. Above this span the advantage of prestressing increases.

Shell construction in laminated timber produces a very light structure. A timber hyperbolic paraboloid shell 18 m square will be about 50 mm thick and weigh approximately 25 kg/m^2; a comparable concrete shell would be about 63 mm thick weighing about 150 kg/m^2. Laminated timber domes and vaults show similar weight advantages. Elliptical shell domes have been constructed with diameters over 45 m and barrel vaults over spans of up to 30 m.

Grid structures in flat or curved form are particularly useful over very wide spans where the advantage of rigidity and light weight are important. Single-layer flat grids are economical up to 15 to 24 m span. Braced barrel vaults have the advantage over reinforced concrete shell vaults of considerable reduction in weight and of rapid erection on the site. They require no shuttering and may be partially or completely prefabricated. For spans above about 18 m, they are about 10 per cent cheaper than reinforced concrete shells or even 20 per cent in some cases. By the use of folded grids and double-layer flat grids, spans of over 90 m can be covered economically. Where the circular or polygonal plan form is acceptable, braced domes of various types can be used over spans of up to 90 to 120 m. As explained on page 303 aluminium alloys are advantageously employed in grid structures.

The relative advantages of steel, aluminium alloys, concrete and timber are discussed on pages 301-304. The metals are produced not only as 'I', channel and other common sections in hot-rolled steel or extruded aluminium alloy, but also as tubes, hollow sections and, in the case of steel, as cold-formed sections produced from thin strip steel.

The manufacture of cold-formed steel sections and the design considerations involved in their use are described on page 203. Although they are used for two- or three-storey structural frames as described in chapter 5, they are used to greatest advantage in roof structures where a low dead/live load ratio is important. For spans up to about 15 m the weight of cold-formed steel construction per m^2 of floor area is about 40 to 60 per cent less than that of similar hot rolled steel construction, the greater savings being over the shorter spans. This is largely due to the fact that when hot-rolled sections

---

[1] See also page 304 on the use of timber in this context.

are used for short-span structures, sections for the various members cannot always be obtained small enough to match the minimum area required for stress resistance, as the choice is limited by the smallest sections rolled. In addition, the wide variety of shapes which can be cold-formed makes it possible to avoid eccentric loading of the members and the use of gusset plates, and this further decreases the weight of the structure.

As far as cost is concerned, beyond very small spans the increased cost of fabrication, and sometimes of the material, is likely to make the price equal to or more than that of ordinary hot-rolled construction, but against this must be placed the advantages of the comparative ease with which sections of different shapes can be produced, the simple fixing of roof claddings and ease and speed of erection. A light lifting pole only is required for erection purposes because the fabricated units are much lighter than those in hot-rolled construction. Savings in erection costs are given as around 15 to 25 per cent for trusses and 10 per cent for purlins and bracings. Provided that detailing is simple, construction in cold-formed sections can be very economical when used on standard units or with a high degree of repetition. Over medium spans with light loading, cold-formed lattice beams show savings over hot-rolled lattice beams although these savings are not so great as in the case of trusses. Because of the efficient distribution of metal in cold-formed sections and the advantages shown in their repetitive use, considerable economies result when they are used as purlins, particularly as zed sections, in conjunction with main roof members of hot-rolled steel.

Roof structures constructed from welded steel tubes will be considerably lighter than similar structures constructed in ordinary steel sections. The weight of a tubular framework per m^2 of floor area may be from 25 to 40 per cent less than that of a hot-rolled structure for spans up to about 15 m. As in the case of cold-formed sections the greater savings are made in the shorter spans. Actual savings will depend upon the simplicity of detailing. Although there will be a saving in weight by the use of tubular construction, a reduction in cost compared with that of normal steel construction is unlikely unless simple detailing is used and gussets are reduced or eliminated. However, if these points are watched and there is considerable repetition, tube frameworks will generally be cheaper in overall cost, or at least competitive with hot-rolled construction.

Efficient large-span construction can be obtained by using both tubes and hot-rolled sections together: the latter as tension members and the former as compression members. This is because tubes provide a more efficient section in compression, and are designed to a lower basic tension stress than, for example, hot-rolled angles. Erection takes rather longer than with ordinary steel frames in spite of the fact that tubular frames are lighter and stiffer in lifting. This is largely because erectors cannot walk as easily along tubular members, so that in carrying out fixing operations they cannot so readily work on the structure itself.

If welding instead of bolting of normal steel sections is employed there will be an overall saving in the weight of the structure, but not necessarily a saving in cost unless there is considerable repetition of the work and simple detailing. This is because the saving in weight is offset by the increased cost of fabrication. In order to reduce the fabricating cost to a minimum, welding should be restricted to the shop or to the floor on the site. Where possible, any connections required to be made when the structure is erected in position should be bolted.

Speed and economy in erection

As already indicated, the cost of erection in terms of the speed and facility with which it may be carried out is an important factor in the overall cost of a roof construction. Metal and timber structures are generally quicker to erect that those in *in situ* concrete. Reference has already been made to the ease with which structures fabricated from cold formed sections are erected. The use of precast concrete, which can be assembled fairly quickly, reduces the time disadvantage of concrete. The erection time of wide-span concrete structures can be reduced by the use of prestressing with precast concrete elements.

The precasting of concrete roof frames from 6 to 40 m or so in span, the large ones being cast in three to four units, bolted together on the site, is common practice. By this means they can be easily and quickly erected by the use of simple derricks or mobile cranes. The erection of large rigid frames cast in three or four sections is carried out with the use of a timber trestle, either on a track or moved by crane into each frame position, which supports the separate units until the joints between them are bolted up. Three-pin frames in lattice construction, particularly in aluminium, may be erected in two halves which are swung over into position using the base connections as hinges. In the case of large span concrete structures, erection can be speeded up and a great amount of expensive shuttering avoided by the application of post-tensioning to small precast units. These can be stressed into a single element and then hoisted into position by cranes or hydraulic jacks.

Although rigid concrete frames of various types are those most commonly carried out in precast construction, beyond certain heights it may sometimes be cheaper to precast small shell vaults and domes on the ground and then hoist them into position. 7.50 m square shell concrete domes weighing 13.70 tonnes each have been precast on site and lifted into position by a pair of cranes. Lightweight concrete shell barrel vaults over 18 m long and weighing 20 tonnes each have similarly been cast and hoisted into position. In each case a number of such units was cast for the job. In the case of the domes twelve, and in the case of the barrel vaults, six units were required.

In each of these examples the weight of the units was within the lifting capacity of a crane or pair of cranes, but in certain circumstances it may be cheaper than other forms of construction to cast very much larger units and lift by hydraulic jacks. Groups of prestressed concrete barrel vaults linked together by end frames into units measuring 56 m x 33 m overall and weighing about 1500 tonnes have been cast on the floor of a building, with all internal roof finishings and external roof coverings applied at ground level, and then raised nearly 15 m by means of hydraulic jacks. Two pairs of jacks were positioned at each of the four supporting column positions, the columns being built up on the jacks with specially designed interlocking concrete blocks placed in position at the completion of each lift by the jack.

Grid structures of all types are fabricated on the site as a whole, or in large sections from prefabricated units, and are then lifted into position usually by crane, or by hydraulic jack in the case of very large structures raised as a whole. Steel lattice grids more than 90 m sq. and 3 m deep, weighing over 1000 tonnes have been lifted in a similar manner to lift slab construction in reinforced concrete[1].

CONSTRUCTIONAL METHODS

The basic principles and economic applications of roof structures have been described in the preceding pages. It now remains to describe the constructional methods adopted in the application of different materials to the various types of structure.

Beams

These are constructed in timber, steel and concrete.

Timber

Glued and laminated beams   These are commonly called 'glulam' beams and may be constructed of laminations arranged vertically or horizontally (figure 205 *A*). Horizontal lamination allows a more economical use of timber, particularly for larger section beams, since the depth of the beam may be formed of multiple laminates and is not limited by the width of any given plank. Furthermore, horizontal lamination facilitates shaping (*B*). The thickness of each laminate is usually about 50 mm for straight members, reducing to 16 mm according to the radius of curve for shaped members. The shapes of beams may be varied to give sloping upper surfaces, slight cambers and sections of varying depth. Beams much over 18 m in length are not usually formed in glulam construction. Beams may be secured to masonry or be supported on solid or laminated timber columns and secured by metal angles, plates and bolts as shown at (*F*).

Web beams   For spans of up to 12 or 15 m a more economical distribution of material is obtained by the combined use of solid timber flanges and webs of plywood (*C*). Such beams may be stiffened by using a number of webs to form a hollow box section and additional cross sectional area may be added to the flanges between the webs to accommodate larger bending moments (*D*). Web stiffeners are required and at points of high shear stress

[1] For a full discussion of the economic aspect of single-storey construction, see 'Economics of Framed Structures', by Leonard R. Creasy, BSc (Eng), MICE, *Proceedings of the Institution of Civil Engineers*, Vol 12, March 1959. See also *Factory Building Studies* no. 7, 'Structural Frameworks for Single-Storey Factory Buildings', HMSO 1960.

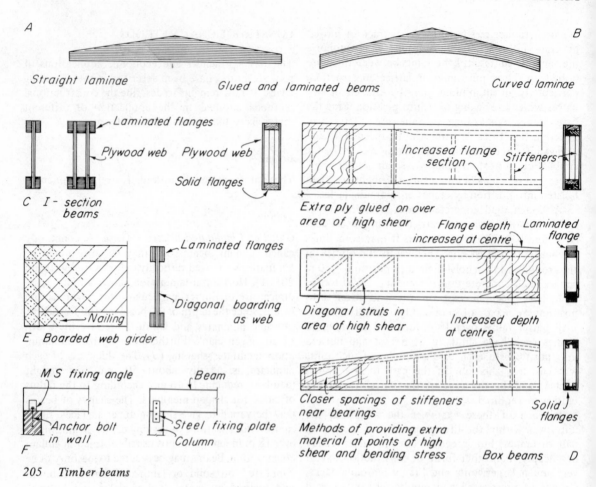

A

Straight laminae

Glued and laminated beams

B

Curved laminae

Laminated flanges

Plywood web   Plywood web

Solid flanges

C   I - section
    beams

Increased flange
section   Stiffeners

Extra ply glued on over
area of high shear

Flange depth   Laminated
increased at centre   flange

Laminated flanges

Diagonal boarding
as web

Nailing

E   Boarded web girder

Diagonal struts in
area of high shear

Increased depth
at centre

M S fixing angle   Beam

Anchor bolt
in wall   Steel fixing plate

Column

F

205   Timber beams

Closer spacings of stiffeners
near bearings

Methods of providing extra
material at points of high
shear and bending stress

Solid
flanges

Box beams   D

these may be more closely spaced or alternatives may be adopted as shown. For spans over 18 m and up to 30 m the use of laminated glued and nailed beams of I-section will produce a more economical structure than solid rectangular laminated beams. The webs and flanges are formed of boards approximately 25 mm thick fixed together with glue and carefully calculated nailing (*E*). This system enables short pieces of timber to be used economically, each flange being composed of a number of overlapping lengths and the web formed from diagonal boarding braced where necessary by the addition of vertical stiffeners glued and nailed to it.

Steel
British Standard Universal beams may be used where head room is critical and minimum beam depth is

desirable. Reference should be made to chapter 5 where the merits of mild and high-tensile steel sections and castellated beams are discussed (page 187. See also page 307).

Concrete
For spans much over 6 to 9 m reinforced concrete is not likely to be economical for beams, but wide spans can be covered economically if the beams are prestressed (see page 303).

Trusses and lattice girders

Timber
Girders and trusses may be fabricated with connectors, nails or glues and the applications of these techniques are illustrated in Part 1.

340

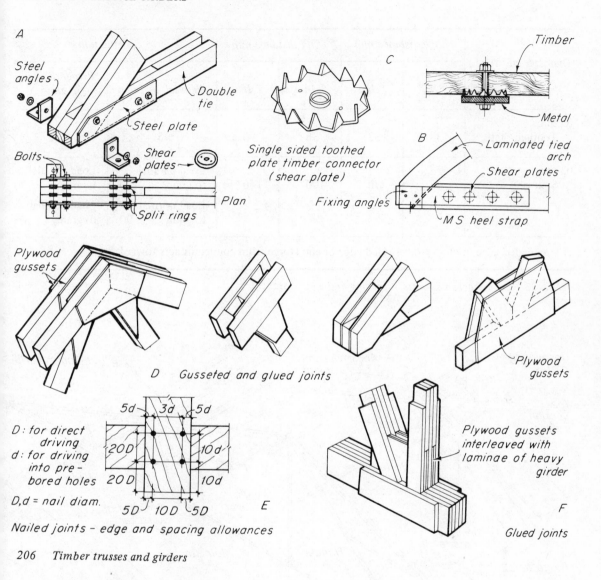

A

Steel angles

Double tie

Steel plate

Bolts

Shear plates

Plan

Split rings

C

Single sided toothed plate timber connector (shear plate)

Timber

Metal

B

Laminated tied arch

Shear plates

Fixing angles

MS heel strap

Plywood gussets

D   Gusseted and glued joints

Plywood gussets

D: for direct driving
d: for driving into pre-bored holes

D,d = nail diam.

5d   3d   5d

20D   10d'

20D   10d

5D   10D   5D

E

Nailed joints – edge and spacing allowances

Plywood gussets interleaved with laminae of heavy girder

F

Glued joints

206   Timber trusses and girders

The two commonly used connectors are the toothed plate and split ring, illustrated in figure 138, Part 1. Toothed connectors are useful for joining sawn timber. Split-ring connectors carry greater loads than toothed connectors but require accurately machined grooves cut in the timber and the timbers are usually faced on four sides and prepared in the shop.

As shown in Part 1 (see page 181) gussets give greater fixing areas with all methods of jointing and permit members to butt against each other rather than overlap. When steel plate is used for this purpose (figure 206 A), or for heel straps (B), with connectored joints a shear plate connector is used as shown in (A) and (C), the plate side being in contact with the steel.

In order to permit the development of the full allowable load at each connector minimum spacings and distances from the bolt centre to the edges and ends of the members being joined must be ensured. These vary with the type and size of connector. Typical distances are given in table 23. For smaller distances the maximum allowable load must be reduced.

Diameter and type	Edge distance mm		End distance mm		Spacing mm	
	A	B	A	B	Parallel to grain	Perpendicular to grain
63 mm Toothed plate	38	38	38	90	95	75
63 mm Split ring	45	70	100	140	Load parallel to grain 170 Load normal to grain 90	Load parallel to grain 90 Load normal to grain 108

A = unloaded edge or end   B = loaded edge or end (force from connector acts towards it)

Table 23   *Connectored joints – spacing and edge distances*

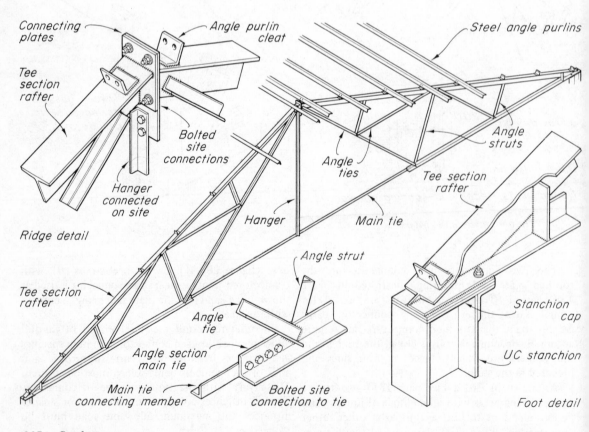

Connecting plates
Angle purlin cleat
Steel angle purlins
Tee section rafter
Bolted site connections
Hanger connected on site
Ridge detail
Angle struts
Angle ties
Tee section rafter
Hanger
Main tie
Tee section rafter
Angle strut
Angle tie
Angle tie
Angle section main tie
Main tie connecting member
Bolted site connection to tie
Stanchion cap
UC stanchion
Foot detail

207   *Steel trusses*

Nailing is sufficient for the relatively lightly loaded trussed purlin and roof truss shown in Part 1 but it is not efficient enough to be used alone in heavy structures, although it may be used with glue (figure 205 *E*). As with bolted joints minimum spacings and edge distances must be provided (figure 206 *E*).

When glued joints are adopted for trusses and girders of heavier types than those illustrated in Part 1 greater gluing surface can be achieved by the use of thick ply gussets sandwiched between double lapped members as shown in figure 206 *D* or interleaved with the laminae of heavy built-up or glulam members as at (*F*).

Girders up to about 45 m and trusses to about 60 to 75 m may be constructed in timber. The upper chords of girders may be curved, parallel or inclined, and are usually laminated in the larger spans.

Figure 208 indicates the relative efficiency of the types of connection used in timber construction.

Relative efficiency of connections	% Efficiency	
	Single lap	Double lap
Nails in prebored holes	4.22	8.44
Screws in prebored holes	2.82	5.65
Bolts - 1 at 25 mm diam.	21.30	21.30
Toothed connector - 64 mm with 13 mm bolt	21.10	19.10
Split ring connector - 64 mm diam.	38.50	—
Urea formaldehyde glue	46.40	112.50
Efficiency = $\dfrac{\text{Max. working load in joint}}{\text{Max. working load in member}}$		

208   Relative efficiency of connections

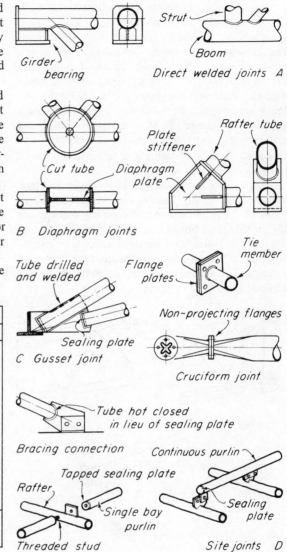

Direct welded joints A

B   Diaphragm joints

C   Gusset joint

Cruciform joint

Site joints   D

209   Welded tube connections

Steel

Girders fabricated from hot-rolled sections are described and illustrated on pages 191-192. Conventional steel trusses are described in Part 1, page 196. These may be bolted or welded complete in the fabricating shop or, if too large for transport, are made in two halves with the necessary connecting plates and provision for attaching connecting members as shown in the welded example in figure 207. Assembly, by means of bolts, is carried out at ground level prior to hoisting to position. Trusses and girders fabricated from welded tubes possess the advantage of lightness and stiffness. This form of construction is discussed on page 338, where the need for simplicity of detailing and, as far as possible, the avoidance of gussets is emphasized. Typical joints are shown in figure 209.

343

The jointing of members by direct welding is shown at (A) (see also figure 210 C). This is facilitated by the use of rectangular hollow sections and these are often used for the major members of a truss or girder for this reason. The use of diaphragms at the junction of tubes of equal or near equal diameter facilitates the jointing (B). When gussetted joints are adopted the tubes may be cut and welded to them as shown at (C). In large trusses and girders some site joints are always necessary and examples of purlin, bracing and tie joints are shown in (D).

Light lattice beams formed from angles and flats are used for lightly loaded roof structures up to spans of 12 m. Such beams frequently incorporate high-tensile steel bottom chords (figure 210 A). The use of cold-rolled steel sections welded or riveted together in lattice beams is discussed on page 338. Examples of these beams are illustrated in figures 154 and 210 B and in figure 114, Part 1.

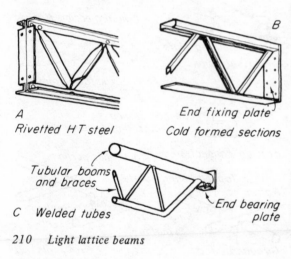

A
Rivetted H T steel

B
End fixing plate
Cold formed sections

Tubular booms and braces

C Welded tubes

End bearing plate

210   Light lattice beams

Aluminium

Aluminium and steel trusses and girders are generally similar in form. As explained on page 302 care must be taken to prevent local buckling of members in compression, and while typical aluminium sections will be used some members may need to be doubled up to produce adequate stiffness.

Concrete

Modern light concrete trusses have resulted from the introduction of prestressing which enables small members to be drawn together by the stressing cables, so that the tension members of the trusses are pre-compressed (see page 233). Some typical examples are shown in figure 211. (A) shows a built-up truss with the bottom boom tensioned by cables. (B) shows a large concrete bowstring girder and (C) a lattice beam formed in a similar way from precast elements stressed together. (D) shows an example of a concrete beam with exposed stressing cables, forming a trussed beam.

Rigid frames

Timber

The techniques used in fabrication of timber beams, trusses and girders are also applicable to the construction of rigid frames. Lattice frames are built up with bolted or glued joints on the same lines as trusses and girders. Figure 212 shows examples of rigid frames constructed in hollow box, solid laminated, and built-up 'I' form. Where glued laminations are employed in frames with curved angles or in arch ribs (B, C, E) the laminates must be thin enough to bend easily. The ratio of the radius of curvature to the thickness of the laminates must be not less than 100, and ·150 is normal practice. The cost of fabrication rises as the laminates get thinner, therefore sharp curves should be avoided where possible. Metal fittings are employed to connect the members and details of connections will vary according to the degree of rigidity required at the joints. Base connections are formed by cleats (A, D) or by locating the feet of frames in metal shoes bolted to concrete foundations (C, E). The degree of rigidity provided by a connection will vary with the size of the frame and the nature of the connection. That shown in (G) and the similar example in (E) provide considerable flexibility. Those shown in (A), (C) and (F) and the box connection in (E) will be stiff or rigid connections. In larger frames hinged metal bearings may be bolted to the feet of the frames (H). Crown joints are either bolted through the apex (C, D) or secured by splice plates (B, K) to provide a rigid connection, or are provided with a hinged bearing (E, J). Boxed plywood frames are used for spans up to about 18 m, glulam frames up to about 24 m and lattice and built-up I section frames up to about 45 m. Arch ribs are constructed in glulam up to spans of more than 60 m.

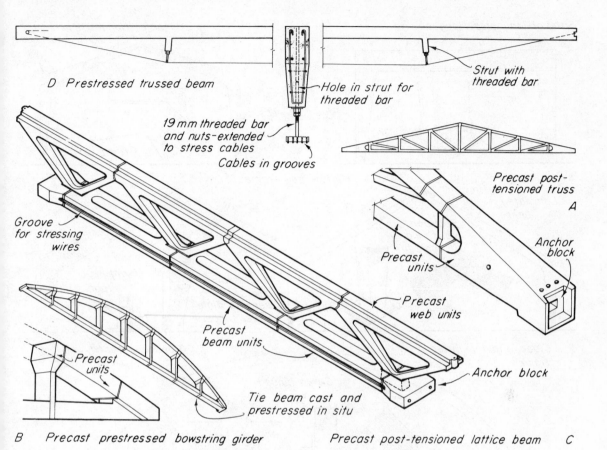

D Prestressed trussed beam

Hole in strut for threaded bar

Strut with threaded bar

19 mm threaded bar and nuts-extended to stress cables

Cables in grooves

Precast post-tensioned truss

A

Groove for stressing wires

Precast units

Anchor block

Precast web units

Precast beam units

Precast units

Anchor block

Tie beam cast and prestressed in situ

B   Precast prestressed bowstring girder

Precast post-tensioned lattice beam   C

211   Prestressed concrete trusses and girders

Steel

Rigid frames constructed in steelwork may be of welded solid web construction or of lattice construction. Some typical forms are shown in figure 213. The lattice form is commonly employed in the construction of large-span frames. A number of lightweight prefabricated frames have been developed for short to medium spans which are of welded angle, tube or cold-rolled steel construction. Figure 213 A shows an example of a welded steel solid web fixed portal. Where, as in this case, the foundations contribute to the stability of the structure, the feet of the frames must be rigidly fixed to the concrete bases. The illustration shows one method involving embedding the foot in the concrete. Another method uses a heavy gusseted base bolted to the concrete slab. The welded tubular steel three-pin frame shown at (B) is a space frame in form, the triangular cross-section giving great rigidity to a structure constructed of quite small tubes. The nature of the junction between frame and closely spaced purlins (C) gives a rigid joint and provides lateral rigidity to the whole structure. It also results in considerable membrane action so that the whole of the roof structure will act to some extent as a lattice grid. The crown and base connections for this frame are shown in detail.

Crown and base joints vary in complexity and rigidity. In all of the hinge connections shown it will be seen that the frame is free to rotate (one permits rotation in all directions) but is held in position. Both crown hinges (D) and base hinges (E, F) can be formed as rockers or true pin joints. It should be noted that rocker base hinges must incorporate long bolts as shown in order to resist vertical tension and horizontal forces. At a crown rocker a bolt is required to resist shear forces.

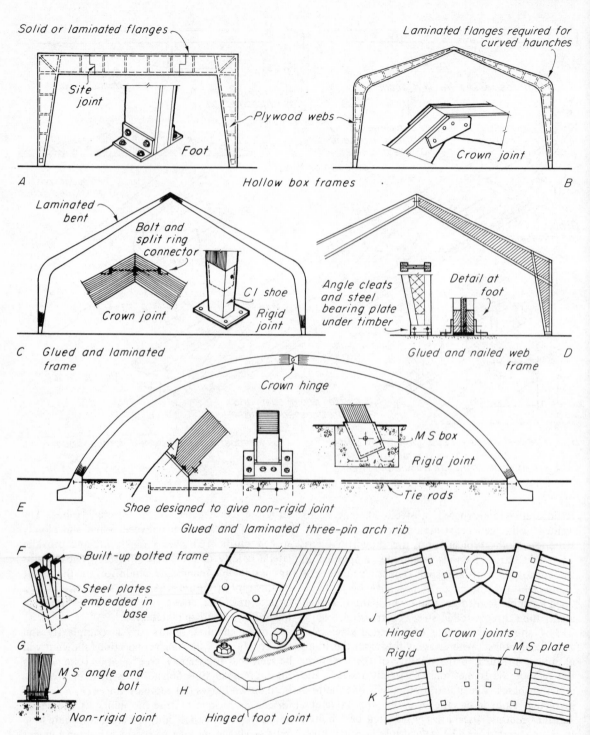

Solid or laminated flanges

Site joint

Foot

Laminated flanges required for curved haunches

Plywood webs

Crown joint

A   Hollow box frames   B

Laminated bent

Bolt and split ring connector

Crown joint

C I shoe

Rigid joint

C   Glued and laminated frame

Angle cleats and steel bearing plate under timber

Detail at foot

Glued and nailed web frame   D

Crown hinge

M S box

Rigid joint

Tie rods

E   Shoe designed to give non-rigid joint

Glued and laminated three-pin arch rib

F   Built-up bolted frame

Steel plates embedded in base

J

Hinged   Crown joints

Rigid

M S plate

G

M S angle and bolt

H

Non-rigid joint

Hinged foot joint

K

212   Timber rigid frames

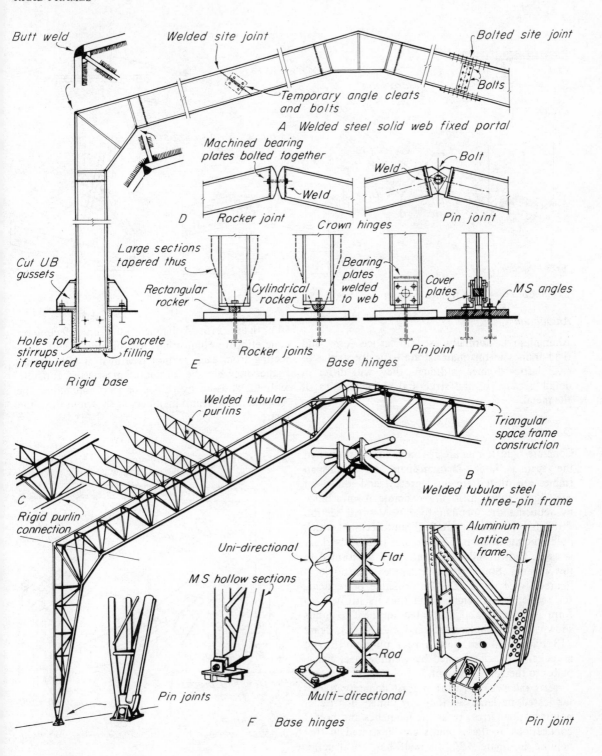

Butt weld

Welded site joint

Bolted site joint

Bolts

Temporary angle cleats
and bolts

A   Welded steel solid web fixed portal

Machined bearing
plates bolted together

Bolt

Weld

Weld

D   Rocker joint

Crown hinges

Pin joint

Cut UB
gussets

Large sections
tapered thus

Bearing
plates
welded
to web

Cover
plates

MS angles

Rectangular
rocker

Cylindrical
rocker

Holes for
stirrups
if required

Concrete
filling

Rigid base

E

Rocker joints

Base hinges

Pin joint

Welded tubular
purlins

Triangular
space frame
construction

B
Welded tubular steel
three-pin frame

C
Rigid purlin
connection

Uni-directional

Flat

Aluminium
lattice
frame

MS hollow sections

Rod

Pin joints

Multi-directional

F   Base hinges

Pin joint

213   Metal rigid frames

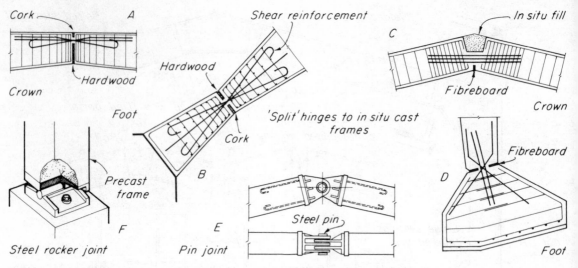

214   *Concrete rigid frames – hinge joints*

Aluminium

Aluminium is normally used in lattice form, and rigid frames in this material are basically similar to steel lattice frames, although they will differ in detail because of the structural characteristics of the metal.

Concrete

Concrete rigid frames are generally cast *in situ* when the span is large. Medium-span and small-span frames are most frequently precast and joined on site. 'Split' joints in *in situ* concrete form a hinge by reducing the member to a very small section through which passes 'bundled' reinforcement which holds the members in position and resists any shear stresses. Crown and base joints may be formed in this manner. Some typical examples are shown in figure 214 *A, B, C, D*. The principle of reducing the section of a member, and thus its stiffness, to form a hinge is also illustrated in the two joints shown in the tubular metal member in (*F*), figure 213. Figure 214 also shows a metal pin joint (*E*) and a rocker joint (*F*) formed by welding steel flange plates to the reinforcement.

As explained in Part 1 precast concrete frames fabricated in large sections are joined on site at points of low stress or at the haunches and bolted connections at these points are illustrated in that volume in figure 152. In multi-bay construction adjacent frames may be bolted together as shown at

(*A*), figure 215. A haunch junction effected by concreting-in connecting reinforcement left projecting from each section is shown at (*B*). Precast frames may also be formed of prestressed concrete sections, stressed by cables or rods connecting the 'column' and 'beam' members as shown in the same figure.

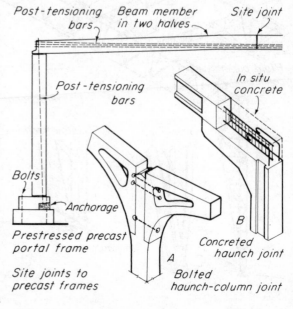

215   *Concrete rigid frames – site joints*

348

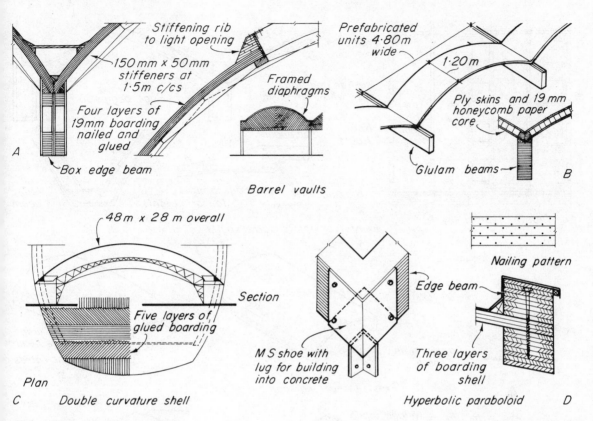

A Box edge beam

Stiffening rib to light opening

150 mm × 50 mm stiffeners at 1·5m c/cs

Four layers of 19mm boarding nailed and glued

Framed diaphragms

Prefabricated units 4·80m wide

1·20 m

Ply skins and 19 mm honeycomb paper core

Glulam beams B

Barrel vaults

48m × 28 m overall

Five layers of glued boarding

Section

Plan

C *Double curvature shell*

M S shoe with lug for building into concrete

Edge beam

Nailing pattern

Three layers of boarding shell

Hyperbolic paraboloid D

216   *Timber shell construction*

Shells

Timber

Some details typical of timber shell construction are shown in figure 216. Site constructed barrel vaults use layers of boarding nailed and glued together. (*A*) shows details of a 30 m span barrel vault the board membrane of which is stiffened by ribs at 1.5 m centres. Small barrels lend themselves to pre-fabrication and light forms of construction. (*B*) shows a shell prefabricated in 1.20 m units, con-structed as a sandwich panel of ply and paper honey-comb core. Edge beams to barrels may be of built-up box form or laminated form as shown.

Timber shell domes are normally built up with boards glued and nailed or screwed together, with laminated edge beams on the lines shown in (*C*). In hyperbolic paraboloids, constructed in the same way, the majority of layers of boarding were, in early examples, laid parallel to the edge beams and at right-angles to each other with some laid diagonally

but it has been found that greater rigidity results from laying the majority diagonally. A detail of the edge beam and shoe bearing of a hyperbolic parabo-loid is shown at (*D*).

Concrete

A typical reinforcement layout for a long barrel vault is shown in figure 217 *A*, and it will be seen that the bars follow the lines of principle shear and tension stresses in the corners and towards the edge beams. Typical edge beam details for long barrels and north-light shells are also shown (*B, C*). It can be seen that the upstand to the gutter at the base of a north-light shell forms its edge beam and also supports struts in the plane of the glazing which, by reducing its span, permits the top edge beam to be quite small.

Reference has been made earlier to the possi-bility of prestressing long barrel shells and alternative ways of doing this are shown at (*D*). When the edge

349

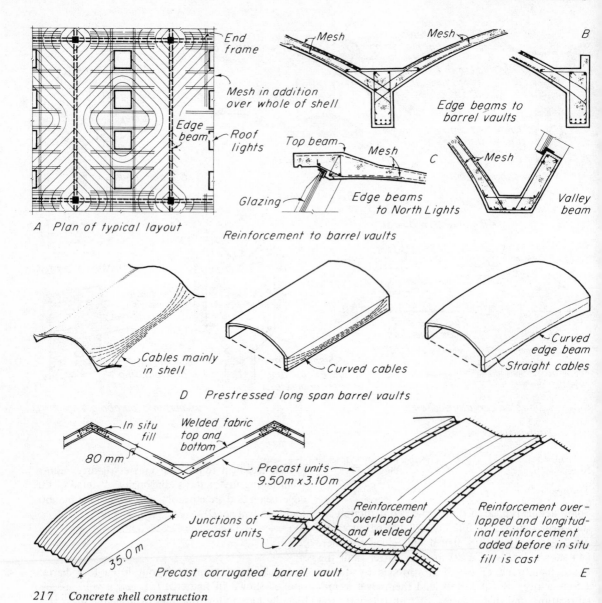

A Plan of typical layout

End frame

Mesh in addition over whole of shell

Edge beam

Roof lights

Mesh

Mesh

Edge beams to barrel vaults

B

Top beam

Mesh

C

Mesh

Glazing

Edge beams to North Lights

Valley beam

Reinforcement to barrel vaults

Cables mainly in shell

Curved cables

Curved edge beam

Straight cables

D Prestressed long span barrel vaults

In situ fill

Welded fabric top and bottom

80 mm

Precast units 9.50m x 3.10m

Junctions of precast units

Reinforcement overlapped and welded

Reinforcement overlapped and longitudinal reinforcement added before in situ fill is cast

35.0 m

Precast corrugated barrel vault

E

217 Concrete shell construction

beams are very shallow the post-tensioning cables are placed in the lower part of the shell. They are, however, placed more easily within edge beams. A double curvature barrel gives the advantage of straight cables. The reasons for curving cables or curving the beams are given on page 231.

Corrugated barrels, such as that shown in figure 217 E, may be constructed on a framework of ribs of tubes, angles or timber shaped to the curve of the

shell, braced apart and covered with hessian. Concrete is applied to the hessian in two or more coats, the hessian sagging between the supports under the weight of the first coat. In the case of large shells the concrete is applied by means of a cement gun. For small shells up to about 12 m span no reinforcement is used and the thickness of the shell varies from 19 to 38 mm. The width of the corrugations would be from 900 mm to 1.80 m and the depth about one-

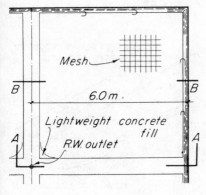

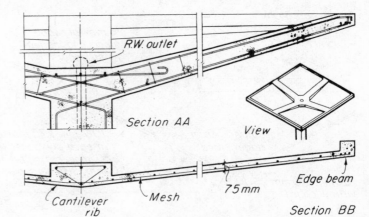

Part plan

218   Hyperbolic paraboloid shell

fifth of the width[1]. For large span vaults the structure is cast on formwork with angular corrugations in the form of folds. Alternatively it may be precast in sections and formed into a monolithic structure on the site as shown at (E).

Details of a cantilever hyperbolic paraboloid are shown in figure 218. The fact that a hyperbolic paraboloid surface can be developed by two groups of mutually perpendicular straight lines has a practical advantage. The shuttering for this type of concrete shell is simple, since it consists wholly of straight timbers. The straight supporting bearers are placed at the appropriate varying angles parallel to the edge beams and are covered with straight planks at right-angles to them to produce the final doubly-curved surface on which the concrete is cast (see figure 258). As all the stresses in these shells are direct, they may be as thin as 38 to 50 mm for spans of 30 to 40 m. In practice the minimum thickness of a concrete shell depends largely on the method employed to place the concrete and slabs as thin as this makes guniting necessary.

Due to the thinness of reinforced concrete shells it is generally necessary to provide an insulating lining and this can often be conveniently achieved by using insulating building boards as permanent shuttering. Alternatively, the interior of the shell may be sprayed with vermiculite or asbestos fibre, these methods having the advantage of allowing inspection of the underside of the concrete. Insulation may be provided externally by such materials as woodwool, cork or vermiculite screed and in this position has the advantage of protecting the concrete from solar heat and consequent thermal expansion.

Plastics

Shells, as noted on page 304, are appropriate structural forms in which to use plastics but, although much research and development work has been carried out in the structural use of plastics their application in this field has not developed to the extent of that in the field of claddings[1].

Folded slabs

Timber

Folded slabs may be constructed in timber as (i) framed panels, (ii) hollow stressed skin panels, (iii) laminated board panels. These are illustrated in figure 219. It will be seen that in each case the folds are stiffened either by additional members or by joining the edge members of adjacent slabs. The framed panels shown (A) have a non-structural covering of strawboard slabs and are joined at the hips. At (B) is an example of a folded slab using 16 mm plywood stiffened on the underside by ribs. The top members are splay cut and are bolted together, and the ends of adjacent panels are bolted together over the supporting beams which are at 8 m centres. In the valleys, support and fixing to

[1]   For greater details of this form of construction, see *Proceedings of Institution of Civil Engineers*, Part III, Aug 1953: 'Corrugated Concrete Shell Roofs', by J. Hardresse de Warrene Waller and Alan Clift Aston.

[2]   For some examples see article on 'Glass Fibre Reinforced Plastics' in the *Architects Journal* 4 April 1973.

351

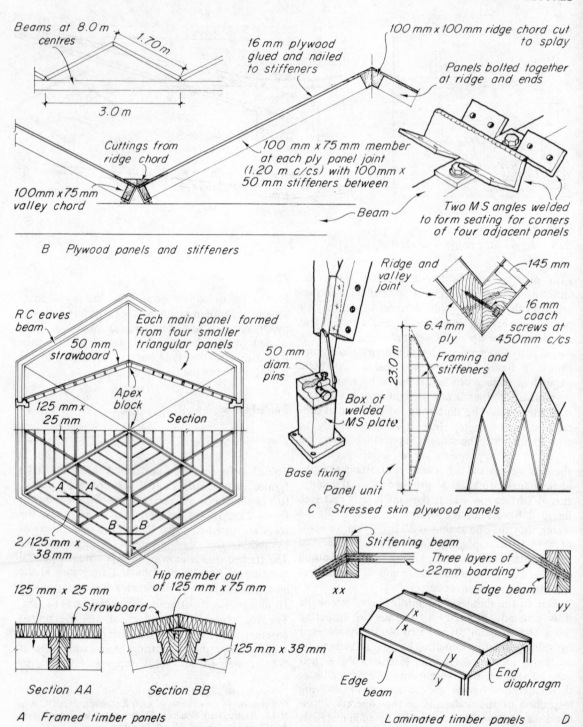

Beams at 8.0 m centres

1.70 m

3.0 m

16 mm plywood glued and nailed to stiffeners

100 mm x 100 mm ridge chord cut to splay

Panels bolted together at ridge and ends

Cuttings from ridge chord

100 mm x 75 mm member at each ply panel joint (1.20 m c/cs) with 100 mm x 50 mm stiffeners between

100 mm x 75 mm valley chord

Beam

Two M S angles welded to form seating for corners of four adjacent panels

B   Plywood panels and stiffeners

R C eaves beam

50 mm strawboard

Each main panel formed from four smaller triangular panels

125 mm x 25 mm

Apex block

Section

A   A

B   B

2/125 mm x 38 mm

Hip member out of 125 mm x 75 mm

125 mm x 25 mm

Strawboard

Section AA

Section BB

125 mm x 38 mm

A   Framed timber panels

Ridge and valley joint

145 mm

50 mm diam. pins

16 mm coach screws at 450 mm c/cs

6.4 mm ply

Framing and stiffeners

23.0 m

Box of welded MS plate

Base fixing

Panel unit

C   Stressed skin plywood panels

Stiffening beam

Three layers of 22mm boarding

Edge beam

xx

yy

Edge beam

End diaphragm

Laminated timber panels   D

219   Timber folded slabs

352

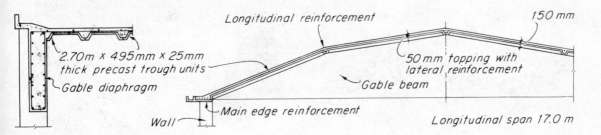

220   Precast concrete folded slab

beams is given by 450 mm long brackets formed from steel angles welded toe to toe. The stressed skin panels of thin ply (*C*) are coach screwed at the folds, thus uniting adjacent edge members. When the slabs are built up with layers of boarding as at (*D*) they must be stiffened at the folds with laminated beams.

Concrete

Concrete folded slabs may be cast *in situ* or be precast. Precasting may be carried out in two ways:

(i) by casting full size panels and uniting a pair at the valley joint before hoisting, the ridge joint then being formed with *in situ* concrete uniting projecting reinforcing bars as in the vault in figure 217 *E*;

(ii) by using smaller precast trough units spanning from fold to fold as shown in figure 220, steel being placed at the longitudinal folds and mesh and shear reinforcement on the top before a structural topping is cast to form a composite construction.

Plastics

Because of their geometrical form folded slabs, like shells, are suitable forms for the structural application of plastics but the remarks made under shells apply here.

Space frames

Timber

The main problem in applying timber to space frame[1] construction is one of connection between timber members in several planes. One solution is to use three lattice girders and connect them by metal lugs at the nodes, as shown at (*A*), figure 221. An alternative is shown at (*B*) using steel lugs bolted to the ends of each member by means of which they are fixed to pressed or cast metal 'multi-directional' connectors.

Steel

Steel space frames may be constructed of cold-rolled sections, angles or tubes riveted, bolted or welded together or to suitably shaped gusset plates or connectors. Tubes are very suitable since they may be more easily joined at any angle, and due to their better performance in compression will produce lighter structures, particularly over large spans. Figure 222 *A, B, C* shows three examples of steel space frame construction in angles and tubes together with details of joints. The types of connectors shown in figure 224 may also be used in the fabrication of space frames instead of welding.

---

[1] For definition of this term as used here see page 326.

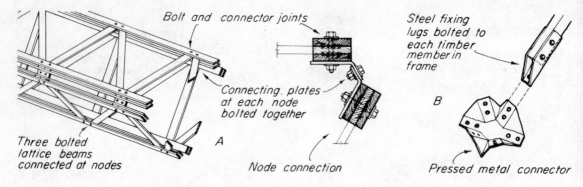

Bolt and connector joints

Connecting. plates
at each node
bolted together

A

Three bolted
lattice beams
connected at nodes

Node connection

Steel fixing
lugs bolted to
each timber
member in
frame

B

Pressed metal connector

221   Timber space frame

Aluminium

As stated earlier, the direct stressing of members and the inherent resistance to torsion in space frames makes them suited to construction in aluminium. Rods or round and rectangular tubes are commonly used, often together with specially extruded sections, as shown in figure 222 *E*. Connectors of various types are also used into which the tubes fit, one of which is illustrated at (*D*). Other suitable connectors are shown in figure 224.

Concrete

The development of precasting and prestressing techniques has made possible the use of concrete members of comparatively light weight and small cross sectional area which may be used in the construction of space frames. The members may be connected together with site bolting and grouting in of reinforcement *in situ*. Some details of a concrete roof constructed of space frames of triangular section closely spaced and carrying light decking are shown at (*F*), figure 222, and indicate the general arrangement of the members and joints.

354

Single-layer grids

Steel

Steel flat and folded grids are constructed of I-beams welded together at the intersections. One element may be continuous, to which the intersecting elements are welded to produce structural continuity in both directions. Alternatively, the beams are 'halved' to each other before welding to facilitate positioning of the beams (figure 223 *A*).

Aluminium

The structural limitations of aluminium mentioned earlier preclude its economic use in single layer flat grids.

Concrete

Concrete grids have commonly been cast *in situ* and this remains the likely method for heavily loaded grids. Precast 'ferro-cement' permanent shuttering[1] has been used to facilitate the accurate casting of the ribs, as shown in figure 223 *B*, but glass fibre reinforced plastic pans are now widely used for this purpose (see page 417).

For lightly loaded roofs of long span a lighter structure can be obtained with post-tensioned precast concrete units as shown at (*C*).

[1]   See footnote page 326.

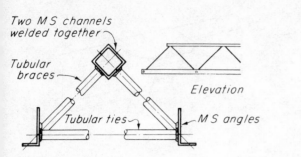

Two MS channels welded together

Tubular braces

Tubular ties

MS angles

Elevation

A   Space frame of welded steel

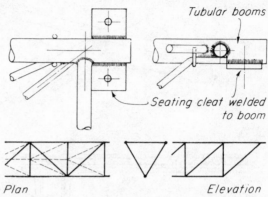

Tubular booms

Seating cleat welded to boom

Plan

Elevation

B   Tubular steel space frame

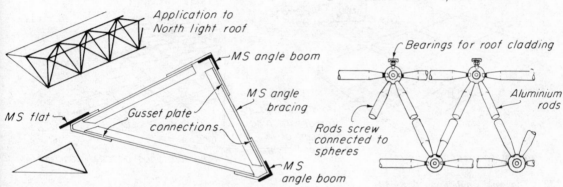

Application to North light roof

MS angle boom

MS angle bracing

MS flat

Gusset plate connections

MS angle boom

C   Steel space frame with gusset plates

Bearings for roof cladding

Aluminium rods

Rods screw connected to spheres

D   Aluminium space frame with node connectors

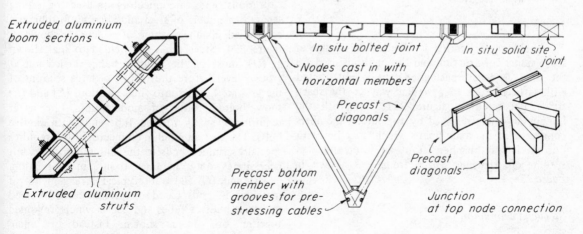

Extruded aluminium boom sections

Extruded aluminium struts

In situ bolted joint

In situ solid site joint

Nodes cast in with horizontal members

Precast diagonals

Precast bottom member with grooves for pre-stressing cables

Precast diagonals

Junction at top node connection

E   Aluminium space frame

Precast concrete space frame   F

222   Space frames

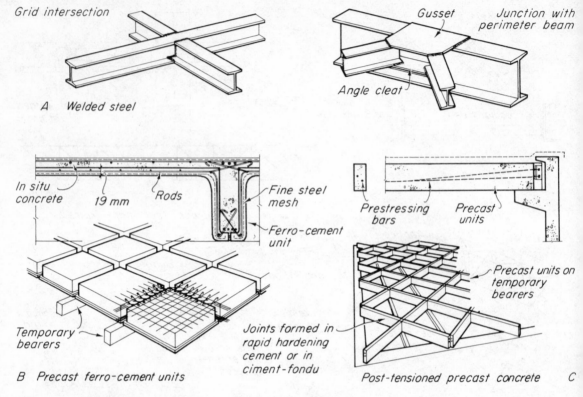

Grid intersection

A   Welded steel

Gusset   Junction with perimeter beam

Angle cleat

In situ concrete   19 mm   Rods   Fine steel mesh

Ferro-cement unit

Prestressing bars   Precast units

Precast units on temporary bearers

Temporary bearers

Joints formed in rapid hardening cement or in ciment-fondu

B   Precast ferro-cement units

Post-tensioned precast concrete   C

223   *Diagonal single layer flat grids*

Double-layer flat grids

Timber

The problem of jointing is identical with that in space frame construction and the methods used are the same. Two-way space grids may be constructed with lattice beams in a similar way to the space frame illustrated in figure 221, but with the horizontal lattice replaced by top lateral members connected at the node points which, together with bottom lateral members form an interconnected system of triangular space frames as shown in figure 224 *A*.

Steel and aluminium

Double-layer grids are constructed in steel and aluminium in basically the same way as described under space frames but invariably making use of connectors at the nodes.

Various types of connectors are used for joining separate members of steel and aluminium tube and cold-rolled channels, some of which are shown in figure 224. Steel tubes are welded to that shown at (*D*), most of the welding being carried out at ground level before the whole, or large sections of the grid are hoisted into final position. (*E*) and (*F*) show bolted forms of node joints, the latter requiring specially formed tube ends with captive bolts. The cast aluminium connector at (*G*) requires specially crimped ends to the aluminium tubes but necessitates only 'hammer fixing' in fabrication.

(*B*) and (*C*) illustrate typical arrangements of members in grids based on prefabricated inverted pyramids built of tubes and angles which are joined together on site as shown. Instead of 'open' pyramids, sheet aluminium or plastic can be used to form them as shown in figure 225, and they are joined together in a similar way but with node connectors to join them to a top grid of tubes.

356

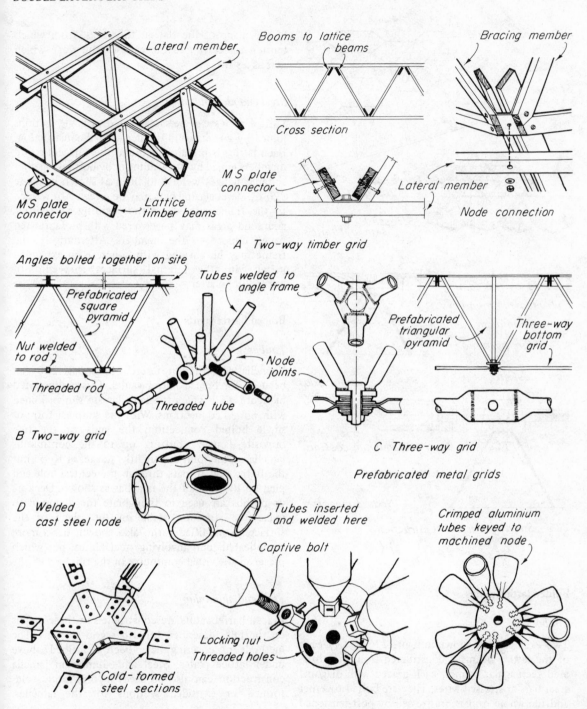

Lateral member

Booms to lattice beams

Bracing member

Cross section

MS plate connector

Lateral member

MS plate connector

Lattice timber beams

Node connection

A Two-way timber grid

Angles bolted together on site

Prefabricated square pyramid

Tubes welded to angle frame

Prefabricated triangular pyramid

Three-way bottom grid

Nut welded to rod

Node joints

Threaded rod

Threaded tube

B Two-way grid

C Three-way grid

Prefabricated metal grids

D Welded cast steel node

Tubes inserted and welded here

Crimped aluminium tubes keyed to machined node

Captive bolt

Locking nut

Threaded holes

Cold-formed steel sections

E   Pressed steel node

F Screw connected node

Key connected node   G

224   Double layer space grids

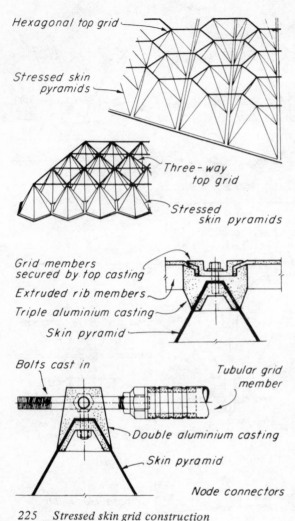

Hexagonal top grid

Stressed skin pyramids

Three-way top grid

Stressed skin pyramids

Grid members secured by top casting

Extruded rib members

Triple aluminium casting

Skin pyramid

Bolts cast in

Tubular grid member

Double aluminium casting

Skin pyramid

Node connectors

225   Stressed skin grid construction

Folded lattice plates

Timber

Two examples are shown in figure 226. In (*A*), each of the lattice beams is constructed from prefabricated rectangular frames, each braced with diagonal steel bars or stressed wires. They are bolted together and the whole roof is prestressed by post-tensioned bars at the eaves. The other example at (*B*) shows a roof with wide plates formed of rafters, trussed to give stiffness to the plates, carrying purlins covered with diagonal boarding. Lattice girders are formed

in the plane of the slab at the eaves, from which tension cables run to the ends of the ridge which acts as a strut.

Steel and aluminium

Lattice plates can be constructed of angle sections joined by gussets or in tubuler welded construction. Each lattice is prefabricated and the adjacent chord members joined by site welding or by bolting together suitable flanges welded to them as shown in figure 226 *C*. Metal glazing bars may be screwed or welded to the framework where roof lighting is required, and solid areas may be covered with prefabricated panels clipped to the members. Alternatively, the frame may be covered with concrete on expanded metal with the internal surfaces sprayed with vermiculite plaster.

Braced barrel vaults

Timber

'Lamella' construction, referred to on page 330, can be used for braced barrel vaults. This is illustrated in figure 227, where the lamellas are shown joined with bolted connections. With the simplest form of single bolted connection, the ends of the two opposite lamellas butting on to the continuous lamella at the joint are slightly staggered to permit the fixing bolt to pass through the central hole and pick up the ends of the lamellas as shown. Decking or purlins are used to triangulate the diamond to make the structure stable in the plane of the surface. A modified form, also shown, has a more complicated joint involving steel plates, but which preserves the visual continuity of the ribs.

Steel and aluminium

Braced barrel vaults are constructed in these metals on the lines described for other grid systems. The methods of covering have been described above under lattice plates. Steel and aluminium lamella construction can also be used, incorporating cold-formed or extruded sections for the lamellas.

Ribbed domes

Two examples are illustrated in figure 228 showing in each case the methods of forming the junction

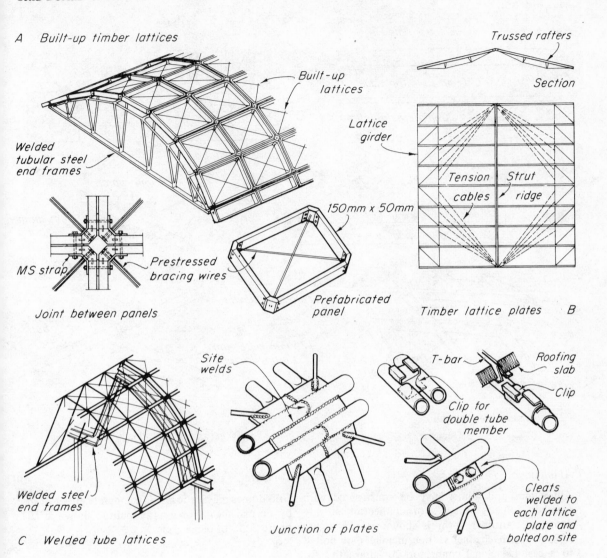

A   Built-up timber lattices

Built-up lattices

Welded tubular steel end frames

Trussed rafters

Section

Lattice girder

Tension cables   Strut ridge

150mm x 50mm

MS strap   Prestressed bracing wires

Joint between panels

Prefabricated panel

Timber lattice plates   B

Site welds

T-bar   Roofing slab

Clip

Clip for double tube member

Welded steel end frames

Junction of plates

Cleats welded to each lattice plate and bolted on site

C   Welded tube lattices

226   Folded lattice plates

at the crown, by connector (*A*) or by compression ring (*B*), and details of the bracing. The timber example shown is for a span of over 90 m and is covered with woodwool slabs. Smaller, lighter ribbed domes may be constructed with laminated ribs covered with plywood or layers of tongued and grooved boarding.

Grid domes

Timber

'Lamella' construction or triangular stiffened ply or

framed panels similar to those in figure 219 may be used for grid domes.

Steel

Tubular steel grid domes may be constructed with node connectors into which the tubes are inserted and welded, similar to those used for double layer grids, but using six-way instead of eight-way connectors to form a three-way grid, as shown at (*A*), figure 229.

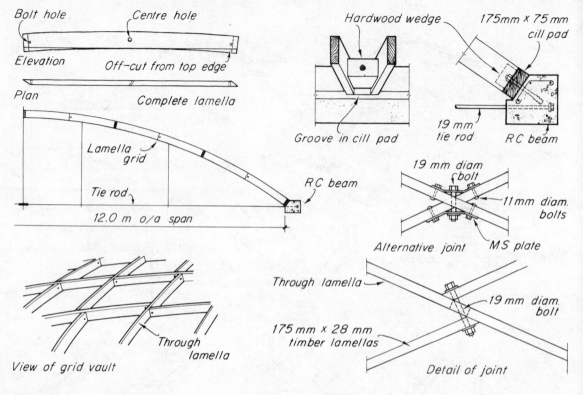

227   Lamella construction

Aluminium

Aluminium grid domes may be constructed with members joined by types of connectors already described. Another form is shown at (B), figure 229, in which channel section members are bolted to a cast hexagonal connector. An alternative for two-way grids using rectangular aluminium tubes cross-halved to each other is also shown (C). The tubes are extruded or built up from extruded sections according to the size of the dome.

The use of sheet material by means of which the grid members may be formed, and are strengthened against buckling by the membrane action of the sheets, has been described. One system (D), uses prefabricated diamond-shaped sheet aluminium panels each with a tube cross strut, three of which form a basic hexagon when joined together. When totally fabricated the struts form an external hexagonal grid. A similar but double-layer system is shown at (E).

Concrete

Grid domes may be formed in concrete either *in situ* by normal means or by the method shown at (B), figure 223, for plane grids.

Plastics

Aluminium is not the only material which can be used in sheet form for structures of this type. Sheet steel would be satisfactory. But as explained earlier in this chapter, materials which lack great stiffness, such as aluminium or plastics, can most effectively be used in this way. Complete pyramidal hexagonal units with flanged base edges may be formed of plastic which are bolted together at the flanges and connected at their nodes to a top grid of tubes. Alternatively the units may be formed of flanged triangular panels bolted together and to node connectors as at (F) to form a geodesic dome.

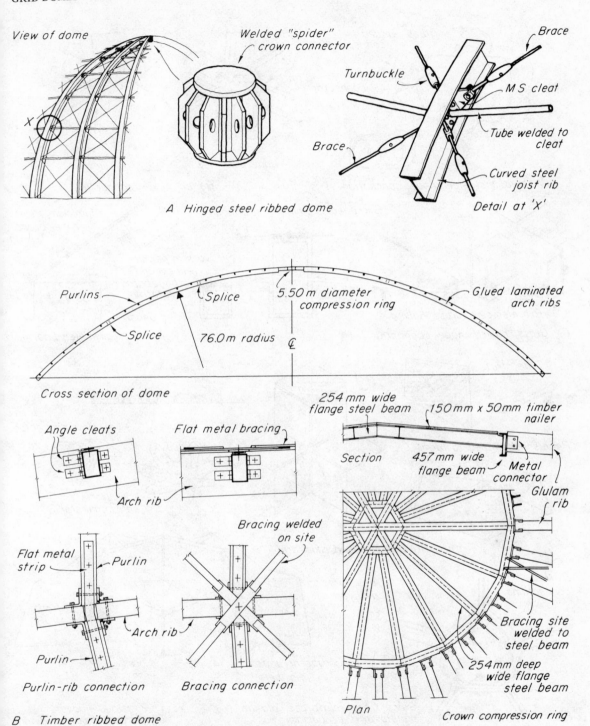

View of dome

Welded "spider" crown connector

Brace

Turnbuckle

M S cleat

Brace

Tube welded to cleat

Curved steel joist rib

A Hinged steel ribbed dome

Detail at 'x'

Purlins

Splice

5.50 m diameter compression ring

Glued laminated arch ribs

Splice

76.0 m radius

Cross section of dome

Angle cleats

Flat metal bracing

254 mm wide flange steel beam

150 mm x 50 mm timber nailer

Section

457 mm wide flange beam

Metal connector

Arch rib

Glulam rib

Flat metal strip

Purlin

Bracing welded on site

Arch rib

Purlin

Bracing site welded to steel beam

Purlin-rib connection

Bracing connection

254 mm deep wide flange steel beam

B Timber ribbed dome

Plan

Crown compression ring

228 Ribbed domes

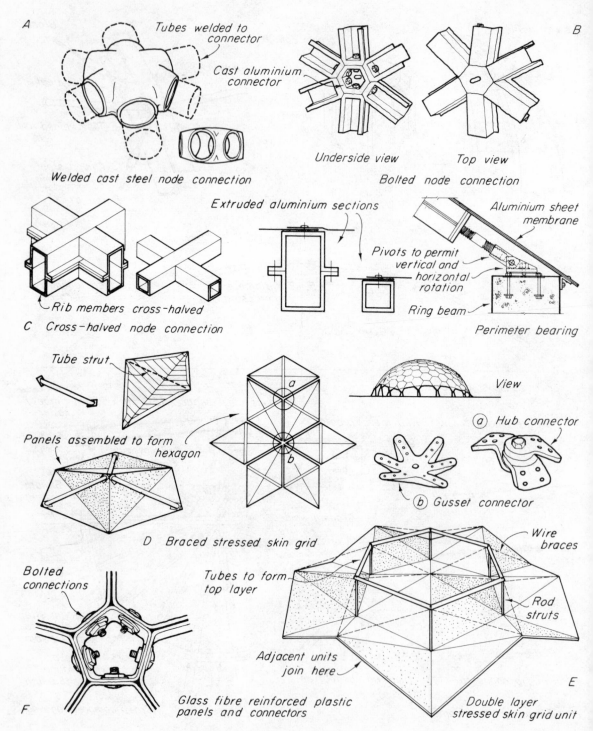

A

Tubes welded to connector

Cast aluminium connector

Welded cast steel node connection

B

Underside view   Top view

Bolted node connection

Extruded aluminium sections

Aluminium sheet membrane

Pivots to permit vertical and horizontal rotation

Ring beam

Rib members cross-halved

C   Cross-halved node connection

Perimeter bearing

Tube strut

Panels assembled to form hexagon

a

b

View

ⓐ Hub connector

ⓑ Gusset connector

D   Braced stressed skin grid

Bolted connections

Tubes to form top layer

Wire braces

Rod struts

Adjacent units join here

Glass fibre reinforced plastic panels and connectors

F

Double layer stressed skin grid unit

E

229   Grid domes

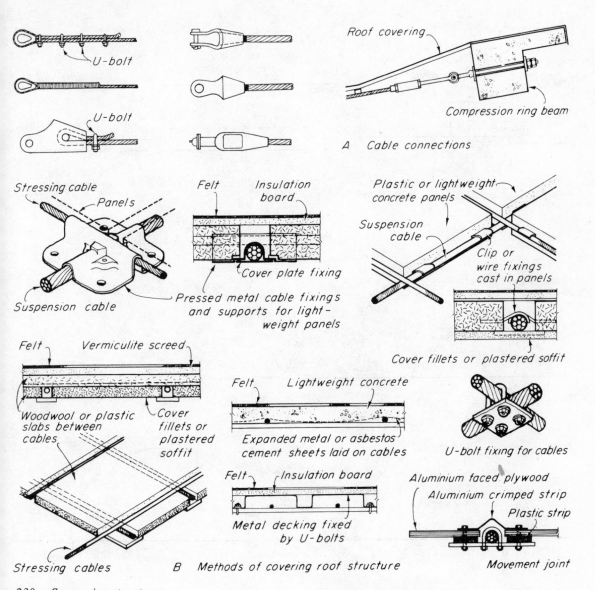

A   Cable connections

Stressing cable

Panels

Suspension cable

Felt   Insulation board

Cover plate fixing

Pressed metal cable fixings and supports for light-weight panels

Plastic or lightweight concrete panels

Suspension cable

Clip or wire fixings cast in panels

Cover fillets or plastered soffit

Felt   Vermiculite screed

Woodwool or plastic slabs between cables

Cover fillets or plastered soffit

Felt   Lightweight concrete

Expanded metal or asbestos cement sheets laid on cables

Felt   Insulation board

Metal decking fixed by U-bolts

U-bolt fixing for cables

Aluminium faced plywood
Aluminium crimped strip
Plastic strip

Movement joint

Stressing cables

B   Methods of covering roof structure

230   Suspension structures

Tension structures

As indicated earlier in this chapter these fall into two categories, suspension and air-supported structures, the former using cables and the latter a membrane.

Suspension structure

Details of this form of construction vary widely.

Various means are adopted to connect the cables to supports and common methods are illustrated in figure 230 *A*. The two sets of cables are secured at their crossings either by U-bolts or by connectors designed to fulfil both this function and that of supporting the cladding elements. Both are illustrated in (*B*).

Cable structures can be clad with light metal sheets such as corrugated aluminium, tongued and grooved timber boards, plywood panels, lightweight

363

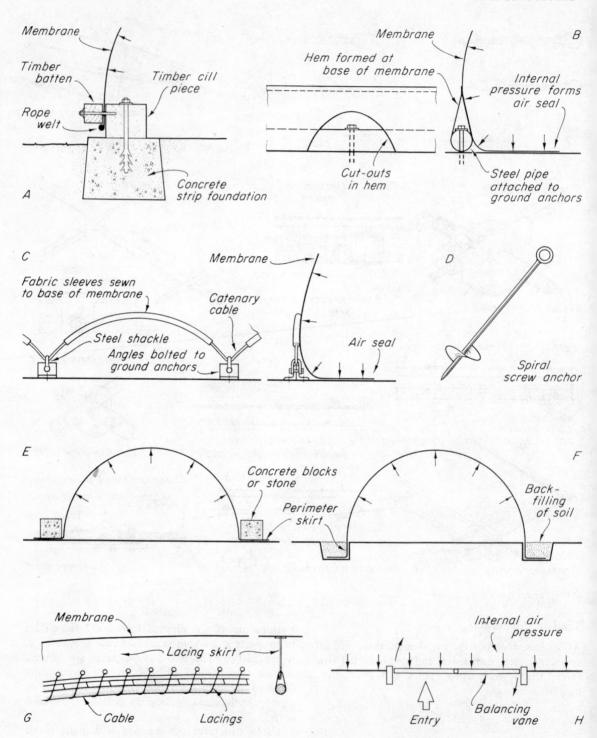

Membrane

Timber
batten

Timber cill
piece

Rope
welt

Concrete
strip foundation

A

Membrane

B

Hem formed at
base of membrane

Internal
pressure forms
air seal

Cut-outs
in hem

Steel pipe
attached to
ground anchors

C

Membrane

Fabric sleeves sewn
to base of membrane

Catenary
cable

Steel shackle
Angles bolted to
ground anchors

Air seal

D

Spiral
screw anchor

E

Concrete blocks
or stone

Perimeter
skirt

F

Back-
filling
of soil

Membrane

Lacing skirt

Internal air
pressure

Cable   Lacings

G

Entry

Balancing
vane

H

231   Air stabilised structures

slabs or panels, 'gunite' concrete sprayed on to expanded metal lathing or with sprayed-on plastic of a suitable type. Some of these methods are shown in figure 230 B.

Temperature changes affect the length of the cables and some provision for movement must be made in the cladding, similar to that shown at the joint between the plywood panels.

Air stabilised structure

Membranes For air supported structures coated fabrics, plastic films, woven metallic fabrics or metallic foils may be used, the first being most commonly employed. These consist of a terylene or nylon fabric coated on one or both sides with a plasticised elastomer such as neoprene or vinyl.

Jointing methods depend on the membrane material used and consist of sewing through a double folded seam, cementing or welding. The first method, although necessitating sealing of the stitches, is the cheapest but produces joints only about 75 per cent as strong as those made by welding or cementing.

The tubes in air inflated tube structures are normally in the form of a circular woven fabric sleeve with an inner airtight elastomer lining and an outer weather resistant coating of similar material. Small tubes requiring very high pressures necessitate high performance standards for the fabric, joints and sealing, since failure of the tube results in a dangerous explosion.

Anchorages Direct or positive anchorage to the ground is used for structures which are to be in position for any length of time. The most reliable form is that in which the membrane is secured throughout its perimeter to a continuous concrete foundation. The bottom edge of the membrane is finished with a rope welt and is clamped to a continuous timber cill piece by coach screws or bolts passing thorough steel angles or timber battens as shown in figure 231 A. The cill piece is rag-bolted at intervals to the concrete strip foundation.

Alternative methods of positive anchorage use pipes or cables, accommodated in fabric sleeves or hems attached to the membrane, which are secured to anchors at intervals round the perimeter. Figure 231 B shows the first type in which sections of pipe are inserted in a hem round the base of the

membrane and are attached to ground anchors at cut-outs in the hem at approximately 1 m intervals. The second type is shown at (C). In this a rope or cable passes through catenary shaped fabric sleeves sewn to the base of the membrane and is secured at intervals to the ground anchors.

These two forms of anchorage do not require a continuous concrete ground anchor but can be used in conjunction with steel spiral screw anchors (D), expanding head anchors installed in previously bored holes or bored piles, placed at the appropriate distances apart and of a depth sufficient to provide the necessary resistance to the membrane uplift.

For structures of a temporary nature ballast anchorages may be used. These may be formed in a number of ways (i) with stones or concrete slabs placed on a perimeter skirt laid flat on the ground (figure 231 E), (ii) with soil backfilling on to a skirt placed in a perimeter trench (F) (iii) with a split ballast tube at the base of the membrane laced up after filling with gravel, sand or soil. Water may also be used as ballast but has the disadvantage that at points of high local lift-up forces it may flow away leaving no anchorage at the point where it is most required.

Cables and cable networks Cables used to divide air supported membranes into smaller elements (see page 334) can cause chafing and wear on the membrane as the two move relative to each other. This problem can be overcome by placing the cables internally, the membrane being attached to them by dropped skirts from the membrane and lacings as shown in figure 231 G.

Access openings As stated on page 332 large openings require air locks or air curtains. These are not essential for small openings but, in order to overcome the problem of the force required on a normal door to overcome the pressure differential, a door counterbalanced by a vane should be used as shown in figure 231 H. The vane, reacting to the same internal pressure as the door, provides a counterbalance and makes opening an easy operation. Where traffic is heavy such a door would result in high air loss because it would be open for too great a length of time. In such circumstances a revolving door should be used. These are not suitable for escape purposes and when adopted some alternative must be provided for use in emergency.

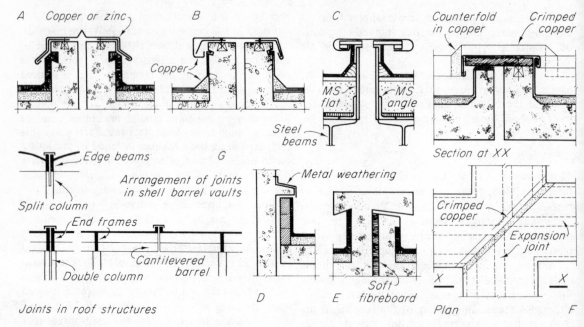

A   Copper or zinc
B
C
Copper
MS flat   MS angle
Steel beams
Counterfold in copper   Crimped copper
Section at XX

Edge beams
G
Split column
Arrangement of joints in shell barrel vaults
Metal weathering
Crimped copper
Expansion joint

End frames
Double column   Cantilevered barrel
Soft fibreboard
D   E

Joints in roof structures
Plan   F

232   Expansion joints

MOVEMENT CONTROL

Settlement movement

Where settlement joints are provided in the structure, as described in chapter 5, the joint in the roof slab must be designed to permit relative movement. Methods of forming and weatherproofing the joint will in principle be similar to those adopted for expansion joints (see figure 232 *D*).

Thermal movement

Thermal insulation placed on top of the roof slab may be sufficient to keep movement to a negligible amount in small buildings, but in large buildings, although movement will be reduced, expansion joints will usually be required. Long parapets should always be provided with expansion joints. Typical details of expansion joints in roofs are shown in figure 232, *A* to *E*. Extensive roofs will require joints in both directions and a method of forming a four-way junction at an intersection is shown at (*F*).

In multi-bay barrel vault roofs expansion joints are made between edge beams and, in continuous barrels, between end frames using double or split

columns as shown in figure 232 *G*. Alternatively, the joint in the length of the barrel can be made between cantilevered sections of the barrel as shown, to avoid the use of double or split columns. Provision for movement in long beams to framed roof structures is made by means of hinged, roller and rocker bearings.

When a long roof slab is supported on loadbearing masonry walls, damage to the walls may be caused by the outward thrust at the ends of the roof unless (i) continuous expansion joints are formed in the roof and walls, or (ii) expansion joints are formed in the roof and a sliding bearing on the wall is incorporated. Crackling in top floor part-itions caused by roof movement (figure 139 *C*) can be avoided by isolating the top of the partition from the roof slab. Plaster should not in this case be carried over the junction of walls or partition with the roof slab.[1]

Reference is made in chapter 5, page 234, to the magnitude of thermal movements and to constructional details of expansion joints generally.

[1] See *Principles of Modern Building*, Volume 1, 3rd Edition, chapter 2, and *Building Research Station Digest* 12 (First series).

10 Fire protection

Fire protection is the protection of the occupants, contents and structure of a building from the risks associated with fire. The subject is of vital importance and is so linked with the design and construction of buildings that it is essential to have an understanding of the factors which influence the nature and form of fire protection and of the principles which are the basis of the various regulations.[1] It is for this reason that it is discussed in this chapter in relatively broad terms, involving as it does considerations of planning as well as of construction, rather than in terms merely of the constructional aspects of protection which, in fact, are conditioned by, and cannot be separated from, planning aspects.

Fires are almost always the result of negligence. The rate of growth, ultimate severity and nature of the risks involved in the event of fire depend largely on the use to which the building is being put. For example, a theatre or concert hall, because of the large number of people accommodated, involves a high life risk even though the combustible contents may be low. But a large warehouse storing much combustible material involves a considerable risk of extensive damage to structure and contents but a low risk to occupants because their number is likely to be small. The degree of damage finally sustained will be influenced by the structure, its effectiveness in confining the fire and its ability to remain stable both at the seat of the fire and remote from it.

It is the purpose of fire protection, therefore, to protect life, goods and activities within a building. These aims are achieved by inhibiting the combustion of the materials from which the building is constructed and by preventing the spread of fire within the building and between buildings. In addition, by ensuring that the elements of construction fulfil their functions for a sufficient length of time during a fire, the occupants are able to escape and the fire brigade is given time to deal effectively with the outbreak and thus limit the total damage. Protective measures involve suitable forms of construction, suitable planning of the building internally and in relation to adjacent buildings, and satisfactory planning and construction of the means of escape.

GROWTH OF FIRE IN BUILDINGS

The origin of a fire is usually the result of negligence, ranging from direct acts such as lighted cigarettes left burning to more indirect causes such as poor installation and maintenance of electrical wiring and the ineffective control of vermin.

The growth of a fire depends on the amount and disposition of combustible material within the building, either in the form of unfixed materials or parts of the fabric such as wall and ceiling linings, which will contribute to the fire[2]. At the beginning of a fire materials near the source of ignition receive heat and their temperatures rise until, at a certain point, inflammable gases are given off. Ignition of the material then occurs and it begins to produce heat instead of merely receiving it. This 'primary' fire then preheats the remaining combustible material and raises its temperature to ignition point. After this a spark or, for example, a flame burning along a floor board, will in a moment start an intense fire involving the whole of the contents. The time taken for this instantaneous spread of the fire, or 'flash-over' as it is termed, to occur depends on the proximity of combustible materials to the source of the fire and on the presence or otherwise of an adequate supply of oxygen. In full-scale tests on domestic living rooms this 'flash-over' occurred after about fifteen minutes[2]. At an early stage the structure at the seat of the fire becomes involved and the strength of some non-combustible materials such as steel, for example, is adversely affected[3].

The further spread of fire occurs by the usual methods of heat propagation, that is, by conduction, convection and radiation, together with the process of flame spread across the surfaces of combustible materials.

[1] This chapter seeks to establish basic principles in a confused situation. Where references are made to Regulations, Codes of Practice or other published documents these are for the purpose of giving point to the principles.

[2] See 'Combustible materials', page 371.

[3] 'Studies of the Growth of Fire', *Fire Protection Association Journal*, Reprint no. 1.

Depending on the construction and design of the building, heat can be conducted through walls causing temperatures high enough to ignite spontaneously any combustible materials stacked on the side remote from the fire.

Any unprotected flue-like apertures such as stair-wells, lift shafts, or light-wells may permit the spread of fire by the passage of convection currents and flying brands.

Heat can be radiated to ignite combustible materials at some distance from the source. In this way fire is spread to other areas as well as to nearby buildings.

The spread of fire is more rapid where combustible linings are used, and particularly where air spaces exist behind them as this provides two surfaces over which flame can spread.

The later stages in the progress of a fire are mostly affected by the behaviour of the particular structure involved. The deformation of supporting columns and beams in a framed building structure may cause apertures in enclosing walls and floors through which direct flame spread can occur. Ultimately, the complete collapse of a building may happen as a result of the weakening of the structure and the gutting of all but the non-combustible parts such as brick walls. It should be noted that structures with a high degree of continuity can be particularly hazardous in this respect, since the collapse of one component can produce far-reaching effects throughout the rest of the building.

The hazards associated with fire may be considered in order of importance as
Personal : the hazard to the occupants of the building
Damage : the hazard to the structure and contents
Exposure : the hazard due to the spread of fire to other buildings[1].

The hazards to occupants are due to the following factors:
Reduction of oxygen   This is due to the consumption of oxygen by the fire and is accompanied by toxic or asphyxiating gases evolved by the fire, particularly carbon monoxide. An associated additional hazard is smoke which results from incomplete combustion. Staircases providing means of escape and corridors giving access to them, where exposed to any particular risk, should be protected by fire-resisting partitions and self-closing fire-resisting doors. Wherever possible staircases should be ventilated to the open air in all storeys.

The degree of hazard to occupants is influenced by the distance between points of escape, size and number of exits and stairs, and the existence or otherwise of a sprinkler system.

Increase in temperature   Breathing is difficult above a temperature of 149°C, and since this temperature will be reached well in advance of the path of the fire, it is essential that automatic alarms be provided, designed to operate at given temperatures (49°C − 70°C).

Spread of flame   The risk here is of burning by physical contact with flame and should be minimized by enclosing escape routes with non-combustible materials, or materials of low flame spread.

The practical requirements in dealing with these hazards are discussed in more detail in later sections of this chapter.

FIRE-GRADING

The term 'fire-grading' has a two-fold application. (a) It is applied to the classification or grading of the elements of structure of buildings in terms of their degree of resistance to fire. (b) With a broader meaning, it is applied to the classification of buildings according to the purpose for which they are used, that is, according to occupancy, and according to the fire resistance of the elements of which they are constructed.

This grading of buildings is considered from two points of view, firstly in terms of damage and exposure hazard, for which the protection is mainly provided by structural precautions, and secondly in terms of personal hazard, for which protection is provided primarily by easy means of escape. The first is considered here and the second later under the sections on fire escape.

The severity of a fire depends largely upon the amount, nature and distribution of combustible material in a building. Thus in determining the requisite degree of fire protection it is necessary to take into account the use and size of a given building, and by this means to assess the probable amount and type of combustible material which would contribute to a fire.

---

[1] See *Post-War Building Study* no. 20, 'Fire Grading of Buildings, Part I', HMSO.

Grade of occupancy	Low fire load	Moderate fire load	High fire load
Fire load kJ/m²	Not exceeding 9495 (18 991 on limited isolated areas)	9495 to 18 991 (not exceeding 37 982 on limited isolated areas)	18 991 to 37 982 (not exceeding 75 964 on limited isolated areas)
Building types	Flats, offices, restaurants, hotels, hospitals, schools, museums, public libraries	Retail shops, eg footwear, clothing, furniture, groceries. Factories and workshops generally	Warehouses, etc, used for bulk storage of materials of non-hazardous nature
Equivalent severity of fire in hours of standard test	1	2	4

Table 24   *Fire load grading*

Fire load

The assessment of the severity of a fire due to the combustible materials in a building is made by reference to what is known as the 'fire load', which is the amount of heat, expressed in kJ, which would be generated per square metre of floor area of a compartment of the building by the complete combustion of its contents and any combustible parts of the building. The fire load is determined by multiplying the weight of all the combustible materials by their calorific values and dividing by the area of the floor; it is based on the assumption that the materials are uniformly distributed over the whole area of the floor.

The calorific value is the property of a material which indicates the amount of heat which will be generated by a particular quantity of that material and it governs the ultimate severity of a fire. Thus the maximum heat is evolved from materials having highest calorific values, eg

bitumen	35 355 kJ/kg
cork	16 747 kJ/kg
paper	16 262 kJ/kg
petrol	46 520 kJ/kg
rubber	39 542 kJ/kg
wood	18 508 kJ/kg[1]

An office, therefore, with 2.5 kg/m² of combustible furniture and papers would have a fire load of 4065.50 kJ/m² to 4627.00 kJ/m².

The fire load is used as a means of grading occupancies, and the grading set out by the Joint Committee on Fire Grading of Buildings[2] is shown in Table 24.

Materials of the same calorific value can give rise to differences in fire risk according to ease of ignition, rate of burning and whether or not, for example, they are explosive or emit dangerous fumes. These would constitute exceptional risks. In addition, certain processes such as paint spraying with inflammable materials, or the application of heat to combustible materials, also constitute exceptional risks[3]. Any building in which these risks are likely to arise or in which the fire load is greater than the maximum for High Fire Load grading must be considered separately.

Fire resistance

It has been stated earlier in this chapter that protective measures against fire include means for limiting the spread of fire together with the provision of structural elements capable of ful-

[1] See *Post-War Building Study* No. 20, Appendix III, for list of calorific values.

[2] See *Post-War Building Study* No. 20.

[3] See *Post-War Building Study* No. 20, appendices I and II for lists of abnormal materials and occupancies.

filling their functions during a fire without the risk of collapse. The first, to some extent[1], and the second entirely, depend upon the use of elements of structure[2] of an appropriate degree of fire resistance.

The term 'fire resistance' used in connection with fire protection has the precise meaning defined in BS 476[3]. It is applied to elements of structure, not to a material. Fire resistance depends on the way in which materials are used in an element and not solely on whether they are combustible or not. This standard lays down tests for establishing the fire resistance of various elements of structure by means of which different forms of construction may be graded according to the length of time during which they will function satisfactorily under the action of the standard test fire (see page 372).

Measurements have been taken of the severity of fires caused by different fire loads. This information has enabled the degrees of severity of fires due to known fire loads to be expressed in terms of periods of exposure to the standard test fire.

For design purposes the equivalent severities shown in table 24 were adopted by the Joint Committee on Fire Grading of Buildings.

This means that if a building is to contain a fire load of 14243 kJ/m^2 then walls, floors and other elements of construction having a fire resistance of 2 hrs would resist the effects of fire without collapse or penetration of the fire even if all the material within it burned.

The determination of the necessary fire resistance of every structure on the assumption that it must withstand the complete burn-out of the contents would pay no regard to the effect of other means of protection, such as fire-fighting and automatic sprinklers, for example, and would render impracticable many sound forms of construction and make the cost of the structure high. Lesser degrees of resistance relative to fire load become practicable, however, by assuming a rational combination of all methods of fire protection and by having regard to the fact that in smaller buildings escape and fire fighting is easier, and fires can more easily be brought under control with less chance of the structure collapsing.

In practice, therefore, the necessary fire resistance of a building (or any part of it if subdivided by suitable walls and/or floors into compartments) in any particular case is determined with reference to types of users, fire load, the areas of floors and the height and cubic capacity of the building.

For convenience the required fire resistance for the elements of structure determined in this way with reference to the size and use of different building types is given in building regulations which tabulate this information to assist designers[4].

MATERIALS IN RELATION TO FIRE

Effect of fire on materials and structures

Non-combustible materials

These are materials which if decomposed by heat will do so endo-thermically, that is, with the absorption of heat or, if they oxidize, do so with negligible evolution of heat. Also included are those materials which require a temperature beyond the range of most fires before they react in any way.

Non-combustible materials do not contribute to the growth of a fire but are damaged when the temperature is reached where decomposition, fusion or significant loss of strength occurs. When incorpor-

---

[1] Other factors in limiting the spread of fire are the use of non-combustible linings and planning considerations.

[2] These are defined in the London Building (Constructional) By-Laws Clause 1.03 as:
(i) Any floor, beam, column or hanger; and any load bearing wall or other load bearing member.
(ii) Any partition or wall which separates parts of a building used for different purposes or tenanted by different persons, other than any partition or wall separating any part of a building used for office purposes from any other part of that building used for office purposes on the same floor; and
(iii) Any staircase including landings and supports other than a secondary staircase between not more than two adjacent storeys.
Regulation E1 of the Building Regulations 1972 includes the following parts of the building fabric in the definition of elements of structure:
Floors (including compartment floors), any load bearing wall, any part of a structural frame (members forming the roof structure only are *not* included) any separating or compartment wall, any independent beam or column, any gallery and any structure enclosing a protected shaft.

[3] BS 476 : Part I, 1953 : *Fire Tests on Building Materials and Structures* − 'a relative term used to designate that property by virtue of which an element of structure as a whole functions satisfactorily for a specified period whilst subjected to prescribed heat influence and load'.

[4] London Building (Constructional) By-law Clause 11.05 and Table 9, Building Regulation E5: Table A, part 1 and part 2.

ated in the structure the loss of strength during a fire may be such that they no longer maintain the integrity of the structure. Examples of non-combustible materials are metal, stone, glass, concrete, clay products, gypsum products and asbestos products.

Apart from marble and gypsum, which liberate free lime under severe heat, the majority of these materials do not decompose chemically under the action of fire. However, certain natural stones, concrete, gypsum and asbestos products decompose by losing their water of crystallization, and in so doing acquire pronounced fire endurance. Although metal and glass suffer negligible decomposition, these materials lose considerable strength at high temperatures and glass and some metals such as aluminium and lead, fuse or soften under heat. Asbestos cement is liable to shatter in intense heat and will disintegrate when struck by water during a fire.

Steel   It should be noted that steel loses strength and rigidity above a temperature of 299°C and at 427 – 482°C, a temperature well within the normal range of building fires, there is a loss of strength of up to 80 per cent. This potential weakness, together with the expansion which takes place under heat, means that in the early stages of a fire unprotected steelwork will bend, buckle and expand. Deformation of a supporting steel member or frame causes walls and floors to fall away and leaves the fire free to spread into other areas which might not have been affected had the structure been maintained. This is an example which underlines the fact that non-combustible materials are not necessarily fire-resisting and this may be further emphasized by considering timber, which although combustible will, if of adequate section, fulfil its structural function longer than mild steel.

Aluminium   The poor performance of aluminium structures in this respect should also be noted. Aluminium has a much lower critical temperature than steel and for elements under load this has been given as 204°C as against 500°C for steel[1]. No reference is made in the by-laws to methods of protecting structural work in aluminium alloy from fire. It would appear that this material is only suitable for buildings of low fire risk or where the solution to the structural problem transcends the fire risk as, for example, in the case of hangars.

Concrete   The behaviour of concrete under action of fire depends largely upon the type of aggregate used. Flint gravel expands greatly and causes spalling of the concrete. Other stones, apart from limestone, behave similarly to a lesser degree, but crushed clay brick and slag do not cause spalling. Provided spalling does not occur, disintegration is slow. Thus additional protection can be given to reinforcement or steel members by increasing the concrete cover so long as non-spalling concrete is used. If a spalling type is used the cover tends to break and fall away.

In terms of fire resistance, aggregates are classified as follows, those in Class 1 being the non-spalling types:

Class 1   Foamed slag, pumice, blast furnace slag, crushed brick and burnt clay products, well-burned clinker, crushed limestone; (expanded slag, expanded clay and sintered pulverised fuel ash are also included in the London Building (Constructional) By-Laws. Pelletted fly ash is included in the Building Regulations).

Class 2   Flint, gravel, granite and all crushed natural stones other than limestone.

At temperatures higher than those required to produce spalling the free lime in the cement is converted into quicklime after which, if the concrete is exposed to water, or even moist air, the lime slakes and in expanding causes complete disintegration of the concrete.

Combustible materials

These are materials which, within the temperature range associated with fires, will combine exothermically with oxygen. That is to say, in their reaction with oxygen considerable heat is evolved and they flame or glow.

Such materials, whether forming part of the structure or the contents of the building, are responsible for the growth of a fire and its ultimate severity. Examples of such materials are wood or wood products, vegetable products, animal products and manufactured products such as fibre-board and strawboard. Within this classification are flammable materials which ignite readily and react vigorously, producing rapid flame spread. Examples of these materials are volatile liquids (petroleum distillates), certain plastics and certain paints based on nitrocellulose products.

[1]   International Convention for the Safety of Life at Sea, 1948. MOT recommendation.

371

Timber Although timber is a combustible material it will, as pointed out earlier, function as a structural member for a longer period than one of metal provided it is of adequate section. If of sufficient size timber is extremely difficult to burn. Some species such as teak, iroko, jarrah and others are highly resistant to fire.

Wood boarding up to 9.5 mm thick ignites relatively easily and continues to burn. But timbers about 150 mm thick will char in depth and this inhibits rapid combustion of the wood beneath. Test have indicated that for any given cross-sectional area and load, a beam having a square cross section will have the longest fire endurance, that is, the time elapsing before collapse occurs.

The degree of combustibility of timber can be reduced by treatment with a suitable fire-retardant. Pressure impregnation gives better results than brush applications. Timber should be worked to finished sizes before impregnation as treated timber is difficult to work.

The following characteristics of combustible materials will influence the precautions necessary in providing adequate protection.

Ignitability A measure of the ease of ignition expressed as the minimum temperature at which the material ignites under given atmospheric conditions, eg wood, wood products and cellulose materials 221 - 298°C, plastics 260 - 482°C, synthetics of nitro-cellulose origin upwards of 138°C, bitumen upwards of 65.5°C, petrol distillates 204 - 482°C.

Flammability The property of a combustible material which determines the severity of flame and flame spread. It is related to volatility and the vigour with which volatile gases react with oxygen. In addition to actual volatile liquids which may be stored within a building, many organic natural and synthetic materials exhibit this property, particularly when well dispersed as in fabrics.

Calorific content The rate at which heat is generated in the reaction between all combustible materials and oxygen is related directly to the flammability of a material, and the ease with which the reaction occurs is related to the ignitability of the material. This has been discussed briefly on page 369.

It will be obvious from what has been said regarding the effect of fire upon materials and structures that in the design of buildings it is

essential to have some means of assessing the likely behaviour in fire of various materials and forms of construction, particularly at a time when new materials and combinations of new materials to form structural elements are constantly being introduced. BS 476: Part 1: 1953, provides means for grading or classifying both materials and structures for this purpose. This has been referred to briefly already with regard to elements of structure and will now be considered generally.

Fire-resistance grading

The fire-resistance grading of elements of structure is determined by tests carried out in accordance with BS 476 by the Department of Scientific and Industrial Research and Fire Offices' Committee Joint Fire Research Organization. The tests are applied to elements of structure whether composed of one or more materials. The elements are graded according to the length of time during which, while exposed to the heat of special furnaces[1], they satisfy certain conditions laid down in the Standard.

It must be understood that the grading is applied to the structural element as a whole and to a precise specification of that element. What may appear to be minor changes in the details of construction of an element may result in a great change in its fire resistance[2].

The notional periods of fire resistance ascribed to elements of structure comprising various materials are given in Schedule II, Tables A to G, and Schedule III (Fire retardant materials) of the London Building (Constructional) By-laws, 1972, and Schedule 8, parts 1 to 8, of the Building Regulations, 1972. It should be borne in mind, however, that other parts of the by-laws, in controlling the details of construction, may make provisions in excess of the requirements of these Tables. For example, London By-law 8.02 lays down minimum solid covers to steel beams and columns in external walls and for steelwork likely to be adversely affected by moisture from adjoining earth, which are greater than those required for the shorter periods of fire

[1] These are fully illustrated in *National Building Studies Research Paper* no. 12, 'Investigations on Building Fires'.

[2] BRS Digest 106, 'Fire: Materials and Structure', gives as an example the fire resistance of a 102.5 mm brick wall which may be increased from just less than 2 hr to 6 hr by the application of 13 mm gypsum perlite plaster on each side.

372

resistance (see footnote page 198). The Building Regulations and the London By-laws both refer to the relevant Codes of Practice for the design of structures in various materials and these should be consulted in conjunction with the Schedules referred to above in designing for a given standard of fire resistance.

Classification of combustible materials

Some materials are obviously combustible. Others are not obviously so because they burn so slowly. Yet in a fire they could contribute to its severity. Further, wall and ceiling linings, which present large surfaces, provide an easy means for the spread of fire when constructed of combustible materials. Particularly is this so where air spaces or cavities exist behind such linings, so permitting the rapid and undetected spread of fire. For practical design purposes some method is required to determine whether or not a material is combustible, and also the ease with which flame is likely to spread over its surface. BS 476: Part 1: 1953 specifies two tests in this connection: (a) combustibility; (b) surface spread of flame. The first decides whether or not a material will burn or contribute to a fire, although no degree of combustibility is defined; the second compares the rate and extent of flame spread along the surfaces of different materials, and classifies the results from 1 to 4. Building Regulation E15 adds a higher class, 0, and these are defined as

Class 0 (i) Entirely non combustible or
(ii) The surface, or (if bonded to a substrate) the surface and substrate have an index (1) not exceeding 12 and sub index (1_1) not exceeding 6 in accordance with BS 467: Part 6: 1968. Special requirements apply to plastic facings with softening points below $120°C$ and reference should be made to BS 2782: 1970 for details of test criteria

Class 1 Surfaces of very low flame spread
Class 2 Surfaces of low flame spread
Class 3 Surfaces of medium flame spread
Class 4 Surfaces of rapid flame spread.

The results of these tests and the results of fire-resistance tests under the same standard, should not be confused. A material which satisfies the combustibility test might fail under the conditions for fire-resistance test; a building-board having Class 1 flame spread classification may or may not make a siginficant contribution to the fire resistance of a composite structure in which it is used. Table 25 shows the surface spread of flame classifications of a number of common lining materials[1].

It will be noticed that the classification of a material may be altered by the application of intumescent paint or fire-retardant solution, but too great a dependence should not be placed on these treatments. Some materials achieving a particular surface classification in this way may, in fact, under certain conditions aid the development of a fire to a greater extent than those achieving the same classification without treatment. This has been shown by the results of fire tests in rooms fully lined with various materials[2].

These tests also indicate that in terms of aiding the development of fire there is not a great difference between this type of Class 1 lining and Class 3 linings. In view of this, therefore, when Class 1 linings must be used it is probably advisable to limit the choice to those achieving this class without surface treatment, as is required by the *Department of Education and Science Building Bulletin* no.7, 'Fire and the Design of Schools'. Further, the behaviour of these paints and solutions during the use of a building is not fully known, and it is possible that their effectiveness may be reduced by heat and condensation, for example. For this reason insurance companies prefer inherently low surface flame spread materials if non-combustible alternatives are out of the question for reasons of economy, and this is reflected in the premiums charged.

The fact that materials of similar surface spread

[1] Fire note no. 9 *Surface spread of flame* – tests on building products by F.C. Adams and Barbara F.W. Rogowski, HMSO (reprinted 1968) gives information on tests and manufacturers under the following classes:

Class 0 (or Grade A)	Wood and wood based material Non combustible substrate Miscellaneous	
Classes 1, 2 & 3	Wood and wood based material	Timber Plywood Particle board Hardboard Fibre insulating board Woodwool
	Plastics Non combustible substrate Miscellaneous	

[2] See BRS Digest 106.

Class 0	Class 2
Asbestos insulating board Asbestos cement sheets PVC/steel laminates 600 680 720 density flame retardant polyurethane finished flooring grade chipboard Special high density asbestos cement silica autoclaved lining board PVC faced asbestos cement board	Synthetic resin bonded paper and fabric sheets Standard hardboard with certain decorative treatments Compressed straw slabs with painted distemper finish

Class 1	Class 3
Plasterboard Woodwool slabs Metal faced (including all edges) plywood Flame retardant hardboard and medium hardboard Flame retardant insulation board Flame retardant hardboard with wood grain veneering Standard hardboard with intumescent paint Compressed straw slabs with surface treatment of vermiculite and wood chippings Compressed straw slabs plastered Hardwood with surface impregnation treatments Glass fibre reinforced plastic sheets to BS 4154: 1967 with flame retardant additives Resin bonded chipboard with proprietary finishes 600 680 720 flame retardant chipboard Asbestos cellulose flexible lining board Melamine impregnated wood veneer laminate (can be class 1 on order) Melamine faced plastic laminate — flame retardant grade Melamine faced hardboard Compressed asbestos clad wood fibre board PVC faced asbestos cement board Flameproofed decorative veneers on plywood backings Composite boards of urethane foam / faced both sides with plasterboard PVC/steel laminates	Timber and plywood weighing more than 416.25 kg/m Chipboard Compressed straw slabs with manilla or impregnated cardboard covering Glass fibre reinforced sheets Standard hardboard Medium hardboard Fibre insulating board with certain decorative treatments

	Class 4
	Plywood and timber weighing less than 416.25 kg/m Fibre insulating boards Compressed straw slabs with one or both faces embossed with PVC copolymer foil 0.25 mm thickness Acrylic sheets (polymethyl methacrylate)

Table 25   *Surface spread of flame classification*

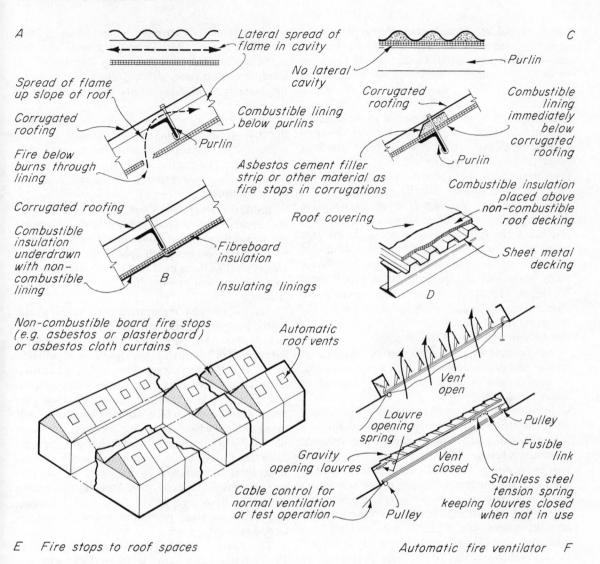

A

Lateral spread of flame in cavity

No lateral cavity

Spread of flame up slope of roof

Corrugated roofing

Fire below burns through lining

Purlin

Combustible lining below purlins

C

Purlin

Corrugated roofing

Combustible lining immediately below corrugated roofing

Asbestos cement filler strip or other material as fire stops in corrugations

Corrugated roofing

Combustible insulation underdrawn with non-combustible lining

B

Fibreboard insulation

Insulating linings

Roof covering

Combustible insulation placed above non-combustible roof decking

Sheet metal decking

D

Non-combustible board fire stops (e.g. asbestos or plasterboard) or asbestos cloth curtains

Automatic roof vents

Vent open

Louvre opening spring

Pulley

Fusible link

Gravity opening louvres

Vent closed

Stainless steel tension spring keeping louvres closed when not in use

Cable control for normal ventilation or test operation

Pulley

E   Fire stops to roof spaces

Automatic fire ventilator   F

233   Fire protection to single storey buildings

The use of combustible materials

of flame classification can, under the same conditions, assist the growth of fire to a different extent has resulted in the development of the Fire Propagation Test as specified in BS 476: Part 6.

Furthermore, certain types of insulating materials such as foamed plastics which melt at temperatures lower than those used in the surface spread of flame test are difficult to classify by that test and the fire hazard of such materials may better be determined by the Fire Propagation Test which is more discriminating amongst materials of low fire hazard than the surface spread of flame test.

As has previously been mentioned, combustible linings are potentially more dangerous when there are air spaces behind them. This condition arises when these materials are used as a wall lining on battens, as a cladding to timber or metal studs in partitions or as false ceilings to mask beams and services. It often occurs in light factory-type buildings where the inner insulating lining is separated from the outer cladding by the structural framing supporting both (figure 233 A). The hazard is particularly great in the case of roofs, since the

375

lateral spread of flame causes the main fire to spread by radiation and by the fall of burning pieces of the lining itself. Linings should therefore be of Class 1 spread of flame and non-combustible or underdrawn with a non-combustible sheeting if combustible, as at (*B*). To avoid continuous air spaces, the lining should be placed directly below the roof covering, and fire breaks or non-combustible infillings provided where corrugated or other profiled sheeting is used, as at (*C*). Further fire breaks can be formed in roof spaces by enclosing roof trusses with fire-resisting building boards. Where a decking is used, combustible insulating linings should be placed above the deck and below the roof covering to avoid the creation of air spaces, as at (*D*). However, advantage should be taken of non-combustible linings by placing them below the deck in order to protect it from fire below. To avoid continuous air spaces above false ceilings formed with combustible linings, the lining should be placed directly below any downstand beams tight against the soffits.

An example of good all-round performance is that of wood slab decking, which is self-spanning, of good insulation value and Class 1 spread of flame classification.

In addition to the fire hazard of insulating lining materials, roof coverings may contribute to a fire. Although regulations require non-combustible coverings generally, they permit the use of combustible coverings on a non-combustible base or in circumstances where the building in question is well away from any adjacent property[1]. Combustible coverings used are mastic asphalt, bitumen felts of both mineral and vegetable base and bitumen protected metal. Mastic asphalt can be considered as of low fire risk due to the large proportion of inert material in it, but bitumen burns easily and should only be used directly on a non-combustible and fire-resistant decking or layer, since collapse of the deck would spread fire into the building from outside and fire to the roof from the inside. In the form of protected corrugated metal sheeting, bitumen coverings are particularly hazardous when combined with combustible linings, as was shown in the Jaguar Car Factory fire in 1957. One of the chief factors contributing to the spread of that fire was the fall of flaming bitumen and lining material on to the factory floor well in advance of the main fire.

The introduction of asbestos into the make-up of bituminous roof coverings considerably reduces their fire hazard. Both asbestos based felts and asbestos shrouded protected metal sheeting afford good protection since, although the bitumen burns, the blanket of asbestos resists the penetration of fire.

In addition to wood wool already mentioned, the following insulating materials may be used without fire hazard: asbestos insulating board, plasterboard, vermiculite, glass fibre and mineral wool (the last two without a bitumen binder or paper covering).

BS 476: Part 3: 1958, specifies a test to assess the capacity of a roof construction to prevent the penetration of fire from outside and the degree to which the covering will spread fire. The test is less severe than those used to establish fire resistance and the resulting roof gradings are not called fire resisting.

Translucent and transparent corrugated plastic sheets are combustible, but the behaviour of different types varies. Special requirements exist governing the use of transparent and translucent panels in roofs. The Building Regulations 1972 3rd amendment has introduced a concession in respect of glass and rigid PVC sheets which cannot be designated under BS 476 but which can be classified as self extinguishing when tested by method 508 A of BS 2782: 1970, whereby such sheets may be used more than 6 m from a boundary and less than 6 m from a boundary in the cases of roofs to garages, conservatories or outbuildings not exceeding 40 m^2 in floor area or for roofs over balconies, verandahs, open car ports, covered ways and detached swimming pools.

The London Building (Constructional) By-laws 1972 (6.05) allow wire laminate PVC which is classified AA (without suffix) generally and also describes conditions in which translucent plastic rooflights and glazing not classified AA may be used whether in the form of panels in the same plane as the main roof, as domelights or as tiles. The main constraints as to dimension are:

(a) Panels in same plane as and secured to adjoining sheets of non-combustible material: area not to exceed 3m^2; the distance between panels not to be less than their widths and lengths and no panel to be nearer to edges of roof than 900 mm

---

[1] London Building (Constructional) By-laws 1972, clause 6.03. The Building Regulations 1972, Clause E17.

(b) If a dome light: not to exceed 1.80 m in any direction; lights not to be nearer to each other than largest dimension and not nearer to edges of roof than 900 mm

(c) Translucent tiles not to exceed 1 m² in any one plane of roof.

Roof lights and glazing may be of translucent plastic material classified not less than CA (without suffix) in positions where plain glass would conform with the provisions of the by-laws.

The roof of a conservatory not exceeding 45 m² attached to a single dwelling may be of rigid PVC sheeting classified as self extinguishing by method 508A of BS 2728: 1970.

By-law 6.06 limits the aggregate area of openings in roofs to half the area of the roof surfaces but excludes: dwelling houses, all buildings where the whole of the glazing is of wired glass in metal frames and if the glazing is classified AA (without suffix). Translucent plastics are taken as glazing in a roof.

FACTORS AFFECTING DESIGN

General design and planning

The broad approach to planning for fire protection is to design the elements of construction to withstand the action of fire for a given period dependent on the size and use of the building, to compartmentalize the building so as to isolate the fire within a given section or area, to separate specific risks within the building and generally to prevent the uncontrolled spread of fire from its source to other parts of the building. Further, a building must be planned to allow the occupants to escape by their own unaided efforts. Suitable separation must be provided to prevent fire, hot gases and smoke from spreading rapidly by means of common spaces such as corridors, staircases and lift shafts, thus trapping the occupants and causing panic.

Separation

The structural elements used to prevent horizontal and vertical spread of fire are the walls and floors, together with any structure necessary to support them, all of which must be of adequate fire resistance. Any openings within these elements of separation must be protected in a manner which does not nullify the effect of that element during a fire, whilst affording access during normal use. Such protection may be given by self-closing, fire-resisting doors or steel doors or roller shutters held open by fusible links and arranged to close automatically in the event of fire. Glazing must be fire resistant where light is required to penetrate the separating elements.

It has been found in practice that where fires occur in large undivided spaces within a building, the greater intensity of heat and volume of smoke generated prevents fire-fighters from attacking the fire at its source. They are driven out of the building, where their hoses become less effective. It is generally considered that 7080 m³ is the maximum volume of any one compartment which could reasonably be tolerated from a fire-fighting standpoint. This figure will vary in practice according to the use to which a building is put and according to such circumstances as the availability of fire-fighting services and the provision of automatic sprinklers.

Further precautions must be taken to prevent the spread of fire between compartments by way of external enclosing surfaces common to them that is, roofs and external walls. For this reason separating walls are often required to extend above adjacent roofs to prevent lateral spread of flame, although in certain types of building lateral spread of fire is considered to be adequately countered by interposing a non-combustible firebreak between adjacent roof spaces extending up to the underside of a non-combustible roof covering. It has been considered essential that the horizontal separation afforded by the floors should not be curtailed at the outer perimeter of a building in a manner which would allow fire to spread vertically between compartments or storeys. Thus regulations limit the proportion of openings and impose dimensional constraints on their sizes and relationships to each other[1]. Figures 234 and 235 show the basic issues affecting the design of external openings as between

[1] London By-law 6.14, clauses 2, 3 and 4. Building Regulations 1972, clause E7 and schedule 9.
Results of an investigation carried out by the Fire Research Station on various types of framing elements and panels indicated that 'a reduction in the fire resistance of the under window panels from ½ hr to nil did not significantly increase the hazard of flames from the ground floor room igniting the contents of the room above it, even with the maximum fire load of 7596 kJ/m². It is also shown that a projecting element is of little use as a firestop (*Fire Research Annual Report*, 1958 (HMSO)).

6·13 (1) Certain domestic, office and garage buildings
exempt from provisions of 6·14 (2 and 3)

General provisions of the London
(Constructional) Bylaws, 1972

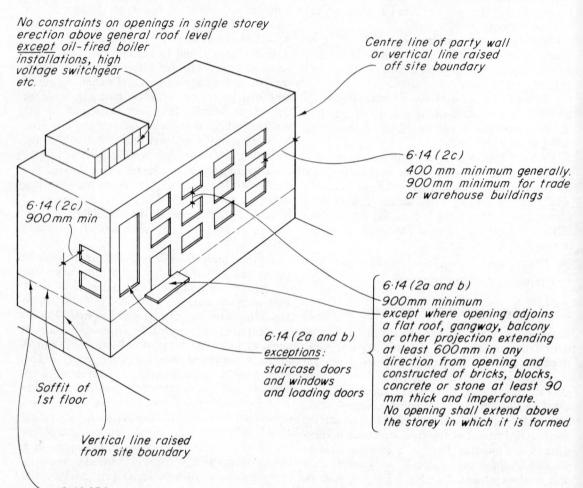

No constraints on openings in single storey
erection above general roof level
except oil-fired boiler
installations, high
voltage switchgear
etc.

Centre line of party wall
or vertical line raised
off site boundary

6·14 (2c)
400 mm minimum generally.
900mm minimum for trade
or warehouse buildings

6·14 (2c)
900mm min

6·14 (2a and b)
900mm minimum
except where opening adjoins
a flat roof, gangway, balcony
or other projection extending
at least 600mm in any
direction from opening and
constructed of bricks, blocks,
concrete or stone at least 90
mm thick and imperforate.
No opening shall extend above
the storey in which it is formed

6·14 (2a and b)
exceptions:
staircase doors
and windows
and loading doors

Soffit of
1st floor

Vertical line raised
from site boundary

6·14 (3)   Generally: the total area of openings in an external enclosure above
the soffit of the 1st floor shall not exceed 50% of the total area of the enclosure (above
the soffit of the 1st floor). All glazing within wall thickness to be deemed to be an opening

234   Openings in external walls and relationship to boundaries

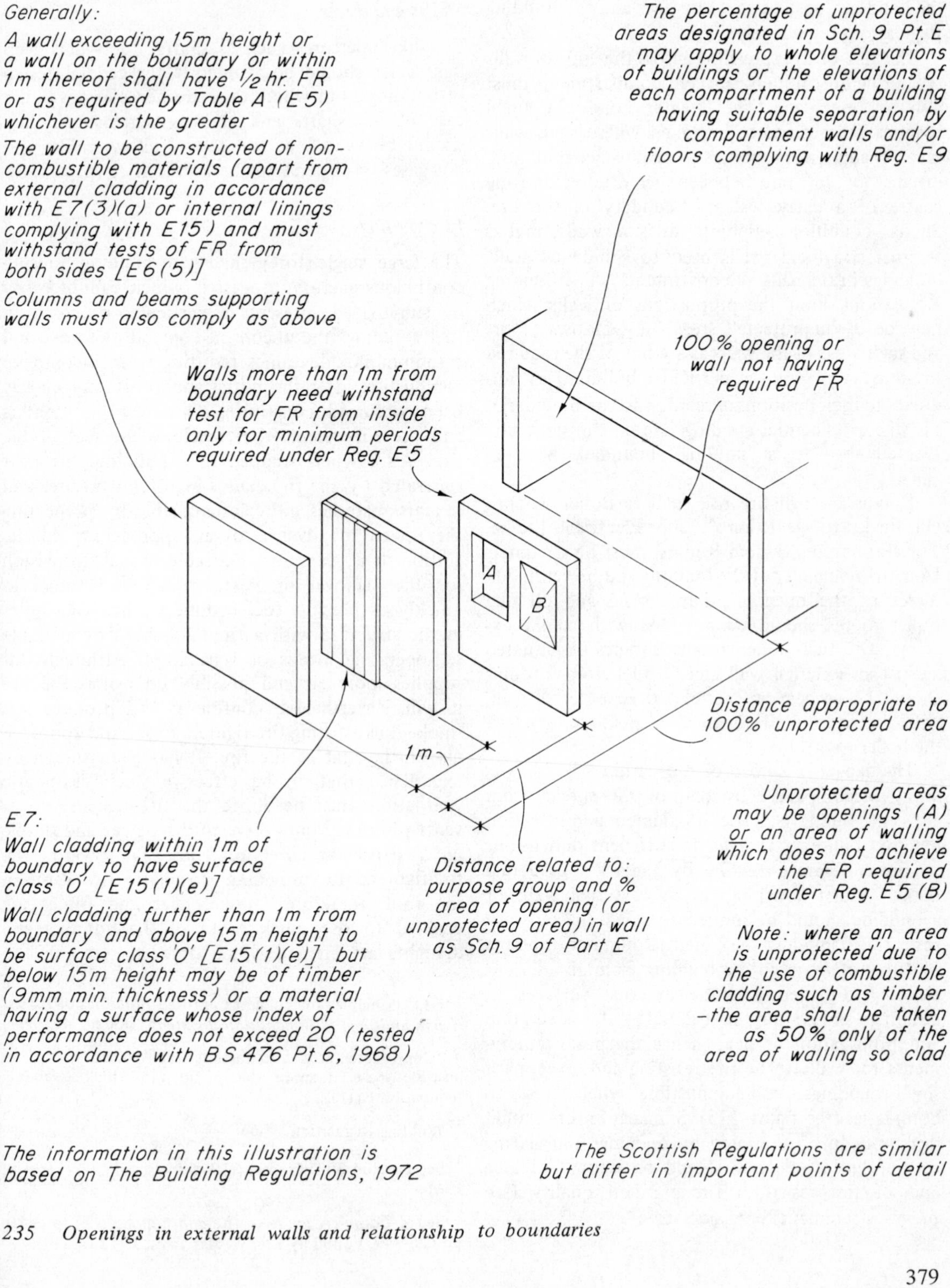

Generally:

A wall exceeding 15m height or a wall on the boundary or within 1m thereof shall have ½ hr. FR or as required by Table A (E5) whichever is the greater

The wall to be constructed of non-combustible materials (apart from external cladding in accordance with E7(3)(a) or internal linings complying with E15) and must withstand tests of FR from both sides [E6(5)]

Columns and beams supporting walls must also comply as above

Walls more than 1m from boundary need withstand test for FR from inside only for minimum periods required under Reg. E5

The percentage of unprotected areas designated in Sch. 9 Pt. E may apply to whole elevations of buildings or the elevations of each compartment of a building having suitable separation by compartment walls and/or floors complying with Reg. E9

100% opening or wall not having required FR

Distance appropriate to 100% unprotected area

1m

E7:

Wall cladding within 1m of boundary to have surface class 'O' [E15(1)(e)]

Wall cladding further than 1m from boundary and above 15m height to be surface class 'O' [E15(1)(e)] but below 15m height may be of timber (9mm min. thickness) or a material having a surface whose index of performance does not exceed 20 (tested in accordance with BS 476 Pt.6, 1968)

Distance related to: purpose group and % area of opening (or unprotected area) in wall as Sch. 9 of Part E

Unprotected areas may be openings (A) or an area of walling which does not achieve the FR required under Reg. E5 (B)

Note: where an area is 'unprotected' due to the use of combustible cladding such as timber – the area shall be taken as 50% only of the area of walling so clad

The information in this illustration is based on The Building Regulations, 1972

The Scottish Regulations are similar but differ in important points of detail

235   *Openings in external walls and relationship to boundaries*

the London Constructional By-laws and the Building Regulations respectively.

The risk of fire spreading across the ends of walls separating horizontally adjacent compartments must also be minimized. The Greater London Council requires precautions to be taken where such walls separate adjacent properties by requiring a minimum distance of 900 mm between any window and the centre of a party wall or boundary of the site. Similar conditions apply to division walls, that is internal fire break walls used to sub-divide trade buildings into smaller compartments[1]. The Building Regulations limit the proportions of walls which may be of 'unprotected areas' (ie windows, doors and such areas of the walls as are not of the required fire-resistance[2] for the particular building) by reference to their positions in relation to the boundaries or, where residential buildings are on the same site, by reference to a 'notional' boundary between them[3].

Rooms of high fire risk such as Boiler Rooms, Oil Fuel Storage Rooms[4] and Electrical Intake, Transformer and Switch Rooms, must be separated from the remainder of the building and be ventilated direct to the open air. Large solid and oil fuel boiler rooms should be provided with smoke extracts. Oil fuel boiler rooms should be situated against an external wall and oil fuel storage rooms should be as near as possible to an external wall, with access from the open air if possible, or from the boiler room.

The danger of spread of fire within a building is always accompanied by that of the spread from adjacent buildings. The fundamental protection between buildings is *space* in sufficient dimensions to prevent spread of fire by radiation or actual flame contact. Openings in the external walls of a building should be limited to resist the outward spread of flame and the potential number of points of entry of flame into a building from an adjacent building. Furthermore, the external surfaces of buildings should not contribute to the generation and propagation of heat, hence the basic requirements for walls to be fire resisting and, except for small buildings, non-combustible when close to boundaries (see figure 235). Similarly, roofs should not be unduly penetrated by openings and glazing or translucent sheeting should be limited in area and of fire resist- or fire-retardant quality (see previous comments on page 376)[5].

Shafts and ducts

Flue-like apertures such as shafts, ducts and deep light-wells, should be avoided in the general design of buildings, but where they are necessary as in the case of lift shafts and staircase enclosures, they should be vented at the top to allow smoke and hot gases to disperse to the atmosphere (page 388).

Fire venting

The large single-storey shed-type building housing continuous factory processes, which cannot easily be sub-divided, presents a particular problem. In such buildings the unconfined spread of smoke and carbon-monoxide fumes resulting from incomplete combustion after the initial supply of oxygen has been consumed, constitutes a hazard to fire-fighters and prevents them from reaching the seat of the fire. Also, when trapped in a building, the heat generated by the fire causes high temperatures and renders materials more inflammable by preheating them well in advance of an approaching conflagration. It is, therefore, desirable to make provision for the removal of heat, smoke and fumes as quickly as possible (often done by breaking holes in the structure with a fireman's axe) by a simple self-operating means of ventilation. Although this supplies more air and possibly intensifies the fire it will, nevertheless, confine it and produce less smoke thus assisting fire fighters to see and approach nearer the seat of the fire. It has been shown by experience that to be effective, such automatic ventilation must be above the fire because cross-ventilation may only serve to drive heat and smoke in a particular direction, possibly towards the fire-fighters. By providing automatic fire vents in the roof, therefore, smoke, heat and fumes are enabled to rise quickly out of the building and draughts are created which draw air towards the

[1] GLC Principles for guidance entitled *Buildings of Excess Height and/or Additional Cubical Extent* (Doc no. 4117).

[2] 'Unprotected area' also includes walls faced with combustible material more than 1.00 mm thick (Building Regulation E1(1)).

[3] Building Regulation E7(5).

[4] See London By-laws, part XIII. See also CP 3002: Part 1: 1961.

[5] See *A Complete guide to fire and buildings*, chapter 10: pp 136-149. Edited by Eric W. Marchant, 1972 (MTP).

fire, thus helping to contain it[1]. The effectiveness of roof vents is enhanced by subdividing the roof space above the ties of lattice trusses into compartments by fire-resisting curtains or non-combustible board fire stops fixed within the roof space. They should extend down to at least the level of the tie members and automatic fire vents should be placed within the bays thus created (see figure 233).

At present no standard of fire venting is laid down, but research into the question of the desirable area of vents and their arrangement, together with fire curtains for buildings of different types, is in progress. It has been suggested, however, that exhaust ventilation should be provided with an effective free area of from 0.5 per cent up to 5 per cent of the total floor area, depending on the fire risk, height and area of the building in question.

A method of providing automatic fire ventilation in such a manner that normal ventilation requirements are also fulfilled is shown in figure 233.

Building regulations contain specific sections devoted to fire-resisting construction but it should be noted that other sections contain requirements which affect the fire protection design of a building. These are sections concerned with the stability of walls, design of steel and reinforced concrete frames, chimneys and fireplaces, roofs and space about buildings.

Access for firefighting

When buildings are of excessive height or cubical extent special precautions must be taken in the provision of division walls to reduce the size of the compartments (page 377). Such buildings must be carefully sited to allow the heaviest fire-fighting units to approach close to all elevations. To provide adequate accessibility for these appliances the Greater London Council requires a portion of every building to abut upon a thoroughfare or open space not less than 12 m wide[2].

When vehicle access is required to a building or part of a building not abutting on a street it should be not less than 2.55 m wide and 3.45 m high to permit the entry of fire-fighting appliances.

In the case of tall buildings external access by ladders can only serve a limited purpose, since floors above 30 m are beyond the reach of most fire brigade ladders. Further, when a tall building is designed on a 'podium' of much larger area two or three storeys high, no external access for fire appliances is possible for any floor, so that internal

access must be designed as an integral part of the building. This is provided in the form of at least one lobby approach staircase, regarded for fire-fighting purposes as an extension of the street. The staircase must be separated from the accommodation on each floor by an enclosed lobby in which all the necessary fire-fighting equipment is installed. Both staircase and lobby must be sited next to an external wall and be provided with adequate ventilation to ensure freedom from smoke logging (see figure 236).

The enclosing walls to staircase and lobby must be constructed to have twice the standard fire resistance for the building. In trade buildings and certain office buildings this may result in a 4 hour standard of construction being required. No openings are permitted in the enclosure other than those for ventilation on to the street and those giving access to the building. The latter must be provided with self-closing fire-resisting doors; in addition, at basement levels and in certain other circumstances, further protection by steel roller shutters is required. As well as such a staircase, a firemen's lift must be provided situated where possible within the staircase lobby or if not, within the staircase or a separate enclosure adjacent having the same degree of fire resistance[3]. (See footnote [2], page 383). This lift need not travel to the top floor. (For details of firemen's lift see *MBC : Environment and Services*, chapter 17).

Means of escape

The object in providing means of escape is to permit unobstructed egress from within a building by way of definite escape routes (exit ways, corridors and stairs) to a street or an open space or to an adjoining building or roof from which access to the street may be obtained.

---

[1] Fire note No. 5. *Fire Venting in single storey buildings*, HMSO.

[2] Buildings exceeding   7 100 m^3: at least $1/6$ perimeter
Buildings exceeding   28 400 m^3: at least $1/4$ perimeter
Buildings exceeding   56 800 m^3: at least $1/2$ perimeter
Buildings exceeding   85 200 m^3: at least $3/4$ perimeter
Buildings exceeding 113 600 m^3: on an island site

[3] GLC Principles for guidance entitled *Buildings of excess height and/or additional cubical extent* (Doc. no. 4117), gives guidance on the requirements of the Council in respect of these classes of buildings.

The primary danger to occupants is that staircases and corridors leading to them from rooms or compartments may be filled with hot gases and smoke, trapping the occupants and causing panic. It is, therefore, important that such escape routes should be enclosed with adequate fire-resisting enclosures. In addition they should be separated by fire-resisting doors planned in strategic positions to prevent the spread of smoke and fire from storey to storey via lift shafts and staircases, and to keep the latter free from smoke when used as an escape route.

Smoke removal from staircases is normally by openable windows at each storey or landing level. Where this is not practicable CP 3 recommends a permanent vent at the top of the staircase of at least 5 per cent of the area of the enclosure. This method of smoke control has come into question due to the variabilities of pressures caused by wind and internal/external pressure differentials and more recently the idea of pressurising the staircase tower against smoke entry has been tested with some measure of success. Tests carried out by the Fire Research Station have led to the following tentative conclusions:

A properly designed system to pressurise a staircase continuously or as activated by smoke detectors, can keep it free of smoke so long as the doors are kept shut. A pressure of 13.01 N/m^2 can prevent entry of smoke through door gaps. Tests on a two storey staircase indicated that a volume of air equal to that of the stairwell per minute is required to clear smoke in about 5 minutes. This might be reduced if air is admitted at more than one point. Pressurisation can overcome adverse pressures developed in a fire and those due to adverse weather conditions and it increases the fire-resistance time for doors. Short period opening of doors does not worsen the smoke conditions in the staircase. Escape routes leading to pressurised staircases should be well sealed to minimise loss of pressure.

The air for pressurisation is supplied preferably by a combination of conventional ventilation system and pressurisation system or by individual fans provided to force air into each escape circulation section which, if not located next to an exterior wall, would have short lengths of duct to connect each fan with atmosphere directly[1]. Systems using long lengths of ducting are not recommended as there is danger of circulating smoke if a main duct is damaged, allowing smoke to penetrate and affect the whole installation. Furthermore the ducts would

have to have equal fire resistance to that of the escape routes they serve.

Another system which has been shown to work effectively is the 'Automatic Air Flow Control' system[2]. This system involves the use of smoke detectors which are connected to the retaining mechanisms of top heavy centre pivot hung windows which open inwards. When the windows are freed in response to the smoke detectors they swing in at the top to form large louvres which allow external air to enter at the bottoms of openings and circulate from floor to ceiling and to leave at the tops of the openings, thus clearing smoke rapidly and assisting fire fighters.

The requirements of building regulations in respect of means of escape vary with the size, construction, use and height of the building in question, the critical heights of floors being 6 m and 12.6 m. In the former case it is dangerous to jump from a building; in the latter case the standard 15.2 m wheeled escape ladder, available in most areas, is at its maximum extension. For smaller buildings with few occupants, approach lobbies with two sets of self-closing fire-resisting doors giving access on to one staircase, may provide adequate protection to means of escape. In large buildings alternative means of escape are required, either (i) to an adjoining building via a roof access which must be properly separated from the downward going flights of the staircase, or (ii) directly to the open air at ground level by means of secondary staircases properly protected and enclosed with fire-resisting walls. In buildings over 30 m high the GLC requires an enclosed stair at each end: in an office building, not more than 7.5 m, and in a residential building, not more than 10.5 m from the end wall. One of these may be the lobby-approach 'fireman's stair'. The general escape requirements of the Greater London Council are given in table 26 and in figure 236. See footnote [2] page 386. Other useful references are given in the note on the facing page.

[1] See Technical Study by L.R. Leworthy, *Architect's Journal* 2.4.1969.

[2] 'Automatic airflow control for escape route smoke movement in multi-storey flats' by J. Wilkinson, *Fire* 62 no. 772 1969. 'The Worthing "AAC" System – Automatic Airflow Control system for escape routes in new multi-storey blocks of flats' : Symposium no. 4, *Movement of Smoke on Escape routes in Buildings* (HMSO).

In residential buildings alternative means of escape may often be provided by open balconies. This permits greater distances between escape stairs, since there is less danger from smoke concentration[1]. Escape from tall residential blocks has been provided by means of close planning round a single fire-resisting staircase, direct access to it from each flat being through an individual ventilated lobby to each flat or through a common cross-ventilated lobby. This method has, however, received some criticism from Fire Authorities, and the Code referred to above recommends some modification of this system to meet the objections[2].

Buildings for public entertainment are required to have comprehensive escape provisions. The number and positions of exits and stairways are related to the sizes of audiences and the particular fire risk involved, such as the use of theatre scenery or film projection. Special regulations[3] cover access to and means of escape from projection and rewinding rooms for cinemas and halls where films may be shown. These require that projection rooms shall have direct access to the open air. Secondary means of escape must be provided to both projection and rewinding rooms which can be via one or other of these rooms. Both must be well ventilated.

Large single-storey buildings must be provided with perimeter exits at sufficiently frequent intervals and the gangways leading to them must at all times be kept clear of materials and other obstructions. Special care is necessary for buildings used for the storage of inflammable materials such as celluloid, petrol, oil and spirit.

Solid and oil fuel boiler rooms and electrical intake, transformer and switch rooms, if large, must be provided with secondary means of escape. Oil fuel boiler rooms and transformer rooms should be approached only from the open air, but where complete separation is impossible approach from the remainder of the building must be through a ventilated lobby. Oil fuel storage rooms should be approached from the open air if possible, or from the boiler room.

---

[1] CP 3 : chapter IV : 1971, *Precautions against Fire. Part I: Flats and Maisonettes (in blocks over two storeys)*, while maintaining the need for alternative means of escape to maisonettes, suggests that this is not necessary for flats provided the living room and bedroom doors open directly on to the entrance lobby of the flat with their doors nearer the flat entrance door than those to the kitchen and living room.

[2] The Code suggests that on each floor the stairs should open, via a ventilated lobby, on to 'a place of safety', that is a lobby or corridor which need not be ventilated to the open air and which may contain the lift. No doorway should be further than 4.5 m away from the 'place of safety'. The requirement that a fire lift must open on to a ventilated lobby has been waived and it is recommended that the lift should now open on to an enclosed hall provided there is a smoke-stop door not more than 4.5 m away leading to a staircase.

[3] The Cinematograph Regulations, 1955, 1958 and 1965.

Note (i)   CP 3 chapter IV: Part I: 1971 gives general recommendations in 'Precautions against Fire in Flats and Maisonettes (in blocks over two storeys)'. Part II deals with shops and departmental stores and Part 3 with office buildings

(ii)   'The Factories Act 1961' — Means of escape in case of fire, deals with factories and the appendices give particulars of highly inflammable materials and examples of trades where there may be a special fire and/or smoke risk

(iii)   'Fire and the Design of Schools', *DES Building Bulletin* no. 7, 4th edition January 1971 gives guidance on the design of schools so that they may satisfy Regulation 51 of the Standards for School Premises Regulations 1959

(iv)   'Planning for Fire Safety in Buildings' series 2: Basic Reference list on Fire for Architects, Surveyors and Builders, published by the Fire Protection Association

(v)   Means of Escape from Fire and assistance to fire service: Building Standards (Scotland) regulations 1963. Explanatory memorandum Part 5, analyses problems of personal safety, nature of hazards and basic principles of protection and gives requirements of individual regulations

(vi)   'Fire Protection', L.R. Leworthy in *Specification* 1973: gives a concise description of the major issues under the headings: Fire protection: Passive Defence, Means of Escape, Active Defence, etc.

New buildings (other than dwellings with no floors above 6 m) require means of escape as follows:

Class	Maximum floor height	Type of escape	Max travel distance	Notes on escape
Class 1 Office or Trade buildings up to 92.9 m² (max. 1st floor area) with timber floors and plastered ceilings or other suitably protected ceilings. **Class 2** ditto, ditto, up to 185.8 m² (max. 1st floor area) with non-combustible floors. **Class 3** ditto ditto up to 278.7 m² (ditto) ditto				
1a 2a, 3a	up to 12.8 m ,,	A,D or E ,,	18.3 m 30.5 m*	Buildings in more than one occupation on top floor require lobbies at that level with D. * 18.3 m of this distance to be within protected corridor
1b 2b, 3b	1 above 12.8 m ,,	B or C ,,	18.3 m 30.5 m*	
1c 2c	2 above 12.8 m ,,	B or C ,,	18.3 m 30.5 m*	Upward escape to be screened at lower of 2 top floors with B.
1d 2d, 3c	2 or more over 12.8 m ,,	C ,,	18.3 m 30.5 m*	Direct access to roof (or other independent alternative) to serve all floors above 12.8 m
Class 4 Small Office or Trade building in single occupation with no floor above 12.8 m				
4	up to 12.8 m	F	18.3 m	Building of limited height and size not requiring staircase enclosure or lobby approach
Class 5 Office or Trade building with 2 staircases. (Of larger area than 1-4 or with different floor construction).				
5	Below 24.4 m	No. 2 as D, or No 2 as E, or D with E or G	30.5 m† (max 61 m apart) †	† 18.3 m and (36.6 m) where timber floor construction Shops, Stores, etc: Floors above 24.4 m to have PS/C to street Escalators in enclosed compartments with lobbies giving on to street may be acceptable
Class 6 One storey Office or Trade buildings of unlimited floor area				
6	—	Exits to open air	18.3 m	Where inflammable materials or liquids are used or stored, at least 2 exits to be provided
Class 7 Dwellings occupied by one family (part use for business permissible), single staircase				
7a 7b	Up to 12.8 m 1 or more above 12.8 m	— Alternative escape to floors above 12.8 m	— —	Minor protective works where special hazards exist Escape by balcony to next building or to open S/C acceptable
Class 8 Blocks of Flats and/or Maisonettes with one staircase				
8a	Up to 12.8 m	H, J or K	30.5 m	Max 6 dwellings per floor (4 if timber floors). Dwellings to have Entrance halls. Fire Brigade access to bedrooms not off hall, and to upper level of maisonettes unless hall is protected and lower rooms have SCFRDs
8a	1 or more above 12.8 m	H, J or K	30.5 m	Alternative escape for dwellings over 12.8 m to roof or balcony giving access to adjacent building Note: Balcony access flats with 1 floor above 12.8 m may have auxiliary S/C to balcony below as alternative means of escape
8b	Below 24.4 m	L	—	All habitable rooms to be off entrance hall of dwelling. Lobby ventilation: min area 25 per cent cross sectional area of lobby or 2.8 m²
8b	Above 24.4 m	L	—	Ditto. ditto. Fire Brigade access strip required at ground level. Also consult Sect 20, 1939 Act
8c	Below 24.4 m	M	—	Where it is necessary to pass another dwelling to reach S/C. As 8b, and Fire Brigade access strip to enable any part of balcony to be reached
8c	Above 24.4 m	M	—	Open balconies on both sides to floors above 24.4 m giving access to common staircase
Class 9 Blocks of Flats and/or Maisonettes with two staircases				
9	—	No. 2 as J	27.45m†† (max 54.9 m apart)	†† Distance measured from dwelling entrance to S/C Means of escape from all bedroom floors as class 8 generally. NB Max travel limit is not applicable when open balcony access is used

S/C = staircase; SCFRD = self-closing fire-resisting door; PS/C = protected staircase

Table 26 *Means of escape in case of fire (reference letters in column 3 relate to figure 236)*

Classes I to VI new office or trade buildings

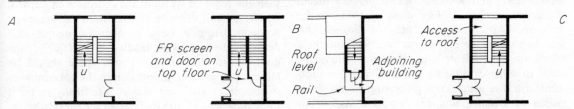

A — P/SC with access to roof and to roof of next building

B — P/SC with screened access to roof and ingress to next building

C — P/SC with protected lobbies or corridors on all floors below top and direct access to roof and into next building for floors above 12·8m

[Note: all P/SC's, lobbies and corridors require ½ hr. FR walls and self-closing doors]

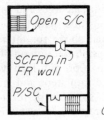

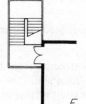

G — Open S/C within protected compartment

F — S/C continued to roof and spandrel to each flight above gd. floor filled with FR partition with SCFR door and FR glazing

E — External S/C with access to street

D — P/SC with protected lobbies or corridors on all floors below top floor. No access to roof

Classes VII to IX residential buildings

Buildings over 30·5m high

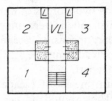

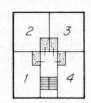

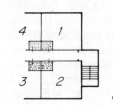

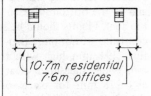

[10·7m residential 7·6m offices]

L — Flats or maisonettes one in line from ventilated lobby or open balcony

H — Flats or maisonettes off common protected S/C

J — Flats or maisonettes off common protected lobby or corridor

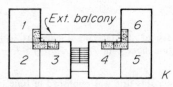

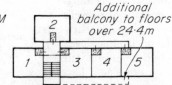

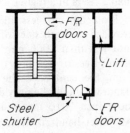

K — Flats or maisonettes off external balcony leading to open or partially open S/C

M — Flats or maisonettes more than one in line from ventilated lobby or open balcony

Fireman's staircase, lobby and lift

Means of escape requirements are often rather complicated but the GLC Code of practice *Means of Escape in Case of Fire* provides guidance for designers. This covers the design of buildings other than those used for public entertainment and those which are very tall or large. Reference should be made to other documents for the detailed requirements for fire escape and protection in the latter classes of building and for requirements outside the Greater London area[1].

FIRE-RESISTING CONSTRUCTION

Design and construction associated with means of escape

The following information extracted from the GLC Code of Practice, 'Means of Escape in Case of Fire', gives a desirable standard for fire-resisting constructional detail for all elements related to a means of escape[2].

Staircases

Staircases should be placed next to an outer wall and should be entered from any floor level in the direction of the flow towards the exit from the building. Landings should be arranged at the top and at the bottom of each flight equal in width to the stair. Flights should be straight without winders, consisting of not more than 16 risers, each not more than 191 mm high with treads not less than 254 mm wide, measured clear of nosings. (152 mm and 278 mm for places of assembly and not less than 3 risers in a flight).

Protection not less than 914 mm high should be provided on both sides of all staircases and landings (not less than 1066 mm when next to open wells or in places of assembly), measured perpendicularly from the centre of the treads or landing level. Handrails should be provided on both sides of a stair more than 1016 mm wide, and on one side if less. An outer handrail should continue round all landings. The width of stairs and landings, measured between finished surfaces of walls or the inner side of balustrades, should not be decreased by any projection other than the handrail which should not project more than 76 mm. The clear width of exit doorways from staircases, measured when the

door or doors are fully open, should not be less than the required width of the staircases concerned. Where an exit from a ground floor or basement accommodating a large number of people also delivers into a staircase landing or exit lobby, the doorway to the street or open space should be increased in width in proportion to the total number of people on all floors. Unless specifically required to be otherwise by the By-laws, or by considerations applicable to public buildings, internal staircases including landings and floors within the staircase enclosure, are permitted to be of wood construction, having not less than a half-hour fire resistance.

External staircases[3], including balconies and gangways, should be of non-combustible materials with the stairs arranged as described above for internal stairs, and of similar dimensions. Protection on both sides is required not less than 1066 mm high with handrails on both sides. Where iron steps and landings are used, they may be solid or perforated, provided openings are not more than 13 mm wide, with 32 mm deep solid nosings. Where open risers are provided nosings should overlap the back of treads below by 25 mm.

Spiral staircases may form secondary means of escape for small numbers of people if they are of non-combustible material not less than 1524 mm in diameter and not exceeding 9.14 m in height.

Step ladders not exceeding two storeys of limited height and not steeper than 60 degrees may be accepted as upward secondary means of escape for not more than 30 persons. Treads must be 127 mm wide not more than 203 mm apart.

Vertical ladders may form means of escape for small numbers of people. They should be not less than 458 mm wide fixed 102 mm clear of the wall face with strings carried up 1066 mm to form hand

[1] A useful summary of the relevant documents will be found in Information library in *The Architect's Journal, 18. 3. 70.*

[2] A new Code under this title was published in June 1974 (GLC Publication 7168 0573 1). The information given here was extracted from an earlier publication and should be checked with the requirements of the new Code.

[3] External staircases are open to attack by fire from adjacent doors and windows and may become unsafe under certain weather conditions. It is preferable not to use them for other than two-storey buildings.

grips. Intermediate landings should be provided if the ladder is more than 6.09 m high and if the height exceeds 9.14 m suitable guards must be provided to form an enclosure about the ladder, as shown in figure 185.

Ramps which afford a means of escape should be protected at the sides and have a gradient of not more than 1 in 10. Steps should not be introduced.

Escape routes over roofs should be protected with railings, balustrades or parapets not less than 1066 mm high.

Enclosures

The enclosures to protected stairs, landings, corridors, passages, lobbies and doorway recesses within the staircase enclosures and exits from staircases should have a standard of fire resistance of not less than one half-hour. Wood or linings of rapid flame spread are not permitted, nor cavities behind linings, within the protected enclosures of a staircase forming the only means of escape. Borrowed lights, fanlights and other glazing, where permitted in enclosures, should be of fire-resisting construction.

Doors

Doors to protected staircases and corridors should be hung to open in the direction of exit and to swing clear of steps, landings, passageways and the public way; they should open the full width with sufficient clearance between two sets of doors.

Doors to enclosed staircases and to external staircases should be of solid timber not less than 44 mm finished thickness or have a fire resistance of not less than one half-hour. They should be properly framed together and the door frames bedded solid[1]. Panelled doors in existing buildings may be retained (i) where the thickness of the stiles is not less than 44 mm, if the panels on the face of the door away from the staircase are entirely covered with asbestos millboard or asbestos wallboard not less than 5 mm thick or with other approved material, (ii) where the thickness of the stiles is less than 44 mm, if the panels on the face of the door away from the staircase are made up solid flush with stiles and rails and that face of the door is protected all over with plasterboard or asbestos wall board not less than 5 mm thick. In existing buildings converted to residential accommodation, panelled doors with stiles not less than 32 mm thick may be retained if protected in the panels by asbestos wallboard not less than 5 mm thick.

Doors to protected or to external staircases should be rendered self-closing and the use of steel bushed rising-butt hinges may be permitted for this purpose in certain cases. Sliding doors where permitted should be self-closing.

Except in factories, exit doors to the street may open inwards provided they are fixed open whilst the building is occupied.

Glazing over, in or at the sides of doors which are required to be fire-resisting, should be also fire-resisting and fixed shut, and may, where appropriate, be of glass bricks.

Doors at the head of staircases giving access to roofs should be glazed in the upper panels.

Doors in factories, warehouses and buildings or portions of buildings involving special hazards must conform to the standards of fire resistance required under the relevant Acts and By-laws.

Revolving doors will not be accepted in escape doorways. Where it is desired to instal such a door, an emergency exit with ordinary hinged doors should also be provided adjacent to the revolving door. This exit should be adequately indicated.

Doors affording access to external escape staircases, balconies and gangways, should be fastened with simple fastenings, easily operated from inside without a key. Exit doors to and from staircases and to places of assembly should be fitted with automatic bolts.

Windows

For purposes of rescue a reasonable number of windows on floors above the ground floor, facing a street or open space to which fire appliances have access, should be made to open at cill level or within 305 mm of the cill level (where a cill ledge is provided). The opening portions should be not less than 838 mm high by 382 mm wide in the clear.

Windows giving access to external escape stairs should be fitted with simple fastenings easily operated from inside without a key, as required for doors serving the same purpose.

[1] Department of Education and Science Building Bulletin no.7, *Fire and the Design of Schools,* Part II makes a distinction between a 'smoke-stop' door to be used to prevent the passage of smoke to escape stairs and corridors and 'fire check' doors to be used to provide a barrier to fire. The requirements of fire resistance and frame detail applying to these doors are as described in Code of Practice 3: chapter IV: Parts 2 and 3: 1968

All glazing vertically under or within 1.83 m of an external staircase should be fire resisting and fixed shut except for that portion statutorily required for the purposes of ventilation or when required for means of escape to the staircase.

Lifts

The motor chamber should be fully enclosed with non-combustible materials and separated from the lift shaft except for openings necessary for the passage.of wires and cables.

In enclosed lift shafts, a smoke outlet to the open-air should be formed, at or near the head of the shaft, or other smoke outlets provided. The smoke outlet should be not less than 0.09 m^2 in area, fitted with an openwork metal grille or widely-spaced louvres.

Gates, doors and shutters to lifts should be provided with automatic control to ensure that they cannot be opened, except the door opposite the lift cage when it is at rest at floor level.

Buildings with one staircase   In blocks of flats or maisonettes the lift shaft will not be permitted within the staircase enclosure. The shaft should be wholly enclosed in fire-resisting materials not less than 76 mm thick and with solid wood doors or steel-shielded gates.

In buildings of other classes under 12.8 m in height, where the motor chamber is at the bottom of the shaft, the lift shaft may be within the stair-case enclosure if protected by solid fire-resisting enclosures and solid wood doors or steel-shielded gates. When the motor chamber is at the head of the shaft the enclosure to the lift may be of metal grilles with collapsible lattice gates at openings.

Buildings with two or more staircases   Lifts may be within the stair enclosure provided that each stair is available for escape for all occupants. Such lifts may be enclosed with metal grilles and collapsible lattice gates irrespective of the position of the motor chamber.

Residential buildings with independent lift shaft The lift shaft should be wholly enclosed in fire-resisting material and with solid wood doors or steel-shielded gates.

It should be noted that although in some cases solid enclosures to a lift shaft are not essential, these may become necessary in order to complete the separation between basement and street exit and to preserve the screened access above the 12.8 level.

Design and construction associated with the limitation of the spread of fire

Walls

The function of walls in providing fire protection and the regulations relating to them have been discussed already. Traditionally fire resisting walls were constructed of non-combustible materials. However, when distances from boundaries are sufficiently great and when, as in the case of the *Department of Education and Science Building Bulletin* no. 7, the requirements for escape are stringent, the use of combustible materials and light claddings is permitted. In some cases, however, the use of light external claddings and, in particular, curtain walling, which are functionally appropriate to framed buildings, present a number of practical difficulties.

Curtain walling   The constructional aspects of curtain walling have been considered in chapter 4. In considering the system from the point of view of fire protection it is necessary to have regard to the requirements of building regulations as far as external walls are concerned. These have been discussed on page 377 and it will be clear that any system of frame and panel walling such as the curtain wall must act as a whole in fulfilling these requirements, but at the same time be light and economical of space if it is to be successful. Clearly no system which employs unprotected steel or aluminium framing can meet the by-law requirements for fire resisting external walls, however fire resistant the panel infillings may be, since these metals have no acknowledged fire resistance. Furthermore, the provision in London By-law 6. 14 (4c) that glazing or glass in the thickness of a wall shall be considered an opening has presented difficulties in the development of the curtain wall. However, By-law 6.09 describes Class IIA enclosures as being constructed entirely of non-combustible materials and an external face of glass and By-law 6.10 describes a non-loadbearing enclosure which includes an independent backing wall or other independent non-combustible construction positioned not more than 100 mm from its inner face and capable of resisting the action of fire for 1 hour (see figure 237 A). These enclosures are limited to certain building types (including offices, flats, single storey garages and roof erections) and have certain dimensional constraints in relation to other buildings and streets. By-law 6.14 allows openings in external enclosures not exceeding a total area of half the total area of the enclosure above the first floor level but 6.14 (5) allows unlimited openings where a Class IIA enclosure is or would be permitted.

The *DES Building Bulletin* no. 7 adopts a similar approach to the Building Regulations in relating openings and areas of unprotected construction to distances from boundaries.

The details shown at (*A*) and (*B*) in figure 237 show two ways of providing a Class IIB enclosure and the necessary separation between windows in adjacent storeys[1]. In (*A*) the external glazed curtain walling is an independent facing to a non-combustible backing wall of not less than 1 hour fire resistance, the glazing being supported in a metal framing, secured to the structure. In (*B*) the spandrel is self supporting, non-combustible and secured directly to the structure. In the Building Regulations the spandrel would constitute fire resisting construction and the window an 'unprotected area'[2].

Detail (*C*) shows a system using prestressed concrete mullions in conjunction with fire-resistant panels forming the fire-resistant wall and having an independent facing of glass or other suitable cladding material held in a metal frame. This system has a fire resistance of 2 hours.

The following points should be borne in mind when selecting or designing a curtain wall system of this type:

(i) the panel and the frame must act together as one fire-resisting element and be tested as a whole under BS 476; (ii) fixings must be protected from fire and should not conduct heat to a vulnerable material. They should be designed to allow for exceptional expansion in a fire.

Openings in walls   Doors and windows require to be fire resisting to retain the integrity of fire separation or compartment walls. Windows of limited pane size in metal frames of melting point not lower than 900°C attain a half hour fire resistance and 1 hour when wired glass is used[3]. Glass blocks properly set in recesses and reinforced with light wire mesh every third horizontal joint attain half hour fire resistance with panels limited to 2.40 m maximum width or height. Fire-resisting glazing must, in most cases, be fixed shut.

A door and its frame form one unit or element and fire tests are made on this unit. Timber doors require deep rebates to prevent flame penetration between the door and frame. A steel frame is necessary if a steel or composite door is to develop its full fire resistance (figure 239). Although not complying strictly with BS 476 for ½ hr fire resistance the 38 mm finished thickness solid timber door is accepted as the standard 'fire check' door of the ½ hr type. BS 459: Part 3: 1951, 'Fire check flush doors and frames', specifies the construction for flush doors with ½ hr and 1 hr resistance. These, together with the appropriate frames are described in detail and illustrated in *MBC:Components and Finishes,* chapter 3. Alternatives are shown in figure 238.

Where a fire resistance of 1 hr or more is required doors are usually of steel or steel and asbestos or, on occasion, steel-encased timber. Uninsulated iron and steel doors and shutters are effective in preventing fire spread for up to 2 hrs although they transmit a lot of·radiant heat. Doors and shutters fitted on each side of an opening are accredited with double the fire resistance of one door only. Any doors provided in addition to steel shutters must be arranged so that they do not interfere with the normal operation of the shutters whether open or closed. Doors should not be placed between double shutters.

Typical details of steel doors and shutters are shown in figure 239. These embody the requirements of the Greater London Council and Fire Offices Committee regarding the construction, together with the main requirements regarding the openings in which they are set[4]. Unless special consent is obtained openings in division and party walls in the Greater London area must not exceed 2.13 m width nor 2.43 m height, except when the doors or shutters are not less than 610 mm apart when the height may be up to 6.59 m. A number of openings should not aggregate more than half the length of the wall in any storey in which they are formed.

---

[1]   London By-laws 6:10 and 6.11

[2]   The Building Regulations 1972, clause E7. The Building Standards (Scotland) (Consolidation) Regulations 1971, clause D17

[3]   London By-laws, Schedule 11 Table F 'Glazing'

[4]   See London Building Acts (Amendment) Act 1939, Section 21; LCC Steel Roller Shutter Regulations, 1939; Rules of the Fire Offices Committee. The dimensions given in figure 239 are rounded off metric equivalents of the imperial dimensions given in these documents.

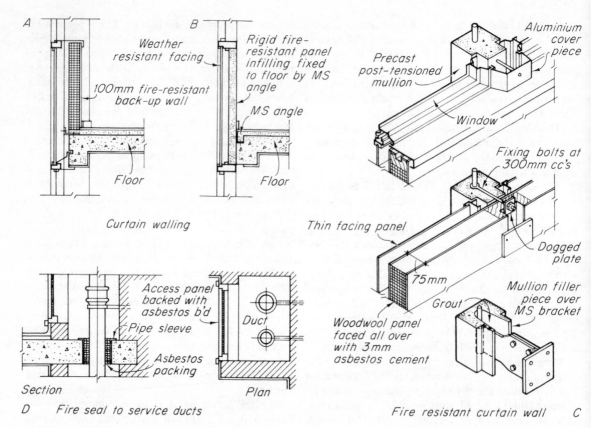

A Weather resistant facing

100mm fire-resistant back-up wall

Floor

B Rigid fire-resistant panel infilling fixed to floor by MS angle

MS angle

Floor

Curtain walling

Access panel backed with asbestos b'd

Pipe sleeve

Duct

Asbestos packing

Section

Plan

D Fire seal to service ducts

Aluminium cover piece

Precast post-tensioned mullion

Window

Fixing bolts at 300mm cc's

Thin facing panel

Dogged plate

75mm

Grout

Mullion filler piece over MS bracket

Woodwool panel faced all over with 3mm asbestos cement

Fire resistant curtain wall *C*

237 Limitation of spread of fire

Special automatic fire doors are required in factories where division walls are penetrated by conveyors. These are generally of the 'garrotting' type comprising a fire door which is released by the action of fire on a fusible link to fall across the track of the conveyor. These doors, or dampers, may have to be double in some cases. Alternatively, protection may be provided by a tunnel incorporating a water drencher system automatically operated in the event of fire.

Floors

Floors of timber joists, boarding and plaster ceilings normally attain ½ hr fire resistance. However, both the London By-laws and the Building Regulations in England and Scotland give up to 1 hr fire resistance to such floors when the ceilings are finished with gypsum or vermiculite plaster on suitable backings or where asbestos insulating board and glass fibre or mineral wool quilts are incorporated in

390

the ceiling. Floor boarding laid on asbestos insulating board with ceilings of the above materials or sprayed asbestos also attains 1 hr fire resistance. The Scottish Building Regulations also describe a timber floor construction with metal lathing and 38 mm sprayed asbestos which attains 2 hr fire resistance.

DES Building Bulletin no. 7 provides for combustible floors in schools up to four storeys in height (except floors or landings within stair enclosures or over boiler rooms)[1]. Floors must have a fire resistance of a ½ hr up to four storeys. For five storeys and over the floors must have a resistance of 1 hr and be non-combustible. The stringent escape requirements of this Bulletin makes the use of combustible materials permissible.

In certain types of industrial buildings with vertical forms of processing, requiring considerable

[1] See Table VI, *DES Building Bulletin* no.7

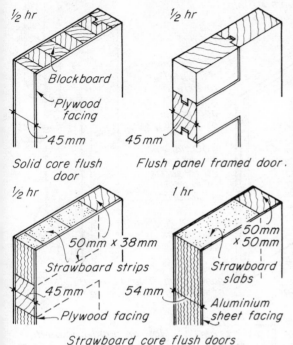

238   Fire resisting timber doors

perforation of the floors to permit the passage of hopper, conveyors and chutes, it becomes impracticable to limit the spread of fire by fire-resisting floor divisions. Thus, provided the area of each floor is sufficiently small and means of escape is good, timber floors can be used, with non-combustible fire-resisting floors limited to the enclosed areas round the vertical circulation of lifts and staircases.

Openings in floors   Apart from lifts and staircases which are enclosed throughout their height, openings in floors may occur where vertical services communicate with successive storeys. Where these are contained in ducts the latter should be sealed off at each floor level with a non-combustible filling to give a fire resistance equal to that of the floor. This prevents them acting as flues to spread fire vertically and may be accomplished by arranging for the floors to close the vertical shaft, openings of the correct dimension to accommodate services being left when the floor is formed. Pipes passing through slabs can be encircled by a ferrule set in the floor and sealed with asbestos to allow for expansion movement (figure 237 *D*). The duct enclosures are built off the floor slabs.

As an alternative to this method the opening in the floor slab may be the full area of the duct. Subsequent to the installation of the service pipes, this may be filled in tightly round the pipes with a suitable non-combustible material. These can be fairly easily removed and replaced when repairs to the services are necessary. Such fillings are asbestos fibre or slag wool with retarded hemi-hydrate plaster, foamed slag concrete or vermiculite concrete, all of which can be laid on asbestos board or expanded metal. Open ducts, or chutes, passing through fire division floors, should be provided with double steel dampers with fusible links and counterweights.

Roofs

The problems of roof construction have already been discussed in the earlier sections of this chapter (see pages 374, 376, 380).

Cavities

The need to avoid continuous cavities in wall and roof construction when formed by the application of combustible linings has been referred to on page 374. Cavities may also be formed in framed structures behind claddings and in curtain walling. In all cases the possibility of the cavity acting as a flue for the passage of flames and smoke in the same way as a duct must always be considered. The two-storey balloon-framed timber wall in which the vertical studs run through the two floors is an example of a cavity arising from the structure used. It requires fire-stops at first floor level, as described on page 124 of Part 1. Current building control requires fire stops within cavity construction at the junctions of elements[1].

Fireplaces and flues

The construction of fireplaces and flues must obviously be designed to contain the fire and prevent the spread of heat by conduction or radiation to combustible parts of the structure, and to ensure that where there is any possibility of the accidental

---

[1] Building Regulation E14 deals with fire-stopping between junctions of elements and in cavities within elements in which combustible material is exposed (lower than Class 0). The Regulation does, however, permit the insertion of combustible *filling* into a cavity.

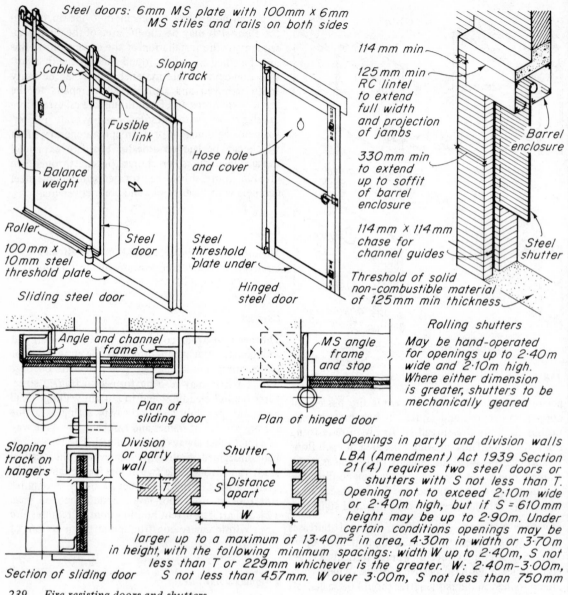

Steel doors: 6mm MS plate with 100mm x 6mm MS stiles and rails on both sides

Cable

Sloping track

Fusible link

Balance weight

Roller

100mm x 10mm steel threshold plate

Steel door

Sliding steel door

Hose hole and cover

Steel threshold plate under

Hinged steel door

114 mm min

125 mm min RC lintel to extend full width and projection of jambs

330 mm min to extend up to soffit of barrel enclosure

114 mm x 114 mm chase for channel guides

Threshold of solid non-combustible material of 125mm min thickness

Barrel enclosure

Steel shutter

Angle and channel frame

Plan of sliding door

Sloping track on hangers

Section of sliding door

Division or party wall

MS angle frame and stop

Plan of hinged door

Shutter

T

S Distance apart

W

Rolling shutters

May be hand-operated for openings up to 2·40m wide and 2·10m high. Where either dimension is greater, shutters to be mechanically geared

Openings in party and division walls

LBA (Amendment) Act 1939 Section 21(4) requires two steel doors or shutters with S not less than T. Opening not to exceed 2·10m wide or 2·40m high, but if S = 610mm height may be up to 2·90m. Under certain conditions openings may be larger up to a maximum of 13·40m² in area, 4·30m in width or 3·70m in height, with the following minimum spacings: width W up to 2·40m, S not less than T or 229mm whichever is the greater. W: 2·40m-3·00m, S not less than 457mm. W over 3·00m, S not less than 750mm

239   Fire resisting doors and shutters

fall of hot embers a non-combustible surface is provided to receive them. Building regulations lay down certain thicknesses and dispositions of non-combustible materials, limit the presence of combustible materials near a fireplace or flue and require that all joints in flue linings shall be properly tight against the passage of smoke or flame. Reference should be made to chapters 8 and 9 of Part 1, where these matters are discussed.

Fire protection for external steelwork

Current building regulations in the United Kingdom require all elements of structure of a given building to be of the same fire resistance grading irrespective of their position in relation to the building enclosure.

The growing tendency for designers to maximise floor space, free of structure, coupled with design philosophies which engender direct expression of structural form has led to a number of developments in the protection of steelwork which may be situated outside the building facade. In this situation it is argued that structural columns and beams are to a large extent protected by the main walling and are clearly not subject to all round attack by fire as are internal columns and, therefore, a reduced fire resistance period should logically be required of them.

In these circumstances the main aim is to keep the temperature of the structural steelwork below the critical temperature of 550°C at which temperature the yield stress reduces to the permissible working stress, resulting in the collapse of loaded structures if the limiting design criterion has been stress as against buckling.

The methods mostly used or investigated to date in order to keep structural steelwork cool (ie below the critical temperature) have been: protective coverings (lightweight concrete, intumescent coatints), solidity of section (massive steel sections, concrete filling of hollow steel sections), water cooling (water spraying, water filled hollow sections) and flame shielding. These methods are briefly described below[1].

Protective coverings

Vermiculite concrete casing is formed of exfoliated vermiculite with a Portland cement matrix reinforced with wire mesh. This has good properties of dimensional stability under conditions of shrinkage, creep, changes of temperature or structural deformation and is frost resistant and weather resistant. An excellent finish can be achieved by plastering techniques.

Thicknesses for hollow protection are 13 mm for up to 1 hr, 25 mm for 2 hrs and 57 mm for 4 hrs fire resistance. For profiled protection, 13 mm for ½ hr, 19 mm for 1 hr and 32 mm for 2 hrs fire resistance.

Intumescent coating systems are based on intumescent paints or mastics which expand on heating to provide a cellular insulating layer, about 50 mm thick, between the fire and the substrate. Coatings can produce up to 1½ hrs fire resistance and are available in a wide range of colours. They are especially useful for the protection of decorative metal structures where preservation of the surface configuration is essential. The main disadvantages appear to be high cost in the case of 1½ hr fire resistance and the need for vigilance and maintenance of the coatings throughout the building's life and during successive ownerships.

Solidity of section

Massive steel sections have inherently great thermal capacity particularly when of low specific surface. Strength in high temperatures is related to their perimeter to section area ratio and a fire resistance period of about 50 minutes can be achieved with a ratio of about 40 at a temperature of 550°C.

Solid steel billets can be more economical in first costs than rolled steel sections or built up sections of similar strength.

Concrete filled sections are hollow steel sections filled with high strength concrete which results in increased load carrying capacity and increased fire resistance. Tests on a nominal 300 mm concrete filled square hollow steel section with wall thicknesses of 10 to 16 mm showed increases from 15 minutes fire resistance to 45 minutes infilled with ordinary concrete and 50 minutes infilled with high strength concrete.

Water cooling

Water spraying   Cooling of steelwork by the transfer of heat from the steel to water may be achieved by spraying directly upon the hot metal or by arranging water to run within a hollow member which has suitable holes in the walls to allow steam to escape. These holes would also act as overflows to enable a given depth of water to be maintained within the member. The heat transfer is maximum within the temperature range 100-150°C and foaming agents can be released to keep the wetted areas wet for longer periods. It is theoretically estimated that the steel would remain below 550°C so long as at least 10 per cent of its surface were to remain wet. This system is in a developmental stage and full sized tests have yet to be carried out. Architectural problems arise in the positioning of pipework.

Water filling   Systems filled with water are of two basically different forms: (a) non-replenishment and (b) replenishment.

[1] An excellent review of these methods and descriptions of buildings in which they have been incorporated is given in the Technical Study by Gordon Cooke, of Pell Frischmann & Partners, entitled 'New methods of fire protection for external steelwork' in the *Architect's Journal* 28.8.74

The non-replenishable systems are simply hollow members filled with water. The natural circulation of the heated water may be aided by the incorporation of an inner tube which promotes the circulation flow up the sides and down the centre of the member. A pressure release valve is provided to each member. The system improves fire resistance from about 15 minutes up to 30 to 45 minutes.

The replenishable systems can achieve high fire resistance so long as the source of water supply is maintained. The systems depend on natural circulation and are usually replenished from a header tank or from the mains. Differential circulation as between horizontal members, in which convective flow is minimal, and vertical members, in which there is maximum flow, can be reduced by arranging the structure on a diagonal elevational pattern or lattice.

Potassium carbonate may be used to prevent bursting stresses due to freezing and potassium nitrate inhibits corrosion in water cooled systems.

Flame shielding

In this method structural members can be planned to occur opposite solid sections of fire-resisting external walling which mask them from flame attack or, alternatively, specially designed shields may be employed which intercept flames.

Examples include metal shields fixed to background structure or to the actual structural members they protect, using minimal attachments to minimise conduction of heat. These systems depend upon an accurate prediction of flame pattern in a wide range of climatic conditions and winds of variable force and direction.

Firefighting equipment

This heading covers various forms of hand extinguishers and types of fixed installation which may be used or come into operation to contain the fire until the arrival of the fire brigade, together with special installations provided for the fire brigade in large and high buildings. For all these reference should be made to chapter 16, 'Firefighting equipment', in *MBC : Environment and Services*.

11   Temporary Works

The subject of temporary constructions which are often a necessary part of the total building process has been introduced in Part 1 and some aspects of timbering for excavations and of formwork have been covered. In this chapter these will be considered further and the subjects of shoring and scaffolding will be taken up.

SHORING

Shoring is the means of providing temporary support to structures that are in an unsafe condition till such time as they have been made stable, or to structures which might become unstable by reason of work being carried out on or near them, such as the underpinning of foundations.

Timber has always been used in the past and is still the most commonly used material for shoring, although effective flying and raking shores can be constructed with tubular scaffolding. But so many struts, ties and couplers are necessary to maintain rigidity that timber is usually found to be quicker and more economical of labour. Steel stanchions and, particularly, steel needles are often used for dead shoring. Tubular scaffolding is less frequently used for this purpose.

Classification   There are three general systems of shoring, (i) raking shores, (ii) horizontal or flying shores, (iii) dead or vertical shores.

Raking shores

These consist of inclined timbers called rakers placed with one end resting against the face of a defective wall, the other upon the ground (see figure 240A). The most convenient and best angle for practical purposes is 60 degrees, but this may vary up to 75 degrees. The angle is often determined in urban areas by the width of the footway. On tall buildings these shores are fixed in systems of two or more timbers placed in the same vertical plane, inclined at different angles to support the building at varying levels, as shown in figure 240 A.

The purpose of a raking shore is to prevent the over-turning of a wall — not, in the case of a tilting or bulging wall, to force it back.

A wall-piece, consisting of a 50 or 75 mm deal, is fixed to the wall by wall-hooks driven in the joints of the brickwork. This receives the heads of the rakers and distributes their thrusts over a larger area of wall (figure 240). For a single raker the width of the wall-piece is usually about 225 mm. In a system of rakers the width should be the same as that of the rakers. The wall-piece should be in one piece throughout the system but, if owing to the length, it is necessary to join two pieces they should be halved and securely spiked as shown in figure 240 C. The length of the longitudinal bevelled halved joint should be six times the thickness of the plate.

To form an abutment for the head of the raker, a needle, consisting of a piece of 100 mm x 100 mm timber about 330 m long, and cut as shown in (B), is passed through a mortice in the wall-piece, and projects into the wall at least 115 mm, a half-brick being taken out to receive it. The function of the needle is to resist the thrust of the raker and prevent it slipping on the wall-piece and to transmit the thrust through the wall-piece to the wall.

The following considerations determine the position of the needles. The head of a raker should only be placed where there is something such as a floor or roof at the back of the wall to resist the thrust, otherwise the wall is liable to bulge inwards at that point. Near the top of a building there is the possibility of the head of the wall being pushed off if there is insufficient weight above the raker. The centre line of rakers should, therefore, pass through the centre of the bed of any wall-plates in the wall (A). If the joists should be parallel to the wall, the produced centre lines of the floor, wall and raker should meet at a point, since it may be assumed that the resultant of the dead weight of the wall and any thrust from the floor passes approximately through this point as shown for the bottom rakers, figure 240 A. The needle should be placed so that the pressure exerted by it takes place along the centre line of the raker, the upper end of which should be notched to receive the needle, thus obviating any tendency to lateral movement. The shoulder bearing on the needle should not be less than 75 mm wide. The needle is supported at its top

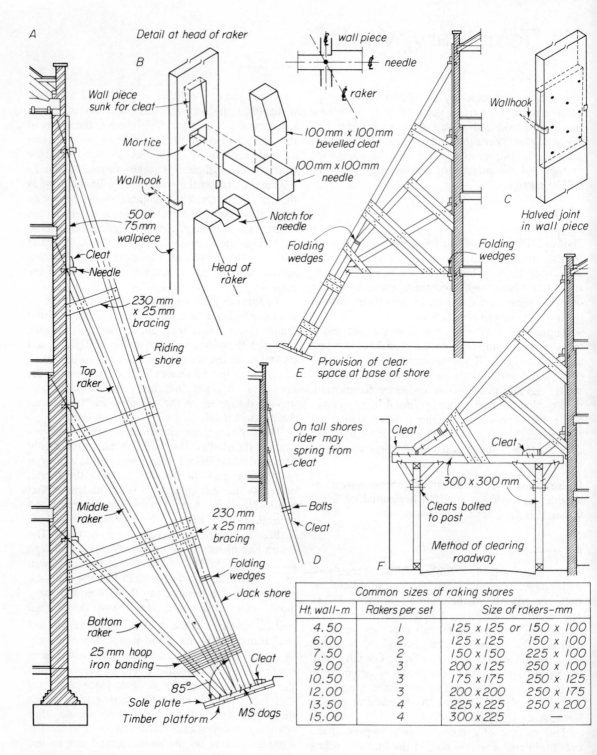

A

Detail at head of raker

B

Wall piece sunk for cleat

Mortice

Wallhook

50 or 75 mm wallpiece

Cleat

Needle

230 mm x 25 mm bracing

Riding shore

Top raker

Middle raker

230 mm x 25 mm bracing

Folding wedges

Jack shore

Bottom raker

25 mm hoop iron banding

85°

Sole plate

Timber platform

Cleat

MS dogs

wall piece

needle

raker

100 mm x 100 mm bevelled cleat

100 mm x 100 mm needle

Notch for needle

Head of raker

Folding wedges

E   Provision of clear space at base of shore

Wallhook

C   Halved joint in wall piece

Folding wedges

On tall shores rider may spring from cleat

Bolts

Cleat

D

Cleat

Cleat

300 x 300 mm

Cleats bolted to post

Method of clearing roadway

F

Common sizes of raking shores		
Ht. wall-m	Rakers per set	Size of rakers-mm
4.50	1	125 x 125 or 150 x 100
6.00	2	125 x 125 150 x 100
7.50	2	150 x 150 225 x 100
9.00	3	200 x 125 250 x 100
10.50	3	175 x 175 250 x 125
12.00	3	200 x 200 250 x 175
13.50	4	225 x 225 250 x 200
15.00	4	300 x 225 —

240   *Raking shores*

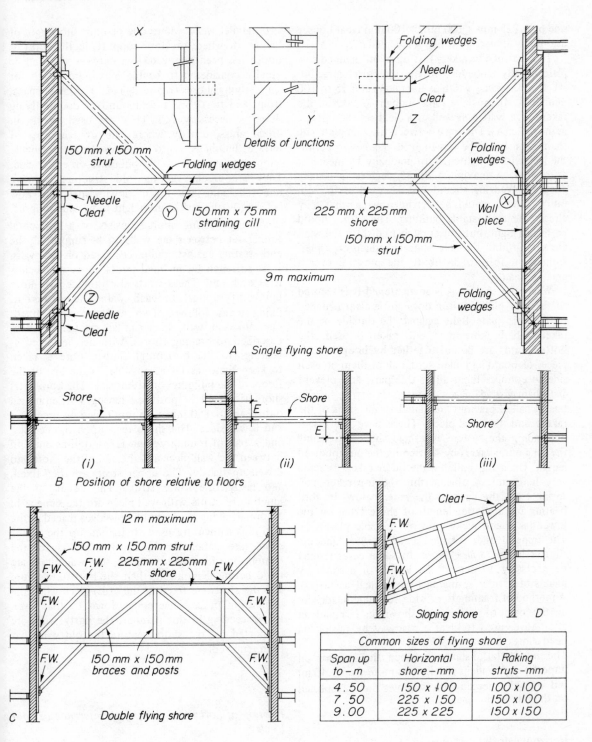

X

Y

Z

Details of junctions

Folding wedges

Needle

Cleat

150 mm x 150 mm
strut

Folding wedges

Needle
Cleat

Y 150 mm x 75 mm
straining cill

225 mm x 225 mm
shore

Folding
wedges

Wall
piece
X

150 mm x 150 mm
strut

9 m maximum

Z

Needle

Cleat

Folding
wedges

A   Single flying shore

Shore

Shore

E

E

Shore

(i)

(ii)

(iii)

B   Position of shore relative to floors

12 m maximum

150 mm x 150 mm strut

F.W.   225 mm x 225mm
shore

F.W.

F.W.

F.W.

F.W.

F.W.

150 mm x 150 mm
braces and posts

C   Double flying shore

Cleat

F.W.

F.W.

Sloping shore   D

Common sizes of flying shore		
Span up to – m	Horizontal shore – mm	Raking struts – mm
4.50	150 x 100	100 x 100
7.50	225 x 150	150 x 100
9.00	225 x 225	150 x 150

241   *Flying shores*

side by a 225 mm x 100 mm x 100 mm cleat housed into the wall-piece to which it is nailed.

The feet of the rakers rest upon an inclined sole plate usually embedded in the ground (figure 240 A). It is the same width as the rakers and 75 to 100 mm thick. It must be long enough to take all the rakers in a system as well as a cleat on the outside as shown at (A). The angle between the sole plate and the rakers should be less than 90 degrees to permit the latter to be tightened up gradually by means of a crowbar. A maximum of 85 degrees is usually adopted. Wedging should not be used as the vibration caused would be detrimental to what may already be an unstable building. The shore should be forced tight, but not enough to disturb the wall. On soft ground the sole plate is bedded on a platform of timber to distribute the pressure over a greater area (A).

When the shore has been tightened it is secured to the sole plate by iron dogs and a cleat is nailed to the sole plate tight against the outside of the raker. Where more than one raker is used, the bottom ends are bound together by hoop iron or pieces of boarding nailed across all of them on each side to connect them all at this part, and prevent disturbance or damage. At intervals in the height, boards called bracings are nailed to the sides of the rakers and the wall-piece. These have the effect of binding the system together, and of stiffening the long outer rakers by shortening the unsupported length. On a tall building the outer rakers become very long and to obviate this the top raker may spring from the back of the raker below, its foot bearing on a shorter length of shore lying on the lower raker and picking up on the sole plate (A). The upper length is called a *riding shore* and the shorter length a *jack shore*. In some cases it may be necessary to extend the length of a shore by means of a rider springing from a cleat as in (D). A method of framing to provide adequate headroom at the base of a shore is shown in (E) and an alternative, used to clear a roadway by raising on dead shores, is shown in (F).

The horizontal distance between the systems on unperforated walls is usually not more than 2.40 m; but on walls pierced with windows they are placed on the intervening piers.

Horizontal or flying shores

These are used to provide temporary support to two parallel walls, where one or both show signs of failure, or where previous support, in the form of floors, has been removed. 9 m between the walls is usually considered to be the maximum length for single flying shores (figure 241 A). For larger spans, from 9 m to 12 m, a compound or double flying shore is necessary (C). They are used mostly in urban areas, usually where one of a number of terrace buildings is to be removed, to provide temporary support to the buildings on either side. They are erected as the old structure is being removed, and are taken down when the new building is of a sufficient height to provide support.

A single flying shore consists of a horizontal timber set between the walls to be supported, the ends resting against wall-pieces fixed on the walls. It is stiffened by inclined struts above and below it at each end. These struts also provide two more points of support to each wall. The method of fixing is as follows. Two wall-pieces are fixed, one on each wall, in a similar manner to those described for raking shores, with needles fixed as bearings to the horizontal timber. Care is taken to keep these, as far as possible, in the line of the floors of the buildings on either side. The horizontal shore is placed in position, having a straining sill out of about 150 mm x 75 mm nailed on the upper and lower sides. The shore rests upon the needles, and a pair of folding wedges is inserted at one end between the wall-piece and the end of the shore and driven up tightly. The upper struts are then fixed, and lastly, the lower struts. The details at the junction of struts with wall-piece are the same as in a normal raking shore. Folding wedges placed in the positions shown are used to tighten up the whole shore (see details X, Y, Z). In the case of the demolition of a terrace building, the wall-pieces are fixed before demolition, then the horizontal shore and the struts are fixed in the order given when the demolition process has come down to that level. By proceeding in this manner, the party walls are supported by the shores before the old work has been removed.

Position of horizontal shore with reference to floor levels

The three conditions of floor positions that are likely to be met in practice are shown in (B):

398

(i) Floors on each side at the same level. The centre line of shore should coincide with centre line of joists, whether these are parallel with or normal to the walls being supported

(ii) Floors at different levels with the same run of joists on each side. The horizontal shore is placed halfway between the two floor levels

(iii) Floors at different levels with different runs of joists on each side. In this case, the wall which is not supported laterally by a floor bearing on it is the weaker of the two. The horizontal shore should therefore be placed in line with the floor joists which run parallel with this wall.

Where, in (ii) and (iii), the shore is positioned between floors, the wall-piece should be stiff enough to transmit any thrust to the floors.

Where walls are of different thicknesses or in different states of repair and with differing floor levels, this must also be taken into consideration in deciding the best position for the shore.

Double flying shores are framed up as shown in (C). They are erected in basically the same manner as single shores.

Where one building is higher than another it may be necessary to erect a raking shore upon a flying shore or sloping shores may be suitable (D).

Horizontal flying shores are usually erected at 3 to 4.5 m intervals on plan. Where considered necessary, horizontal struts are introduced between the shores to act as lateral bracing.

Vertical or dead shores

Shores placed vertically are termed dead shores. They are used for temporarily supporting the upper parts of walls, the lower parts of which are required to be removed, either in the process of underpinning or reinstatement during repair, or for the purpose of making large openings in the lower parts. Where a dead shore immediately under the wall is not convenient a system of dead shores is used, comprising a pair of shores supporting a horizontal beam. The wall is then carried by the beam. If, for example, the lower part of a building is to be removed in order to form a large opening, the procedure would be on the lines illustrated in figure 242 A.

The whole of the floors, the roof, and any other load bearing on the wall are supported by a system of strutting to relieve the wall of all weight normally taken by it. This system of strutting should be firmly supported by a sole piece on the solid ground below the lowest floor. The sole piece should be bedded continuously in mortar and be sufficiently stiff to distribute the weight over its whole length.

Perforations are next made in the wall a short distance above the line of the top of the beam that will ultimately support the wall. Through the holes horizontal beams called needles are inserted, consisting of balk timbers or steel beams. These should not be placed a greater distance apart than 1.80 m in unperforated walls, but when there are windows the needles must be placed under the piers.

The needles are supported by dead shores of timber or steel stanchions, one under each end of the needle. The dead shores rest at their lower ends on sleepers which are horizontal balks of timber properly bedded in mortar for their whole length. It is essential that the sleepers should be bedded on the solid ground, not on the crown of vaults or any other voids. Should there be voids of any kind the work must be solidly strutted below. Pairs of hardwood wedges are placed between the dead shores and the sleepers. Before these are driven up tightly, a bed of cement mortar should be placed on the top of the needle at the point where it passes through the wall to ensure a proper and solid bearing of the wall on the needle. When the wedges are driven home the whole is allowed a few days to set. Lateral bracing is often provided by nailing thick boards diagonally to the faces of the shores.

Where it is not possible to place the inner shores in position in one piece, these must be in two sections. The lower halves are placed first with a transom laid across them. The upper halves are then placed on the transom directly over the lower sections and under the needle at its upper end.

It is essential that the needles and dead shores should have an ample margin of strength, to avoid settlement of the wall through deflection of the needle or compression of the shore or sole piece. The needles, shores and sleepers and any transoms must all be well dogged together before any brickwork is removed. All window openings must be strutted to prevent deformation taking place. In ordinary small windows this consists of an upright against each reveal, with two or three struts between, cut long and driven up tightly as shown in figure 242 A. In large openings a stronger framing

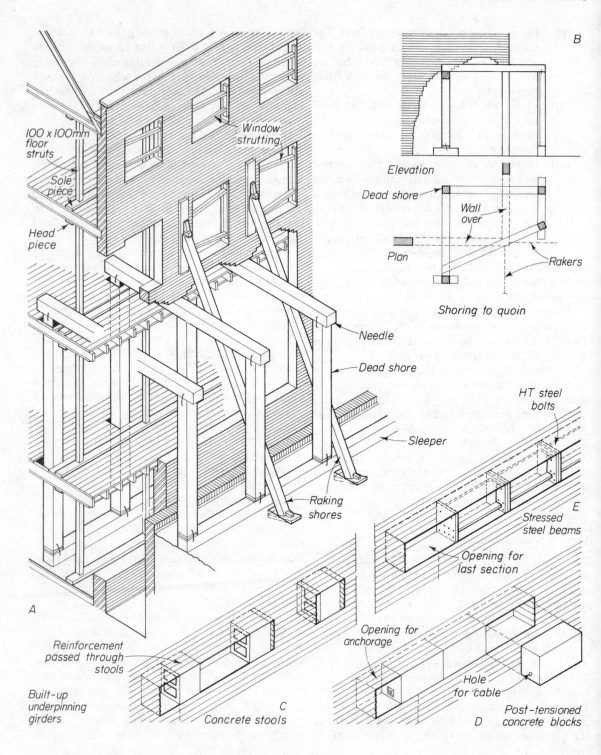

100 x 100mm floor struts

Sole piece

Head piece

Window strutting

B

Elevation

Dead shore

Wall over

Plan

Rakers

Shoring to quoin

Needle

Dead shore

Sleeper

Raking shores

HT steel bolts

E

Stressed steel beams

Opening for last section

A

Reinforcement passed through stools

Built-up underpinning girders

Concrete stools

C

Opening for anchorage

Hole for cable

Post-tensioned concrete blocks

D

242   *Dead shores*

is necessary and any arches will require support by a turning piece, or centre[1], made to fit, with the reveals strutted as before.

If the building is old or at all defective, raking shores are imperative, but it is wise in most circumstances to use them to steady the building during the progress of the works. These are fixed against the piers between the windows and close beside the dead shores as shown.

When all the shores are fixed in position, the two end piers are built, or if the supports are to be stanchions these are erected, the minimum amount of existing wall being taken away to allow for this work, after which the remainder of the wall is removed. The new beam is then raised and fixed, and the brickwork above filled in to the underside of the old work. The new brickwork should be built in cement mortar to avoid settlement in the work.

A week at least should be allowed for the new work to set before any of the shoring is struck. The needles should be eased and removed first, then the strutting from the windows, the floor strutting inside, and, lastly, the raking shores. About two days should be allowed between each of these operations in order that the work may take its bearing gradually on the new supports.

Great care is required in carrying out these operations on a corner building. The needling would be constructed to suit the special requirements of the job and the angle of the building should always be supported by raking shores on each face (figure 242 B).

The lengthy procedure described above can often be avoided by the use of 'stools' or post-tensioned sectional concrete or steel beams (figure 242 C, D, E), inserted in the wall in sections, the lower part of the wall not being removed until the beams are completed. These, and the method of insertion, are described under 'Underpinning' on page 113.

TIMBERING FOR EXCAVATIONS

This subject is introduced in Part 1 where the support required for shallow trenches, that is those not exceeding 1.20 m in depth, is described.

Trenches

Considerable care must be taken in the cutting of trenches and in the selection of walings and struts of adequate dimensions, particularly if the trenches must remain open for some length of time, since, as pointed out in Part 1, the pressures on them can be high.

When trenches over 1.20 m in depth are required in very soft soils, runners should be employed. Runners are sawn timbers 50 to 75 mm thick, usually about 225 mm wide and up to 6 m long. The lower ends are bevelled and shaped to give a cutting edge and are often shod with steel or hoop iron. If the upper soil is reasonably firm, the first stage of the excavation can be supported by poling boards and the following procedure, illustrated in figure 243 A, is adopted. Two 75 mm x 50 mm continuous guides are fixed to the top walings and struts, between which the runners are driven as far as is possible without damage to the head. The soil between is then excavated, care being taken not to remove soil within 300 mm of the toe of the runners, after which driving proceeds again. At approximately 1.20 m intervals, frames of walings and struts are inserted, the lower walings acting as guides to the runners and keeping them in a vertical plane as they are driven. If the soil at the top is not firm enough to permit excavation for poling boards, the top frame of walings and struts is fixed at the surface and continuous guides are firmly framed up about 600 to 900 mm above the ground level to permit driving of the runners. When the excavation is deeper than the maximum length of runner, a further stage of runners must be driven inside the first as indicated in the section, continuous guides being fixed to one of the frames above.

Timbering to a deep trench in firmer soil is shown in figure 243 B.

In deep trenches in loose soils, the withdrawal of the timbers must be very carefully done to prevent the collapse of the trench. Frequently the poling boards at the concrete level must be left in position. For the upper sections, the wall is carried up between the cross struts to a height above the poling boards in each section, as shown in (C). Short struts are then placed on each side of the wall at intermediate positions between the main struts. These subsidiary struts bear on short upright plates

[1] See Part 1, chapter 11

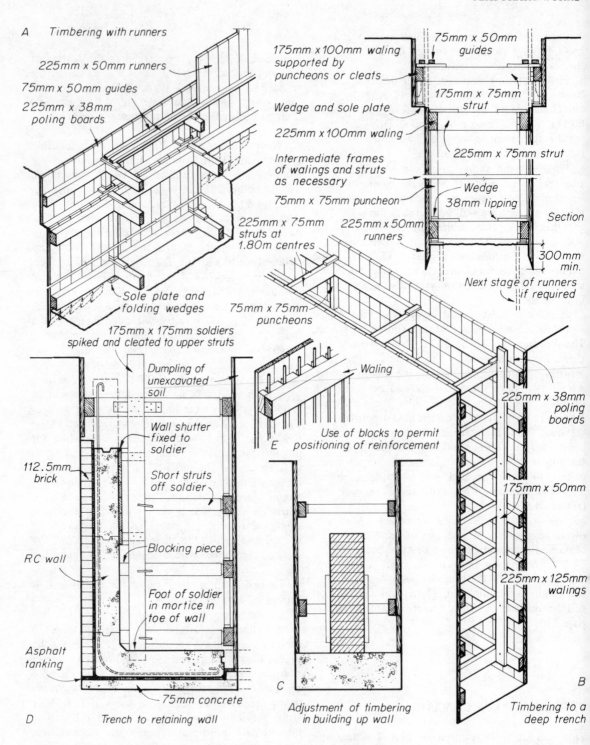

A   Timbering with runners

225mm x 50mm runners

75mm x 50mm guides

225mm x 38mm
poling boards

175mm x 100mm waling
supported by
puncheons or cleats

Wedge and sole plate

225mm x 100mm waling

Intermediate frames
of walings and struts
as necessary

75mm x 75mm puncheon

225mm x 75mm
struts at
1.80m centres

225mm x 50mm
runners

75mm x 50mm
guides

175mm x 75mm
strut

225mm x 75mm strut

Wedge
38mm lipping

Section

300mm
min.

Next stage of runners
if required

75mm x 75mm
puncheons

Waling

Use of blocks to permit
positioning of reinforcement

E

175mm x 175mm soldiers
spiked and cleated to upper struts

Dumpling of
unexcavated
soil

Wall shutter
fixed to
soldier

Short struts
off soldier

Blocking piece

Foot of soldier
in mortice in
toe of wall

112.5mm
brick

RC wall

Asphalt
tanking

75mm concrete

D   Trench to retaining wall

C

Adjustment of timbering
in building up wall

225mm x 38mm
poling
boards

175mm x 50mm

225mm x 125mm
walings

B

Timbering to a
deep trench

243   Timbering to trenches

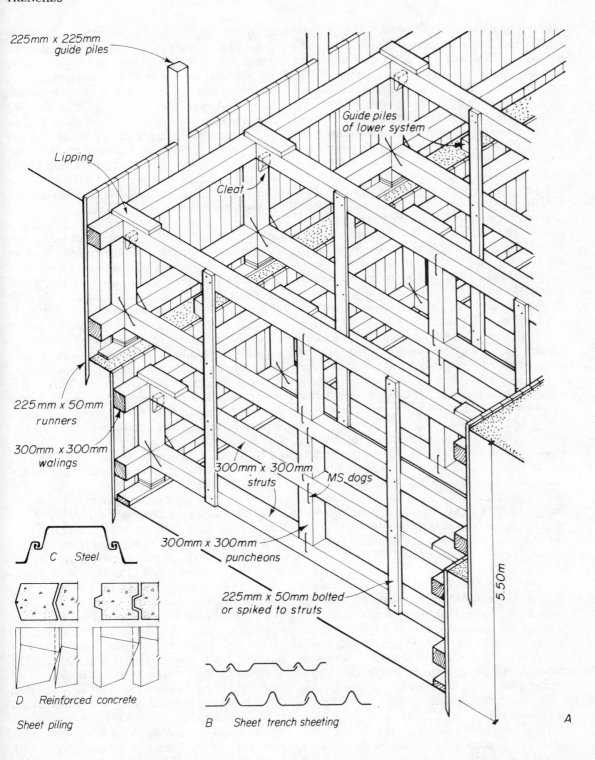

225mm x 225mm
guide piles

Guide piles
of lower system

Lipping

Cleat

225mm x 50mm
runners

300mm x 300mm
walings

300mm x 300mm
struts

MS dogs

300mm x 300mm
puncheons

225mm x 50mm bolted
or spiked to struts

5.50m

C   Steel

D   Reinforced concrete

Sheet piling

B   Sheet trench sheeting

A

244   Large trenches in poor ground

against the wall and care must be taken to place them exactly opposite each other on the two sides of the wall so that the latter is not subjected to bending stresses. The main struts are then removed and the spaces in the run of the wall completed. This process is continued until the wall is above the surface level. Polings and struts are then removed in sections and the earth at once filled in to the foot of the next row of polings, the process being continued until the whole of the timbering is removed.

A method of working adopted when a tanked reinforced concrete retaining wall is to be constructed is shown in figure 243 D. The trench, which will generally be the full width of the toe, is covered with 75 to 100 mm of concrete, the lower boards are removed and the brick protective skin to the asphalt tanking is built up as far as the soil will permit. (In very loose soils it may be necessary to use precast concrete walings or precast concrete sheet piling which will be left permanently in position.) After the asphalt is laid on the concrete blinding and against the lower part of the brick skin, the toe to the wall is cast with mortices formed in the top face to take vertical posts or soldiers. These are later erected with their feet in the mortices and are spiked and cleated to the upper horizontal struts. To permit the pouring of the first lift of wall, the lower struts are removed and replaced by shorter struts bearing on the soldiers. The shuttering to the face of the wall is built up in panels and these are fixed to the soldiers. This process of building up the brick skin, asphalting and pouring subsequent lifts of concrete, is then repeated to the top of the wall. As the wall rises and the upper cross-struts are removed, the lower part of the soldiers can be blocked off the completed lower parts of the wall.

When no external tanking is to be applied and the wall is cast directly against the soil, the position of the vertical reinforcing bars will be close to the poling boards or sheeting and the walings will obstruct them. In these circumstances the walings are blocked off the poling boards by short blocking pieces with spaces between them at the bar centres to permit the bars to pass through them between the walings and the poling boards, as shown at (E).

Large continuous trenches, if made in bad ground, are generally timbered as shown in figure 244 A. At intervals guide piles are driven in, to which

walings are bolted to act as lower guides to an upper stage of runners about 3 m long, inserted between the piles; continuous guides are fixed to the piles 600 to 900 mm above the ground. The runners are driven a short distance into the ground, the soil between the two systems of piles being then taken out to within 300 mm of the bottom of the runners, which are again driven in and the process repeated. After excavation of the first stage, walings, consisting of whole timbers, are placed in position and strutted apart, the struts being also of balk timber. Short strips of board called lippings, or lips, are nailed to the ends of the heavy struts to facilitate handling and fixing, or, alternatively, temporary props are used until the struts are wedged up. Long struts are given intermediate support by short uprights or puncheons, secured to them by dogs. Puncheons are also placed between the waling pieces as each fresh one is inserted. A fresh system of piles and runners is next driven slightly in advance of that to the first stage and the ground excavated as before. Provision must be made for changing the positions of the upper struts in order to permit the driving of runners in the lower stages.

Large trenches in firm ground are timbered in a similar way, using ordinary poling boards, but if the width exceeds 9 m it is cheaper to adopt a system of raking shores instead of horizontal struts (see figure 246).

The method illustrated in figure 245 may be employed where the ground is loose and water-logged. By this method as much of the soil is taken out as is possible without the sides of the excavation falling in, the depth depending on the soil conditions. If the soil will permit, the first stage is supported by poling boards or sheeting, walings and struts, if not, runners must be used, driven as described above. The excavation is continued by lining the trench with 225 mm x 50 mm runners about 2.75 or 3 m in length. These are waled and strutted. Between each runner and waling a wedge or 'page' is inserted to prevent any slip of the soil behind the runners. The method of proceeding with the excavation is as follows: The wedges securing one runner are eased and about 300 mm of soil is removed in front of the runner, the runner being dropped as the ground is removed. It is then re-wedged. Each runner is successively treated in this manner until the whole system has been lowered the necessary amount. As already explained the feet of the runners must at all times

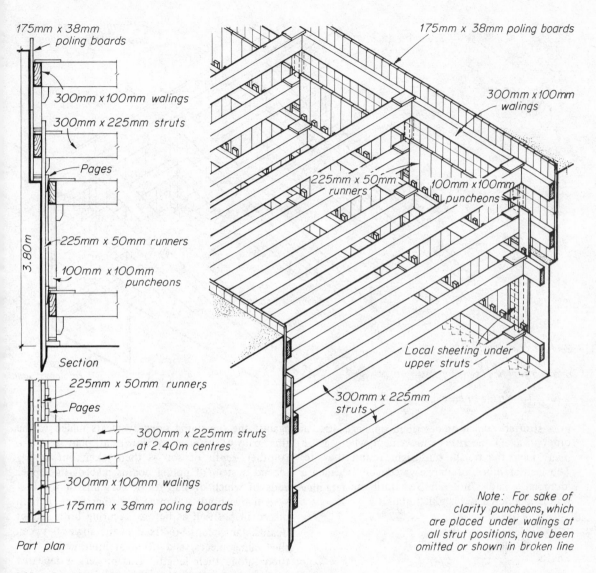

175mm x 38mm poling boards

300mm x 100mm walings

300mm x 225mm struts

Pages

225mm x 50mm runners

100mm x 100mm puncheons

3.80m

Section

175mm x 38mm poling boards

300mm x 100mm walings

225mm x 50mm runners

100mm x 100mm puncheons

Local sheeting under upper struts

300mm x 225mm struts

225mm x 50mm runners

Pages

300mm x 225mm struts at 2.40m centres

300mm x 100mm walings

175mm x 38mm poling boards

Part plan

Note: For sake of clarity puncheons, which are placed under walings at all strut positions, have been omitted or shown in broken line

245   Large trenches in waterlogged soft ground

be kept about 300 mm in the ground, for if any portion of the side of the excavation is exposed, the soil is likely to fall out leaving the back of the runners unsupported, and causing the whole system to collapse. It will be noticed that it is not possible to drive runners under each of the upper pair of struts. Horizontal sheeting in short lengths is therefore introduced behind the two adjacent end runners of each panel to bridge this gap.

Figure 246 shows a method of timbering employed where the ground has to be excavated for a basement and a trench sunk for a retaining wall to support the soil outside the building. In this case the basement may be excavated first. For the first stage about 900 mm of soil is removed; the top broken line shows the soil left in at this stage. The top row of poling boards and walings and the top system of shores are then fixed. The soil indicated by the broken line is then removed and the next system of timbering and shores is fixed, after which the trench for the retaining wall is excavated and timbered. When a deep excavation is near buildings or streets the basement area is often not at first excavated, the wall being built

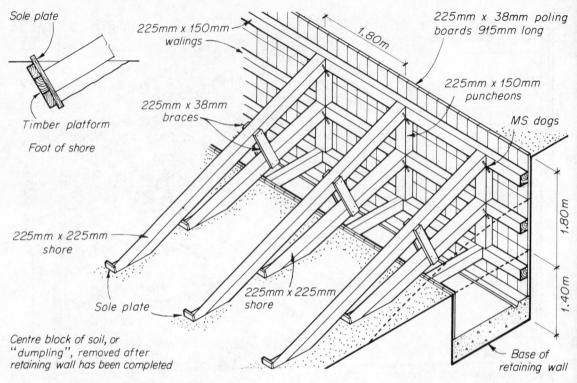

Sole plate

225mm x 150mm
walings

225mm x 38mm poling
boards 915mm long

1.80m

Timber platform

Foot of shore

225mm x 38mm
braces

225mm x 150mm
puncheons

MS dogs

225mm x 225mm
shore

1.80m

Sole plate

225mm x 225mm
shore

1.40m

Centre block of soil, or
"dumpling", removed after
retaining wall has been completed

Base of
retaining wall

246   *Timbering to basement excavation*

in a strutted trench as described on page 404, in order to avoid the risk of movement when, as on many jobs, the trench runs right round the site. The central block of unexcavated soil is called a 'dumpling', and is not removed until the retaining wall is completed and is functioning.

Shafts

It is often necessary to sink shafts for foundations. These are made from 1.20 m square and upwards, the former being the smallest size a man can work in without difficulty.

Shafts from 1.20 to 2.75 m square are timbered as shown in figure 247. One method is shown at (A). In ordinary soils the earth is first excavated to a depth of at least 900 mm and in firm soils 1.80 m. The sides of the excavation are then lined with poling boards strutted apart by frames of horizontal walings, a pair of which are placed in position against two opposite sides, and strutted apart by another pair driven tightly between. The latter are held

against the remaining sides by cleats nailed to the first pair of walings as shown at plan at (BB). Another depth of soil is then taken out, and a second system of poling boards placed, the upper ends of which overlap the lower ends of the first system by about 300 mm. A further frame is then placed in position as before, securing both sets of boards. Puncheons are fixed in the angles between the waling pieces, and often at intermediate positions along their length. This process is repeated until the required depth is obtained.

If the depth is great the timbering must be supported to prevent it sliding down on the removal of the earth from its lower end. Where this is necessary the upper end is left projecting about 900 mm above the ground level, and two substantial beams are laid across the excavation and project about a metre on either side to obtain a good bearing on solid ground. Upright vertical timbers are notched over these (plan at AA) and spiked to the face of the walings below. The whole is thus tied together. This is sometimes supplemented by similar timbers at the bottom of the shaft. These lower timbers are fixed

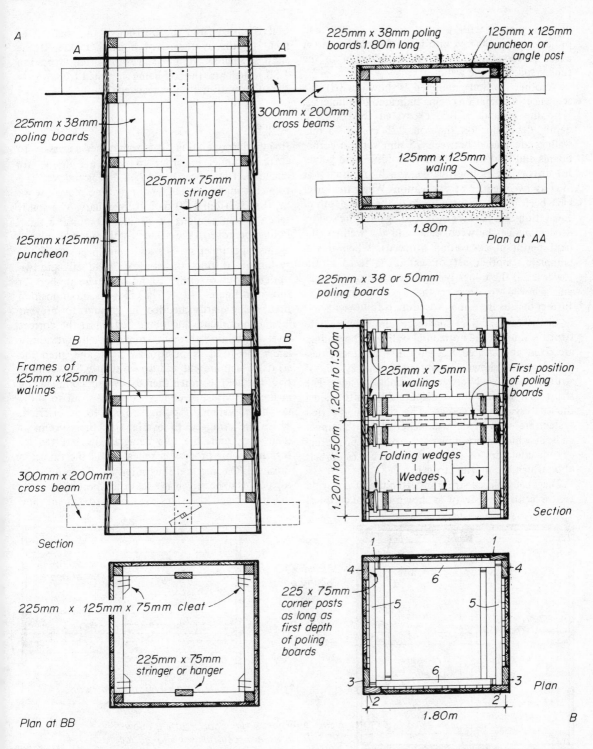

A

A — A

225mm x 38mm poling boards

225mm x 75mm stringer

125mm x 125mm puncheon

B — B

Frames of 125mm x 125mm walings

300mm x 200mm cross beam

Section

225mm x 125mm x 75mm cleat

225mm x 75mm stringer or hanger

Plan at BB

225mm x 38mm poling boards 1.80m long

125mm x 125mm puncheon or angle post

300mm x 200mm cross beams

125mm x 125mm waling

1.80m

Plan at AA

225mm x 38 or 50mm poling boards

1.20m to 1.50m

1.20m to 1.50m

1.20m to 1.50m

225mm x 75mm walings

First position of poling boards

Folding wedges

Wedges

Section

225 x 75mm corner posts as long as first depth of poling boards

1.80m

Plan

B

247   Timbering to shafts

in two pieces, with a scarf in the centre, and project about 900 mm into both sides of the pit. A chain is sometimes employed in addition to the timber spiked to the walings.

Another method employed is shown at (B), the sequence of operations being indicated by numerals. Here the ground is first excavated to a suitable depth, depending on the soil. 225 mm x 75 mm walings are placed between 75 mm vertical corner boards and wedged in position to form rigid guides for driving the poling boards, which are usually as long as two depths of excavation. When the latter have been driven to the bottom of the first depth of excavation and have been forced against the soil by driving wedges between them and the walings, the next depth of excavation proceeds. When this is complete, another set of walings is fixed at the lower level, the upper polings are eased one by one, and allowed to drop to the lower position. For further depths the same procedure is followed.

Shafts over 2.75 m square require intermediate struts to support the horizontal walings. The walings are fixed and cleated to each other as already described and as shown in figure 248. One system of struts is then fixed between two opposite sides. The struts that support the remaining sides butt against the first system as shown. The struts in the first system are supported by puncheons, on the upper ends of which short timbers are placed, projecting beyond the sides of the strut. These act as corbels upon which the shorter struts bear.

Diagonally placed struts are often used to provide a clear central hoistway as shown by broken lines.

Soil is raised from the bottom of the shaft, if of great depth, by means of hoisting tackle. If the shaft is shallow, timber stages are often erected in 1.80 m heights, the soil being shovelled from one to the other until the top is reached.

Tunnels

In building operations it is sometimes necessary to form a tunnel in order to construct drains, for example. The process is carried out as follows. The tunnel is made just large enough for a man to work in, that is, from 1.20 to 2.10 m square. The soil is taken out in sections of about 900 mm at a time. Poling boards of the same length are then placed against the upper surface and kept in position by a system of strutting, consisting of a head, cill, and two uprights, out of either round or square timbers, as shown in figure 249 A. The cill is placed in position first, being partly bedded in ground to prevent lateral movement, and being bedded at the correct level by boning through from the cills previously laid. The head is positioned next and then the struts, which are cut and driven tightly between the two. The next section is then excavated, commencing at the top with just enough soil taken out to allow the next system of poling boards to be inserted. These are arranged to overlap the first system at their back end, the two being finally strutted up together. This process is repeated till the tunnel is finished. The method illustrated at (B) permits the top boards to be driven as excavation proceeds. If the soil is bad and the sides are liable to fall in, they

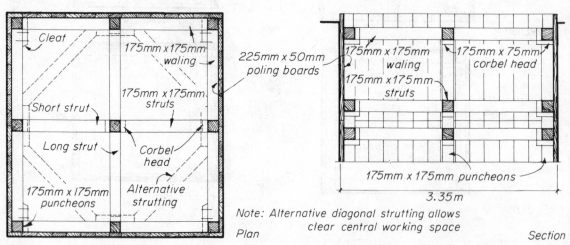

Plan

Note: Alternative diagonal strutting allows clear central working space

Section

248   Timbering to shafts

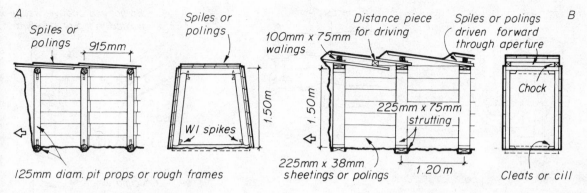

249   Tunnelling

must also be lined with poling boards which are kept in place by the uprights.

Large spikes, similar in shape to floor brads, are driven into the head and cill to secure the struts. The heads are left projecting to permit easy withdrawal. Wood cleats are often used in place of these.

Sheet piling

A form of steel sheet piling, called trench sheeting, is often used instead of timber runners, and the strutting can then be more widely spaced. Light trench sheeting consists of corrugated sheets of steel about 350 to 400 mm wide which are driven by compressed air or petrol hammers so that the edges overlap (figure 244 *B*). Withdrawal is by crane, usually with a shackle fixed through a hole in the pile or with pulling tongs or grips which grip the faces of the sheets. Other types of heavier steel sheet piling with interlocking edges, and precast reinforced concrete sheet piles, are used for heavy or permanent works, such as for retaining walls and coffer dams, as well as for excavation work (*C* and *D*). These are driven by drop hammer, diesel hammer or double-acting steam or compressed air hammer. Alternatively, a system which vibrates the pile into the ground can be used which drives at a very high rate with much less noise or one in which the piles are driven by a multi-ram hydraulic driver, the reaction for which is obtained from other partly driven piles in the group being driven by the machine. This method is vibrationless and is also quiet in operation.

These piles are withdrawn by large extracting grips used in conjunction with a crane, a double acting hammer or hydraulic jacks. Normally only steel sheet piling and timber piling when used in temporary works, is extracted, and then generally only when the piles are less than about 12 m in length. Piles over this length require a very large extracting force and it is sometimes cheaper to leave the pile in position rather than withdraw it.

SCAFFOLDING

Temporary erections, constructed to support a number of platforms at different heights to enable workmen to reach their work and to permit the raising of materials, are termed scaffolds (figure 251).

Tubular metal scaffolding is in almost universal use, although timber scaffolding may still occasionally be used in rural districts.

Tubular scaffolding

This has considerable advantages over timber. The small diameter and the standard lengths simplify storage and transport; if overloaded it does not suddenly break like timber, but gives ample warning by bending. Its adaptability to any purpose required on a building job, such as storage racks for timber or any other material, or the framing for temporary buildings and sheds, constitutes a valuable asset.

The tubing employed is 38 mm internal diameter weld-less steel steam tubes, no. 6 gauge, 5.20 kg/m run, or light alloy tubing weighing only 1.5 kg/m run. The standard length of the unit is 5.50 m, but shorter and longer lengths can be obtained. The shorter lengths in most common use are 1.80 m, 3.60 m and 4.30 m. The weight of a standard length of steel tube is 29 kg and the external diameter is nominally 50 mm; it can thus be easily handled by the average man.

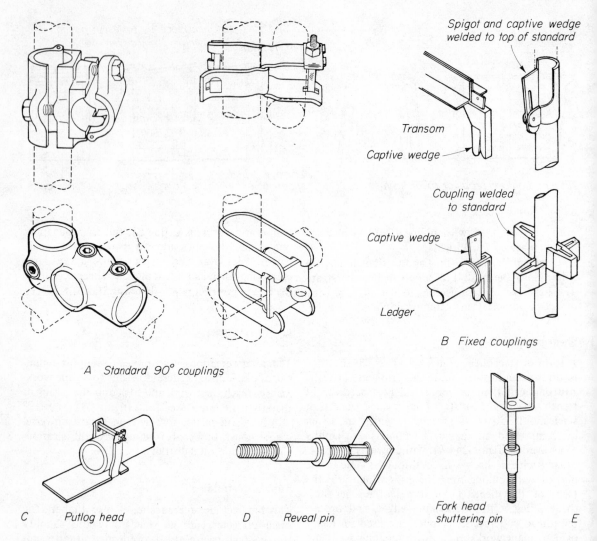

Spigot and captive wedge welded to top of standard

Transom

Captive wedge

Coupling welded to standard

Captive wedge

Ledger

B   Fixed couplings

A   Standard 90° couplings

C   Putlog head

D   Reveal pin

Fork head shuttering pin

E

250   Tubular metal scaffolding – couplings

Some forms of the standard couplings used to frame up the tubes are shown in figure 250 A. Some scaffolding systems incorporate fixed couplers the parts of which are integral with the different members as shown in (B). Square or circular base plates are provided with a central pin that fits into the base of the standards. These are sufficient to take weight of the scaffold on ordinary firm ground; for soft ground or over cellars or pavement lights stout planking is used to distribute the pressure. The base plates can be spiked to the planking, holes usually being provided in them for this purpose. Adjustable bases are available with a range of height of a few centimetres and the tubing itself can be extended to any length, by means of end to end couplers.

Two types of scaffold are used (i) the putlog or bricklayer's scaffold and (ii) the independent scaffold.

In both types a frame of vertical tubes, called standards, and horizontal tubes, called ledgers, is erected about 1.20 to 1.50 m from the line of the building, the standards being about 2.40 m apart and the ledgers at vertical distances of about 2 m. Diagonal braces are applied to the outside of the frame to provide rigidity.

410

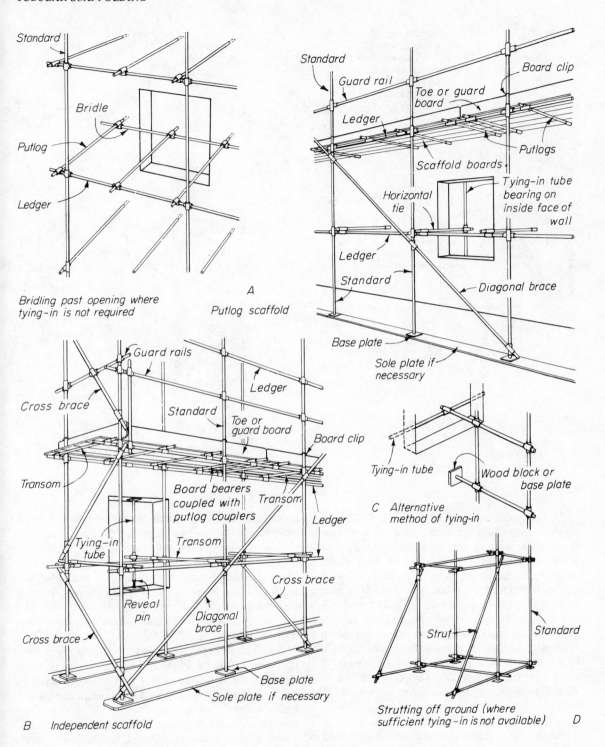

Standard

Bridle

Putlog

Ledger

Bridling past opening where tying-in is not required

A

Standard

Guard rail

Toe or guard board

Ledger

Board clip

Putlogs

Scaffold boards

Horizontal tie

Tying-in tube bearing on inside face of wall

Ledger

Standard

Diagonal brace

Base plate

Sole plate if necessary

Putlog scaffold

Guard rails

Cross brace

Ledger

Standard

Toe or guard board

Board clip

Transom

Board bearers coupled with putlog couplers

Transom

Ledger

Tying-in tube

Transom

Reveal pin

Diagonal brace

Cross brace

Cross brace

Base plate

Sole plate if necessary

B   *Independent scaffold*

Tying-in tube

Wood block or base plate

C   *Alternative method of tying-in*

Strut

Standard

Strutting off ground (where sufficient tying-in is not available)   D

251   *Tubular metal scaffolds*

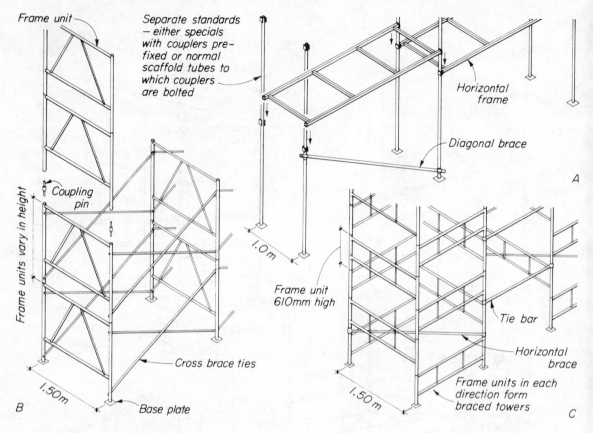

Frame unit

Separate standards
– either specials
with couplers pre-
fixed or normal
scaffold tubes to
which couplers
are bolted

Horizontal
frame

Diagonal brace

A

Frame units vary in height

Coupling
pin

Cross brace ties

1.0 m

Frame unit
610mm high

Tie bar

Horizontal
brace

Frame units in each
direction form
braced towers

B

1.50m

Base plate

1.50 m

C

252   Metal scaffolding frames

In the putlog scaffold (figure 251 *A*) the cross members or putlogs, which carry the scaffold boards forming the working platforms, bear one end on the scaffold frame and the other on the wall which is being built, the putlog tube being flattened at one end and driven into a joint. Alternatively, a putlog head may be used (figure 250 *C*). The scaffold is tied back to the wall at suitable openings as shown in (*A*), or as in (*B*) where the tie is coupled to a vertical or horizontal tying-in tube secured tightly between the reveals of an opening by means of a reveal pin (figure 250 *D*). This consists of a bolt with ferrule which fits the bore of a tube. The bolt may be rotated in the end bearing plate by a spanner to act as a light screw jack.

As the name implies, an independent scaffold is self-supporting and has an inner frame erected about 150 mm from the building face (figure 251 *B*). The

cross members, here called transoms, bear on and are coupled to the two frames. For tying back to the wall a tying-in tube with a reveal pin as shown in (*B*) or the alternative method shown in (*C*) must be adopted. Where insufficient tying-in is available the scaffold must be strutted at ground level (*D*). In this form of scaffold lateral rigidity must also be ensured by the use of cross braces running from frame to frame.

Scaffold boards are fixed to the putlogs or transoms by board clips and guard boards to the standards in the same way (*A, B*). Guard rails at normal handrail height should be fixed to the outer frame and at the ends.

On high buildings double or treble tube standards, linked by couplers, are required for the lower part when heavy loads must be supported.

Metal scaffolding frames

In order to reduce the multiplicity of couplings and coupling operations in the framing of independent scaffolds, ledgers and transoms in some systems are prefabricated into horizontal frames of circular or rectangular hollow steel sections (figure 252 A). These may rapidly be secured to separate standards by lugs welded to the corners of the frame which fit into special separate couplers or into lugs bolted or welded to the standards at the appropriate intervals. Diagonal bracing is applied as in normal scaffolds.

Other systems prefabricate frames in the vertical plane as shown in figure 252 B, C. In (B) the basic frame consists of two standards spaced the normal scaffold width apart by cross tubes, the height of the frames varying from 900 mm to 1.80 m. These are built up ladder style using coupling pins or by slotting directly one into another. Bracing is bolted to the standards. In (C) the frames are of a single, smaller standard dimension and are built-up in both directions with coupling pins to form 'towers'. Diagonal bracing is required only in the horizontal plane as shown and the 'towers' may be tied together by longitudinal tie-bars. These last two systems are most useful in forming short lengths of scaffold or individual access towers where operations take place at the top level[1].

Gantries

These are structures erected primarily to facilitate the loading and unloading of material, and for its storage during building operations. They consist of an elevated staging erected in front of buildings in the course of erection, designed to act as unloading platforms. They extend usually from the face of the intended structure to a short distance from the edge of the kerb, covering the footways. A gangway is provided under the staging for the convenience of the public. Some form of hoisting tackle is provided for raising material from lorries on to the platform. The use of the tower crane reduces the value of the gantry for this particular purpose, but it is still useful on sites where it is essential to have room outside the site for the storage of materials and for agents' huts.

Gantries are constructed of tubular scaffolding or of steel, as shown in figure 253. The uprights here consist of light steel beam sections and these are connected by channels at their bottom, which act as sleepers. Their upper ends are connected to a light lattice beam bolted on the face of the columns and resting on cleats. The two frames are connected with cross frames bolted to them. The various parts are standardized, which ensures simplicity and rapidity in erection, and enables them to be used many times. The inner frames of both timber and steel gantries frequently have to be taken down to a basement level for a bearing. In this case either longer standards are employed, or a subsidiary frame is erected to support the upper frame, being cleated to the upper lengths of the uprights by fish-plates through the web.

If the gantry is over the public way, it must be double-boarded to prevent dust, rubbish or water falling upon pedestrians, or be under-decked with corrugated iron sheeting.

FORMWORK

The subject of formwork has been introduced briefly in Part 1 where that for a simple slab is described. As indicated there concrete must be given form by casting it in a mould. These moulds are known as formwork or shuttering[2]. Reference has already been made in chapter 5 to the fact that the cost of the formwork may be as much as one-third or more of the total cost of the concrete work as a whole, and to the effect that the nature of the formwork can have on its cost. The formwork for any job must be considered at the design stage. Economy is more likely to be achieved if it is designed and worked out in detail before work commences on the site, taking into consideration the nature of the elements to be cast and the methods of handling likely to be used on the site. For example, handling by crane makes possible the use of much larger sizes of wall and floor shutters than if manually handled.

The general requirements governing the design and construction of formwork are as follows:

[1] The regulations governing the construction and maintenance of scaffolds are to be found in the Construction (Working Places) Regulations, 1966, (HMSO)

[2] *Formwork* is a general term which covers all types of mould for cast *in situ* concrete. The word *shuttering* is correctly applied only to the flat panels which are fixed together to make the complete formwork. Parts of the formwork such as column and beam boxes are called *forms*. Boxes for precast concrete are called *moulds*

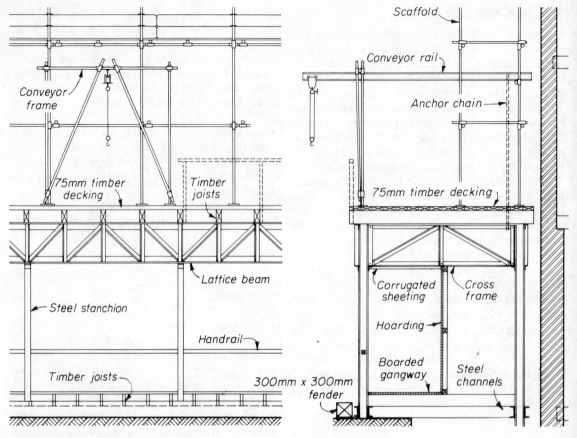

253   Steel gantry

(i) It should be strong enough to bear the weight of the wet concrete and all incidental working loads and it should be rigid enough to prevent excessive deflection during the placing of the concrete.

(ii) The joints should be tight enough to prevent the loss of fine material from the concrete.

(iii) It should be so designed and constructed that erection and stripping is orderly and simple and all units are of such a size that they can be easily handled. It should be possible for the side forms to be removed before the soffit shuttering is struck.

(iv) If the concrete is to be fair-face the formwork in actual contact with the concrete should be so arranged and jointed that the resulting concrete has a good appearance.

In horizontal work the formwork must support its own weight, the weight of the wet concrete and reinforcement placed upon it, and the weight of men and any transporting equipment which is being used for the work. The formwork for vertical work must resist the pressure of the wet concrete pushing it outwards, and wind pressure. The outward pressure of the concrete depends upon its stiffness, the depth of concrete placed at one time and the way in which it is consolidated and will increase with increased wetness of mix and an increased height of concrete placed. When tamping by vibrators is used, board joints must be tight, any wedges must be nailed and the whole must be sufficiently braced to prevent any movement.

The formwork must be designed and constructed so that it may easily be removed or 'stripped' without damage to the formwork itself or to the hardened concrete. To facilitate this, nailing in timber forms should be kept to a minimum. Erection should be such that the formwork can be struck in the following order: One side of columns, sides of beams, bottoms of slabs and beams, the remaining sides of columns (see table 28, p. 425).

Formwork may be constructed of any suitable

material. Timber was once always used for this purpose and, by its nature, still has the advantage of the flexibility of forms which may be produced by its use. Plywood is widely used in place of boards for the working faces. On *in situ* work timber forms are usually unfit for further use after four to six times, although they may then be cut up and parts used in the construction of other forms. If reasonable care is taken in fixing and striking, plywood will give about thirty to forty re-uses.

In precast work, particularly in factory production, up to twenty re-uses are possible with timber forms. Steel shutters are available, generally in standard units of suitable sizes, and are designed to eliminate timber. They can be quickly erected and dismantled and can be used a greater number of times than timber forms, the actual number of times depending on the way in which they are handled and dismantled. In factory produced precast work as many as 100 re-uses are possible.

Timber formwork

Timber should be sound and well seasoned and may be dressed on all four sides, on one side and one edge, or on one side and two edges. Timber dressed on all four sides is uniform in size and therefore more easily adapted for different purposes. The advantages arising from this often make it more economical to use than timber dressed in the other ways.

Timber of any one size should be dressed to a uniform thickness so that each piece will match up, particularly in the case of boarding. Close, square-edge boarding is most commonly used but tongued and grooved boarding gives best results, particularly for face work. When a good finish is required, used and new boards should not be incorporated in the same panel. Plywood used instead of boards or as a lining to forms should be resin-bonded external grade. The thicknesses of the timbers will depend on loads to be carried and on the available supply. The latter is generally the governing factor as any ordinary size can be used by adjusting the spacing of supports.

As mentioned earlier, nailing should be kept to a minimum and, where used, the nail heads should not be quite driven home, as this makes it easier to draw them with a claw hammer or nail bar. Bolts and wedges are preferable to nailing, but are more costly.

Column forms

In the case of rectangular columns, the forms consist of four shutters or panels made up of boards nailed securely to cleats and held together by a series of yokes or clamps to form the column casing, as shown in figures 254 *A* and 255. The thickness of the boards varies between 25 and 50 mm depending on the size of the column and the spacing of the yokes. In practice it is usually the thickness of the board which dictates the number and spacing of the yokes, which are more closely spaced at the bottom where the outward pressure is greatest, than at the top. This is shown in figure 254.

The yokes may be made up of 75 or 100 mm x 100 mm timber, with 13 or 16 mm diameter bolts to clamp together the two longer pieces, the other two being tightened against the shutter by pairs of wedges driven between the bolts (figures 254 *B* and 255). Where convenient, when the column is very tall and large it is usual to erect and strut three sides only of the box and to fix the fourth side in 900 mm sections as concreting proceeds. This facilitates hand tamping of the concrete after each pouring and avoids the provision of hand holes at the bottom of the column form, although it is still essential to make certain that the bottom is quite clear of rubbish before the concrete is deposited.

Adjustable steel yokes or clamps as shown in figure 254 *A* are now widely used, as they reduce the amount of timber required, have a long life and are quickly fitted and dismantled.

Circular column forms are built up from narrow vertical boards, called 'staves', shaped to the correct curve and fixed to shaped yokes of various forms. The latter are in two halves secured by bolts or steel clamps (figure 254 *C*).

Beam forms

Beam sides are generally built up of 25 mm boards nailed to 100 mm x 25 mm battens at 600 or 760 mm centres. The bottom should be thicker, about 50 mm, and where practicable should be made from a single width of board (see figure 254 *D*). In erection, the bottoms are first placed in position between the column boxes and are supported by props, to the top of which are fixed cross pieces or headtrees about 380 mm longer than the width of the beam. These are braced to the props by pairs of struts. The beam bottom is carefully levelled by

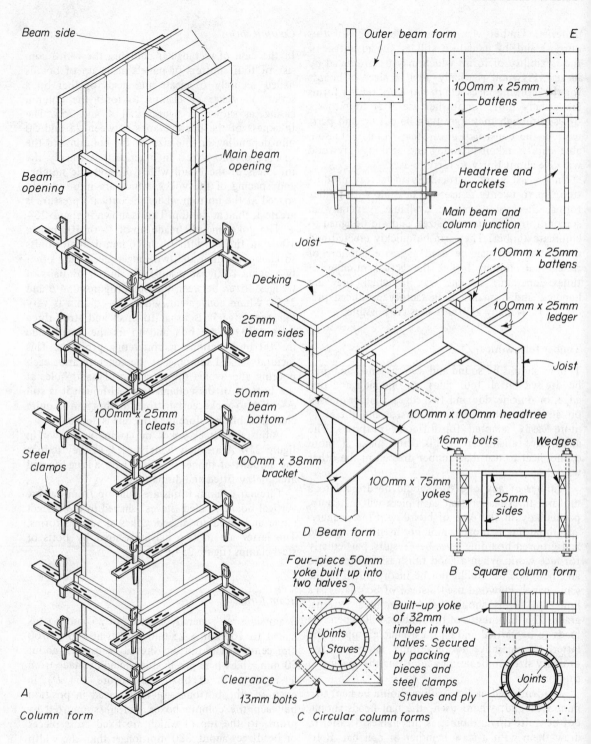

Beam side

Main beam opening

Beam opening

Steel clamps

100mm x 25mm cleats

A

Column form

E

Outer beam form

100mm x 25mm battens

Headtree and brackets

Main beam and column junction

Joist

Decking

25mm beam sides

50mm beam bottom

100mm x 38mm bracket

100mm x 25mm battens

100mm x 25mm ledger

Joist

100mm x 100mm headtree

100mm x 75mm yokes

D Beam form

16mm bolts

Wedges

25mm sides

B Square column form

Four-piece 50mm yoke built up into two halves

Joints

Staves

Clearance

13mm bolts

Built-up yoke of 32mm timber in two halves. Secured by packing pieces and steel clamps

Staves and ply

Joints

C Circular column forms

254 Formwork for columns and beams

means of wedges placed between the lower ends of the props and the cill pieces on which they rest. The bottom board is usually given a slight camber of about 6 mm in 3 m of length to allow for deflection. The sides are then placed in position and nailed to the edges of the bottom and secured to the column boxes. Stops or stop boards with wedges are sometimes preferred to nailing the sides to the edges of the bottom. The sides are braced apart to the correct width by temporary strainers.

The free side of an outer beam is braced at the top by struts off the ends of the headtrees, which are extended for this purpose (figure 255).

To avoid construction on scaffolding, whenever possible beam forms should be made at ground level and hoisted into position.

When secondary beams bear on main beams, openings for them are cut in the sides of the main beam shutter in the correct position and a bearer nailed on to receive the soffit board of the secondary beam form, the construction of which is carried out as described above (figure 255). Junctions of beams and columns are formed in the same way as shown in figure 254 A, E. A 100 mm x 25 mm ledger or runner is nailed to the battens of the beam sides to form a bearing for the slab joists. The tops of the ledgers are set at such a level that the slab decking will bear on the top of the sides (figures 254 D and 255).

Floor shutters

The construction of a simple floor shutter for a slab bearing on walls is illustrated and described in Part 1 and as indicated there it is made up of boards or 'decking' on which the concrete is placed, supported by joists, ledgers and props. Shuttering related to beams is illustrated in figure 255 and the description given in Part 1 will here be extended in relation to this. 25 or 32 mm thick decking is generally used, thinner decking necessitating a closer and less economic spacing of the joists. The boards should run the length of the floor panels so that they do not require cutting into short lengths, and so that the joists span across the shortest dimension. The usual size of joist is 150 mm x 50 mm, but may range from 100 mm x 50 mm to 250 mm x 75 mm. Long narrow areas of floor shutter may be made up in a number of smaller panels consisting of a number of joists or cleats carrying the decking (figure 255). These may be handled easily and facilitate re-use. Plywood, rather than boards, is widely used for

floor shutters, and very large panels consisting of plywood on joists with framing round all four edges can be used when a crane is being employed on the site for handling purposes.

The actual thickness of the decking and the sizes of the joists used will depend upon the loads the forms must carry, the spacing and the span of the joists and the maximum deflection of the shuttering which may have been specified. Boarding used for formwork is usually one standard size on a job, 150 mm x 32 mm being the most common size. This facilitates the ordering of material and reduces wastage.

The joists are supported on ledgers fixed to the beam sides. To minimize the size of the joists intermediate ledgers are sometimes used, carried on props bearing on the floor below (figure 200 C, Part 1). Folding wedges are placed between the bottom of the joists and the ledger to permit final adjustment of height and to facilitate removal. The size of joists and props should be such that ledgers and props do not have to be too closely spaced, causing excessive obstruction of working space. Adjustable steel props may be used which are simple to install and avoid the use of wedges (see figure 259 B). Tubular steel scaffolding with forkheads (figure 250 E) may also be used for the same purpose. The latter is particularly suitable where the height from the floor or ground is great.

Adequate cross bracing of all supports is essential in order to avoid movement and failure.

Rectangular-grid floors and roofs (figures 144 and 223) present a special problem because of the large number of intersecting beams or ribs to be formed. This is usually solved by the use of square box forms or pans of metal or glass-fibre reinforced plastic which are in the form of deep trays with projecting horizontal edges or lips. The pans are laid on temporary skeleton formwork with the edges touching to form the soffits of the ribs. The depth of the pans vary according to the required depth of the ribs.

The metal or plastic pans can be re-used a great number of times but, as an alternative, stout, stiffened cardboard boxes can be used as expendable forms. The use of precast 'ferro-cement' units for this purpose, which serve as permanent shuttering, is referred to on page 354.

Wall forms

Boarding used for wall shuttering varies from 25 to 50 mm in thickness; for heights up to about three

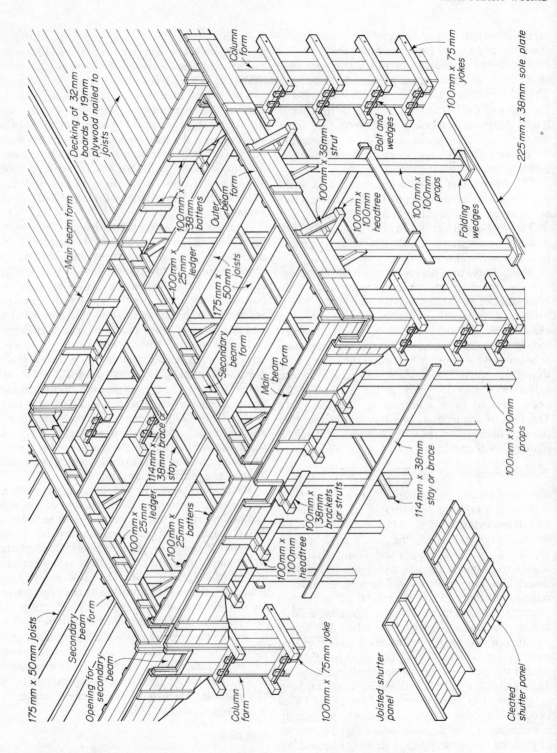

Decking of 32mm boards or 19mm plywood nailed to joists

Column form

100mm x 75mm yokes

225mm x 38mm sole plate

Main beam form

100mm x 38mm battens

100mm x 25mm ledger

Outer beam form

100mm x 38mm strut

Bolt and wedges

100mm x 100mm headtree

100mm x 100mm props

Folding wedges

175mm x 50mm joists

Secondary beam form

Main beam form

100mm x 25mm ledger

114mm x 38mm brace or stay

100mm x 25mm battens

100mm x 38mm brackets or struts

100mm x 100mm headtree

114mm x 38mm stay or brace

100mm x 100mm props

175mm x 50mm joists

Secondary beam form

Opening for secondary beam

Column form

100mm x 75mm yoke

Joisted shutter panel

Cleated shutter panel

255   *Formwork for floors*

418

metres 38 mm boarding is usually used. Alternatively 16 mm plywood is commonly used for this purpose. The boards are fixed to 100 mm x 50 mm studs, known as soldiers, at 760 mm spacings, and horizontal walings are fixed to the studs at intervals. The shutters are supported by struts, the bottoms of which are secured by chocks fixed to a sole plate or sleeper of timber (figure 256 A). If well strutted the two shutters are kept the thickness of the wall apart by timber cross-pieces nailed to the tops of the posts. Spacing at the bottom, as in the case of columns, is provided by a 'kicker' of concrete about 50 mm high and the exact width of the wall, which is cast on top of the foundations or of a concrete floor. With thin walls it is an advantage first to erect one side of the formwork to the full height of the wall, and then to fix the reinforcement to the full height, followed by the formwork for the second side, which may be erected to the full height immediately or in lifts. If hand compacting of the concrete is to be used, the second side should be erected in successive lifts of 600 or 900 mm in height; if vibration is to be used and the thickness of the wall is sufficient, the second side may be erected to the full height before concrete is placed. An alternative method in the case of hand compacting is to erect both sides together as the work proceeds, one lift at a time.

To avoid the use of large studs and an excessive amount of strutting, spacers and wire ties are used, the spacers holding the two shutters the correct distance apart, and the ties resisting the outward pressure of the concrete when poured (figure 256 A, B). The wires are passed through the boarding and round the walings on each side. When the wall is cast in lifts the spacers may be of timber, which are raised as the concreting proceeds; but when the wall is cast in one operation, the removal of the spacers is difficult, so that concrete spacers are used and are left in position. When the formwork is struck the protruding ends of the wires must be cut back to at least 13 mm below the face of the wall and the holes carefully filled. As wire ties are likely to cause rust stains at the points where they are cut back, bolts are used as an alternative, being well greased or fitted with sleeves to enable them to be drawn out from the concrete when the formwork is struck (C). A number of proprietary ties are available which secure the formwork without wire or spacers (D). Several systems of clamps which dispense with the need for ties altogether are available

for the construction of thin walls and a typical example is shown at (E).

Wall forms are frequently made up in panels about 1.80 m x 600 mm, a size which can be easily handled and stripped, and are often constructed with their supporting studs in such a way that the panels, after the first lift has been poured and has hardened, can be supported on the walling below which has already been cast (F).

Metal faced plywood stiffened with small steel angles can be used for wall and floor panels; the metal facing protects the plywood and increases the life of the panels.

Stair forms

These are constructed on the lines shown in figure 257. The shutter is carried on cross joists and raking ledgers. The risers, which are fixed after the reinforcement is placed, are 38 to 50 mm thick and bevelled at the bottom to permit the whole of the tread face to be trowelled. The outer ends are carried by a cut string and the wall ends by hangers secured to a board fixed to, or strutted against, the wall face. The treads are left open to permit concreting.

Shell barrel vault forms

The radii of these vaults in most cases permits plywood decking to be bent cold round the curve. An indication of the radii appropriate to various ply thicknesses is given in table 27.

Thickness in mm	Approximate minimum radius in metres	
	Across grain	Parallel to grain
10	0.900	1.40
13	1.80	2.40
16	2.40	3.00
19	3.00	3.60

Table 27   *Bending of plywood for forms*

The decking is supported by straight joists, spaced at intervals appropriate to the thickness of the plywood (figure 258 A). For a 75 to 100 mm thick barrel these would be 635 mm for 13 mm, 735 mm for 16 mm, and 900 mm for 19 mm ply.

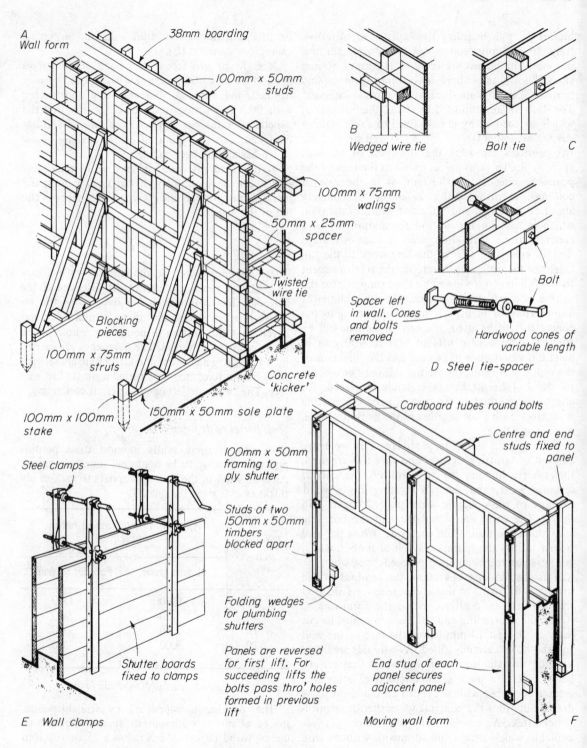

A Wall form

38mm boarding

100mm x 50mm studs

100mm x 75mm walings

50mm x 25mm spacer

Twisted wire tie

Blocking pieces

100mm x 75mm struts

100mm x 100mm stake

Concrete 'kicker'

150mm x 50mm sole plate

B Wedged wire tie

Bolt tie C

Bolt

Spacer left in wall. Cones and bolts removed

Hardwood cones of variable length

D Steel tie-spacer

Cardboard tubes round bolts

Centre and end studs fixed to panel

100mm x 50mm framing to ply shutter

Studs of two 150mm x 50mm timbers blocked apart

Folding wedges for plumbing shutters

Panels are reversed for first lift. For succeeding lifts the bolts pass thro' holes formed in previous lift

End stud of each panel secures adjacent panel

Steel clamps

Shutter boards fixed to clamps

E   Wall clamps

Moving wall form F

256   Formwork for walls

420

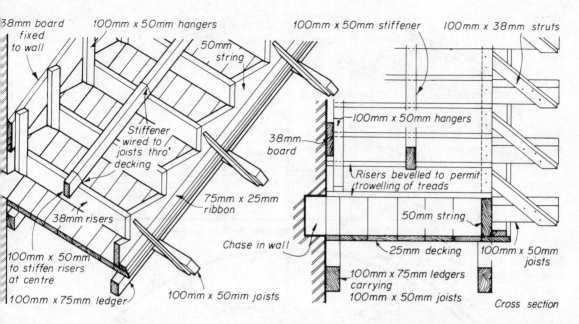

38mm board fixed to wall

100mm x 50mm hangers

50mm string

100mm x 50mm stiffener

100mm x 38mm struts

Stiffener wired to joists thro' decking

100mm x 50mm hangers

38mm board

Risers bevelled to permit trowelling of treads

75mm x 25mm ribbon

38mm risers

Chase in wall

50mm string

100mm x 50mm to stiffen risers at centre

100mm x 75mm ledger

100mm x 50mm joists

25mm decking

100mm x 50mm joists

100mm x 75mm ledgers carrying 100mm x 50mm joists

Cross section

257   Formwork for stairs

The joists bear on curved ledgers. For smaller radii curves, where boards or plywood cannot be bent, shaped joists, carried by longitudinal ledgers, are used to carry longitudinal decking (*B*). In each case the formwork is carried on a system of braced shores. End diaphragms are deep thin beams, and are formed in the same way as thin walls, the whole formwork being supported by braced shoring. Edge beams are formed as normal beams (*A*).

Shell dome forms

Narrow board decking is used for these, carried on curved radiating built-up ribs, some of which stop short of the crown to avoid jointing difficulties. The boarding is laid in 'panels' at different angles to minimize curvature and taper cutting. Noggings are inserted between the ribs where required to take the ends of the boards. With small radius domes it may be necessary to use two layers of thin boarding (10 mm) in order to bend it to the required curve. The built-up ribs are carried by a braced framework bearing on ledgers supported, usually, by braced scaffolding.

Doubly curved shell forms

As explained in chapter 9, the hyperbolic paraboloid form can be developed from straight lines. The formwork can, therefore, be made up entirely from straight members. In principle, therefore, the construction is the same as for floor slabs, with decking, joists, ledgers and props. The edge beam forms are framed up, braced and supported as normal outer floor beams as shown in figure 258 *C*.

Steel formwork

This type of formwork is designed to eliminate timber and to be quickly erected and dismantled, and being in panel units the lower panels can be released individually and used at a higher level. Many systems are available, generally consisting of panels made of steel sheet on light steel angle framing similar to that shown in figure 259. The sizes of the panels for floor and wall shutters are usually about 900 mm x 600 mm or 600 mm square, and narrow width units and strips for making up dimensions are available. Special panels for circular work may be obtained. Steel forms are most commonly used for wall and floor shutters (*A*, *B*) although they can be used for columns, particularly if the latter are large (*D*). Telescopic steel beam units (*B*) can be used for the support of slab formwork; they are easily removed and reduce the number of supporting props required.

Adjustable steel beam clamps (*C*) provide head-

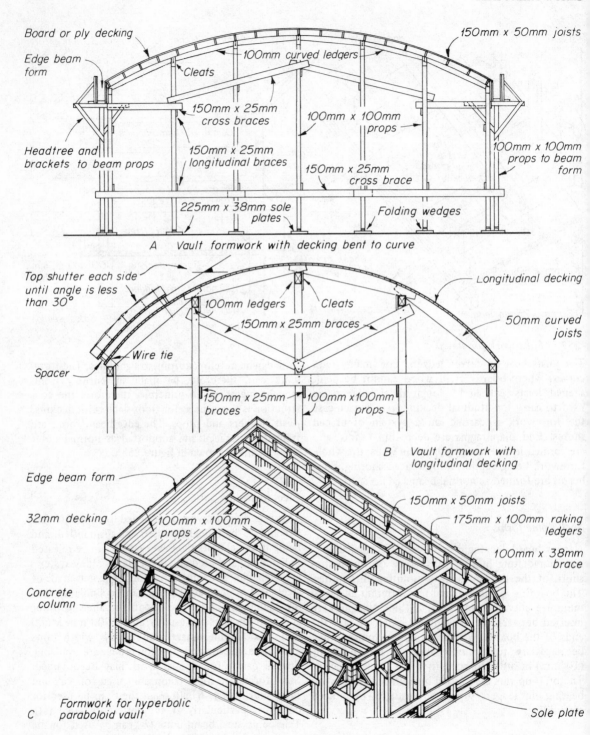

Board or ply decking

Edge beam form

Cleats

100mm curved ledgers

150mm x 50mm joists

150mm x 25mm cross braces

100mm x 100mm props

Headtree and brackets to beam props

150mm x 25mm longitudinal braces

150mm x 25mm cross brace

100mm x 100mm props to beam form

225mm x 38mm sole plates

Folding wedges

A   Vault formwork with decking bent to curve

Top shutter each side until angle is less than 30°

100mm ledgers

Cleats

Longitudinal decking

150mm x 25mm braces

50mm curved joists

Wire tie

Spacer

150mm x 25mm braces

100mm x100mm props

B   Vault formwork with longitudinal decking

Edge beam form

32mm decking

100mm x 100mm props

150mm x 50mm joists

175mm x 100mm raking ledgers

100mm x 38mm brace

Concrete column

C   Formwork for hyperbolic paraboloid vault

Sole plate

258   Formwork for shell vaults

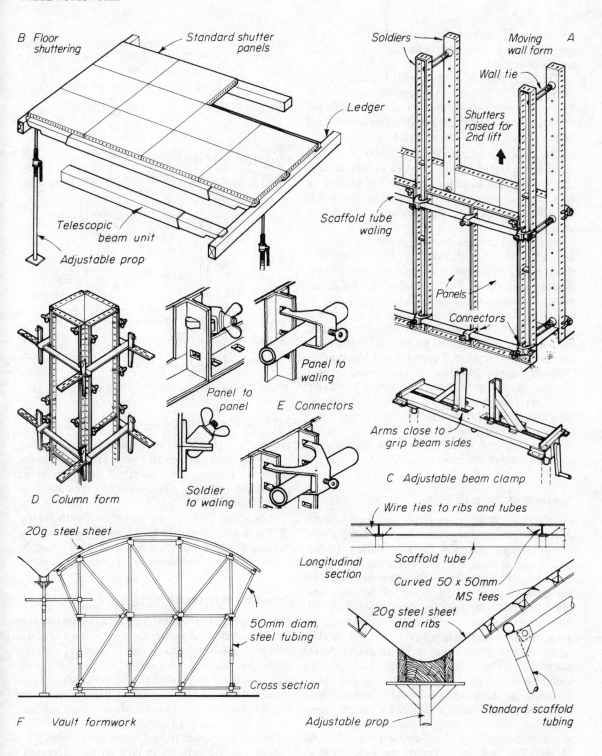

B Floor shuttering
Standard shutter panels
Ledger
Telescopic beam unit
Adjustable prop

Soldiers
Moving wall form A
Wall tie
Shutters raised for 2nd lift
Scaffold tube waling
Panels
Connectors

Panel to panel
Panel to waling
E Connectors
Soldier to waling
D Column form

Arms close to grip beam sides
C Adjustable beam clamp

Wire ties to ribs and tubes
Longitudinal section
Scaffold tube
Curved 50 x 50mm MS tees
20g steel sheet and ribs
Standard scaffold tubing

20g steel sheet
50mm diam. steel tubing
Cross section
F Vault formwork
Adjustable prop

259 Steel formwork

tree and strut support (see figure 255) to bottom and sides of beams, with adjustment for varying widths of beams.

The angle framing to the panels is perforated to permit them to be connected together or to walings by various types of connectors (E). These simple means of fastening and dismantling permit the rapid raising of wall forms as shown at (A).

This type of formwork can be re-used a great many times, but if roughly handled needs considerable maintenance in straightening and in welding up broken and cracked edges. This must be put against the savings arising from the greater number of re-uses.

Steel forms can also be used in the construction of shell vaults (see figure 259 F). Longitudinal shutters, of thin sheet steel stiffened by ribs, are supported on curved T-sections carried by a framework of metal scaffold tubes. Adjustable props, or jack bolts fitted to the heads of the props, provide vertical adjustment. Curved scaffold tubes, carrying flexible steel sheets secured by special shutter clips, may be used instead of T-sections and longitudinal stiffened shutters. Mobile scaffolds running on rails may be employed and this is a useful method on very long vaults, but necessitates a clear barrel soffit. All stiffening ribs and frames to the vault must, therefore, be above the curved shell. Unless the vault is long, striking and re-erection of the shuttering is generally quicker and more accurate.

Slip forms or sliding shutters

For the rapid construction of constant section walls it is possible to use a continuously rising form, usually known as a slip form or sliding shutter. By this means work may proceed continuously, the shutter rising from 150 to 300 mm per hour depending upon the rate of hardening of the concrete, since the cast concrete very rapidly becomes self-supporting. The form is about 900 mm or 1.20 m deep, fixed to and held apart by timber or steel frames or yokes, as shown in figure 260 A, B. On top of each yoke is fixed a hydraulic jack, through which passes a high tensile steel jacking rod, about 25 mm in diameter, which is cast into the wall as it rises. The jack contains a ram and a pair of upper and lower jaws which can grip the jacking rod and it works in cycles, each cycle giving a rise of about 25 mm. The jack works against the lower jaws to raise the yoke and the form with it. When the

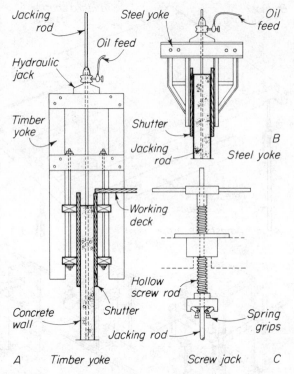

A   Timber yoke   Screw jack   C

260   Sliding shutters

pressure is released, the upper jaws grip the rod and the lower jaws are released and raised under the action of a spring. An alternative to the hydraulic jack is the manually operated screw jack which is also illustrated (C).

A working deck is constructed level with the top of the form, from which is usually suspended a hanging scaffold from which the concrete may be inspected and rubbed down as it leaves the shutters.

Treatment of formwork

The nature and treatment of the working faces of the formwork, that is, the faces in contact with the concrete, will affect the finished surface of the concrete. All working faces should always be treated with mould oil to prevent the concrete adhering to them and thus reduce the risk of damage when the formwork is stripped. In cases where a good key will be required on the final surface or where it is desired ultimately to expose the aggregate on the surface, a retarding liquid may be applied to the formwork. This prevents the setting of the cement at the surface, so that when the formwork is struck, the concrete face may be brushed down with stiff brushes to form a rough surface of exposed agg-regate. Alternatively, aggregate transfer may be used. This consists of sticking selected aggregates to shutter liners with a suitable water-soluble adhesive. On stripping the shutters the aggregate is transferred to the concrete to which it is, by then, bonded. To produce a good smooth face, the formwork may be lined with plywood, hardboard or plastic sheeting. Where a patterned surface is required, patterned tough rubber sheet or expanded plastic is used as a lining or glass reinforced plastic forms may be used. Very deep patterning may be produced by the first two methods because the linings can be pulled away reasonably easily from the set concrete[1].

Permanent shuttering of precast concrete may be used to provide the final finished face. Concrete pipes may be used for circular columns and concrete slabs for walls (see 'Facings', chapter 4). Wood wool slabs used for thermal insulation to a wall may be used as an inside permanent shutter.

[1] For the special requirements of 'board-marked' concrete finish and for other finishes referred to here see *Guide to Exposed Concrete Finishes* by Michael Gage, Architectural Press, 1972

Striking times for formwork from CP 110 are given in table 28. These are intended as a general guide only. Actual times will vary on each job according to the size of member, type of structure and day to day weather conditions.

Formwork	Surface temperature of concrete	
	16°C	7°C
Vertical formwork to columns, walls and large beams	9 hours	12 hours
Soffit formwork to slabs	4 days	7 days
Props to slabs	11 days	14 days
Soffit formwork to beams	8 days	14 days
Props to beams	15 days	21 days

Note: These periods are for ordinary Portland Cement concrete. In cold weather the above periods should be increased. For soffit formwork increase the periods for 7°C by half a day for each day on which the concrete temp-erature was generally between 2°C and 7°C and by a whole day for each day on which the concrete temperature was below 2°C.

Table 28   *Minimum times for striking formwork*

Index

427

435